T0136695

FLORA ZAMBESIACA

Flora terrarum Zambesii aquis conjunctarum

VOLUME EIGHT: PART SIX

FLORA ZAMBESIACA

MOZAMBIQUE

MALAWI, ZAMBIA, ZIMBABWE

BOTSWANA

VOLUME EIGHT: PART SIX

Edited by
J.R. TIMBERLAKE & E.S. MARTINS

on behalf of the Editorial Board:

J.R. TIMBERLAKE
Royal Botanic Gardens, Kew

M.A. DINIZ
*Centro de Botânica, Instituto de Investigação
Científica Tropical, Lisboa*

D.P. SIMPSON
Royal Botanic Gardens, Kew

Published by the Royal Botanic Gardens, Kew
for the Flora Zambesiaca Managing Committee
2015

ROYAL BOTANIC GARDENS

First published in 2015 by
Royal Botanic Gardens, Kew,
Richmond, Surrey, TW9 3AB, UK
www.kew.org

Distributed on behalf of the Royal Botanic Gardens, Kew in North America by the
University of Chicago Press, 1427 East 60th Street, Chicago, IL 60637, USA

ISBN 978-1-84246-413-7
eISBN 978-1-84246-518-9

British Library Cataloguing in Publication Data
A catalogue record for this book is available from the British Library

Typesetting by Christine Beard
Publishing, Design and Photography
Royal Botanic Gardens, Kew

Printed in the UK by Marston Book Services Ltd
Printed in the USA by The University of Chicago Press

For information or to purchase all Kew titles please visit
www.kewbooks.com or email publishing@kew.org

Kew's mission is to inspire and deliver science-based plant conservation worldwide,
enhancing the quality of life.

Kew receives half of its running costs from Government through the Department for
Environment, Food and Rural Affairs (Defra). All other funding needed to support Kew's
vital work comes from members, foundations, donors and commercial activities including
book sales.

CONTENTS

FAMILIES INCLUDED IN VOLUME 8, PART 6

ACANTHACEAE (Part 2)

LIST OF NEW NAMES PUBLISHED IN THIS PART

126. ACANTHACEAE (Part 2)

by Iain Darbyshire, Kaj Vollesen and Ensermu Kelbessa

Tribes **Barlerieae** and **Justicieae** (genera 24–42)

Acanthaceae Part 2 covers cultivated species and genera 24–42. The key to all native genera is given in Part 1 (vol. 8.5).

Cultivated Acanthaceae in the Flora Zambesiaca region

Over 50 species of Acanthaceae have been reported in the literature or in herbarium collections as having been cultivated in the Flora region. Of these, a number are native species which have been brought into cultivation locally; these are treated in the native species accounts. Here we restrict ourselves to treating the most widely cultivated ornamentals which have not, to our knowledge, become naturalised in the Flora region.

We have relied on the collections in the Harare and Kew herbaria but also to a large extent on Maroyi's Preliminary Checklist of Introduced and Naturalized Plant Species in Zimbabwe (Kirkia **18**: 177–247, 2006). Species are listed alphabetically, with only one herbarium specimen cited per country.

Anisacanthus quadrifarius (Vahl) Nees
Anisacanthus wrightii sensu Maroyi in Kirkia **18**: 193 (2006), not (Torr.) A. Gray
Soft-wooded shrub to 1 m; stems thinly hairy in two bands; cystoliths present. Leaves narrowly ovate-elliptic; apex acuminate; base cuneate. Flowers in loose racemiform subglabrous cymes, sometimes aggregated into large panicles; bracts and bracteoles minute. Calyx 5–10 mm long, deeply divided into 5 linear-lanceolate cuspidate lobes. Corolla bright red, to 4.5 cm long (along upper lip), with 5 spreading to recoiled lobes. Stamens 2, spreading out between lobes; anthers bithecous, thecae slightly unequal, at same height, rounded.
Zimbabwe. Harare, 27.xi.1973, *Biegel* 4430 (SRGH).
Native of Mexico. Occasionally cultivated elsewhere (South Africa, Egypt, Europe). Maroyi (2006) lists this as *A. wrightii*, a species with more densely hairy stems and inflorescences and a shorter calyx.

Barleria cristata L. —Maroyi in Kirkia **18**: 193 (2006). —Darbyshire in F.T.E.A., Acanthaceae: 353 (2010). Philippine Violet.
Subshrub; cystoliths present. Leaves elliptic, ovate or lanceolate, base and apex attenuate or acute, surfaces strigose mainly on veins beneath. Flowers held in contracted 2–5-flowered unilateral cymes, or single-flowered in upper leaf axils; bracteoles linear-lanceolate, 8.5–16 mm long, apex spine-tipped. Calyx conspicuous, initially green or tinged purple, turning scarious, anterior and posterior lobes ovate-trullate, the former 14.5–19 × 6.5–8.5 mm, the latter somewhat longer, margins spinulose-toothed; lateral lobes lanceolate, 6–8 mm long. Corolla blue, mauve or white, tube cylindrical towards base, campanulate above attachment point of stamens, 31–43 mm long; limb of 5 lobes, abaxial lobe splitting from tube before remaining 4 lobes, each 11–22 mm long. Stamens 2, exserted, bithecous, thecae at equal height; staminodes 3, lateral pair with small antherodes.
Zambia. C: Lusaka, Mt Makulu, fl. 27.ii.1968, *Fanshawe* A8 (K, NDO). **Zimbabwe**. C: Harare, Golden Stairs Nursery, fl. 17.iv.1975, *Biegel* 4974 (K, SRGH).
Native of tropical and subtropical Asia. Widely and commonly cultivated in the tropics.

Crossandra nilotica Oliv. —Vollesen in F.T.E.A., Acanthaceae **1**: 139 (2008).

Crossandra massaica sensu Maroyi in Kirkia **18**: 193 (2006), not Mildbr.

Shrubby herb to 0.5 m; cystoliths absent. Leaves dark green, glossy, ovate to elliptic, to 10 × 5 cm; apex acute to acuminate; base decurrent. Flowers in dense terminal pedunculate racemoid cymes to 5 cm long; bracts large, pale green, ovate to elliptic, to 2.5 cm long; bracteoles linear-lanceolate, to 1 cm long. Calyx divided into 5 unequal sepals, thickened and horny at base, dorsal 2-veined and 2-toothed, ventral and lateral smaller, 1-veined and 1-toothed, to 1 cm long. Corolla bright red, to 5 cm long, split dorsally to give a 5-lobed lower lip and no upper lip. Stamens 4, sessile, included in corolla tube; anthers monothecous, rounded to finely apiculate.

Zimbabwe. C: Harare, 17.xi.1982, *Biegel* 5895 (SRGH).

Native of eastern tropical Africa. Occasionally cultivated in the tropics and also grown as a pot plant in Europe and the USA.

Crossandra nilotica Oliv. has stalked capitates glands and long eglandular hairs on the bracts while *C. massaica* Mildbr. has long eglandular hairs but no capitate glands. The Harare specimens clearly have glands and are therefore *C. nilotica*.

Eranthemum pulchellum Andrews. —Maroyi in Kirkia **18**: 193 (2006). —Darbyshire *et al.* in F.T.E.A., Acanthaceae **2**: 730 (2010). Blue Sage.

Shrub; cystoliths present. Leaves ovate or elliptic, base and apex attenuate, lateral veins and scalariform tertiary veins prominent beneath. Flowers held in dense spikes, terminal and in upper axils; bracts showy, imbricate, white or silverish with green pinnate-reticulate venation, ± elliptic, 14–20 mm long. Calyx of 5 ± equal lanceolate lobes shorter than bracts. Corolla blue, salver-shaped, tube narrowly cylindrical, curved, to c.20 mm long; limb of 5 subequal spreading lobes to c.10 mm long. Stamens 2, exserted, bithecous, thecae at equal height; staminodes 2, minute.

Zimbabwe. C: *Phipps* 2164 (SRGH, cited by Maroyi). **Malawi.** S: Zomba Botanic Garden, fl. n.d., *Salubeni* 817 (K, SRGH).

Native of the Indian subcontinent. Widely cultivated in the tropics.

Graptophyllum pictum (L.) Griff. —Maroyi in Kirkia **18**: 193 (2006). —Darbyshire *et al.* in F.T.E.A., Acanthaceae **2**: 730 (2010).

Shrub; cystoliths present. Leaves variegated, either green and yellow or dark green and pink, elliptic, base cuneate to attenuate, apex shortly acuminate, glabrous. Flowers held in a short terminal thyrse; bracts and bracteoles inconspicuous, lanceolate, 2–3 mm long. Calyx of 5 ± equal lanceolate lobes c.4 mm long. Corolla deep reddish-purple, 35–40 mm long, tube gradually expanded to mouth, curved, to c.25 mm long; limb 2-lipped, upper lip 2-lobed, lower lip divided into 3 oblong lobes. Stamens 2, exserted, bithecous, thecae at equal height; staminodes 2.

Zambia. C: Lusaka, fl. 1.x.1994, *Bingham* 10157 (K). **Zimbabwe.** C: Harare, st. 13.iii.1975, *Biegel* 4938 (K, SRGH).

Native of the Indo-Pacific region, though its exact native distribution apparently unknown. Very widely cultivated in the tropics.

The Zimbabwe specimen cited lacks flowers and fruits making confirmation of its identity difficult; we have had to accept Biegel's identification.

The genus *Graptophyllum* has an unusual distribution, the species all being from SE Asia and Australasia with the exception of *G. glandulosum* Turrill, native to the Cameroon Highlands.

Justicia adhatodioides (Nees) V.A.W. Graham

Duvernoia adhatodioides Nees. —Maroyi in Kirkia **18**: 193 (2006).

Shrub to 2 m; cystoliths present. Leaves ovate, to 10 cm long, apex subacute to broadly rounded, base attenuate. Flowers in pedunculate racemoid cymes from upper axils; bracts green, obovate, to 1 cm long, puberulous; bracteoles linear-lanceolate to 7 mm long. Calyx 5–7 mm long, with 5 broadly triangular acute lobes to 1 mm long. Corolla white with purple streaks on lower lip, to 3 cm long (along upper lip), with a broad hooded 2-toothed upper lip and 3-lobed lower lip with large spreading middle lobe and 2 narrow deflexed lateral lobes. Stamens 2, held under upper lip; anthers bithecous, thecae at different height, lower with small white basal appendage.

Zimbabwe. C: Harare, 10.xii.1955, *Wild* 4718 (K, SRGH).
Native of South Africa. This extremely attractive shrub has been cultivated in and around Harare since 1951, but there are no records of it being cultivated elsewhere.

Justicia aurea Schtdl. —Maroyi in Kirkia **18**: 193 (2006). Yellow Jacobinia.
Shrubby herb to 2 m; cystoliths present. Leaves ovate, to 30 × 15 cm, apex acuminate, base cuneate to truncate. Flowers in large pedunculate terminal racemoid cymes to 30 cm long; bracts green, middle and upper linear-lanceolate, to 2.5 cm long; bracteoles linear-lanceolate, to 1.5 cm long. Calyx deeply divided into 5 lanceolate acute puberulous lobes to 1 cm long. Corolla yellow to orange, to 6 cm long (along upper lip), with a hooded 2-lobed upper lip and 3-lobed lower lip with strongly curled spreading lobes. Stamens 2, held under upper lip; anthers bithecous, thecae at slightly different height, lower theca apiculate.
Zambia. S: Livingstone, 30.iv.1972, *Fanshawe* D14 (K). **Zimbabwe**. C: *Biegel* 5260 (SRGH, not seen). **Mozambique**. M: Maputo, 20.vii.1971, *Balsinhas* 1920 (K).
Native of tropical America. Widely cultivated in the Old World tropics.

Justicia brandegeeana Wassh. & L.B. Sm. —Maroyi in Kirkia **18**: 193 (2006). — Darbyshire *et al.* in F.T.E.A., Acanthaceae **2**: 731 (2010). Shrimp Plant.
Beloperone comosa Nees, *Beloperone guttata* Brandeg.
Perennial or shrubby herb to 1 m; cystoliths present. Leaves ovate, apex acute, base cuneate. Flowers in subsessile or shortly pedunculate racemoid cymes from upper axils; bracts imbricate, to 2 cm long, red to purple or green, ovate, puberulous and ciliate; bracteoles similar to bracts. Calyx deeply divided into 5 lanceolate acute puberulous lobes to 5 mm long. Corolla white with reddish purple stripes on lower lip or whole lower lip red, to 3 cm long (along upper lip), with a hooded 2-lobed upper lip and spreading 3-lobed lower lip. Stamens 2, held under upper lip; anthers bithecous, thecae at different height, both with flat white basal appendage.
Zimbabwe. C: Harare, 6.iv.1969, *Biegel* 3157 (K, SRGH).
Native of tropical America. Widely cultivated in all tropical and subtropical regions and as a pot plant in temperate regions.

Justicia carnea Lindl. —Maroyi in Kirkia **18**: 193 (2006). Pink Jacobinia.
Shrubby herb to 1.5 m; cystoliths present. Leaves ovate, apex acute, base cuneate. Flowers in shortly pedunculate ovoid terminal racemoid cymes to 6 cm long; bracts green, ovate, to 2 cm long; bracteoles lanceolate, to 1.2 cm long. Calyx deeply divided into 5 lanceolate acute puberulous lobes to 1 cm long. Corolla rose-pink, to 6 cm long (along upper lip), with a hooded 2-lobed upper lip and recoiled 3-lobed lower lip. Stamens 2, held under upper lip; anthers bithecous, thecae at different height, lower theca apiculate.
Zambia. W: Kitwe, 15.i.1969, *Fanshawe* R9 (SRGH). **Zimbabwe**. C: Harare, 12.xi.1975, *Biegel* 5176 (K, SRGH).
Native of tropical America. Widely cultivated in the Old World tropics.

Justicia fulvicoma Cham. & Schtdl.
Maroyi (Kirkia **18**: 193, 2006) records this as cultivated in Zimbabwe based on *Biegel* 4936 (SRGH, not seen). There are no other records of this species – a Mexican endemic – ever having been in cultivation anywhere.

Justicia spicigera Schtdl. —Maroyi in Kirkia **18**: 194 (2006). —Darbyshire *et al.* in F.T.E.A., Acanthaceae **2**: 731 (2010).
Justicia oblongata sensu Maroyi in Kirkia **18**: 193 (2006), non Link & Otto
Shrubby herb to 1 m; cystoliths present. Leaves ovate to elliptic, apex acute, base attenuate. Flowers in open axillary cymes from upper axils; bracts and bracteoles minute. Calyx deeply divided into 5 linear glabrous lobes to 2 mm long. Corolla orange to orange-red or mauve to pink, to 5 cm long (along upper lip), with slightly hooded 2-lobed upper lip and recoiled 3-lobed lower lip. Stamens 2, held under upper lip; anthers bithecous with thecae almost at same height, lower theca apiculate.

Zimbabwe. C: Goromonzi, 8.ii.1973, *Biegel* 4215 (K, SRGH).
Native of Central and South America. Widely cultivated in the Old World tropics.
Maroyi (2006: 193) cites this collection as *Justicia oblongata* Link & Otto, which is a
quite different species; the specimen has been misidentified.

Mackaya bella Harv. —Maroyi in Kirkia **18**: 194 (2006). —Darbyshire *et al.* in F.T.E.A.,
Acanthaceae **2**: 731 (2010).

Shrub to 4 m; cystoliths present. Leaves elliptic, coarsely toothed, apex acuminate, base cuneate
to attenuate. Flowers in elongated terminal racemoid cymes, either 1 or 2 per node; bracts and
bracteoles inconspicuous. Calyx with 5 linear cuspidate lobes to 5 mm long. Corolla lilac or pale
mauve with darker lines in throat, to 6 cm long, with narrow basal tube, campanulate throat and
5 erect to spreading subequal lobes. Stamens 2, included in throat; anthers bithecous, thecae
linear, at same height, apiculate at base.

Zimbabwe. C: Harare, 27.ix.1981, *Biegel* 5867 (K, SRGH).
Native of South Africa. Cultivated in eastern and southern Africa.

Megaskepasma erythrochlamys Lindau. —Maroyi in Kirkia **18**: 194 (2006).

Robust shrub to 3 m; cystoliths present. Leaves ovate-elliptic, to 50 cm long, coarsely toothed,
apex acuminate, base attenuate. Flowers in elongated terminal racemoid cymes; bracts reddish
purple, ovate to 3 × 2 cm. Calyx of 5 lanceolate cuspidate lobes to 1.5 cm long. Corolla white,
turning pale pink, to 5.5 cm long, tube to 2 cm long, lobes in upper lip erect, hooded, to 3.5 cm
long, those in lower lip deflexed, twisted, to 3.5 cm long. Stamens 2, exserted; anthers bithecous,
thecae linear, one slightly longer, apiculate at base.

Zambia. C: University of Zambia Farm, 15 km S of Lusaka, 20.iv.1979, *Critchett* 19/79
(K). **Zimbabwe**. C: Harare, 13.iii.1975, *Biegel* 4939 (K, SRGH).
Native of Venezuela. Widely cultivated in the tropics and in greenhouses elsewhere.

Odontonema callistachyum (Schltdl. & Cham.) Kuntze. —Daniel in Contrib. Univ.
Mich. Herb. **20**: 154 (1995). —Maroyi in Kirkia **18**: 194 (2006). Purple Fire Spike.

Perennial herb or shrub; cystoliths present. Leaves petiolate, ± elliptic, base attenuate, apex
acuminate, largely glabrous except along veins beneath. Flowers held in dense to more lax
terminal thyrse, racemose, sometimes with 2 lateral branches from base, shortly pubescent on
rachis, bracts, pedicels and sometimes calyx; bracts and bracteoles inconspicuous, lanceolate
or subulate, 1–9 mm long; flowers pedicellate. Calyx of 5 ± equal subulate lobes 2–6 mm long.
Corolla pink or mauve, 17–30 mm long, tube slender, to 20 mm long, basal tube cylindrical,
throat narrowly expanded; limb 2-lipped, upper lip 4.5–11 mm long, shortly 2-lobed, lower lip
divided into 3 lobes, 4–12 mm long. Stamens 2, exserted or included (flowers heterostylous),
bithecous, thecae at equal height; staminodes 2, minute.

Zambia. W: Mufulira, fl. 12.vi.1973, *Fanshawe* M16 (K, NDO). **Zimbabwe**. C: Harare,
Marlborough Nursery, fl. 17.vii.1978, *Biegel* 5606 (K, SRGH). **Malawi**. S: Limbe, fl.& fr.
17.vii.1978, *Emtage* s.n. (K).
Native of Mexico, Belize and Guatemala. Widely cultivated in the tropics.

Odontonema cuspidatum (Nees) Kuntze. —Daniel in Contrib. Univ. Mich. Herb. **20**:
160 (1995). —Darbyshire *et al.* in F.T.E.A., Acanthaceae **2**: 732 (2010). Fire Spike.
Mottled Toothedthread or Cardinal's Guard.

Odontonema strictum sensu Maroyi in Kirkia **18**: 194 (2006), non (Nees) Kuntze.

Shrub resembling *O. callistachyum* but inflorescence rachis, bracts, etc. less densely pubescent.
Corolla red, 21–35 mm long, tube to 27 mm long, upper lip 2–5 mm long, lower lip with lobes
2.5–6 mm long.

Zambia. C: Chilanga, Mt Makulu Research Station, fl. xii.1978, *Critchett* 13/79
(K). **Zimbabwe**. C: Harare, fl. 16.i.1979, *Biegel* 5650 (K, SRGH). **Malawi**. S: Zomba,
Government Hostel, fl. 20.ii.1979, *Brummitt* 15415 (K, MAL). **Mozambique**. M:
Maputo, Tunduru Botanic Garden, fl. viii.2011 (Darbyshire sight record).

Native of Mexico and possibly the West Indies. Widely cultivated in the tropics. African specimens of this ornamental have often been named *Odontonema strictum* (Nees) Kuntze (= *O. tubaeforme* (Bertol.) Kuntze), a closely related species from Central America, but it appears that *O. cuspidatum* is the correct name for the widely cultivated species.

Pachystachys lutea Nees. —Maroyi in Kirkia **18**: 194 (2006).

Shrub to 1 m; cystoliths present. Leaves narrowly ovate-elliptic; apex acuminate; base decurrent, with two small auriculate lobes. Flowers in a dense terminal racemiform cyme; bracts conspicuous, imbricate, orange yellow, ovate, to 2.5 cm long; bracteoles similar, to 1.5 cm long. Calyx c.5 mm long, deeply divided into 5 lanceolate lobes. Corolla white, to 6.5 cm long (along upper lip), with a hooded 2-lobed upper lip and spreading to recurved 3-lobed lower lip. Stamens 2, held under upper lip; anthers bithecous, green, thecae at same height, rounded.

Zimbabwe. C: Harare, 23.x.1979, *Biegel* 5702 (SRGH).

Native of Peru. Occasionally cultivated elsewhere (Kenya, New Caledonia, USA, Bolivia).

Petalidium bracteatum Oberm. —Maroyi in Kirkia **18**: 194 (2006).

Shrub to 2 m tall; cystoliths present. Leaves ovate to elliptic or cordiform, to 5 cm long, entire; apex subacute to rounded and apiculate; base attenuate to cordate. Flowers in axillary racemoid cymes to 7 cm long; bracteoles large, membranaceous, white or purple-tinged, enclosing calyx and corolla tube, to 3 cm long. Corolla bright red; tube to 2 cm long, widening upwards; lobes c.1.5 × 0.8 cm, 2 in upper lip erect, middle lobe in lower lip deflexed, 2 lateral lobes in lower lip vertical. Stamens 4, erect, held under upper lip; anthers bithecous, oblong, rounded at apex, with short spurs at base.

Zimbabwe. N: Makonde Dist., Banket, i.1975, *Scott* s.n. (SRGH).

Native to Angola and Namibia. There are no records of this species in cultivation elsewhere.

Pseuderanthemum carruthersii (Seem.) Guill.

Soft-wooded shrub to 1.5 m tall; cystoliths present. Leaves ovate to elliptic, to 17 cm long, often variegated, grossly and irregularly crenate; apex acute base attenuate. Flowers in cymules aggregated into racemoid or paniculate cymes; bracts minute. Calyx 3–5 mm long, divided almost to base into 5 narrowly triangular lobes. Corolla white to pink or pale purple; tube 1–1.3 cm long; lobes c.1 × 0.5 cm, 2 in upper lip erect, 3 in lower lip deflexed. Stamens 2, erect, held under upper lip; anthers bithecous, oblong, rounded.

Zambia. S: Livingstone, 30.iv.1972, *Fanshawe* E14 (K). Mozambique. M: Maputo, 3.iv.1972, *Balsinhas* 2401 (K).

Native of the South Pacific. Widely cultivated throughout the tropics.

Vollesen in Flora Zambesiaca 8(5): 136 (2013) mentions three species of *Ruellia* as being cultivated in Zambia and Zimbabwe. Maroyi (2006: 194) also mentions three species, but the names used only partly overlap. A little cleaning up is clearly needed.

Ruellia brevifolia (Pohl) Ezcurra. —Vollesen in F.Z. **8**(5): 136 (2013). Red Ruellia.
Ruellia graecizans Backer. —Maroyi in Kirkia **18**: 194 (2006).

Zambia. C: Lusaka, 12.iii.1973, *Fanshawe* X15 (K). Zimbabwe. C: Harare, 12.iii.1974, *Biegel* 4505 (K, SRGH).

Ruellia elegans Poir. —Maroyi in Kirkia **18**: 194 (2006). Scarlet Ruellia.
Ruellia formosa Andr. —Vollesen in F.Z. **8**(5): 136 (2013).

Zimbabwe. C: Harare, 26.ii.1975, *Biegel* 4928 (K, SRGH).

Ruellia portellae Hook. f. —Maroyi in Kirkia **18**: 194 (2006). —Vollesen in F.Z. **8**(5): 136 (2013).

Zimbabwe. C: Harare, 12.iii.1974, *Biegel* 4504 (K, SRGH).

Ruttya fruticosa Lindau. —Maroyi in Kirkia **18**: 194 (2006). —Vollesen in F.T.E.A., Acanthaceae **2**: 489 (2010). Jammy Mouth.

Shrub to 3 m tall; cystoliths present. Leaves ovate to elliptic, to 10 cm long, entire, apex acute to rounded, base decurrent. Flowers in 4–8-flowered axillary dichasia; lower bracts foliaceous, upper minute. Calyx 3–6 mm long, with 5 narrowly triangular lobes. Corolla orange red to bright red (rarely yellow), a large glossy black nectariferous patch on base of lower lip; tube to 1.2 cm long, lobes to 2.5 × 1.2 cm, 2 in upper lip erect, 3 in lower lip strongly deflexed. Stamens 2, erect, held under upper lip; anthers monothecous, ellipsoid, rounded.

    Zambia. W: Kitwe, 12.xi.1968, *Fanshawe* M8 (SRGH). **Zimbabwe**. C: Harare, 14.viii.1969, *Biegel* 3168 (SRGH). **Malawi**. S: Blantyre, 5.xi.1981, *Patel* 1778 (SRGH).

    Native of eastern Africa. Widely cultivated elsewhere in Africa.

Sanchezia parvibracteata Hook. f.

    Sanchezia nobilis sensu Darbyshire *et al.* in F.T.E.A., Acanthaceae **2**: 732 (2010) in part, non Hook. f.

    Sanchezia oblonga sensu Maroyi in Kirkia **18**: 194 (2006), non Ruiz & Pav.

Shrub; cystoliths present; stems usually reddish. Leaves with midrib and lateral veins yellow, contrasting with bright green surface, (oblong-)elliptic, base cuneate-attenuate, margin serrate or crenate, apex acuminate, glabrous. Inflorescence terminal, spiciform but often few-branched towards base, 1–3 sessile flowers per fertile bract, spikes often 1-sided; bracts conspicuous, usually red, ovate (excluding those subtending inflorescence branches), 10–22 cm long, apex obtuse, glabrous. Calyx of 5 narrowly oblong or spathulate lobes, extending beyond bracts. Corolla yellow or yellow-orange, to c.40 mm long, tube gradually expanded upwards, pubescent externally; limb c.4 mm long, equally 5-lobed, lobes revolute at anthesis. Stamens 2, exserted, bithecous, thecae at equal height, hairy; staminodes 2, linear.

    Zimbabwe. C: Harare, Forest Nurseries, Enterprise Road, fl. 12.vi.1975, *Biegel* 5131 (K, SRGH). **Malawi**. C: Kasungu Dist., Kasungu main street, fl. 3.vi.1989, *Brummitt* 18320 (K, MAL).

    Native range unclear but probably from northern South America. Described from cultivation and commonly cultivated in the tropics (but see note).

    There is considerable confusion over the correct name(s) to be applied to commonly cultivated species of this genus. It is quite possible that either too many species are currently recognised within *Sanchezia* and/or that there has been some hybridization within the cultivated stock. Certainly some of the characters used to separate the species seem to be inconsistent or rather insignificant when the full range of variation is taken into account. We have not found the synopsis of the genus by Leonard & Smith in Rhodora **66**: 315 (1964) to be particularly useful in resolving this problem. Cultivated African specimens have most often been named *S. nobilis* in the past. The type of *S. nobilis* (not seen when the F.T.E.A. account was written), from cultivated material originally derived from Ecuador, has a sparsely hairy corolla and large, shortly hairy bracts. Of the two taxa present in the Flora Zambesiaca region and elsewhere in Africa, one has smaller, glabrous bracts and a hairy corolla and the other has large, glabrous bracts and a glabrous corolla. Thus it seems that, although all three are clearly close, the name *S. nobilis* is not applicable to either of the species commonly cultivated in Africa. Maroyi (2006) named *Biegel* 5131 from Zimbabwe as *S. oblonga* Ruiz & Pav., a further species in the *S. nobilis* complex. However, whilst that species shares the pubescent corolla and glabrous bracts, it has considerably larger bracts than the Biegel collection and so is not considered to be the correct name. Instead, we tentatively name the Biegel collection and matching material as *S. parvibracteata*, described from cultivated material. It should be noted, however, that typical *S. parvibracteata* can have even smaller bracts than the African specimens seen, though with some overlap. See also discussion in Staples & Herbst, A Tropical Garden Flora: 92 (2005).

Sanchezia speciosa Leonard

Sanchezia nobilis sensu Darbyshire *et al.* in F.T.E.A., Acanthaceae 2: 732 (2010) in part, non Hook. f.

Shrub resembling *S. parvibracteata* but bracts larger, 27–41 mm long, longer than and largely enclosing the calyx. Corolla glabrous externally; inflorescence often unbranched, spiciform. Leaves with veins often whitish or pale yellow.

Mozambique. M: Maputo, Jardim Vasco da Gama, fl. 11.x.1972, *Balsinhas* 2442 (K).

Native range unclear but probably northern South America or possibly the Caribbean. Described from cultivation, and fairly widely cultivated in the tropics.

This species was described from cultivated material on Cuba; it is unclear as to how it relates to wild species in the neotropics. It is similar to *S. oblonga* Ruiz & Pav. except that it has a glabrous corolla.

Strobilanthes anisophylla (Hook.) T. Anderson. —Maroyi in Kirkia **18**: 194 (2006).

Strobilanthes persicifolia (Lindl.) J.R.I. Wood in Kew Bull. **64**: 27 (2009).

Shrub or subshrub; cystoliths present. Leaves strongly anisophyllous or pairs subequal, larger of each pair lanceolate or oblong-lanceolate, 5–19 cm long, base cuneate, margin shallowly serrate or subentire, apex long-acuminate, glabrous. Inflorescence axillary, of few-flowered pedunculate heads, sometimes branched, axis glabrous; bracts and bracteoles small, elliptic, green. Calyx of 5 linear or narrowly oblong-elliptic lobes, glandular-pubescent. Corolla 25–28 mm long, pale blue to mauve, glandular- and eglandular-pubescent, basal tube narrowly cylindrical, abruptly expanded into ventricose throat, limb of 5 short subregular lobes; stamens 4, didynamous, all included.

Native of India and Bhutan (rather rare in the wild). Commonly cultivated in the tropics and in glasshouses in temperate regions.

Two forms are recognized, both cultivated in the Flora region.

a) forma **anisophylla**

Leaves strongly anisophyllous, smaller leaf of each pair minute and soon deciduous so that leaves can appear alternate; leafy stems often zigzagged.

Zambia. C: Mt Makulu Research Station, Chilanga, fl. 21.v.1979, *Critchett* 21/79 (K). **Zimbabwe.** C: Biegel 5896 (SRGH, not seen).

b) forma **isophylla** (Nees) J.R.I. Wood

Strobilanthes isophylla (Nees) T. Anderson. —Maroyi in Kirkia **18**: 194 (2006).

Strobilanthes persicifolia (Lindl.) J.R.I. Wood forma *isophylla* (Nees) J.R.I. Wood in Kew Bull. **64**: 28 (2009).

Leaves equal to only slightly unequal; leafy stems not markedly zigzagged.

Zambia. C: Lusaka, fl. 12.iii.1973, *Fanshawe* V15 (K, NDO). **Zimbabwe.** C: Harare, Burncoose, Enterprise road, fl. 15.vi.1976, *Biegel* 5319 (K, SRGH).

Wood (2009) remarks that all the wild plants are strongly anisophyllous and forma *isophylla* appears always to be cultivated.

The descriptions for the three *Strobilanthes* species are adapted from longer descriptions provided by J.R.I. Wood, to whom we are most grateful.

Strobilanthes auriculatus Nees var. **dyeriana** (Mast.) J.R.I. Wood. Aluminium Plant.

Strobilanthes dyerianus Mast. —Maroyi in Kirkia **18**: 194 (2006).

Shrub or subshrub; cystoliths present. Leaves variegated, dark green with deep pink or purple blotching above when young, becoming whitish on older leaves, elliptic, 12–30 cm long, attenuate and usually with sessile auriculate base, margin serrulate, apex acuminate, upper surface sparsely pubescent. Inflorescence a pedunculate axillary or terminal spike, subtended by a series of foliose bracts resembling reduced leaves; floral bracts imbricate, obovate, apex acute to emarginate, sometimes reflexed, surface densely pilose and/or pubescent including glandular hairs. Calyx with 5 linear-spathulate lobes, posterior 3 lobes often united in lower half, as densely hairy as bracts. Corolla 25–38 mm long, blue to pale purple, pubescent; tube curved, basal tube narrowly cylindrical, throat markedly ventricose, limb of 5 short subregular reflexed lobes. Stamens 4, didynamous, all included.

Zimbabwe. C: *Biegel* 4300 (SRGH, not seen). Native of Burma and Thailand; the species as a whole more widespread in Pakistan, India, Nepal, China, Burma, Thailand, Malaysia. Commonly cultivated in the tropics. This is a widely cultivated variety of *S. auriculatus* distinguished primarily by its variegated leaves.

Strobilanthes hamiltoniana (Steud.) Bosser & Heine. —Maroyi in Kirkia **18**: 194 (2006).

Perennial herb or shrub; cystoliths present. Leaf pairs somewhat unequal or subequal, blade elliptic, 5–19 cm long, base cuneate or uppermost leaves rounded, margin serrate, apex acuminate, surfaces glabrous, lateral veins prominent beneath. Inflorescence a ± large, lax terminal panicle, ultimate branches wiry, spreading at acute angles, glabrous; bracts and bracteoles < 5 mm long, oblong-elliptic or obovate, green, often faling early. Flowers shortly pedicellate. Calyx of 5 equal linear-oblong lobes with apex obtuse or emarginate. Corolla 30–40 mm long, glabrous, pale mauve to rose-purple, basal tube narrowly cylindrical, throat markedly ventricose, limb of 5 short subregular lobes, the lower 3 protruding; stamens 4, didynamous, all included.

Zambia. C: Lusaka, fl. 1982, *Sculsby* s.n. (K). **Zimbabwe**. C: Harare, Kia Ora nurseries, fl. 15.vi.1976, *Biegel* 5322 (K, SRGH). **Malawi**. C: Karonga Dist., Kayelekera camp, 30 km W of Karonga, fl. 12.vi.1989, *Brummitt* 18515 (K, MAL).

Native of India, Nepal, Bhutan, China and Burma. Commonly cultivated elsewhere.

Thunbergia battiscombei Turrill. —Maroyi in Kirkia **18**: 194 (2006). —Vollesen in F.T.E.A., Acanthaceae **1**: 58 (2008).

Perennial herb with several erect stems from woody rootstock; cystoliths absent. Leaves slightly fleshy, ovate to elliptic, to 17 cm long, subentire or with a few large teeth, apex acute to acuminate, base truncate to subcordate. Flowers in many-flowered axillary racemes; bracts minute; each flower clasped by 2 large ovate to elliptic bracteoles to 2.7 cm long, pale green with conspicuous green venation. Calyx with numerous broadly triangular segments. Corolla purple to royal blue; tube white, to 4.5 cm long, lobes 5, subequal, spreading, to 1.5 × 2 cm. Stamens 4, included in throat; anthers bithecous, one theca with long curved spur.

Zimbabwe. C: Harare, 11.xii.1974, *Biegel* 4460 (K, SRGH).

Native of East Africa. Cultivated within its native area.

Maroyi (2006: 194) describes this as an 'evergreen shrub'; it is always a herb.

Thunbergia grandiflora Roxb. —Maroyi in Kirkia **18**: 194 (2006). —Darbyshire *et al.* in F.T.E.A., Acanthaceae **2**: 733 (2010). Bengal Trumpet Vine or Blue Trumpet Vine.

Vigorous herbaceous twiner to 5 m, often totally covering walls, roofs or trees; cystoliths absent. Leaves triangular in outline, to 25 cm long, 7-veined from base, 2–4 large triangular teeth per side, apex acute to acuminate, base cordate. Flowers in pendulous racemoid cymes to 60 cm long; bracts minute; each flower clasped by 2 large obovate greenish bracteoles to 4 cm long with large black urn-like glands. Calyx an undulate puberulous rim. Corolla white or blue; tube to 2.5 cm long, broadly campanulate, lobes 5, subequal, spreading, to 3.5 × 3.5 cm. Stamens 4, included in throat; anthers bithecous, thecae pubescent, with 5 mm long basal spurs.

Zimbabwe. C: Harare, 11.xii.1973, *Biegel* 4459 (K, SRGH). **Malawi**. C: Lilongwe, 5.v.1984, *Patel et al.* 1507 (K, MAL).

Native of India and SE Asia. Widely cultivated elsewhere in tropical Asia and in Africa.

Thunbergia holstii Lindau. —Vollesen in F.T.E.A., Acanthaceae **1**: 40 (2008). — Darbyshire *et al.* in F.T.E.A., Acanthaceae **2**: 733 (2010).

Shrub to 1.5 m tall; cystoliths absent. Leaves ovate to elliptic with undulate margin, to 8 cm long, apex acuminate to rounded, base cuneate to rounded. Flowers solitary in leaf axils; each flower clasped by 2 large pale green ovate to elliptic bracteoles to 3.5 cm long. Calyx with numerous linear segments to 1 cm long. Corolla purple to dark purple or dark blue, tube to 4.5 cm long, lobes subequal, spreading, to 2.5 × 2.5 cm. Stamens 4, included in throat, anthers bithecous, thecae at base with a boss, covered with short stiff glossy setose hairs.

Zimbabwe. C: Goromonzi, 9.ii.1973, *Biegel* 4214 (K, SRGH).
Native of East Africa. Widely cultivated throughout the tropics.
It is thought that plants that have been grown widely in Africa under the names
Thunbergia erecta T. Anderson and *T. affinis* S. Moore all belong to this species,
and probably all originated from Kenyan material. The first confirmed record of
cultivation is from 1919 (*Battiscombe* 943); the label states "Common at Machakos.
Specimen cultivated at Nairobi". It is such an attractive plant that it soon spread and
by the early 1930s there were collections from all East African countries.

Thunbergia mysorensis (Wight) T. Anderson. —Maroyi in Kirkia **18**: 194 (2006). —
Darbyshire *et al.* in F.T.E.A., Acanthaceae **2**: 733 (2010).

Vigorous herbaceous twiner to 6 m, often totally covering walls or roofs; cystoliths absent.
Leaves ovate, to 15 cm long, 3-veined from base, irregularly dentate, apex acuminate to cuspidate,
base truncate. Flowers in pendulous racemoid cymes to 50 cm long; bracts minute, each flower
clasped by 2 large purplish brown ovate bracteoles to 2.5 cm long, partly fused along lower edge,
no large urn-shaped glands. Calyx an undulate glabrous rim. Corolla yellow outside, deeper
yellow inside and lower lip maroon distally, tube to 3 cm long, campanulate, upper lip of 2 erect
and lower of 3 deflexed lobes to 3 cm long. Stamens 4, held under upper lip, anthers bithecous,
thecae densely pubescent, with 5 mm long basal spurs.

Zimbabwe. C: Harare, 1.ii.1970, *Biegel* 3462 (K, SRGH).
Native of India. Occasionally cultivated in Africa.
Maroyi (2006) gives the name Lady's Slippers for this species, a name not mentioned
on any Kew specimen and a bit misleading since it is a long-established name for a
well-known orchid.

Whitfieldia elongata (P. Beauv.) De Wild. & T. Durand was recorded by Maroyi (2006:
194) as cultivated in Harare, Zimbabwe based on *Biegel* 4529 (SRGH).
Native in East Africa and Zambia and is fully treated in Flora Zambesiaca **8**(5): 175
(2013), but no mention was made of it having been cultivated elsewhere in the Flora
area.

24. NEURACANTHUS Nees[1]

Neuracanthus Nees in Wallich, Pl. Asiat. Rar. **3**: 76, 97 (1832); in De Candolle, Prodr.
11: 248 (1847). —Bidgood & Brummitt in Kew Bull. **53**: 1–76 (1998).

Perennial herbs including suffrutices with annual stems from woody rootstocks, subshrubs
or shrubs; cystoliths conspicuous or not. Leaves opposite, entire. Inflorescences axillary or
terminal, spicate to globose, sometimes subtended by bract-like leaves, sometimes with sterile
spinose inflorescences at base of or surrounding fertile inflorescences; bracts in 4 or 6 rows,
imbricate, with 3–7 prominent veins, each subtending a single sessile or subsessile flower.
Calyx 2-lipped, deeply divided between lips, dorsal lip 3-veined, ventral 2-veined, veins each
terminating in a tooth or the teeth fused into a single apex. Corolla puberulous outside in bands
along middle of lobes, inside hairy in a ventral band; basal cylindric tube short; throat broadly
funnel-shaped; lobes spreading to suberect; lower lip of 3 broad triangular lobes, upper lip of
1 shortly bifid lobe. Stamens 4, didynamous, inserted at base of throat, included, ventral pair
inserted below dorsal pair; ventral pair with bithecous anthers, dorsal pair with monothecous
anthers, thecae oblong, hairy. Style glabrous; stigma with a single flattened linear-oblong lobe,
upper lobe missing. Capsule 2–4-seeded, oblong or triangular-ovate, usually glabrous. Seeds
discoid, covered with hygroscopic hairs.

A genus of 30 species in tropical Africa, Madagascar, Arabia and tropical Asia,
especially numerous in NE and E Africa.

---

[1] By Kaj Vollesen

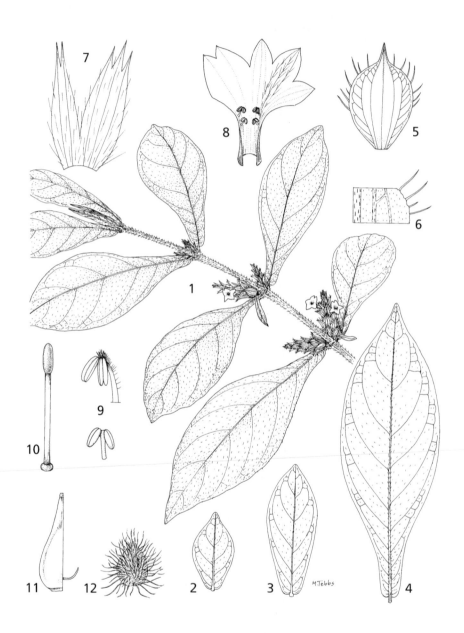

Fig. 8.6.**34**. NEURACANTHUS AFRICANUS. 1, habit (× ²/₃); 2–4, variation in leaf shape and size (× ²/₃); 5, bract (× 4); 6, detail of bract indumentum (× 8); 7, calyx opened up (× 4); 8, corolla opened up (× 4); 9, stamens (× 14); 10, style and stigma (× 14); 11, capsule (× 4); 12, seed (× 4). 1 & 7–10 from *Fanshawe* 8302, 2 & 5–6 from *Goodier* 855, 3 & 11 from *Fulwood* 1, 4 from *Fanshawe* 4530, 12 from *Bidgood* 1222. Drawn by Margaret Tebbs. Reproduced from Flora of Tropical East Africa (2010).

Suffrutescent herb with erect unbranched stems; inflorescences all terminal; lower and middle bracts 1.5–2 × 1–1.5 cm . . . . **2. decorus**
– Perennial herb or subshrub with branched stems, or if unbranched then with lateral inflorescences; lower and middle bracts 0.8–1.2 × 0.5–0.7 cm . . . .**1. africanus**

1. **Neuracanthus africanus** S. Moore in J. Bot. **18**: 37 (1880). —Clarke in F.T.A. **5**: 137 (1899). —Vollesen in F.T.E.A., Acanthaceae **2**: 289 (2010). Type: Mozambique, Lupata, iv.1860, *Kirk* 10 (K holotype). FIGURE 8.6.**34**.

   Neuracanthus africanus var. *limpopoensis* Bidgood & Brummitt in Kew Bull. **53**: 36 (1998). Type: Zimbabwe, Chiredzi Dist., Chipinda Pools, 20.i.1960, *Goodier* 855 (K holotype, SRGH).

Perennial herb, subshrub or shrub, often growing in clumps with several stems from a large woody rootstock; stems erect to decumbent, to 1.25 m long, simple or branched, sericeous to pubescent or sparsely so when young with broad curly multicellular hairs. Mature leaves sparsely curly-pubescent below on midrib, rarely glabrous, glabrous on lamina, with uniformly scattered sparse hairs above, rarely glabrous; petiole 1–5 mm long; lamina ovate to elliptic or subpandurate, largest 3–13(17.5) × 2–6.5 cm; apex acuminate to subacute, base truncate to subcordate; margin often concave above base. Flowers in axillary strobilate spikes, single or 2–3 per axil; peduncle 0–1(3) mm long, apically with a pair of foliaceous bracts to 1.5(2.5) cm long; spike 1–3(6) × 0.5–1 cm; bracts loosely imbricate, in 4 rows, 0.8–1.2 × 0.5–0.7 cm, green with (3)5–7 pale to dark brown veins and reticulation, broadly elliptic to broadly obovate, with conspicuous shoulders to almost wings below the spinescent acuminate tip, subglabrous to sparsely sericeous, with long stiff (rarely curly) hairs to 2 mm on margins, sometimes almost white from dense cystoliths. Calyx subglabrous to sericeous-puberulous on lamina, with few to many long strigose hairs, sometimes only on margins or near apex; lobes ovate or narrowly so, 9–12 × 2–4 mm, similar or dorsal slightly longer and wider, dorsal 3-toothed, ventral 2-toothed, teeth 2–3(4) mm long. Corolla white or white with pink to purple lines in throat, sericeous to puberulous, 8–13 mm long of which tube is 4–6 mm, limb subactinomorphic, 8–12 mm diameter. Larger anthers c.0.5 mm long. Capsule 2-seeded, ovoid, beaked, 8–10 mm long. Seeds cordiform, 4–5 × 3–4 mm.

Zambia. C: Luangwa Dist., Katondwe, fl. 4.ii.1964, *Fanshawe* 8302 (K). S: Mazabuka Dist., 55 km on Mochipapa–Sinazongwe road, fl. 2.iii.1960, *White* 7566 (FHO, K). **Zimbabwe**. N: Guruve Dist., 25 km W of Kanyemba, near Zambezi–Tunsa R. junction, 600 m, fl. 2.ii.1966, *Müller* 343 (K, SRGH). W: Hwange Dist., N escarpment of Lukosi R., 28.ii.1963, *Wild* 6061 (SRGH). S: Gwanda Dist., Bubye Ranch, 550 m, fl.& fr. 3.v.1958, *Drummond* 5548 (K, SRGH). **Malawi**. S: Chikwawa Dist., Livingstone Falls, W bank of Shire R., 100 m, fl. 21.iv.1970, *Brummitt* 10003 (K, MAL). **Mozambique**. N: Quissenga Dist., Lupangua Hill, 25 m, fr. 23.ii.2009, *Clarke* 137 (K, LMA, LMU). T: Moatize Dist., Lupata, fl.& fr. 20.iv.1860, *Kirk* 10 (K). MS: Machaze Dist., Massagena, L margin of Save R., 11.vi.1942, *Torre* 4307 (BR, LISC, LMU, LUA, PRE, WAG). M: Moamba Dist., Moamba, road to Ressano Garcia, fl. 3.xii.1940, *Torre* 2201 (LISC).

Also in Tanzania and South Africa (Limpopo). Mopane woodland and bushland, mixed woodland and bushland on rocky slopes, riverine grassland and thicket, often on heavy clay soils; (25)100–600 m.

Conservation notes: Widespread; not threatened.

Material from the Flora area belongs to subsp. *africanus*. There are two further subspecies from eastern Africa – subsp. *masaicus* (Bidgood & Brummitt) Vollesen from S Kenya and N Tanzania, and subsp. *ruahae* (Bidgood & Brummitt) Vollesen which is endemic to C Tanzania.

Typical material from northern parts of the Flora area with its large pandurate leaves is strikingly different from the small-leaved non-pandurate forms in the south. But a number of collections are intermediate and small-leaved pandurate forms occur in South Africa. If judged against the variation in material of subsp. *africanus* and subsp. *masaicus* from Tanzania, all southern African material can easily be accommodated within a single taxon.

2. **Neuracanthus decorus** S. Moore in J. Bot. **18**: 307 (1880). —Clarke in F.T.A. **5**: 138 (1899). —Vollesen in F.T.E.A., Acanthaceae **2**: 295 (2010). Type: Angola, Huíla, between Lake Ivantala and Quilengues, ii.1860, *Welwitsch* 5057 (BM holotype, G, K, P).

Subsp. **strobilinus** (C.B. Clarke) Bidgood & Brummitt in Kew Bull. **53**: 40 (1998). —Vollesen in F.T.E.A., Acanthaceae **2**: 295 (2010). Type: Malawi, Nyika Plateau, vii.1896, *Whyte* 138 (K holotype).

> *Neuracanthus strobilinus* C.B. Clarke in F.T.A. **5**: 138 (1899).

Perennial herb with a single (rarely 2) erect (rarely decumbent) stem from a creeping rootstock with fleshy roots; stems to 35 cm long, densely pubescent to tomentose when young with broad curly multicellular hairs. Mature leaves subglabrous to sparsely curly-pubescent below on midrib and major veins, with uniformly scattered sparse hairs above; petiole 1–3(6) mm long; lamina elliptic to obovate, largest 6–10.5(16) × 2–4 cm; apex acute to rounded, base cuneate to truncate. Flowers in a single dense strobilate quadrangular terminal spike; peduncle 1–4(8) mm long, no sterile bracts; spike 3–7.5 × 2–3 cm; bracts densely imbricate, in 4 rows, 1.5–2 × 1–1.5 cm, pale green with 5–7 prominent darker veins and reticulation, broadly ovate to obovate to almost orbicular, acute to acuminate, minutely puberulous on veins and with dense long stiff tubercle-based pale brown hairs to 2 mm long on margins (sparser on veins and lamina). Calyx with long pale brown strigose hairs to 2 mm, densest near base and on margins, lobes ovate, 9–14 × 3–4 mm, similar or dorsal slightly longer and wider, dorsal 3-toothed, ventral 2-toothed, either with short (1–3 mm) teeth or divided about halfway down. Corolla white or white with pale pink veins, sericeous outside (puberulous on lobes), 7–10 mm long of which tube is 2–3 mm, limb subactinomorphic, 5–6 mm diameter. Anthers not seen. Capsule 2-seeded, narrowly ovoid, beaked, 9–11 mm long. Seed cordiform, 4–5 × c.3 mm.

Zambia. W: Kitwe, fr. 2.vi.1955, *Fanshawe* 2313 (EA, K). **Malawi**. N: Nyika Plateau, 1525 m, fl.& fr. vii.1896, *Whyte* 138 (K).

Also in D.R. Congo and Tanzania. In woodland; 1100–1550 m.

Conservation notes: Not threatened.

Subsp. *decorus* from Angola has narrower (1.5–2(2.5) cm) spikes and smaller (8–15 × 5–10 mm) bracts with rusty brown hairs. See discussion in Bidgood & Brummitt (1998) on the differences between these two taxa; the material of subsp. *decorus* is all old and poor.

25. CRABBEA Harv.[2]

Crabbea Harv. in London J. Bot. **1**: 27 (1842), conserved name. —Nees in De Candolle, Prodr. **11**: 162 (1847). —Thulin in Nordic J. Bot. **24**: 501–506 (2007).

> *Crabbea* Harv., Gen. S. Afr. Pl.: 276 (1838), rejected name.

Perennial or subshrubby herbs, sometimes acaulescent; cystoliths conspicuous. Leaves opposite, entire to crenate-dentate. Flowers in dense axillary sessile or pedunculate heads composed of several racemoid scorpioid 3–5-flowered cymes on flattened axes, surrounded by numerous large spiny-bristly or entire bracts which get gradually smaller inwards, peduncle with 2 sterile bracts at apex. Calyx glumaceous, deeply divided into 5 subequal or unequal (dorsal sepal much larger than others) sepals. Corolla hairy outside, usually also glandular, rarely glabrous, subactinomorphic (lobes fused higher up in upper lip), in bud the 3-lobed lower lip folded over the 2-lobed upper lip; basal cylindric tube widened into a long cylindric throat; lobes oblong to rounded. Stamens 4, didynamous, inserted at base of throat, included; anthers bithecous, thecae of slightly different lengths, rounded, line of long white hairs on both sides of aperture. Ovary with 2–4 ovules per locule; style glabrous, articulated at base; lower stigma lobe rhomboid-rounded, upper missing or a small tooth. Capsules 4–8-seeded, oblong (rarely ovoid), glossy. Seeds discoid, covered with hygroscopic hairs.

---

[2] By Kaj Vollesen

A genus of 10–15 species from Ethiopia and Somalia through eastern tropical Africa to South Africa. The species are closely related and often difficult to separate.

1. Acaulescent or sub-acaulescent herb with leaves appressed to ground 2
 – Plant with clearly defined aerial stem, leaves not appressed to ground. 3
2. Leaves in 2(3) pairs, less than twice as long as wide; bracts entire, less than 3 times as long as wide . . . . **6.** *kaessneri*
 – Leaves in 3–5 pairs, more than 3 times as long as wide; bracts with bristly-toothed margin, more than 3 times as long as wide . . . . **7.** *zambiana*
3. Corolla pink with yellow gibbose patches in throat; bracts on all inflorescences entire . . . . **2.** *coerulea*
 – Corolla white with yellow gibbose patches in throat; bracts with bristly-toothed margin, rarely some or all inflorescences with entire bracts 4
4. Peduncle with stalked capitate glands; corolla c.30 mm long of which basal tube is c.12 mm long and throat c.18 mm, with scattered capitate glands in bud but without hairs . . . . **3.** *glandulosa*
 – Peduncle without stalked capitate glands; corolla 11–22 mm long of which basal tube is 3–7 mm long and throat 8–15 mm, hairy in bud, usually also glandular outside 5
5. Dorsal calyx lobe 2–3 mm wide, distinctly wider than others; capsule 8–10 mm long. . . . . **1.** *velutina*
 – Dorsal calyx lobe 1–1.5 mm wide, ± same width as others; capsule 10–15 mm long. 6
6. Inflorescence subsessile or with peduncle to 0.5(1) cm long. . . . . **5.** *cirsioides*
 – Inflorescence with peduncle 1–5(15) cm long . . . . **4.** *nana*

1. **Crabbea velutina** S. Moore in J. Bot. **32**: 135 (1894). —Clarke in F.T.A. **5**: 119 (1899). —Vollesen in F.T.E.A., Acanthaceae **2**: 297 (2010). Types: Kenya, Mombasa, 1888, *Gregory* s.n. (BM syntype); Mombasa, 1888, *Taylor* s.n. (BM syntype). FIGURE 8.6.35.
 Crabbea reticulata C.B. Clarke in F.T.A. **5**: 119 (1899). Types: Kenya, "Ukamba", 1893-94, *Scott Elliot* 2309 (K syntype, BM); Tanzania, Bukoba Dist., Karagwe, viii.1893-94, *Scott Elliot* 8147 (K syntype, BM).

Perennial herb with 1 to several erect to decumbent stems from a branched creeping rhizome, roots not fleshy; stems 2–20(40 in decumbent plants) cm long, sparsely to densely strigose-pubescent when young, rarely puberulous. Leaves subglabrous to strigose-pubescent, densest along veins below, often glabrous on lamina, uniformly so above; petiole 0–2 cm long; lamina elliptic to obovate or narrowly so, largest 6–21 × 2–6.5 cm; apex subacute to broadly rounded, base attenuate, often decurrent to stem; margin crenate, sometimes indistinctly. Heads 1–3 cm across; peduncle (0.5)1–9 cm long, from sparsely puberulous with a few scattered long hairs to densely strigose-pubescent; outer bracts green, ovate or narrowly so, 1.5–3.5 × 0.8–1.6 cm, sparsely puberulous to sparsely pubescent, at least along veins, subacute to acuminate, varying from entire to densely bristly-spinose with bristles to 1.2 cm. Calyx puberulous at base and along veins, distinctly ciliate, dorsal lobe distinctly longer and wider, (8)10–17 × 2–3 mm, ventral and lateral 1–3 mm shorter, all acute to acuminate. Corolla white with yellow gibbous patches in throat, 11–22 mm long of which cylindric tube is 3–7 mm and throat 8–15 mm long; lobes 5–11 × 4–9 mm. Filaments 2–4, 3–5 mm long; anthers 1–2 mm long. Capsule 4–8-seeded, oblong, 8–10 mm long. Seeds rhomboid, 2.5–3 × 1.5–2 mm.

Caprivi. E Caprivi, Impalila (Mpilila) Is., 925 m, fl. 13.i.1959, *Killick & Leistner* 3351 (K, PRE). **Botswana**. N: Ngamiland Dist., Kwebe Hills, fl.& fr. 26.xii.1977, *P.A. Smith* 2163 (C, K, LISC, SRGH). **Zambia**. C: Chongwe Dist., Great East Road, 20 km W from Luangwa Bridge, 650 m, fl. 1.i.1973, *Kornaś* 2913 (K). E: Nyimba Dist., Great East Road between Hofmeyr turn-off and Kachalolo, fl.& fr. 12.xii.1958, *Robson* 915 (BM, K, LISC). S: Sinazongwe Dist., Siazwela Village, 650 m, fl. 26.ii.1997, *Luwiika et al.* 509 (K, MO). **Zimbabwe**. N: Kariba Dist., Kariba Nat. Park, 30 km on Makuti–Kariba road,

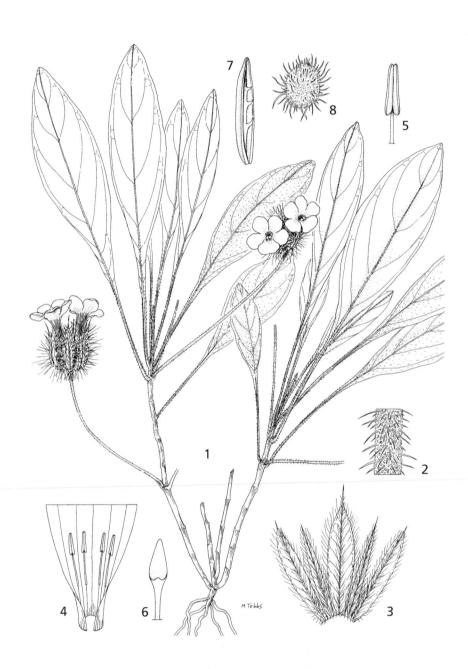

Fig. 8.6.**35**. CRABBEA VELUTINA. 1, habit ($\times$ $^2/_3$); 2, detail of stem indumentum ($\times$ 4); 3, calyx opened up ($\times$ 3); 4, corolla tube opened up with stamens ($\times$ 2); 5, anther ($\times$ 8); 6, apical part of style and stigma ($\times$ 12); 7, capsule ($\times$ 4); 8, seed ($\times$ 6). 1 & 3–7 from *Bidgood & Lovett* 256, 2 from *Kerfoot* 443, 8 from *Verdcourt* 815. Drawn by Margaret Tebbs. Reproduced from Flora of Tropical East Africa (2010).

925 m, fl. 19.ii.1981, *Philcox et al.* 8771 (C, K). W: Hwange Dist., Victoria Falls, 875 m, fl. 8.i.1979, *Mshasha* 154 (K, SRGH). C: Kwekwe Dist., 8 km NE of Kwekwe (Que Que), Sable Park, fl., 13.i.1976, *Chipunga* 96 (K, SRGH). E: Mutare Dist., Matika's Kloof, fl.& fr. 8.i.1950, *Chase* 1876 (BM, K, LISC, SRGH). S: Chiredzi Dist., Chipinda Pools, fl. 19.xi.1959, *Goodier* 724 (K, SRGH). **Malawi.** S: Mangochi Dist., 2 km SW of Monkey Bay, 525 m, fl. 20.ii.1982, *Brummitt et al.* 16006 (K, MAL). **Mozambique.** N: Marrupa Dist., road to Mucuaiaia, near Momopsus, fl. 23.ii.1981, *Nuvunga* 686b (K, LMU). Z: Caia Dist., near mouth of Shire R., Chamo (Shamo), fl. 1863, *Kirk* s.n. (K). T: Moatize Dist., Lower Zambesi R., Lupata, fl.& fr. iv.1860, *Kirk* s.n. (K). MS: Machaze Dist., Madanda Forest, 125 m, fl. 5.xii.1906, *Swynnerton* 1142 (BM).

Also in Sudan, Ethiopia, Somalia, Uganda, Kenya, D.R. Congo, Rwanda, Burundi, Tanzania, Swaziland and South Africa. In a wide range of woodland, wooded grassland, bushland, grassland, riverine forest and thicket, rocky hills, on a wide range of soils, often in disturbed places and in secondary vegetation; 100–1300 m.

Conservation notes: Widespread; not threatened.

It is quite common that the first inflorescence produced has entire bracts and the following have toothed bracts; only rarely do the following inflorescences have entire bracts. However, I have seen a few specimens from Zambia where all inflorescences have entire bracts. They can be distinguished from *C. coerulea* by having puberulous (not strigose-pubescent) peduncles and by the white (not pink) corolla.

2. **Crabbea coerulea** Vollesen, sp. nov. Differs from *C. velutina* in the always entire bracts and the pink (not white) corolla which drops at the faintest touch. The corolla also has a longer (7–10 mm not 3–7 mm) basal tube. Type: Malawi, Karonga Dist., 25 km W of Karonga on Stevenson Road, 3.i.1974, *Pawek* 7739 (K holotype, MAL, MO, SRGH).

Perennial herb with 1 to several erect to decumbent stems from a branched creeping rhizome, roots not fleshy; stems 2–5 cm long, sparsely to densely sericeous-pubescent. Leaves sparsely puberulous to pubescent below along midrib, sparser or subglabrous on lamina, with uniformly scattered hairs above; petiole 0.5–1(2.5) cm long; lamina elliptic to obovate, largest 4–7.5(13) × 1.5–2.5(4) cm; apex subacute to broadly rounded, base attenuate, decurrent; margin indistinctly crenate. Heads 1–2.5 cm across; peduncle 2–7 cm long, sparsely to densely strigose sericeous-pubescent; outer bracts green, ovate-elliptic or narrowly so, 1.7–3 × 0.8–1.4 cm, pubescent or sparsely so along midrib and on edges, sparser on lamina, acute, with entire margin. Calyx glabrous or with a few downwardly directed hairs at base, upwards with long hairs on midrib (a few on lamina), distinctly ciliate; dorsal lobe distinctly longer and wider, 9–14 × 1.5–3(4) mm, ventral and lateral 1–2 mm shorter, all acute or subacute. Corolla dropping easily, pink with yellow gibbous patches in throat, 19–25 mm long of which cylindric tube is 7–10 mm and throat 12–15 mm long; lobes 9–13 × 8–11 mm. Filaments 2–4 and 3–5 mm long; anthers c.2 mm long. Capsule 6-seeded, ovoid, 10–11 mm long. Seeds not seen.

Malawi. N: Karonga Dist., 35 km NW of Karonga on Stevenson road, 925 m, fl. 26.iv.1977, *Pawek* 12700 (K, MAL, MO, SRGH); Chitipa Dist., 30 km W of Karonga, Thulwe Hills, fr. 25.v.1989, *Pope et al.* 2343 (K, MAL).

Not known elsewhere. *Brachystegia* woodland on stony or gravelly soil, often on hillsides; 500–1200 m.

Conservation notes: Known only from nine collections in N Malawi; possibly Near Threatened.

This pretty species is known only from Chitipa, Karonga and Mzimba Districts in N Malawi. It is most easily distinguished from *C. velutina* (which so far has not been collected in N Malawi) by the constantly entire bracts and pink corolla with longer basal tube. The corolla is also said to drop at the faintest touch. It is a species of *Brachystegia* woodland, a habitat where *C. velutina* is very rarely seen.

3. **Crabbea glandulosa** Vollesen, sp. nov. Differs from *C. velutina* in having capitate glandular hairs on apical part of stems and peduncles and in the larger corolla (c.3 cm not 1.1–2.2 cm long), of which the basal tube is c.1.2 cm not 0.3–0.7 cm long, and throat is c.1.8 cm not 0.8–1.5 cm long. Type: Zambia, Muchinga escarpment, Mutumba's Village, 18.iii.1988, *Phiri* 2174 (K holotype, UZL).

Perennial herb with a single stem from an unbranched creeping rhizome, roots not fleshy; stems to 20 cm long, basal part creeping and buried, apical part erect, buried part densely sericeous, aerial part sparsely puberulous with scattered stalked capitate glands towards apex. Leaves sparsely puberulous below on midrib and on edges, with uniformly scattered hairs above; petiole 0–2 cm long; lamina elliptic, largest 10.5–12 × 4–6 cm; apex subacute, base attenuate, decurrent, often to stem; margin crenate. Heads 2–3 cm across; peduncle 0.5–1 cm long, puberulous, sometimes with scattered longer hairs, also with dense stalked capitate glands; outer bracts green, ovate, 2.5–3.5 × 1.3–1.8 cm, subglabrous or with scattered long hairs on midrib and edges, subacute to rounded, densely bristly-spinose with teeth to 7 mm. Calyx with scattered strigose hairs along midrib and on edges in upper part, at base with scattered stalked capitate glands; dorsal lobe distinctly longer but only slightly wider, 15–17 × 2.5 mm, ventral and lateral 2–4 mm shorter, all acuminate. Corolla white with yellow gibbous patches in throat, c.30 mm long of which cylindric tube is c.12 mm and throat is c.18 mm long, with scattered capitate glands but no hairs; lobes c.12 × 11 mm. Filaments c.3 and 5 mm long; anthers c.2.5 mm long. Capsule and seeds not seen.

Zambia. N: Mpika Dist., North Luangwa Nat. Park, 11°38'S 32°04'E, 1200 m, fl. 15.i.1995, *P.P. Smith* 740 (K). C: Mpika Dist., Muchinga escarpment, Mutuma's Village, 750 m, fl. 18.iii.1988, *Phiri* 2174 (K, UZL).

Not known elsewhere. *Brachystegia* woodland on rocky slopes; 750–1200 m.

Conservation notes: Known only from central Zambia; possibly Near Threatened.

Known only from these two collections. Differs most conspicuously from *C. velutina* in the glandular indumentum (most conspicuous on the peduncles) and the larger corolla which has an indumentum of stalked glands only. It is the only species of *Crabbea* which has stalked capitate glands on the vegetative parts.

4. **Crabbea nana** (Nees) Nees in De Candolle, Prodr. **11**: 162 (1847). —Clarke in Fl. Cap. **5**: 38 (1901) in part. Type: South Africa, E Cape, Queenstown, Swart Kei R., n.d., *Zeyher* s.n. (K lectotype, GZU), lectotypified here.

Ruellia nana Nees in Linnaea **15**: 355 (1841).

Crabbea pedunculata C.B. Clarke in Fl. Cap. **5**: 40 (1901). Types: South Africa, KwaZulu-Natal, Inanda, vii.1879, *Wood* 365 (K lectotype, SAM); Krantz Kloof, n.d., *Schlecter* 3210 (B† syntype); KwaZulu-Natal, no locality, 1860, *Sanderson* 466 (K syntype), lectotypified here.

Crabbea galpinii C.B. Clarke in Fl. Cap. **5**: 40 (1901). Type: South Africa, Mpumalanga, Barberton, iv.1890, *Galpin* 1148 (K holotype).

Perennial herb with 1–5 erect stems from a branched creeping rhizome, roots fleshy; stems to 30 cm long, subglabrous to puberulous when young, with scattered to dense strigose hairs. Leaves strigose-pubescent or sparsely so below, densest on midrib and edges, subglabrous above to sparsely and uniformly puberulous; petiole 0–5(10) mm long; lamina narrowly elliptic to narrowly obovate, largest (3)4–8(10.5) × 1–2.5(3) cm; apex subacute to rounded, base attenuate, decurrent, often to stem; margin entire to crenate. Heads 1.5–3 cm across; peduncle 1–5(15) cm long, with scattered to dense long hairs; outer bracts green, narrowly ovate, 2.5–4.5 × 0.8–1.7 cm, subglabrous or with scattered long hairs on midrib and edges, acute to cuspidate, margin entire to densely bristly-spinose with bristles to 6 mm. Calyx with sparse to dense strigose hairs, at apex only or to base along midrib and margins; dorsal lobe distinctly longer but only slightly wider, 13–20 × 1–1.5 mm, ventral and lateral 2–3 mm shorter, all cuspidate. Corolla white with yellow gibbous patches in throat, 12–18 mm long of which cylindric tube is 4–5 mm and throat is 8–13 mm long; lobes 4–6 × 3–5 mm. Filaments c.3 and 4 mm long; anthers c.1.5 mm long. Capsule 4-seeded, oblong, 11–12 mm long. Seeds not seen.

Mozambique. M: Matutuine Dist., Catuane, fr. 29.ix.1983, *Zunguze et al.* 580 (K, LMU);

Namaacha Dist., Goba, Changalane, fl.& fr. 20.x.1983, *Zunguze et al.* 708a (K, LMU).
Also in E South Africa and Swaziland. Alluvial grassland and riverine scrub; 50–100 m.
Conservation notes: Local in Flora area; probably not threatened.

5. **Crabbea cirsioides** (Nees) Nees in De Candolle, Prodr. **11**: 163 (1847). —Bandeira
et al., Fl. Nat. Sul Moçamb.: 190 (2007). Type: South Africa, KwaZulu-Natal, "Katri
River, Beaufort, Mt Chumi", *Ecklon* (not traced).

Ruellia cirsioides Nees in Linnaea **15**: 354 (1841).
Crabbea hirsuta Harv. in Hooker, London J. Bot. **1**: 27 (1842). —Nees in De Candolle,
Prodr. **11**: 163 (1847). —Clarke in Fl. Cap. **5**: 39 (1901). —Mapura & Timberlake, Checklist
Zimb. Vasc. Pl.: 13 (2004). Type: South Africa, KwaZulu-Natal, Durban (Port Natal), n.d.,
Williamson s.n. (K holotype).
Crabbea angustifolia Nees in De Candolle, Prodr. **11**: 163 (1847). —Clarke in Fl. Cap. **5**:
39 (1901). —Setshogo, Checklist Pl. Botswana: 18 (2005). Type: South Africa, Gauteng,
Magaliesberg, n.d., *Burke* 405 (K holotype, PRE, SAM, TCD), see note.
Crabbea ovalifolia Ficalho & Hiern in Trans. Linn. Soc., Bot. **2**: 24, t.6 (1881). Type:
Angola, Ninda R., viii.1878, *Serpa Pinto* 21 (LISU holotype).
Crabbea undulatifolia Engl. in Bot. Jahrb. Syst. **10**: 263 (1889). Type: Namibia, Grootfontein,
ii.1886, *Marloth* 1079 (B† holotype).
Crabbea nana sensu Clarke in F.T.A. **5**: 118 (1900) in part & in Fl. Cap. **5**: 38 (1901) in
part. —Binns, Checklist Herb. Fl. Malawi: 13 (1968), non Nees (1847).
Crabbea robusta C.B. Clarke in Fl. Cap. **5**: 39 (1901). Type: Swaziland, Horo Concession,
xii.1890, *Galpin* 1265 (K holotype).

Perennial herb with 1–5 decumbent to prostrate (rarely erect) simple or branched stems from
a branched creeping rhizome, roots fleshy; stems to 40 cm long, subglabrous to puberulous
when young, with scattered to dense strigose-pubescent hairs. Leaves sparsely puberulous to
sparsely strigose-pubescent below on midrib and edges, occasionally on lateral veins, glabrous
or sparsely puberulous above on midrib; petiole 0–5(10) mm long; lamina lanceolate to elliptic
or elliptic-obovate, largest 7–12 × 0.7–3.3 cm; apex subacute to broadly rounded, base attenuate,
decurrent, often to stem; margin entire to crenate. Heads 1.5–4(5) cm across; peduncle 0–5(10)
mm long, puberulous or densely so; outer bracts green, narrowly ovate, 2.5–4(5) × 0.7–1.5 cm,
puberulous or sparsely so, with scattered longer hairs along midrib and on edges, cuspidate,
margin entire to densely bristly-spinose with bristles to 7 mm. Calyx with sparse to dense strigose
hairs at apex only or to base along midrib and margins, dorsal lobe distinctly longer but not or
only slightly wider, 16–21 × 1–1.5 mm, often with 1–2 teeth per side near apex, ventral and lateral
lobes 2–3 mm shorter, all cuspidate. Corolla white with yellow gibbous patches in throat, 12–15
mm long of which cylindric tube is 4–5 mm and throat is 8–10 mm long; lobes 4–5 × 4–5 mm.
Filaments c.3 and 4 mm long; anthers c.1.5 mm long. Capsule 4–6-seeded, ovoid-oblong, 10–14
mm long. Seeds square in outline, c.3 × 3 mm.

Botswana. SE: South East Dist., Gaborone, 1000 m, fl. 1.i.1974, *Mott* 107b (K). **Zambia**.
C: Lusaka Dist., 25 km SE of Lusaka, Kanyanja, 1300 m, fl. 25.ii.1996, *Bingham & Harder*
10932 (K). **Zimbabwe**. N: Mazoe Dist., 10 km W of Mvurwi (Umwukwes), fl. 10.ii.1982,
Brummitt & Drummond 15842 (K, LISC). W: Matobo Dist., Besna Kobila farm, 1475 m,
fl. ii.1955, *Miller* 2683 (K, SRGH). C: Harare Dist., Ruwa R., fl. 14.ii.1971, *Linley* 613
(K, LISC, SRGH). E: Chipinge Dist., Ngungunyana Forest Res., 1075 m, fl. iii.1966,
Goldsmith 2/66 (K, LISC, SRGH). S: Masvingo Dist., Kyle Nat. Park, Mutunhumushava
kopje, 1075 m, fl.& fr. 8.iv.1971, *Basera* 319 (K, SRGH). **Malawi**. C: Ntchisi Dist., between
Kongwe and Mwera Hill, 1400 m, fl. 21.ii.1959, *Robson & Steele* 1696 (BM, K, LISC). S:
Chikwawa Dist., Manganja Hills, Mbame, 925 m, fl. vii.1861, *Kirk* s.n. (K). **Mozambique**.
M: Namaacha Dist., Mt M'ponduine, fl. 4.iii.2001, *Bolnick* s.n. (K).

Also in Angola, Namibia, South Africa and Swaziland. Grassland, bushland and
woodland on sandy to stony soil, rocky slopes, often on dolomitic rocks; 900–1500
(?1850) m.
Conservation notes: Widespread; not threatened.

Nees (1841) erroneously cites the type of *C. angustifolia* as having been collected by Burchell.

6. **Crabbea kaessneri** S. Moore in J. Bot. **48**: 252 (1910). —Vollesen in F.T.E.A., Acanthaceae **2**: 300 (2010). Type: D.R. Congo, Katanga (Shaba), Tonkoshi, 14.i.1908, *Kässner* 2337 (BM holotype, K).

Acaulescent or sub-acaulescent perennial herb with unbranched (rarely branched) creeping rhizome, roots fleshy; aerial stems absent but with short subterranean stems to 2 cm long, apical part of rhizome and subterranean stems yellowish sericeous-tomentellous. Leaves in 2(3) decussate pairs appressed to ground, yellowish puberulous or sparsely so below on midrib and lateral veins, glabrous on lamina, not ciliate, glabrous or with scattered pale yellow hairs above; petiole 0–1 cm long, yellowish strigose pubescent; lamina ovate to elliptic or broadly so, largest 5–12 × 3.5–8 cm, less than twice as long as wide; apex broadly rounded, rarely obtuse or retuse; base attenuate, decurrent, often to stem; margin subentire to crenate. Heads 1.5–4.5 cm across; peduncle 0–6(10) mm long, densely yellowish strigose-pubescent; outer bracts yellowish green, rarely purplish tinged, ovate to elliptic or broadly so, 2–3.5 × 0.8–2.5 cm, usually less than twice as long as wide, subglabrous to sericeous-pubescent along midrib and on veins, acute to broadly rounded, margin entire to dentate, but never bristly. Calyx apically with sparse to dense strigose hairs, with sparse to dense short capitate glands at base, dorsal lobe longer and much wider, 12–15 × c.2 cm, ventral and lateral lobes 2–4 mm shorter, all acuminate to cuspidate. Corolla white with yellow gibbous patches in throat, 21–32 mm long of which cylindric tube is 7–12 mm and throat is 14–20 mm long; lobes 8–10 × 8–10 mm. Filaments 2–4 and 4–5 mm long; anthers 1.5–2 mm long. Capsule (immature) ellipsoid, c.13 mm long. Seeds not seen.

Zambia. N: Mbala Dist., Kalambo Falls, 1200 m, fl. 9.ii.1965, *Richards* 19631 (K). W: Solwezi Dist., 59 km W of Mutanda, 18 km W of Muheba R., 1400 m, fl. 19.i.1975, *Brummitt et al.* 13855 (K).

Also in D.R. Congo and Tanzania. *Brachystegia* woodland on sandy to stony or loamy soil, rocky outcrops; 1200–1750 m.

Conservation notes: Local in the Flora area, but not threatened.

7. **Crabbea zambiana** Vollesen, sp. nov. Differs from *C. kaessneri* in the larger (14–19 not 5–12 cm long) narrower (more than three times as long as wide) leaves, narrower (more than three times as long as wide) bristly (not entire or dentate) bracts, shorter (dorsal lobe c.11 not 12–15 mm long) calyx and smaller (c.5 × 5 not 8–10 × 8–10 mm) corolla lobes. Type: Zambia, Kaputa Dist., Kalungwishi R., Kundabwika Falls, 17.iv.1989, *Pope et al.* 2165 (K holotype, BR, K, LISC, SRGH, UZL).

Acaulescent or sub-acaulescent perennial herb with an unbranched or branched creeping rhizome, roots fleshy; aerial stems absent or up to 1 cm long, also with short subterranean stems, stems and apical part of rhizome sericeous-pubescent. Leaves in 3–5 decussate pairs appressed to ground, sparsely puberulous below on midrib and margins, glabrous on lamina, glabrous above or with a few hairs on midrib; petiole 0–1.5 cm long; lamina elliptic, largest 14–19 × 4.5–6 cm, more than three times as long as wide; apex acute or subacute, base attenuate, decurrent, often to stem; margin crenate. Heads 1.5–3 cm across; peduncle 0–5 mm long, densely sericeous-puberulous; outer bracts green or with purplish venation, oblong-elliptic, 2.5–3.5 × 0.7–1.2 cm, ± three times as long as wide, sparsely sericeous-puberulous, densest on midrib and veins, acute or subacute, margin with numerous weak bristles to 6 mm. Calyx densely sericeous-strigose all over, also with dense stalked capitate glands towards base, dorsal lobe longer and much wider, c.11 × 2 mm, ventral and lateral c.3 mm shorter, all acuminate. Corolla white with yellow gibbous patches in throat, c.25 mm long of which cylindric tube is c.11 mm and throat is c.14 mm long; lobes c.5 × 5 mm. Filaments c.3 and 5 mm long; anthers c.2 mm long. Capsule and seeds not seen.

Zambia. N: Kaputa Dist., Kalungwishi R., Kundabwika Falls, fl. 17.iv.1989, *Pope et al.* 2165 (BR, K, LISC, SRGH, UZL). C: Mpika Dist., Mutinondo Wilderness Area, 1450 m, fl. 22.ii.2009, *Merrett* 371 (K).

Also in D.R. Congo. *Brachystegia* woodland and dry riverine forest; 1000–1450 m.
Conservation notes: Known only from these two collections, plus several from
Katanga; possibly Vulnerable.

In its acaulescent habit it seems to be closest to *C. kaessneri*, but it differs in much
larger narrower leaves, narrower bristly bracts, smaller calyx and smaller corolla lobes.

26. LEPIDAGATHIS Willd.[3]

Lepidagathis Willd., Sp. Pl., ed.4: 400 (1800). —Nees in De Candolle, Prodr. **11**: 249
(1847).
Volkensiophyton Lindau in Bot. Jahrb. Syst. **20**: 27 (1894).
Lindauea Rendle in J. Bot. **34**: 411, t.362 (1896). —Clarke in F.T.A. **5**: 129 (1899).

Perennial herbs and subshrubs. Stems 4-angular to subterete. Leaves evergreen or deciduous,
opposite-decussate, sessile or base decurrent into petiole, blade often with minute sunken glands
on lower surface; cystoliths numerous, ± conspicuous. Inflorescences of unilateral (scorpioid)
fasciculate to spiciform cymes, axillary and/or terminal, often compounded into a dense
synflorescence, rarely a dichasial thyrse or flowers solitary; bract pairs equal to dimorphic, free,
sterile bracts in unilateral cymes typically imbricate and subtending flowers, fertile bracts and
bracteoles usually adpressed to calyx; bracteoles paired, (sub)equal, free. Calyx divided almost to
base, unequally 5-lobed, posterior lobe broader than remaining lobes, anterior pair sometimes
partially fused, lateral (inner) pair linear-lanceolate. Corolla bilabiate; tube cylindrical in lower
portion, throat campanulate, base of throat ± densely hairy within; upper lip hooded, straight
or arcuate, apex shortly bilobed or shallowly emarginate; lower lip with apex 3-lobed, palate
upraised with a central furrow. Stamens 4, didynamous, filaments arising from near base of
corolla throat; anterior pair of stamens longer, bithecous, thecae at an equal height or offset;
posterior pair shorter, monothecous or bithecous. Ovary bilocular, 1 or 2 ovules per locule; style
filiform; stigma capitate-bilobed or one lobe expanded and spoon-shaped. Capsule compressed,
2-seeded, face then ovate, or 4-seeded, face then oblong-elliptic, sometimes with a short sterile
apical beak, placental base inelastic. Seeds held on retinacula, discoid or flattened-triangular,
clothed in hygroscopic hairs.

A genus of around 100 species with a tropical and subtropical distribution.

Two infrageneric groups are separable in the Flora area on the basis of variation in
anther characteristics, capsule shape, number of seeds and leaf venation. These largely
correspond to Clarke's two subgenera (F.T.A. **5**: 120, 1899) which he separated purely
on whether the posterior stamens have bithecous (*Eulepidagathis*) or monothecous
(*Neuracanthopsis*) anthers. In both groups, the individual cymose inflorescence units
are unilateral, one bract of each pair being sterile; these cymes are often compounded
into complex synflorescences. The third group recognised in the F.T.E.A. treatment
(*Teliostachya*), in which all bracts are fertile and the inflorescence is a terminal thyrse,
is so far not known from the Flora area, though *L. alopecuroides* (Vahl) Griseb. could
possibly occur in the wet riverine forests of NW Zambia.

1. Leaves pinnately veined; posterior (shorter) pair of stamens monothecous or with
    a vestigial second theca, thecae glabrous; capsule face oblong-elliptic, potentially
    4-seeded (Group A, *Neuracanthopsis*). 2
–   Leaves either prominently 3(5)-veined from base, or if linear then sometimes
    only midrib prominent; stamens all bithecous, thecae with short hairs and glands
    dorsally and with cilia along suture; capsule face ovate, 2-seeded (Group B,
    Lepidagathis) 8

---

[3] By Iain Darbyshire

2. Stem and young leaves with dense stellate or dendritic indumentum . . . **2.** *scariosa*
 – Stem and leaves with simple hairs only or glabrous 3
3. Bracts and calyx lobes aristate; anterior pair of calyx lobes fused for 1–2.5 mm
  beyond short calyx tube; posterior calyx lobe margin not inrolled nor partially
  enveloping other lobes; corolla 11.5–14.5 mm long . . . . **1.** *glandulosa*
 – Bracts and calyx lobes not aristate; anterior pair of calyx lobes free above short
  calyx tube; posterior calyx lobe margin ± inrolled, often partially enveloping
  other lobes; corolla 13.5–40 mm long 4
4. Inflorescence fasciculate, axillary; calyx lobes longer than and clearly extending
  beyond bracts. . . . . **6.** *nemorosa*
 – Inflorescence a contracted unilateral spike, sometimes strobilate, axillary and/
  or terminal; calyx lobes subequal in length to and largely hidden behind fertile
  bracts 5
5. Corolla (30)35–40 mm long including lobes of lower lip 12.5–21 mm long;
  capsule 17–20 mm long. . . . . **7.** *macrochila*
 – Corolla 13.5–28 mm long including lobes of lower lip 3–10 mm long; capsule
  7–12 mm long 6
6. Sterile bracts straight, ± adpressed to inflorescence axis, brown-scarious at least
  in distal half, sometimes green below; night-flowering; palate of lower corolla lip
  shortly pubescent. . . . . **5.** *pallescens*
 – Sterile bracts falcate, diverging from inflorescence axis, green at first, only later
  turning brown; day-flowering; palate of lower corolla lip glabrous 7
7. Plant caulescent; leaves ovate, lanceolate or elliptic with apex acute, obtuse
  or subattenuate, largest blade 1.8–9.5 cm long with 3–5 pairs of lateral veins;
  peduncle 0–6 mm long . . . . **3.** *scabra*
 – Plant rosulate; leaves oblong to oblanceolate with apex rounded or emarginate,
  largest blade 9–16 cm long with 6–10 pairs of lateral veins; peduncle 10 mm long
  or more. . . . . **4.** *plantaginea*
8. Plant erect, decumbent or straggling; inflorescence held above ground level at
  leafy axils, often only in distal half of stems (rarely also basal) 9
 – Plant trailing or procumbent, if leafy then stems erect to decumbent and without
  inflorescence at leafy axils; inflorescence held against ground, basal and/or
  axillary on trailing stems. 12
9. Inflorescence indumentum yellow; flowering calyx lobes green throughout,
  lacking a darker acumen . . . . **14.** *sp. A*
 – Inflorescence indumentum white or rarely pale buff; flowering calyx lobes green
  with conspicuous brown or purple acumen. 10
10. Sterile bracts* broadly oblong-elliptic, rounded or somewhat obovate, 5–13.5
  mm wide, surface convex, apex abruptly narrowed into an apiculum or short
  acumen (Fig. **37**, 4); cymes compounded into dense ± globose heads 3–6.5 cm in
  diameter. . . . . **9.** *eriocephala*
 – Sterile bracts (linear-)lanceolate to ovate, surface ± flat, apex gradually narrowed,
  attenuate to long-acuminate; cymes less densely compounded, or if in globose
  heads to 5 cm wide then sterile bracts very narrow, 1.5–4 cm wide 11
11. Inflorescence densely white-plumose; 1(2) unilateral cymes per axil, alternate or
  often in opposite pairs, usually shortly spiciform or triangluar; leaves often lanate
  when young; anther sutures with cilia sparse or absent; plants often robust. . . . .
  . . . . **8.** *andersoniana*
 – Inflorescence white-pilose, not plumose; unilateral cymes several per node,
  fasciculate, together forming globose or hemispheric heads at maturity; leaves
  sparsely to densely pilose when young, not lanate; anther sutures with numerous
  cilia; plants often slender and wiry. . . . . **11.** *randii*

12. Sterile bracts* broadly oblong-elliptic, rounded or somewhat obovate, surface convex, apex abruptly narrowed into an apiculum or short acumen (Fig. 37) . .
. . . . **9.** *eriocephala*
 – Sterile bracts lanceolate, linear-lanceolate or triangular, surface usually flat or only slightly convex, apex more gradually narrowed, attenuate to long-acuminate . . 13
13. Inflorescence axillary, either in distal half of stems only or more evenly distributed, usually at leafy axils, more rarely spaced along largely leafless proximal portion of stems; stems usually trailing or procumbent 14
 – Inflorescence basal, often compounded to form cushions, sometimes also ± densely clustered along leafless proximal portion of stems; leafy stems often erect or decumbent. 15
14. Sterile bracts large and long-acuminate, (17.5)21.5–34 mm long; inflorescences solitary or few in distal portion of leafy stems only; young leaves glabrous or with hairs only along margin and midrib, not lanate . . . . **10.** *longisepala*
 – Sterile bracts shorter, 6–15 mm long; inflorescences often more numerous and evenly distributed throughout or in proximal half of stems only; young leaves often lanate . . . .**12.** *lanatoglabra* [4]
15. Inflorescence appearing brown or purplish, bracteoles and calyx lobes with a prominent linear, usually brown or purple acumen, contrasting with green lower portion, only acuminate portions usually exposed beyond bracts . . . . . **13.** *fischeri*
 – Inflorescence appearing greyish-green, green bracteoles and calyx lobes clearly exposed beyond bracts, with or without a short acumen which is green throughout or at most brown-tipped . . . .**15.** *sp. B*

* The sterile bracts referred to here are mature bracts at a flowering node, not reduced sterile bracts at the very base of inflorescence or sheathing the peduncle.

Group A, 'Neuracanthopsis'

Corolla with upper lip straight or arcuate; lower lip with palate lacking membranous portions, glabrous or shortly pubescent on bosses. Anterior (longer) pair of stamens bithecous, posterior (shorter) pair monothecous, sometimes with vestigial second theca; filaments glabrous or with few eglandular hairs at base; thecae glabrous. Stigma capitate-bilobed or with one expanded, spoon-shaped lobe. Capsule 4-seeded (or 2 by abortion), oblong-ellipsoid; seeds with hygroscopic hairs (green-)brown or yellow.

1. **Lepidagathis glandulosa** A. Rich., Tent. Fl. Abyss. **2**: 147 (1850). —Clarke in F.T.A. **5**: 128 (1899). —Darbyshire in F.T.E.A., Acanthaceae **2**: 305 (2010). Type: Ethiopia, Tigray, Mt Soloda (Scholoda), fl. 27.x.1837, *Schimper* 1: 44 [cited by Richard as 1: 41 in error] (K lectotype, BR, GZU, S), lectotypified by Darbyshire (2010).

Erect, decumbent or straggling, single- to several-stemmed perennial herb or suffrutex, 10–90 cm tall, sometimes rooting at lower nodes; stems 4-angular, crisped-pilose to antrorse-pubescent or largely glabrous. Leaves ovate, lanceolate or elliptic, 3.2–10.5 × 1.2–3.5 cm, base (cuneate-) attenuate to shallowly cordate, apex acute to attenuate, surfaces pilose to antrorse-pubescent at least on margin and veins beneath, rarely glabrous; lateral veins 5–8 pairs; petiole 0–18 mm long. Inflorescence terminal on main shoots and pseudo-axillary on poorly developed lateral branches, cymes unilateral, fasciculate to spiciform, compounded into dense ovoid or globose heads, 1.3–6.5 cm long, subsessile; fertile and sterile bracts subequal, (white-)green with purple apex and main veins or brown-scarious throughout, (linear-)lanceolate, 5.5–12.5 × 0.7–2.5 mm, apex

---

[4] Occasional trailing plants of *L. randii* may key out here; see note under that species to separate this from *L. lanatoglabra* – the differences are quite descriptive and so are better not included in the key.

aristate, shortly pubescent at least on margin where sometimes densely ciliate, with or without a few glandular hairs, with 3–5 subparallel principal veins and ± prominent reticulate tertiary venation; bracteoles as bracts but 1–2 mm wide, 3-veined or midrib only prominent, glandular hairs sometimes more numerous. Calyx as bracteoles but more tardily scarious; lobes aristate, anterior pair fused for 1–2.5 mm, (linear-)lanceolate, 7–12 × 0.8–1.5 mm; posterior lobe ovate to lanceolate, 8.5–13 × 1.5–3 mm, 3–5-veined from base, with prominent reticulate tertiary venation. Corolla 11.5–14.5 mm long, white to mauve, often with purple markings on lower lip, pubescent externally mainly on limb; tube 7–9 mm long; upper lip 3.5–5 mm long; lower lip 4–6 mm long including lobes 2–3.5 mm long. Anterior (longer) stamens with filaments 4.5–7 mm long, thecae 1–1.3 mm long, barely offset; posterior pair with vestigial second theca. Ovary glabrous or with few hairs at apex; style shortly pubescent and/or with minute stalked glands in proximal half; stigma capitate. Capsule 6–7 mm long, largely glabrous; seeds 1.3–1.6 × 1.1–1.4 mm.

Zambia. N: Mansa Dist., Mansa (Fort Rosebery), fl.& fr. 8.v.1964, *Fanshawe* 8547 (K, NDO). W: Kitwe, fl. 18.iii.1955, *Fanshawe* 2150 (BR, EA, K, NDO, SRGH).

Also in Cameroon, D.R. Congo, South Sudan, Ethiopia, Kenya and Tanzania. In miombo woodland, often on rocky hillslopes, and riverside grassland; c.1200–1500 m.

Conservation notes: Assessed as Least Concern in F.T.E.A.; uncommon in the Flora region but widespread and more common in E and NE Africa.

The description covers only the forms of this variable species recorded within the Flora region. Plants with considerably smaller flowers (including the type) are recorded from Ethiopia and Kenya; see F.T.E.A. for further discussion.

2. **Lepidagathis scariosa** Nees in Wallich, Pl. Asiat. Rar. **3**: 95 (1832). —Nees in De Candolle, Prodr. **11**: 251 (1847).—Clarke in F.T.A. **5**: 122 (1899). —Darbyshire in F.T.E.A., Acanthaceae **2**: 307 (2010). Type: India, fl. 1830, *Wallich Cat.* 2354b (K lectotype, K-W, NY, P), lectotypified by Darbyshire (2010).

　　Ruellia aristata Vahl, Symb. Bot. **2**: 73 (1791). Type: Yemen, *Forsskål* 372 (C microfiche 116: III. 3–4 holotype, K photo).

　　Lepidagathis aristata (Vahl) Nees in De Candolle, Prodr. **11**: 251 (1847), illegitimate name, non *L. aristata* Nees in Wallich (1832).

　　Lepidagathis terminalis Nees in De Candolle, Prodr. **11**: 251 (1847). Types: Ethiopia, Tigray, below Sessaquilla, fl. 29.ix.1838, *Schimper* II: 815 (BM, GZU, K, M, S, TUB, WAG syntypes); Sudan, Fazokl (Fazohel), fl. 1837–38, *Kotschy* 482 (K syntype).

　　Volkensiophyton neuracanthoides Lindau in Bot. Jahrb. Syst. **20**: 27 (1894). Type: Tanzania, Kilimanjaro, Lake Chala, fl. 15.vi.1893, *Volkens* 318 (B† holotype, BM, K).

　　Lepidagathis sciaphila S. Moore in J. Bot. **51**: 215 (1913). Types: D.R. Congo, W Kundelungu, fl. 17.v.1908, *Kässner* 2800 (BM, K syntypes); Tanzania, ?Uruwira (Uvira), fl. 10.vii.1908, *Kässner* 3070 (BM syntype).

Perennial herb or subshrub, much-branched, 30–120 cm tall; stems with dense pale buff to golden stellate or dendritic hairs. Leaves ovate(-lanceolate) or oblong-elliptic, 2.8–9 × 1–3.6 cm, base obtuse, shortly attenuate or cuneate, apex attenuate, acute or obtuse, surfaces pale stellate-pubescent, dense beneath particularly on principal veins, hairs often long-armed; lateral veins 4–6 pairs; petiole to 7.5 mm long. Inflorescence terminal, cymes fasciculate to spiciform, compounded into globose or conical heads, 1.5–4.5 cm long, sessile; leafy bracts subtending heads rhombic, ovate or lanceolate, often with an expanded basal portion, stellate pubescent; fertile and sterile floral bracts subequal, brown- to purple-scarious, elliptic(-obovate) to oblong-lanceolate, 4–14 × 1–3.5 mm, apex mucronulate to aristate, with few to numerous pale long-pilose hairs at least on margin, surface also with short glandular and/or eglandular hairs, venation parallel; bracteoles as bracts but (linear-)lanceolate to oblanceolate, 8–17 × 0.5–3 mm, pilose hairs more dense. Calyx purple-tinged at least in distal half, scarious; lobes aristate, densely long-pilose, with short glandular hairs particularly in distal half, anterior pair fused for 2.5–4.5 mm, (linear-)lanceolate, 9–15.5 × 1–2.5 mm; posterior lobe elliptic(-obovate) to oblong-lanceolate, 9–18 × 3–6.5 mm, with 5 or 7 subparallel principal veins. Corolla 19–22 mm long, pink to purple, rarely white, numerous long hairs on lobes, shortly pubescent externally on upper tube and limb; tube 10.5–13 mm long; upper lip arcuate, 7.5–11 mm long; lower lip

8–10 mm long including lobes 4.5–6 mm long. Anterior (longer) pair of stamens with filaments 12–14.5 mm long, declinate, thecae 1.3–2 mm long, barely offset. Ovary pubescent in distal half; style with short hairs and short-stalked glands in proximal half; stigma capitate. Capsule 8.5–10 mm long, shortly pubescent towards apex; seeds 2–3 × 1.8–2.7 mm.

Zambia. N: Kawambwa Dist., Kawambwa, fl.& fr. 16.viii.1973, *Chisumpa* 91 (K, NDO). W: Chingola Dist., fl. 26.viii.1954, *Fanshawe* 1503 (K, NDO). C: Serenje Dist., Luangwa R., fl. 5.vi.1958, *Fanshawe* 4532 (K, NDO, SRGH). E: Chipata Dist., Machinje Hills, fl. 15.v.1965, *Mitchell* 2972 (K). S: Mazabuka Dist., Kafue Gorge (S), fl.& fr. 18.viii.1963, *Robinson* 5606 (EA, K, SRGH). **Zimbabwe**. N: Binga Dist., Mwenda Research Station, fl.& fr. 3.vi.1966, *Grosvenor* 87 (K, SRGH). W: Hwange Dist., Deka, fl.& fr. 16.v.1974, *Gonde* 91/74 (BR, K, SRGH). C: Chegutu Dist., Chegutu (Hartley), fl.& fr. 19.vi.1943, *Hornby* 2258 (K, PRE). E: Chimanimani Dist., Umvumvumvu, fl. v.1923, *Swynnerton* 4002 (K). S: Masvingo Dist., top of Mtunumushava kopje, Mutirikwi (Kyle) Recreational Park, fl.& fr. 21.v.1971 *Ngoni* 85 (K, SRGH). **Mozambique**. MS: Báruè Dist., 40 km from Changara towards Catandica (Vila Gouveia), track on right to Chief Catunguinene (Luenha valley), fl. 28.v.1971, *Torre & Correia* 18701 (LISC, LMU, PRE).

Also in Mali, Benin, Nigeria, Cameroon, D.R. Congo, Sudan, South Sudan, Eritrea, Ethiopia, Somalia, East Africa, Namibia, Angola, Arabia and India. In miombo and mopane woodland and associated grassland, often on rocky hillslopes; 600–1400 m.

Conservation notes: Widespread and common; assessed as Least Concern in F.T.E.A.

A very variable species; the above description covers only forms recorded within the Flora region. These populations often have larger bracts, bracteoles, calyces and capsules than East African material but there is much overlap in all but the lattermost character (in East Africa capsules to only 7.5 mm long). Plants from N Zambia are of the form previously separated as *L. sciaphila* – the leaves are oblong-elliptic to oblong-lanceolate with ± prominent reticulate venation beneath and the stems have a golden-yellow indumentum. Elsewhere the leaves are ovate to ovate-elliptic, usually with inconspicuous tertiary venation, and the stem hairs are grey, buff or pale yellowish. *L. sciaphila* is a good candidate for future subspecific recognition, but *L. scariosa* requires a full review across its range before infraspecific taxa are described.

3. **Lepidagathis scabra** C.B. Clarke in F.T.A. **5**: 129 (1899). —Darbyshire in F.T.E.A., Acanthaceae 2: 309, fig.46 (2010). Type: Tanzania, Lake Chala, fl.& fr. vi.1893, *Volkens* 320 (K lectotype, BM), lectotypified by Darbyshire (2010). FIGURE 8.6.**36**.

Erect, decumbent or prostrate perennial herb or subshrub, 10–50 cm tall; stems pale antrorse-pubescent. Leaves ovate, elliptic or lanceolate, 1.8–9.5 × 0.8–3.6 cm, base cuneate or attenuate, apex acute, obtuse or subattenuate, surfaces pale-pubescent mainly on midrib and veins beneath; lateral veins 3–5 pairs; petiole 0–15 mm long. Inflorescence axillary, contracted unilateral spikes 1.5–4.5 cm long, subsessile or peduncle to 6 mm long; sterile bracts held dorsally in two ± divergent rows, green, subulate-falcate, 11–20 × 1–3 mm, with 3–5 prominent parallel veins; fertile bracts appressed to calyx, oblong-lanceolate, pale and scarious at least towards base, 7–13.5 × 1.5–4 mm, apex acuminate, venation parallel, indumentum of longer ascending to spreading hairs at least on margin, often with shorter hairs elsewhere, lowermost 1–2 fertile bracts of each spike markedly enlarged and green, ovate to obovate, venation palmate; bracteoles linear-lanceolate, 6.5–12.5 × 1–2 mm. Calyx pale-scarious, indumentum as bracts; anterior lobes free, subulate, 8.5–11.5 × 1.5–2.5 mm, margin somewhat inrolled, apex acute to acuminate; posterior lobe oblong-lanceolate, 9–13.5 × 2.5–3.5 mm, margins inrolled and often partially enveloping other lobes. Corolla 13.5–20 mm long, pale blue to mauve or white, with purple markings on lower lip and throat, pubescent externally mainly on lower lip, throat sometimes with short-stalked glands; tube 8–10.5 mm long; upper lip straight, 4–8 mm long; lower lip 6–10 mm long including lobes 5.5–7 mm long. Anterior (longer) stamens with filaments 4–6.5 mm long, glabrous, thecae 1.5–2 mm long, parallel; posterior stamens with vestigial second theca sometimes present. Ovary with or without a few appressed hairs towards apex, with a ring of short hairs at style attachment;

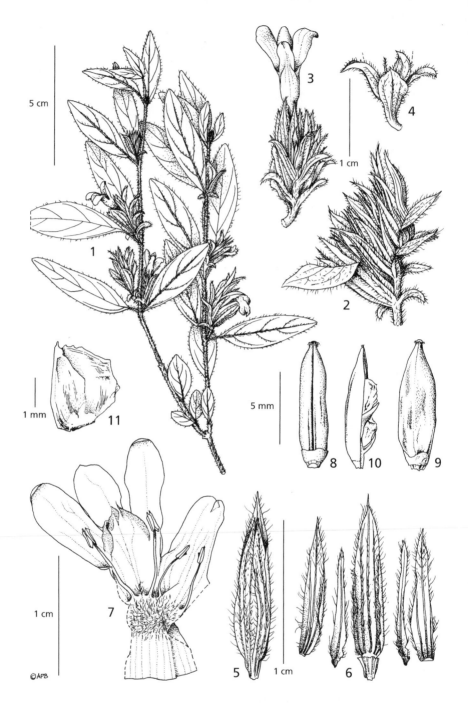

Fig. 8.6.**36**. LEPIDAGATHIS SCABRA. 1, habit; 2, typical elongate inflorescence; 3, inflorescence, more fasciculate variant; 4, base of inflorescence showing enlarged basal fertile bract; 5, calyx, anterior aspect; 6, dissected calyx showing inner surfaces of each lobe, anterior lobes outermost, posterior lobe central, lateral lobes in between; 7, dissected corolla with stamens; 8, capsule, outer surface of single valve; 9, capsule, face view prior to dehiscence; 10, capsule valve after dehiscence; 11, detail of mature seed. 1 from *Richards* 25542, 2, 7–11 from *Bidgood et al.* 2212, 3–6 from *Gillett* 13003. Drawn by Andrew Brown. Reproduced from Flora of Tropical East Africa (2010).

style with spreading hairs and/or minute stalked glands in proximal half; stigma with spoon-shaped lobe. Capsule 7–10 mm long, glabrous or with few appressed to antrorse hairs towards apex; seeds 2.5–3 × 2–2.7 mm.

Botswana. N: Central Dist., c.128 km WNW of Francistown on Maun road, fl.& fr. 29.iv.1957, *Drummond* 5280 (K, SRGH). **Zambia**. S: Livingstone Dist., Victoria Falls, fl.& fr. 24.iii.1984, *Brummitt* 16914 (K, SRGH). **Zimbabwe**. N: Binga Dist., Mwenda Research Station, fl.& fr. 3.vi.1966, *Grosvenor* 91 (K, LISC, SRGH). W: Nyamandhlovu Dist., Nyamandhlovu Pasture Research Station, fl.& fr. iv.1956, *Plowes* 1944 (BR, K, SRGH). S: Mwenezi Dist., Bubi R., near Bubye Ranch homestead, fl.& fr. 3.v.1958, *Drummond* 5550 (K, SRGH).

Also in Ethiopia, Kenya, Tanzania, South Africa and Swaziland. In woodland, typically of mopane, *Kirkia*, *Commiphora* and/or *Acacia*, including in disturbed areas; 500–1100 m.

Conservation notes: Widespread and locally common; assessed as Least Concern in F.T.E.A.

4. **Lepidagathis plantaginea** Mildbr. in Notizbl. Bot. Gart. Berlin-Dahlem **13**: 286 (1936). —Darbyshire in F.T.E.A., Acanthaceae **2**: 312 (2010). Type: Tanzania, Lukuledi, fl. 20.iv.1935, *Schlieben* 6339 (B† holotype, BM, BR, G, HBG, K, M, MA, S).

Rosulate herb with short creeping rhizome. Leaves appressed to ground, 2–5 pairs; blade oblong or oblanceolate, 9–16 × 3.5–5.5 cm, base cuneate-attenuate, apex rounded or shallowly emarginate, glabrous except for sparse hairs on midrib and main veins beneath; lateral veins 6–10 pairs; petiole to 20 mm long. Inflorescence axillary but appearing terminal above rosulate leaves, contracted unilateral spikes 2–6 cm long, sometimes single-branched at base; peduncle 10–42 mm long, crisped-pilose; sterile bracts held in two divergent rows, green, (oblong-)lanceolate, ± falcate, 10–15 × 1.5–4 mm wide, glabrous or margin with short-ascending and/or long-crisped hairs; fertile bracts pale brown-scarious, 10–11.5 × 2.5–3.5 mm, apex acuminate, margin with long straight and crisped white hairs often numerous towards apex, elsewhere glabrous, lowermost fertile bract of each spike somewhat larger and green; bracteoles as fertile bracts but 8–9 × 1.5–2 mm, enveloping calyx. Calyx scarious, indumentum as bracts; anterior lobes free, subulate, 10–11 × 2–2.5 mm, margin somewhat inrolled, apex acuminate; posterior lobe oblong-elliptic, 10.5–12 × 3.5–5 mm, margin strongly inrolled and enveloping the other lobes. Corolla 16–20 mm long, white to pale pink with darker markings on lower lip; throat and limb pubescent externally; tube 10.5–12.5 mm long; upper lip 3.5–4.5 mm long; lower lip 5.5–7.5 mm long including lobes 4–4.5 mm long. Anterior (longer) stamens with filaments 4.5–5 mm long, thecae 1.2–1.5 mm long, parallel; posterior stamens with vestigial second theca. Ovary glabrous except for a ring of minute hairs at style attachment; style shortly and sparsely hairy; stigma not seen. Capsule and seeds not seen.

Mozambique. N: Montepuez Dist., slopes of Mt Matuta, c.5 km S of Rio Messalo (M'salo), near Nantulo, bud 9.iv.1964, *Torre & Paiva* 11831 (LISC, PRE); Mueda Dist., Ngapa to Negomano, old infl. 27.xi.2009, *Luke* 13927 (EA, K).

Also in SE Tanzania. In miombo and *Sterculia–Adansonia* woodland, sometimes on termitaria; 150–400 m.

Conservation notes: Since this species was assessed as Vulnerable (VU D2) in F.T.E.A. it has been recorded from 3 more sites in NE Mozambique, but is still considered threatened owing to the very restricted range with likely loss of habitat.

5. **Lepidagathis pallescens** S. Moore in J. Bot. **18**: 308 (1880). —Clarke in F.T.A. **5**: 127 (1899). —Darbyshire in F.T.E.A., Acanthaceae **2**: 312 (2010). Type: Angola, Pungo Andongo, fl. iii.1857, *Welwitsch* 5084 (BM holotype, K, LISU).

Suffruticose perennial, 1 to few erect stems from a woody base, 5–60 cm tall; stems glabrous, antrorse-pubescent and/or ± spreading pilose. Leaves ovate, (oblong-)lanceolate or elliptic,

3–10.5 × 0.8–3.7 cm, base cuneate or attenuate, apex acute or obtuse, surfaces glabrous or pubescent to pilose; lateral veins 4–7 pairs; petiole 0–7 mm long. Inflorescence (sub)terminal and often also in upper axils, strobilate unilateral spikes, 1.5–9 cm long, sessile or peduncle to 7(18) mm long; bracts ovate, lanceolate or obovate, 9–18 × 2.5–7 mm, apex attenuate to acuminate or rounded to emarginate, venation prominent, parallel towards centre, usually with prominent reticulation towards margin, surface minutely pubescent and/or with long ascending to spreading white hairs sometimes restricted to margin; fertile bracts similar but 3.5–7.5 mm wide; bracteoles (oblong-)lanceolate to oblong-elliptic, 8.5–19 × 1.5–5 mm, apex often narrowed into a linear acumen. Calyx brown-scarious in distal half, pale (yellow-)green below, venation and indumentum as bracts; anterior lobes free, subulate to narrowly oblong-elliptic, 11.5–16 × 1.5–3.7 mm, apex obtuse to attenuate; posterior lobe elliptic, lanceolate or oblanceolate, 12–17 × 2.5–7 mm, margins ± strongly inrolled and partially enveloping other lobes. Night-flowering; corolla (14)19–28 mm long, white or cream, with mauve markings on lower lip, throat and limb pubescent externally; tube (8)11–15 mm long; upper lip straight, (4.5)6–8.5 mm long; lower lip (6)8–13 mm long including lobes (3)6.5–10 mm long, palate shortly pubescent on bosses. Anterior (longer) stamens with filaments 4–6.5 mm long; thecae 1.4–1.7 mm long, parallel; posterior stamens with vestigial second theca. Ovary with few minute hairs at apex; style glabrous or shortly hairy; stigma with spoon-shaped lobe. Capsule 10–12 mm long, glabrous or with few hairs towards apex; seeds ± 3 × 2.5 mm.

a) Subsp. **pallescens**

Leaves elliptic, ovate or lanceolate, 1.2–3.7 cm wide, length:width ratio (leaves at midpoint of stems) 1.9–5:1. Sterile bracts ovate or lanceolate with apex attenuate or acuminate; bracteoles and anterior and posterior calyx lobes with acuminate apices. Corolla 19.5–28 mm long in our region – see note.

Zambia. N: Mbala Dist., above Pine Apple Farm, Inono valley, fl.& fr. 13.iv.1955, *Richards* 5432a (K, LISC). W: Mwinilunga Dist., Mwinilunga, by canning factory, fl. 26.i.1971, *Anton-Smith* in SRGH 213,261 (SRGH). C: Mpika Dist., Kapamba R., Luangwa Valley, bud 20.iii.1968, *Taylor* 64 (SRGH). E: Lundazi Dist., Lusantha, on road to Malawi, bud i.1968, *Anton-Smith* in SRGH 201,751 (SRGH). **Malawi**. N: Karonga Dist., North Rukuru R., fl.& fr. 31.iii.1989, *Steiner* 611 (K, UPS). C: Kasungu Dist., just outside Kasungu Nat. Park on Kasungu road, fl.& fr. 18.iv.1991, *Bidgood & Vollesen* 2195 (CAS, K, MAL). **Mozambique**. N: Marrupa Dist., Marrupa, c.25 km on road to Mecula, fl. 12.ii.1981, *Nuvunga* 514 (K, LISC, LMU). Z: Gurué Dist., 22 km from Nintulo towards Lioma, near junction for Mutuáli, fl. 10.ii.1964, *Torre & Paiva* 10513 (LISC, LUA, MO, PRE, WAG). T: Macanga Dist., Macanga (Furancungo), fl. 17.iii.1966, *Pereira, Sarmento & Marques* 1793 (BR).

Also in Tanzania, D.R. Congo and Angola (see note). In miombo woodland and associated grassland on sandy to rocky soils, *Combretum–Terminalia* woodland, and sandy dambos; 500–1500 m.

Conservation notes: Fairly common in the miombo ecoregion; assessed as Least Concern in F.T.E.A.

As noted in F.T.E.A., the type locality in Angola is isolated from the remainder of the range and the single (not well-preserved) flower on the holotype is only c.10 mm long, considerably smaller than in our region. It is quite possible that two taxa are involved here, but since no other Angolan specimens are known it is difficult to draw firm conclusions.

Some populations from N and W Zambia (e.g. *Bredo* 3945, *Anton-Smith* in SRGH 213,261) and adjacent D.R. Congo have densely long-ciliate bracts and bracteoles; D. Champluvier (BR) has proposed to recognise this form as var. *villosa* but this name has not yet been published.

b) Subsp. **obtusata** I. Darbysh., subsp. nov. Differs from *L. pallescens* sensu stricto in the leaves being narrower and oblong-lanceolate, length : width ratio of leaves at midpoint of the stem 5–8: 1 versus 1.9–5: 1, in the sterile bracts being obovate or obovate-elliptic with a rounded- or emarginate-apiculate apex (vs. ovate or lanceolate with an attenuate or acuminate apex), and in the bracteoles and anterior and posterior calyx lobes lacking acuminate apices, being abruptly narrowed at apex and blunt or at most apiculate. Type: Angola, Bié–Cuando Cubango, Menongue, Caiundo, Capico, near Mission, fl. 2.ii.1960, *Mendes* 2320 (LISC holotype).

Leaves narrowly oblong-lanceolate, length : width ratio (leaves at midpoint of stems) 5–8 : 1. Sterile bracts obovate or obovate-elliptic, apex rounded- or emarginated-apiculate; bracteoles and anterior and posterior calyx lobes lacking attenuate or acuminate apices, more abruptly narrowed and at most minutely apiculate. Corolla c.14 mm long (see note).

Zambia. S: Namwala Dist., W boundary of Kafue Nat. Park, 35 km from southern tip, bud 7.ii.1965, *Mitchell* 25/96 (BR, K, SRGH); Kalomo Dist., Kalomo, bud 10.ii.1965, *Fanshawe* 9169 (K, NDO).

Also in S Angola. In *Julbernardia–Burkea* miombo woodland; also in areas of abandoned cultivation in Angola; 1000–1300 m.

Conservation notes: Very local, known from only 6 collections; it may prove to be more common since S Angola is so under-collected.

This distinctive subspecies is currently known only from the two Zambian specimens cited plus four collections from S Angola – *Gossweiler* 3074 & 3580 (both BM), *Mendes* 2320 (LISC) and *Menezes et al.* 4447 (LISC, SRGH). Only the Mendes specimen has any open corollas and these are rather withered and perhaps not truly representative, though they do appear to be smaller than in subsp. *pallescens* from the FZ and FTEA regions (but see note on type of *L. pallescens* above).

6. **Lepidagathis nemorosa** S. Moore in J. Bot. **48**: 253 (1910). —Darbyshire in F.T.E.A., Acanthaceae 2: 313 (2010). Type: D.R. Congo, Lofoi R., fl. 21.iii.1908, *Kässner* 2655 (BM holotype, K, P).

> *Lepidagathis lindaviana* Buscal. & Muschl. in Bot. Jahrb. Syst. **49**: 494 (1913). —Piscicelli, Nella Regione Laghi Equatoriali: 116 (1913). Type: Zambia, Samfya Dist., Lake Bangweulu (Banguelo), 6.iv.1910, *von Aosta* 1073 (B† holotype).
> *Lepidagathis ringoetii* De Wild. in Repert. Spec. Nov. Regni Veg. **13**: 146 (1914). Type: D.R. Congo, Katanga, Shinsenda, fl. 14.iii.1912, *Ringoet* 511 (BR holotype).
> *Lepidagathis persimilis* S. Moore in J. Bot. **67**: 51 (1929). —Mapaura & Timberlake, Checklist Zimb. Vasc. Pl.: 14 (2004). —Phiri, Checklist Zamb. Vasc. Pl.: 19 (2005). Type: Zimbabwe, Hurungwe Dist., Mwami (Miami), fl. iv.1926, *Rand* 77 (BM holotype).

Suffruticose perennial, few to numerous erect or decumbent branches from a woody rootstock, 10–60 cm tall; stems antrorse-pubescent to crisped- or spreading-pilose, or glabrous except for a line of hairs at nodes. Leaves glossy above, elliptic, narrowly oblong-elliptic to -lanceolate or spathulate, 3.3–9 × 0.9–3 cm, base cuneate to attenuate, apex acute to rounded, glabrous or pubescent mainly on principal veins; lateral veins 4–6 pairs, ± prominent beneath; petiole 0–18 mm long. Inflorescence axillary, usually opposite, fasciculate, subsessile or rarely peduncle to 4 mm long; sterile bracts green or brown with age, oblong, lanceolate or spathulate, 4–9.5 × 1–3 mm, crisped-pubescent at least on margin, parallel venation prominent; fertile bracts and bracteoles equal, together enveloping lower portion of calyx, green in proximal half and (black-)brown in distal half or turning brown-scarious throughout with age, ovate-elliptic, 5–9 × 2.5–4 mm, apex rounded to attenuate, mucronulate, ciliate, sometimes also shortly hairy on surface. Calyx colour as fertile bracts; lobes glabrous or often with short cilia, surface sometimes puberulent in distal half, anterior pair free, oblong-lanceolate, 8–13.5 × 2.5–3.7 mm, margin somewhat inrolled, apex acute to attenuate; posterior lobe oblong-elliptic to -lanceolate, 3.5–5 mm wide, margins strongly inrolled and partially enveloping other lobes. Corolla 20–28 mm long, white to pale blue or mauve, with pink to purple markings on lower lip, pubescent externally on lower lip and throat or rarely glabrous; tube 10–16.5 mm long; upper lip 6–8 mm long; lower lip 9.5–12.5 mm long including lobes 6–8.5 mm long. Anterior (longer) stamens with filaments 7.5–10.5 mm long; thecae 2–2.8 mm long, parallel; posterior stamens with vestigial second theca. Ovary glabrous or with few short hairs at apex; style glabrous; stigma with spoon-shaped lobe. Capsule 10–11 mm long, glabrous or with few hairs at apex; seeds c.2.7 × 2 mm.

Zambia. B: Kaoma Dist., 6.4 km NE of Kaoma on Kasempa road, 2 km S of Kalamba village, fl. 28.ii.1996, *Harder et al.* 3571 (K, MO). N: Mbala Dist., Chisungu bush, fl. 20.iv.1959, *Richards* 11283 (K, SRGH). W: Kasempa Dist., 56 km N of Kasempa on Solwezi road, fl. 22.iii.1961, *Drummond & Rutherford-Smith* 7189 (BR, EA, K, LISC, SRGH). C: Serenje Dist., Serenje, between town and main road, fl. 28.iii.1984, *Brummitt, Chisumpa & Nshingo* 16939 (K, SRGH). E: Nyimba Dist., 4 km W of Kachalola on Great East road, fl. 17.iii.1959, *Robson* 1745 (BM, K, LISC, SRGH). S: Mazabuka Dist., 40 km N of Pemba, near Kanchale village, fl. 11.ii.1969, *White* 6964 (FHO, K). **Zimbabwe**. N: Hurungwe Dist., Zwipani, fl. 19.iii.1958, *Goodier* 550 (BR, EA, K, SRGH). **Malawi**. N: Mzimba Dist., 17.6 km N of Mzambazi, 9.6 km S of Mperembe, fl. 10.iii.1978, *Pawek* 14013 (BR, K, MAL, MO, SRGH, UC).

Also in Tanzania and D.R. Congo. In miombo woodland and degraded open miombo, usually on sandy soils or rocky hillslopes; 500–1500 m.

Conservation notes: A common miombo woodland species, particularly in Zambia; assessed as Least Concern in F.T.E.A.

Fanshawe 9168 from Kalomo closely resembles *L. nemorosa* in bract and calyx shape and dimensions, but differs in having a shortly spiciform inflorescence; no flowers or fruits are present on the K sheet.

7. **Lepidagathis macrochila** Lindau in Warburg, Kunene–Sambesi Exped.: 378 (1903). Type: Angola, Likise on Kusisi R., fl. 16.iii.1900, *Baum* 779 (B† holotype, BM, BR, HBG, K, M, S, W).

Subshrub 60–120 cm tall; stems with buff-coloured antrorse to appressed hairs, dense when young. Leaves sometimes immature at flowering, narrowly elliptic to lanceolate, 3–6.5 × 0.8–2 cm, base cuneate to acute, apex acute, pubescent beneath when young, becoming sparse and largely restricted to principal veins at maturity; lateral veins 4–5 pairs; petiole to 2.5 mm long. Inflorescence axillary and subterminal, strobilate spikes 2–4.5 cm long, sessile or peduncle to 3.5 mm long; bracts and bracteoles subequal, red-brown scarious, sometimes green at base, broadly ovate-elliptic, 4.5–9 × 3.5–5 mm, apex obtuse to shallowly emarginate, apiculate, margin and apex with long pale silky hairs, sometimes also scattered on surface and at base; lowermost sterile bracts often narrower, lanceolate and densely hairy on margin and midrib; fertile bracts and bracteoles together enveloping lower portion of calyx. Calyx colour and indumentum as bracts; anterior lobes free, narrowly oblong-elliptic, 10.5–13.5 × 3.5–5 mm, apex shortly attenuate, margin somewhat inrolled; posterior lobe elliptic, 5.5–8 mm wide, margins strongly inrolled and partially enveloping other lobes, apex acute to obtuse. Corolla (30)35–40 mm long, white to mauve, densely long-pubescent externally on limb and throat; tube 11.5–14.5 mm long; upper lip arcuate, 16.5–19.5 mm long; lower lip 19–26 mm long including lobes 12.5–21 mm long. Anterior (longer) stamens with filaments 17–20.5 mm long, thecae 3–3.5 mm long, subparallel. Ovary with dense tuft of hairs at apex; style pubescent in proximal half; stigma capitate-bilobed. Capsule 17–20 mm long, with apical tuft of hairs or glabrescent; seeds c.4.8 × 3.8 mm.

Zambia. B: Kaoma Dist., c.5 km along road to Luampa Hospital from intersection with Lusaka–Mongu road, fl. 2.iii.1996, *Harder, Zimba & Luwiika* 3637 (K, MO). W: Mwinilunga Dist., 96 km S of Mwinilunga on Kabompo road, fl.& fr. 3.vi.1963, *Loveridge* 745 (BR, K, LISC, SRGH).

Also in Angola. In *Cryptosepalum* forest and thicket on Kalahari sands, miombo woodland, roadsides; 1000–1500 m.

Conservation notes: Localised but frequent in suitable habitats; probably not threatened.

Group B, '**Lepidagathis**'

Corolla with upper lip straight; lower lip with a membranous portion on either side of centre of palate, with an adjacent line of silky hairs (Fig. 8.6.**37**), tube with a dense ring of silky deflexed hairs within below stamen attachment. Stamens all bithecous; filaments with subsessile or short-stalked glands mainly in distal half or glabrous; anther thecae somewhat offset and oblique, with short hairs and subsessile glands dorsally and a line of cilia along suture (sometimes sparse). Stigma capitate, bilobed. Capsule 2-seeded, face ovate; seeds with hygroscopic hairs cream to pale buff.

Species delimitation within this group of fire-adapted woodland and grassland plants is highly problematic. Whilst several seemingly distinct taxa are usually recognizable, all of species 8–16 here are clearly closely related and hybridization/intergrading apparently occurs in parts of their ranges. One of the most notable differences between the taxa is in growth habit and the position and form of the inflorescence. It is quite possible that these are influenced primarily by ecological factors, most notably local variation in fire regime. However, ecological influence is very difficult to infer from the study of herbarium material alone. A full revision of the group, including the species from West Africa and the Indian subcontinent and involving extensive field studies, is required to gain a full understanding of taxon delimitation and suitable taxonomic ranks within this group. In view of this, I have chosen to maintain the generally separable taxa as distinct species here, although it is accepted that they may ultimately be treated as fewer or even a single polymorphic species. The key above should allow identification of most specimens although naming is most easily achieved by comparison to previously named herbarium collections. See F.T.E.A. Acanthaceae 2: 314 (2010) for further discussion on the complex.

8. **Lepidagathis andersoniana** Lindau in Bot. Jahrb. Syst. **20**: 16 (1894); in Engler, Pflanzenw. Ost-Afrika **C**: 368 (1895) as *L. andersonii*. —Clarke in F.T.A. **5**: 126 (1899) in part, excluding *Whyte*s.n. from Chitipa (Fort Hill), Malawi. —Darbyshire in F.T.E.A., Acanthaceae 2: 318, fig.47.1–2 (2010). Types: Tanzania, Ussindje (?Usinge), 7.iii.1892, *Stuhlmann* 3525 (B† syntype); Malawi, Shire Highlands, fl. xii.1881, *Buchanan* 325 (B† syntype, K); Malawi, no locality, fl. 1891, *Buchanan* 774 & 832 (both B† syntypes, BM, K). FIGURE 8.6.**37**, 1–2.

Erect or straggling suffruticose herb, few- to much-branched from a woody base, (15)40–130 cm tall; stems weakly 4-ridged when young, subterete and often stout with age, white(-buff) pilose to lanate when young, later ± glabrescent. Leaves subsessile, often drying black, elliptic, lanceolate or linear, 3.5–16.5 × 0.3–2.5 cm, base cuneate to obtuse, margin entire or minutely toothed, apex acute-apiculate, indumentum as stems when young, hairs persisting on margin and veins beneath or glabrescent; prominently 3(5)-veined from base. Inflorescence axillary in distal portion of stems, 1(2) per axil, often in opposite pairs, each a contracted unilateral spike, 1.5–5.5 cm long; peduncle 0–7 mm long; sterile bracts black- to brown-scarious, (oblong-)ovate to lanceolate, 11–25 × 2.5–7(11) mm, apex long-attenuate to acuminate, straight to somewhat falcate, white- or buff-plumose at least on margin; fertile bracts similar but often somewhat smaller, surface more densely hairy; bracteoles brown-scarious or one of a pair green with dark brown apex, narrowly elliptic to lanceolate, 11–20.5 × 1.5–6 mm, with a linear acumen. Calyx lobes green with dark brown(-purple) acumen, indumentum as fertile bracts, dense; anterior lobes oblong-elliptic to lanceolate, 9–16 × 2–4.5 mm, linear acumen sometimes recurved; posterior lobe ovate to lanceolate, 10.5–17 × 2.5–7.5 mm, venation inconspicuous. Corolla 11–16 mm long, white, grey or pale purple, lobes usually purple, limb with purple markings towards margin and with yellow(-orange) median patch on lower lip, throat and limb silky-hairy externally; tube 6–9 mm long; upper lip 4–6 mm long; lower lip 4.5–7.5 mm long including lobes 1.5–3 mm long. Anterior (longer) stamens with filaments 3–4.5 mm long, thecae 1.4–1.8 mm long, suture cilia short and often sparse. Ovary glabrous; style with long hairs and subsessile glands in proximal half. Capsule 7–8 mm long, glabrous; seeds 4–4.5 × 3–3.5 mm.

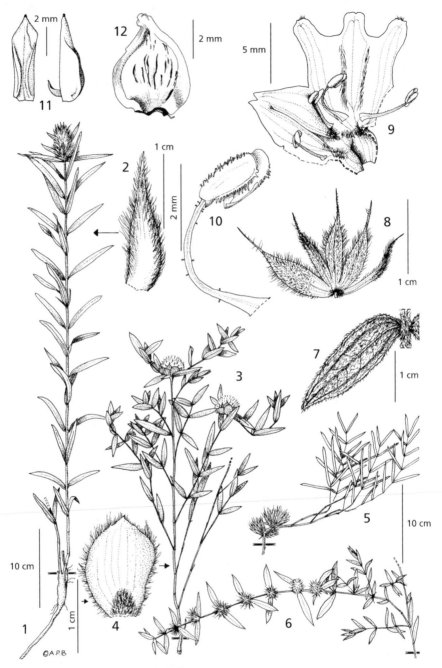

Fig. 8.6.**37**. LEPIDAGATHIS ANDERSONIANA. 1, habit; 2, sterile bract from near base of inflorescence. LEPIDAGATHIS ERIOCEPHALA. 3, habit; 4, sterile bract from near base of inflorescence. LEPIDAGATHIS FISCHERI. 5, habit. LEPIDAGATHIS LANATOGLABRA. 6, habit; 7, abaxial leaf surface showing primary venation; 8, dissected calyx, external surface, posterior lobe outermost left; 9, dissected corolla with stamens; 10, detail of abaxial pair of stamens; 11, capsule valve, lateral and face views; 12, detail of mature seed. 1, 2 from *Milne-Redhead & Taylor* 9598, 3, 4 from *Richards* 13798, 5 from *Bidgood et al.* 7058, 6 from *Kayombo* 1017, 7–12 from *Richards* 13537. Drawn by Andrew Brown. Reproduced from Flora of Tropical East Africa (2010).

Zambia. E: Lundazi Dist., south of Katete R., Luangwa Valley, fl. 27.iv.1969, *Taylor* 309 (SRGH). **Malawi**. S: Blantyre Dist., Michiru Mt, near CDC Estate, fl. 6.iii.1979, *Blackmore & Brummitt* 615 (BM, K, MAL). **Mozambique**. N: Mandimba Dist., 16 km NE of Mandimba border post, fl.& fr. 3.v.1960, *Leach* 9905 (K, LISC, LMA, SRGH). Z: Gurué Dist., near Gurué (Vila Junqueiro), fl. 5.iv.1943, *Torre* 5081 (BR, LISC, LUA, MO, PRE, WAG).

Also in Rwanda, Burundi, D.R. Congo and Tanzania. In miombo woodland, clearings and fire-prone grassland, typically on sandy soils or rocky hillslopes, usually flowering during and immediately after the main rains, Feb–June(July); 550–1300 m.

Conservation notes: Fairly common in both S Malawi and N Mozambique; assessed as Least Concern in F.T.E.A.

9. **Lepidagathis eriocephala** Lindau in Bot. Jahrb. Syst. **30**: 409 (1901). —Darbyshire in F.T.E.A., Acanthaceae **2**: 320 (2010). Type: Tanzania, Usafwa (Usafua), Pungulumo (Bunguluma) Mt, fl. vii.1899, *Goetze* 1083 (B† holotype, BM, BR). FIGURE 8.6.**37**, 3–4.

Lepidagathis andersoniana sensu Clarke in F.T.A. **5**: 126 (1899), in part for *Whyte* s.n. from Chitipa (Fort Hill), Malawi, non Lindau.

Erect to decumbent, rarely prostrate, suffruticose herb, often much-branched from a woody base, 20–80 cm tall; stems 4-ridged, white-pilose when young at least at nodes, later glabrescent. Leaves subsessile, elliptic to narrowly lanceolate, 2.5–7.5 × 0.4–1.3 cm, base cuneate to rounded, margin entire or minutely toothed, apex acute-apiculate, surface glabrous or sparsely white-pilose mainly on margin and midrib beneath; prominently 3-veined from base. Inflorescence axillary in distal portion of stems, rarely basal (see note), of several cymes compounded into dense globose heads 3–6.5 cm in diameter; sterile bracts dark brown-scarious, broadly oblong-elliptic to rounded or somewhat obovate, 11–17 × 5–13.5 mm, external surface convex, apex abruptly narrowed into an apiculum or short acumen, margin white-plumose; fertile bracts similar but oblong to spathulate, 3.5–7 mm wide, apex abruptly acuminate, margin and midrib densely white-plumose; bracteoles as fertile bracts but 1.5–4.5 mm wide. Calyx lobes green with dark brown(-purple) acumen, surface and margin densely white-pilose in proximal half; anterior lobes narrowly oblong-elliptic to lanceolate or oblanceolate, 9.5–11.5 × 1.5–2.5 mm; posterior lobe oblong-elliptic to -lanceolate, 10–13 × 2–4 mm, parallel-veined. Corolla 11.5–15 mm long, white to grey with purple markings on limb, with yellow median patch on lower lip, limb silky-hairy externally; tube 7.5–9.5 mm long; upper lip 4–5.5 mm long; lower lip 4–6 mm long including lobes 1.5–3.5 mm long. Anterior (longer) stamens with filaments 3–4.5 mm long; thecae 1.2–1.5 mm long. Ovary glabrous; style with long hairs and subsessile glands in proximal half. Capsule ± 6.5 mm long, glabrous; only immature seeds seen.

Zambia. E: Chama Dist., Nyika Plateau, Rest House, fr. 27.x.1958, *Robson* 392 (K). **Malawi**. N: Chitipa Dist., Misuku Hills, Kanjera area, fl. 19.iv.1976, *Pawek* 11134 (BR, DSM, EA, K, MAL, MO, SRGH, UC). S: Ntcheu/Mwanza Dist., lower Kirk Range, Chipusiri, fl. 17.iii.1955, *Exell, Mendonça & Wild* 957 (BM, LISC, SRGH). **Mozambique**. N: Lichinga Dist., Posto Zootécnico de Lichinga (Vila Cabral), fl. 28.ii.1964, *Torre & Paiva* 10886 (BR, LISC). T: Tsangano Dist., Vila Mouzinho, Estação Zootécnica, fl. 16.vii.1949, *Barbosa & Carvalho* 3640 (K).

Also in S Tanzania. In fire-prone grassland and miombo woodland, typically on sandy soils or rocky hillslopes, flowering during and immediately after the main rains, Feb–May (occasionally later in July, or Sept–Oct following burning); 1000–1600 m.

Conservation notes: Rather scarce and patchily distributed; assessed as Near Threatened in F.T.E.A.

Separated from other members of the *L. andersoniana* complex by having large, ± globose compounded synflorescences with broad, convex, abruptly narrowed sterile bracts and a dense wooly indumentum. Typical *L. eriocephala* from S Tanzania and N

Malawi is erect or decumbent with the inflorescence held aerially. However, plants from N and W Mozambique and S Malawi can be prostrate with the inflorescences apparantly held against the ground. Some specimens from the Nyika Plateau (Malawi, Zambia, e.g. *Robson* 392) appear to match *L. eriocephala* in terms of bract morphology but have the flowering heads held at or towards the base of the stems, similar to *L. fischeri*; these latter populations may be of hybrid origin or an ecological variant in response to differences in fire regime.

10. **Lepidagathis longisepala** C.B. Clarke in F.T.A. **5**: 125 (1899). Type: Zambia, Urungu, Fwambo, S of Lake Tanganyika, fl. 1896, *Nutt* s.n. (K holotype).

Trailing suffruticose herb, branches 20–75 cm long; stems wiry, 4-ridged, glabrous except for tuft of white hairs at nodes. Leaves subsessile, narrowly elliptic to lanceolate, 2.7–6 × 0.4–0.9 cm, base cuneate to acute, margin entire, apex acute-apiculate, surface glabrous or with a few long hairs along margin and midrib beneath when young; 3-veined from base. Inflorescence axillary in upper half of stem, of several fasciculate unilateral cymes compounded into heads 3.5–4 cm in diameter, 1 to few per stem, subsessile; bracts green with purple towards apex when young but soon brown-scarious; sterile bracts lanceolate, (17.5)21.5–34 × 3.5–6 mm, attenuate, straight to somewhat falcate, margin white-pilose, surface largely glabrous or shortly hairy; fertile bracts subhyaline and pale brown when young, narrowly lanceolate, (16)18–22 × 2.5–3.5 mm, long-acuminate, margin and often midrib white-plumose; bracteoles as fertile bracts but somewhat narrower. Calyx lobes green with dark purple-brown acumen, white plumose in upper half and along margin; anterior lobes lanceolate, (11)15.5–17 × 1.3–3 mm wide, with long linear acumen; posterior lobe as anterior but oblong-lanceolate, (12)16–19.5 × 2.5–4 mm, with 5 or 7 parallel veins. Corolla 11.5–15 mm long, white or pale mauve, palate of lower lip with purple markings towards margin and throat, yellow-orange markings centrally, silky-hairy externally mainly on limb; tube 6.5–8.5 mm long; upper lip 3.5–4.5 mm long; lower lip 5–6.5 mm long including lobes 3 mm long. Anterior (longer) stamens with filaments 3–4 mm long; thecae 1.6–1.8 mm long. Ovary glabrous; style with long hairs and subsessile glands in proximal half. Capsule and seeds not seen.

Zambia. N: Mbala Dist., Lumi R., 3 km from Kawimbe, fl. 31.v.1957, *Richards* 9953 (BR, K, SRGH).

Not known elsewhere. On ironstone rock outcrops and amongst grass; flowering towards end or immediately after main rains, April–May; c.1500 m.

Conservation notes: Endemic to NE Zambia, known only from 3 collections within a highly restricted range; possibly threatened.

This species was separated from closely related taxa by Clarke on account of its long-acuminate calyx lobes, bracts and bracteoles. With the combination of trailing stems and inflorescences held in the leafy axils, it is most likely to be confused with *L. lanatoglabra* or prostrate forms of *L. eriocephala*. It is easily separable from the former by the much larger bracts and much longer silky hairs on the inflorescence, and from the latter by the longer and much more gradually tapered bracts and much longer acumen to the calyx lobes.

11. **Lepidagathis randii** S. Moore in J. Bot. **64**: 307 (1926). Type: Zimbabwe, Mwami (Miami), fl. 8.vi.1926, *Rand* 152 (BM holotype).

Lepidagathis dicomoides Hutch., Botanist Sthn. Afr.: 505 (1946). —Mapaura & Timberlake, Checklist Zimb. Vasc. Pl.: 14 (2004). —Phiri, Checklist Zamb. Vasc. Pl.: 19 (2005). Type: Zambia, 9.6 km SW of Mpika, fl.& fr. 16.vii.1930, *Hutchinson & Gillett* 3779 (K holotype, BM).

Erect, decumbent or rarely trailing suffruticose herb, few- to several-branched from a woody base, 15–70 cm tall; stems often wiry, 4-ridged, glabrous or sparsely to densely pale pilose when young. Leaves subsessile, linear(-lanceolate) or narrowly oblong, 4.5–8.8 × 0.2–1 cm, base cuneate, margin entire, apex acute-apiculate, sparsely to more densely pilose when young, at least along margin, later glabrescent; 3-veined from base but sometimes only midrib prominent. Inflorescence axillary in upper half of stems, rarely also basal, several fasciculate unilateral cymes per axil, together forming dense globose or hemispheric heads 1.8–5 cm in diameter; bracts brown-scarious with darker acumen, sterile bracts (linear-)lanceolate, 10–17 × 1.5–4 mm, apex attenuate or with a linear acumen, margin white-pilose, surface glabrous or shortly hairy; fertile bracts similar but often subhyaline, 11.5–19 × 1.5–3.5 mm, linear acumen more pronounced, surface often more densely hairy at least along midrib; bracteoles as fertile bracts but to 2.7 mm wide, those on exterior of inflorescence sometimes green. Calyx lobes green with dark brown(-purple) acumen, with rather dense mixed short and long ascending white hairs, hairs on margin more spreading; anterior lobes (linear-)lanceolate, 10–16 × 1.3–2.7 mm, acumen linear; posterior lobe 2–3.2 mm wide. Corolla 10–15 mm long, white, grey or pale pink with pink-purple lobes, palate of lower lip with purple markings towards margin, yellow markings centrally, upper lip sometimes with few purple markings, silky-hairy externally mainly on limb; tube 5.5–8.5 mm long; upper lip 3–5 mm long; lower lip 3.5–6.5 mm long including lobes 1–2.5 mm long. Anterior (longer) stamens with filaments 3–5 mm long; thecae 1.2–1.7 mm long. Ovary glabrous or with minute sessile glands in distal half; style with long hairs and subsessile glands in proximal half. Capsule 5.5–7.5 mm long, glabrous; seeds 3–3.7 × 2.5–3 mm.

Zambia. N: Mpika Dist., North Luangwa Nat. Park, fl. 14.v.1994, *P.P. Smith* 608 (K). C: Serenje Dist., road from Kapiri to Serenje, fl.& fr. 9.ix.1967, *Richards* 22303 (K). S: Choma Dist., Siamambo, fl. 17.v.1961, *Fanshawe* 6573 (K, NDO, SRGH). **Zimbabwe**. N: Hurungwe Dist., Zwipani, fl.& fr. 12.x.1957, *Phipps* 787 (BR, EA, K, LMA, SRGH). C: Chegutu Dist., Hartley A Safari Area, fl. 3.vii.1974, *Müller* 2172 (SRGH).

Not known elsewhere. In miombo and *Combretum–Terminalia* woodland, open grassland, roadsides and other disturbed areas, rocky hillslopes, flowering at end and after main rains, Mar–July(Oct); 700–1600 m.

Conservation notes: Locally frequent in E Zambia and N Zimbabwe in suitable habitat; Least Concern.

The typical erect to decumbent form of this species is close to *Lepidagathis andersoniana*, differing principally in having more numerous, fasciculate cymes at each node which, when mature, form a globose or hemispheric synflorescence; *L. andersoniana* has only 1(2) cyme in each axil, which is often more clearly elongate and usually does not form globose or hemispheric heads. *L. randii* is usually more wiry than *L. andersoniana* with narrower leaves, bracts, bracteoles and calyx lobes, although with some overlap with slender forms of *L. andersoniana* from SE Tanzania and N Mozambique. In addition, *L. randii* is never lanate on the leaves and stems when immature, the hairs on the margins of the calyx and bracts/bracteoles are generally shorter and the cilia on the anther sutures are more numerous and evenly distributed.

Some plants of *L. randii* have trailing branches and these can appear very similar to *L. lanatoglabra*. *L. randii* typically has more dense, globose or hemispheric synflorescences, with the bracteoles and calyx lobes having a more pronounced linear acumen with more numerous long ciliate hairs; in *L. lanatoglabra* the long ciliate hairs are usually dense on the widened basal portion of the calyx lobes but sparse or largely absent on the acumen (although they are white-plumose throughout in the types of *L. sparsiceps*).

One plant of *Bingham* 13076 (from C Zambia, 70 km E of Lusaka) is unusual in having a basal synflorescence in addition to those in the upper half of the stems.

12. **Lepidagathis lanatoglabra** C.B. Clarke in F.T.A. 5: 124 (1899). —Darbyshire in F.T.E.A., Acanthaceae 2: 321 (2010). Types: Malawi, Tanganyika Plateau (Chitipa Dist.), fl. vii.1896, *Whyte* s.n. (K syntype); Khondowe to Karonga, fl. vii.1896, *Whyte* s.n. (K syntype); North Nyassa, Songwe and Karongas, fl. vii.1896, *Whyte* s.n. (K syntype). FIGURE 8.6.**37**, 6–12.

Lepidagathis lanatoglabra var. *latifolia* C.B. Clarke in F.T.A. **5**: 125 (1899). Type: Malawi, Tanganyika Plateau, Chitipa (Fort Hill), fl. vii.1896, *Whyte* s.n. (K holotype).
Lepidagathis sparsiceps C.B. Clarke in F.T.A. **5**: 124 (1899). Types: Malawi, Mpata and commencement of Tanganyika Plateau, fl. vii.1896, *Whyte* s.n. (K syntype); Nyika Plateau, fl. vi.1896, *Whyte* s.n. (K syntype); Manganja Hills, fl. iv.1859, *Kirk* s.n. (K syntype).
Lepidagathis nematocephala Lindau in Bot. Jahrb. Syst. **30**: 409 (1901). Type: Tanzania, Usafwa, Pungulumo (Bunguluma) Mt, fl. vii.1899, *Goetze* 1084 (B† holotype, BM, BR).

Prostrate, procumbent or weakly decumbent suffrutex, branching from a woody base, branches to 30–60 cm long; stems slender, 4-ridged, sparsely to densely pale pilose when young or glabrous. Leaves (sub)sessile, linear, lanceolate, ovate or elliptic, 2.7–7 × 0.2–2 cm, base cuneate to rounded, margin entire, apex acute-apiculate, surface glabrous or pale-pilose, often lanate when young; prominently 3-veined from base. Inflorescence axillary either throughout stem or mainly at proximal nodes, held at ground level, cymes fasciculate, 1–3 per axil, together forming heads 1.5–3.7 cm in diameter, subsessile; bracts brown-scarious with darker, sometimes purple-tinged acumen, rarely green when young, sterile bracts lanceolate, 6–15 × 1.5–5 mm, apex attenuate or with a linear acumen, straight to somewhat curved, white-pilose mainly or only on margin; fertile bracts similar but often hyaline when young, 11.5–15 mm long, linear acumen more pronounced, often more densely hairy on the prominent midrib; bracteoles as fertile bracts but 1.5–3 mm wide, those on exterior of inflorescence sometimes green-tinged and with fewer hairs on midrib. Calyx green, paler towards base, with dark purple(-brown) acumen; anterior lobes elliptic-lanceolate, 9–15 × 1.5–3 mm, acumen linear, ± straight, surface with mixed short and long ascending white hairs, with dense long ± spreading hairs on margin except on acumen; posterior lobe ovate to lanceolate, 2–4 mm wide, with 5 or 7 parallel veins ± prominent. Corolla 11–14.5 mm long, white or more rarely blue to violet, palate of lower lip with yellow or orange markings centrally and purple markings towards margin and/or in throat, limb silky-hairy externally; tube 7–8.5 mm long; upper lip 3–4.5 mm long; lower lip 3.5–6 mm long including lobes 1.5–3 mm long. Anterior (longer) stamens with filaments 3.7–4.7 mm long; thecae 1.2–1.7 mm long. Ovary glabrous; style with long hairs and subsessile glands in proximal half. Capsule 6–7.5 mm long, glabrous; seeds 3.3–4 × 2.5–3 mm.

Zambia. N: Nchelenge Dist., Chiengi, L. Mweru, fl. 16.vii.1957, *Whellan* 1391 (K, LISC, SRGH). W: Mwinilunga Dist., between Mwinilunga and Matonchi Farm, fl. 16.v.1986, *Philcox et al.* 10331 (BR, K, SRGH). C: Mpika Dist., Mutinondo Wilderness Area, fl. 14.vi.1998, *P.P. Smith* 1715 (K). E: Chama Dist., 96 km NW of Lundazi, E Luangwa escarpment, fl. 2.vi.1954, *Robinson* 815 (BR, K, SRGH). **Malawi**. N: Karonga Dist., near Kayelekera, 30 km W of Karonga, fl. 6.vi.1989, *Brummitt* 18376 (BR, K, MAL). C: Mchinji Dist., Kalombo village between Bua R. and Mchinji (Fort Manning), fl.& fr. 5.viii.1936, *Burtt* 6201 (BM, BR, K). **Mozambique**. N: Cuamba Dist., 45 km NW of Cuamba (Nova Freixo), fl. 25.v.1961, *Leach & Rutherford-Smith* 11006 (K, LISC, SRGH).

Also in S and W Tanzania and SE D.R. Congo. In miombo woodland and grassland including recently burnt areas, typically on rocky hillslopes or over laterite, also on roadsides, bare or disturbed ground; usually flowering towards end or after main rains, Apr–July(Sept); 900–1700 m.

Conservation notes: Common particularly in N Zambia; assessed as Least Concern in F.T.E.A.

Specimens intermediate between *L. lanatoglabra* and *L. fischeri* are occasionally recorded (e.g. *Brummitt* 11596 from Malawi) and it is quite possible that variation across these taxa will prove to be clinal, but most specimens are easily placed in one or the other.

Plants with rather broad sterile bracts and densely woolly inflorescences from central Malawi (e.g. *Brummitt* 9588) and adjacent Tete Province of Mozambique tend towards *L. eriocephala*; further studies are required to quantify the extent of intergrading. The types of *L. sparsiceps* also have woolly inflorescences but are otherwise a good match for typical *L. lanatoglabra*.

13. **Lepidagathis fischeri** C.B. Clarke in F.T.A. **5**: 123 (1899). —Darbyshire in F.T.E.A., Acanthaceae **2**: 322 (2010). Types: Tanzania, Singida Dist., Unyamwezi, Ussure (Usuri), n.d., *Fischer* 490 (B† holotype); Tanzania, Tabora Dist., 23 km on Ipole to Inyonga road, fl. 27.v.2008, *Bidgood, Leliyo & Vollesen* 7079 (K neotype, CAS, DSM, NHT), neotypified by Darbyshire (2010). FIGURE 8.6.**37**, 5.

   Lepidagathis lindauiana De Wild. in Ann. Mus. Congo, Bot., sér.4 **1**: 145 (1903). Type: D.R. Congo, Lukafu, v.1900, *Verdick* 520 (BR holotype).

   Lepidagathis rogersii Turrill in Bull. Misc. Inform., Kew **1912**: 360 (1912). —Morton in Bot. J. Linn. Soc. **96**: 342, fig.3 (1988). —Phiri, Checklist Zamb. Vasc. Pl.: 19 (2005). Type: D.R. Congo, Sakania, fl. 18.viii.1911, *Rogers* 10032 (K holotype, BM, GR, NBG).

Suffruticose herb with prostrate, decumbent or erect leafy stems branching from a woody base and/or rootstock, leafy stems 5–40 cm long, rarely undeveloped at flowering; stems 4-ridged, sparsely or rarely densely pale pilose when young or glabrous throughout. Leaves (sub)sessile, linear to broadly (ovate-)elliptic, 1.5–13 × 0.15–3 cm, base cuneate to rounded or subcordate, margin entire, apex acute-apiculate, glabrous or sparsely pale-pilose, rarely densely so when young; prominently 3(5)-veined from base, then often with scalariform tertiary veins visible, sometimes only midrib prominent in linear-leaved variants. Inflorescence basal and/or at lower, usually leafless nodes, often compounded into dense cushion-like synflorescences, 3–11.5 cm in diameter, subsessile or on short peduncles with reduced sterile bracts; sterile floral bracts brown-scarious or green to purple-brown when young, lanceolate, 11–25.5 × 2.8–6.5 mm, apex attenuate to acuminate, margin white-pilose to -plumose, surface glabrous or often minutely pubescent; fertile bracts similar but often thinner, 12–22.5 × 2.5–4 mm, acumen often more pronounced, often with long pilose hairs along midrib; bracteoles as fertile bracts but 2–3.3 mm wide, those on exterior of inflorescence often green with a darker acumen, with fewer hairs on midrib. Calyx lobes green, usually with dark purple or brown acumen, surface with short or mixed short and long ascending white hairs, margin with dense long ± spreading hairs, often shorter on acumen; anterior lobes (elliptic-)lanceolate, 10.5–18 × 2–3.7 mm, acumen linear, ± straight; posterior lobe 12–19 × 2.5–5 mm, with 5 or 7 parallel veins ± prominent. Corolla 11.5–18 mm long, white or pale purple, palate of lower lip with yellow to orange markings centrally and with purple markings towards margin and/or in throat, limb silky-hairy externally; tube 5.5–11 mm long; upper lip 3.5–5 mm long; lower lip 4.5–7 mm long including lobes 2.5–3.7 mm long. Anterior (longer) stamens with filaments 3.5–4.5 mm long; thecae 1.3–1.8 mm long. Ovary glabrous; style with few hairs and subsessile glands in lower half. Capsule 8–9 mm long, glabrous; seeds c.5 × 3.5 mm.

Zambia. N: Chinsali Dist., 26 km on Chinsali to Kasama road, fl. 13.iv.1986, *Philcox, Pope & Chisumpa* 9911 (BR, K, SRGH). W: Mufulira, fl. 5.vi.1934, *Eyles* 8200 (BM, K, SRGH). C: Mumbwa Dist., Mumbwa, fl. n.d., *Gairdner* in *Macauley* 828 (K). E: Lundazi Dist., Lundazi, fl. 20.viii.1965, *Fanshawe* 9292 (K, NDO). **Malawi**. N: Rumphi Dist., 1.6 km into Nyika road, fl. 4.vii.1971, *Pawek* 4946 (K, MAL).

Also in Tanzania and D.R. Congo. In miombo woodland and grassland and recently burnt areas, typically on rocky hillslopes or over laterite, also on roadsides, bare or disturbed ground; flowering after main rains or immediately following burning, Apr–Aug(Nov); 1000–1700 m.

Conservation notes: Rather common in miombo woodland in N Zambia; assessed as Least Concern in F.T.E.A.

Several regional forms are recognisable in this variable species (see F.T.E.A.). In our area, a rather distinctive form from the vicinity of Kalambo Falls in NE Zambia has linear leaves and very slender bracts, bracteoles and calyx lobes, each with a long linear acumen.

14. **Lepidagathis sp. A** of F.T.E.A., Acanthaceae **2**: 324 (2010).

Erect, decumbent or spreading suffruticose herb; stems somewhat 4-ridged when young, soon terete, glabrous to pale-yellow pilose. Leaves sessile, ovate, lanceolate or oblong-elliptic, 5.5–7 × 1.3–2.5 cm, base obtuse to subcordate, margin entire, apex acute-apiculate, glabrous or pilose on margin when young; prominently 3-veined from base, with scalariform tertiary venation. Inflorescence axillary, restricted to distal portion of stems or more widespread, cymes fasciculate, compounded into globose heads 3–5 cm in diameter when mature, sessile; sterile bracts green at first but turning brown-scarious, lanceolate, 18–24 × 3–5 mm, apex attenuate, yellow-pilose at least on margin; fertile bracts and bracteoles similar but green with prominent pale midrib, narrowly lanceolate, 18–22 × 1.5–3 mm, gradually narrowed into linear acumen, more densely yellow-pilose throughout. Calyx lobes green throughout, ± densely long yellow pilose; anterior lobes linear-lanceolate, 15.5–19 × c.2.5 mm, acumen linear; posterior lobe lanceolate, 3.5–4.5 mm wide, with 3 or 5 prominent parallel veins. Corolla 13.5–15 mm long, white or cream, sometimes green-veined, palate of lower lip with orange markings centrally and with purple markings towards margin, limb silky-hairy externally; tube 8–10 mm long; upper lip 3.5–5.5 mm long; lower lip 4.5–5.5 mm long including lobes 2.5–3 mm long. Anterior (longer) stamens with filaments 4–4.5 mm long; thecae c.1.5 mm long. Ovary with few sessile glands towards apex; style with long hairs in proximal half. Capsule and seeds not seen.

Zambia. N: Mbala Dist., Kalambo R., path to Sansia Falls, fl. 8.v.1961, *Richards* 15115 (K, LISC); Mbala Dist., no locality, fl. 19.iv.1969, *Richards* 24484 (K).

Also in SW Tanzania. In miombo woodland over gritty soils, flowering towards end of main rains, Apr–May; 1500–1600 m.

Conservation notes: Known from only 3 specimens; assessed as Data Deficient in F.T.E.A. but possibly threatened.

The yellow-green inflorescences are very distinctive but more material of the *L. andersoniana* complex from the Zambia–Tanzania border region is required before conclusions on its status and rank can be reached.

15. **Lepidagathis sp. B** of F.T.E.A., Acanthaceae **2**: 324 (2010).

Pyrophytic herb, leafy stems usually developing during or after flowering, erect to decumbent from a woody rootstock, sometimes rhizomatous; stems 4-ridged, white-pilose or largely glabrous. Leaves subsessile, often immature or absent at flowering, linear to elliptic, 3–6(10) × 0.3–0.7(1.5) cm, base cuneate, margin entire, apex acute-apiculate, surface glabrous or white-pilose; 3-veined from base. Inflorescence basal, cymes fasciculate or shortly spiciform to 4 cm long, usually compounded into cushions 2.5–8 cm in diameter, subsessile or shortly pedunculate, peduncle with triangular scale-like sterile bracts; sterile floral bracts inconspicuous, brown-scarious, triangular to lanceolate, 4.5–11 × 1.5–3.5 mm, apex acute to attenuate, white-pilose at least along margin; fertile bracts similar but pale-hyaline or green when young, sometimes scarious with age, more densely pilose; bracteoles green throughout or one of each pair hyaline and with brown apex, (linear-)lanceolate to elliptic or oblanceolate, 5.5–13 × 1–2.5 mm, apex acute or shortly acuminate, surface densely white-pilose. Calyx lobes green, with or without a short brown acumen, surface densely white-pilose, with minute subsessile glands; anterior lobes lanceolate, elliptic or oblanceolate, 8.5–14.5 × 1.8–4 mm; posterior lobe ovate(-elliptic) to lanceolate, 2.3–4.5 mm wide, with 5 or 7 parallel veins ± prominent. Corolla 13–16 mm long, white, palate of lower lip and/or throat with yellow markings centrally and with brown or purple markings towards margin, limb silky-hairy externally; tube 8–9.5 mm long; upper lip 3.5–5 mm long; lower lip 4.5–7 mm long including lobes 2.5–3.5 mm long. Anterior (longer) stamens with filaments 3–5 mm long; thecae 1.4–2.1 mm long. Ovary glabrous; style with long hairs and subsessile glands in proximal half. Capsule c.7.5 mm long, glabrous. Seeds not seen.

Zambia. W: Solwezi Dist., Solwezi dambo, fl. ix.1962, *Holmes* 1518 (K, NDO).

Also in S and W Tanzania and S D.R. Congo. On recently burnt ground in dambos; c.1300 m.

Conservation notes: Rather local in Flora area; assessed as Data Deficient in F.T.E.A., but perhaps not threatened.

Most likely a further form of *L. fischeri*, but kept separate here as it appears quite discrete in parts of their overlapping ranges (notably in the Southern Highlands of Tanzania), though the distinction appears to break down somewhat in southern Congo.

27. BARLERIA L.[5]

Barleria L., Sp. Pl.: 636 (1753). —Nees in De Candolle, Prodr. **11**: 223 (1847). —Obermeijer in Ann. Transv. Mus. **15**: 127–180 (1933). —M.-J. & K. Balkwill in Kew Bull. **52**: 535–573 (1997); in J. Biogeogr. **25**: 95–110 (1998).

Perennial herbs, shrubs or climbers, sometimes spiny; stems 4-angular or subterete; cystoliths numerous, often occuring in adjacent cells and sometimes appearing to cross, ± conspicuous. Leaves evergreen or deciduous, opposite-decussate, petiolate or sessile, pairs equal or somewhat unequal, margin (sub)entire, rarely toothed or spinose. Inflorescences simple and axillary or compounded into a terminal synflorescence; cymose with monochasial and/or dichasial branching or reduced to a single flower; bract pairs (sub)equal, foliaceous or reduced and/or highly modified; bracteole pairs equal or dimorphic, often highly modified from bracts. Calyx divided almost to base, 4-lobed, anterior and posterior (outer) lobes usually considerably larger and proportionally broader than lateral (inner) lobes, anterior lobe often emarginate or bifid. Corolla usually showy; tube cylindrical throughout or campanulate to funnel-shaped above attachment point of stamens; limb 5-lobed or rarely the adaxial pair fully fused or absent, subregular to zygomorphic (see note). Fertile stamens (in our region) 2, abaxial; filaments arising from various levels within corolla tube, twisted through 180° and crossing near base; anthers bithecous, thecae parallel, muticous, exserted or rarely included within corolla tube; staminodes 2 or 3, adaxial staminode often less well developed than lateral pair or absent, with or without antherodes, these often producing a few pollen grains. Disk cupular. Ovary bilocular, 2 ovules per locule or one aborting very early; style filiform; stigma subcapitate to linear, either with 2 subconfluent lobes, or with 1 lobe much-reduced or absent. Capsule ± compressed laterally, fertile portion rounded to fusiform, with or without a prominent sterile apical beak, placental base inelastic but lateral walls sometimes thin and tearing from thickened flanks on dehiscence. Seeds 2 or 4 per capsule, held on retinacula, usually (ovate- or triangular-)discoid, with surfaces covered in hygroscopic hairs, rarely only subflattened and with unevenly distributed minute hairs.

A genus of c.300 species, mainly in the palaeotropics and subtropics with one species in the neotropics. Most diverse in the woodlands, bushlands and dry rocky terrain of eastern and southern Africa. Many of the species are very striking in flower, yet surprisingly few are widely cultivated; Jex-Blake (Gardening E. Africa, ed.4: 104, 1957) noted "the genus is sadly liable to disease".

The infrageneric classification proposed by M.-J. & K. Balkwill (1997) is followed here with minor modifications. Species are keyed out separately within each section. Sections in *Barleria* are easily separated when mature fruits are available and the key to sections for fruiting material should be used wherever possible. However, as species are more often observed and collected when in flower due to their showy corollas and to the persistence of the calyx lobes which often hide the fruit, a key to sections for flowering material is also given. As flowering material is often required for identification to species, ideal collections would comprise both flowers and fruits.

Notes on descriptions:

Calyx – although comprising a single whorl, the anterior and posterior lobes are typically larger than, and often enclose, the two lateral lobes and so are referred to collectively as the 'outer lobes'. The shape and indumentum of the posterior lobe is only described if differing from that of the anterior lobe.

Corolla – the corolla limb is rather variable in *Barleria* and can be divided into several arrangements:

Subregular, in which the corolla lobes are divided from the tube at ± the same point and are subequal in size, although the adaxial pair are often slightly narrower and the abaxial lobe is often slightly broader than the lateral pair (e.g. Fig. 8.6.**43**).

---

[5] By Iain Darbyshire

2+3 arrangement, in which the adaxial lobe pair form a 2-lobed upper lip and the abaxial and lateral lobes form a 3-lobed lower lip. In some cases in Sect. *Somalia* the adaxial lobe pair are partially fused so that the two lips are quite pronounced. In some species of Sect. *Barleria* the adaxial lobe pair are not partially fused but differ sufficiently from the lateral and abaxial lobes to form a weakly defined 2-lobed upper lip (e.g. Fig. 8.6.**38**).

1+3 arrangement, an extreme form of the 2+3 arrangement in Sect. *Somalia* in which the adaxial pair of lobes fully fuse.

0+3 arrangement, in which the adaxial pair of lobes are essentially absent; in our region this arrangement is restricted to *B. velutina*.

4+1 arrangement, the abaxial corolla lobe splits away from the tube considerably earlier than the lateral and adaxial lobes, the resultant limb being highly zygomorphic with a 1-lobed lower lip and 4-lobed upper lip (e.g. Figs. 8.6.**40,41,44**). This is the most widespread limb arrangement in the genus. In some species, the offset is only slight and so the arrangement is intermediate between 4+1 and subregular.

Cavirostrata-type, in species of Sect. *Cavirostrata* from our region, the lobes split from the tube at ± the same point but the two adaxial lobes are widely divergent to form a highly zygomorphic limb (Fig. 8.6.**42**).

The length of the corolla tube is measured from the base to the point at which the first corolla lobe splits; in species with the 4+1 arrangement, the length by which the abaxial lobe is offset from the remaining lobes (this being an 'open tube') is also recorded. The length of the corolla is measured from the base of the tube to the apex of the lateral lobes when flattened.

Androecium – in our region the lateral staminodes are much reduced in comparison to the fertile stamens pair, but they often bear antherodes which produce some pollen grains, leading to some authors to treat them as stamens. Therefore, although always referred to as staminodes in the descriptions and keys here, the key to genera in Flora Zambesiaca vol. 8.5 includes *Barleria* in both leads to couplet 6 which separates genera with 2 and 4 fertile stamens. In very rare cases, abnormal flowers occur in which all 5 stamens are developed. In some species from outside our region, four fertile stamens are present (M.-J. & K. Balkwill 1997).

Key to sections of Barleria *– fruits present*

1. Seeds not or only partially flattened, surface black with unevenly distributed minute pale grey hairs, glabrescent; capsule potentially 4-seeded*, prominently beaked . . . . .Sect. IV. **Cavirostrata** (p. 91)
   – Seeds discoid (surface often ovate or triangular), clothed in cream, golden, bronze or purple-brown hygroscopic hairs; capsule 2-seeded, or if 4-seeded then lacking a prominent beak 2
2. Capsule (potentially) 4-seeded, lacking a prominent beak, septum woody except for a shallow membranous portion above upper retinacula . . . .
   . . . . .Sect. I. **Barleria** (p. 40)
   – Capsule 2-seeded, septum largely membranous or if largely or wholly woody then capsule with a prominent beak. 3
3. Capsule with septum largely membranous, apex lacking a prominent beak, if beaked then stem and leaf indumentum stellate or dendritic; capsule wall often tearing from thickened flanks at dehiscence. 4
   – Capsule with septum woody throughout or with a shallow membranous portion above retinacula, apex prominently beaked; stem and leaf indumentum never stellate or dendritic; capsule wall not tearing from flanks 5
4. Indumentum of simple hairs; stigma subcapitate or clavate . . . .
   . . . . Sect. II. **Fissimura** (p. 81)

- Indumentum of predominantly stellate or dendritic hairs, often with a long central 'arm'; stigma linear. . . . . Sect. III. **Stellatohirta** (p. 85)
5. Capsule with a shallow membranous portion above retinacula; seeds with wavy (woolly) hygroscopic hairs; plants unarmed . . . . Sect. V. **Somalia** (p. 92)
- Capsule septum woody throughout; seeds with straight hygroscopic hairs; plants usually spiny, rarely unarmed . . . . .Sect. VI. **Prionitis** (p. 109)

Key to sections of Barleria *– flowers only*

1. Corolla yellow, orange, apricot, salmon or ochre-coloured, rarely brick red; anterior calyx lobe lanceolate, ovate or elliptic, to 10.5 mm wide but often less than 5 mm; plant often armed with 2–4(6)-rayed axillary spines clearly differentiated from bracteoles . . . . Sect. VI. **Prionitis** (p. 109)
- Corolla variously white or pink to blue or purple, if rarely yellow (42. *B. calophylloides*) or red (29. *B. repens*) then anterior calyx lobes broadly ovate or cordiform, (5.5)9–29 mm wide; plant unarmed, or if axillary spines present then these paired, being the persistent spinose bracteoles of an aborted or old inflorescence 2
2. Corolla limb with sinus between two adaxial lobes at markedly wider angle than other sinuses (Fig. **42**.9); outer calyx lobes with only palmate or parallel principal veins conspicuous, margins entire . . . . .Sect. IV. **Cavirostrata** (p. 91)
- Corolla limb with sinus between two adaxial lobes not at markedly wider angle than other sinuses, or if so (some species in Sect. *Barleria*, Fig. **39**.1) then outer calyx lobes with conspicuous palmate-reticulate venation and/or toothed or spinose margins 3
3. Indumentum stellate or dendritic, hairs often with long central 'arm'; outer calyx lobes with only palmate or subparallel principal veins conspicuous; flowers held in well-defined globose to cylindrical heads with each axil single-flowered. . . . . .
. . . . .Sect. III. **Stellatohirta** (p. 85)
- Indumentum of simple and/or medifixed (biramous) hairs, or if long-armed stellate hairs present then outer calyx lobes with conspicuous palmate-reticulate venation and flowers held in 1 to several-flowered axillary cymes, sometimes clustered towards stem apex but not in well-defined heads 4
4. Plant with bracteoles spinose or spine-tipped, those of old or aborted inflorescence often persisting as paired axillary spines . . . . Sect. I. **Barleria** (p. 40)
- Plant without spines. 5
5. Corolla tube abruptly funnel-shaped above attachment point of stamens, cylindrical below, limb in 4+1 arrangement; stigma subcapitate or clavate, broader than style; lateral staminodes with well-developed antherodes** . . Sect. II. **Fissimura** (p. 81)
- Corolla form not as above, or if so then stigma linear, ± same width as style; staminodes with or without antherodes. 6
6. Calyx thin and soon-scarious, with ± conspicuous reticulate venation, margin often toothed, and/or lateral staminodes with antherodes present and/or flowers held in conspicuously pedunculate single- or few-flowered cymes widely spreading from or patent to stems . . . . Sect. I. **Barleria** (p. 40)
- Calyx not scarious, only palmate or parallel main veins conspicuous, margin entire; staminodes always lacking antherodes; cymes not as above. . . . .
. . . . Sect. V. **Somalia** (p. 92)

* One or more seeds often abort in all sections but it is easy to determine the potential number of seeds by checking for reduced retinacula and comparing the size of the mature seeds to the size of locule.
** Some specimens of 31. *B. gueinzii* (Sect. *Barleria*), would key out to Sect. *Fissimura* here in the absence of fruits; this species is therefore included in the key to Sect. *Fissimura* (p. 81).

Barleria L. sect. I. **Barleria**, M.-J. Balkwill & K. Balkwill in Kew Bull. **52**: 555 (1997).

Barleriacanthus Oerst. in Vidensk. Meddel. Dansk Naturhist. Foren. Kjobenhavn: 136 (1854), lectotypified by M.-J. & K. Balkwill in Kew Bull. **52**: 555 (1997).

Barleriosiphon Oerst. in Vidensk. Meddel. Dansk Naturhist. Foren. Kjobenhavn: 136 (1854).

Dicranacanthus Oerst. in Vidensk. Meddel. Dansk Naturhist. Foren. Kjobenhavn: 136 (1854), lectotypified by M.-J. & K. Balkwill in Kew Bull. **52**: 555 (1997).

Barleria subgen. *Acanthoidea* sensu Clarke in F.T.A. **5**: 141 (1899) in part, excl. *B. flava.*

Barleria sect. *Eubarleria* subsects. *Pungentes, Aculeatae, Innocuae* and *Heterotrichae* sensu Obermeijer in Ann. Transv. Mus. **15**: 133–138 (1933).

Barleria sect. *Chrysothrix* M. Balkwill in J. Biogeogr. **25**: 110 (1998) in part. —M. & K. Balkwill in Kew Bull. **52**: 558 (1997) in part (see Darbyshire in F.T.E.A., Acanthaceae 2: 329, 2010 for explanatory note).

Plants unarmed or often with spinose bracteoles of aborted or old inflorescences persisting as paired axillary spines. Indumentum simple or stellate; leaves often with broad sessile glands towards base, these sometimes also on bracteoles and calyx. Inflorescences unilateral (scorpioid) or single-flowered cymes, or rarely partially dichasial at maturity; axillary with foliaceous bracts, more rarely compounded into a terminal thyrse or spike with ± modified bracts. Calyx often turning scarious, outer lobes usually with conspicuous palmate-reticulate venation. Corolla white, blue, purple or red, often drying blue with darker venation or blue-black throughout; limb in 4+1, 2+3 or subregular arrangement. Staminodes 3, lateral pair with antherodes well-developed or absent. Stigma linear to capitate, apex ± bilobed. Capsule usually drying black or dark brown, laterally flattened, oblong-fusiform, without a prominent beak; lateral walls usually remaining attached to flanks at dehiscence; septum with a shallow membranous portion above upper retinacula, elsewhere woody. Seeds 4 (or fewer by abortion), discoid with dense matted hygroscopic hairs, often purplish, brown or bronze.

A section of over 100 species, most diverse in southern Africa; the largest section in our region.

Key to Sect. Barleria

1. Indumentum of simple hairs only. 2
– Indumentum including at least some long-armed stellate hairs on stems, leaves and/or calyx (check young parts as stellate base sometimes caducous) 32
2. Cyme conspicuously pedunculate, primary peduncle 8–45 mm long, often held at wide angle from stem and ± arcuate; single-flowered, or if 2 or more flowered then lax 3
– Cyme sessile or with a short and inconspicuous primary peduncle to 5 mm long, or if longer (to 9 mm) then cymes 2 or more flowered and contracted. 7
3. Adaxial corolla lobes 3.8–5 mm long, less than ⅓ of lateral lobe length; outer calyx lobes in fruit sometimes with prominent spinulose teeth along margin of distal half. . . . .**18.** *hydeana*
– Adaxial corolla lobes 9–21 mm long, over ⅔ or subequal to lateral lobe length; outer calyx lobes with margin entire or with minute teeth formed from swollen hair bases. 4
4. Cyme usually 2 or more flowered (in our region); outer calyx lobes appearing striate due to prominent dark parallel main veins; capsule 13–16.5 mm long; corolla tube (narrowly) campanulate above attachment point of stamens; lateral staminodes with well-developed antherodes. . . . . **32.** *obtusa*
– Cyme usually single-flowered (rarely 2-flowered); outer calyx lobes with darker reticulate venation, not appearing striate; capsule 17–23 mm long; corolla tube subcylindrical, barely widened towards mouth; lateral staminodes with antherodes absent or obsolete. 5

5. Inflorescence with glandular hairs (if present) interspersed within dense indumentum of eglandular hairs; bracteoles 2–6.5 mm long, if longer then ovate, apiculate; style with short glandular hairs towards apex . . . . **15.** *aromatica*
   – Inflorescence densely viscid-glandular, eglandular hairs absent or sparse and inconspicuous; bracteoles 5–33 mm long, linear, apex spinulose or bracteoles spinose throughout; style lacking glandular hairs 6
6. Bracteoles 5–13 mm long; corolla tube 26–35 mm long, less than twice as long as abaxial lobe; anthers 3–5 mm long . . . . **16.** *glutinosa*
   – Bracteoles 15–33 mm long; corolla tube 35–65 mm long, 2–3.5 times longer than abaxial lobe; anthers 5–7.5 mm long. . . . .**17.** *molensis*
7. Adaxial pair of corolla lobes absent or obsolete, to 2 mm long, not narrowed towards base; style with short glandular hairs towards apex . . . .**10.** *velutina*
   – Adaxial pair of corolla lobes present, sometimes considerably smaller than lateral and abaxial lobes but always well-developed and narrowed towards base; style lacking glandular hairs. 8
8. Stout suffrutex with erect, largely unbranched stems from woody base; flowers held in dense terminal cylindrical or conical heads, bracts clearly modified from leaves; outer calyx lobes broadly ovate with coarsely and irregularly dentate margin. . . . . **11.** *sceptrum-katanganum*
   – Without above combination of characters 9
9. Plant unarmed 10
   – Bracteoles spinose or prominently spine-tipped; calyx and leaf tips often also spiny. 14
10. Corolla bright red to rose-pink; outer calyx lobes ± broadly ovate with rounded or cordate base, margin (sub)entire, accrescent and turning glossy brown in fruit; stigma subcapitate. . . . . **31.** *repens*
    – Corolla blue, mauve, purple or white; if outer calyx lobes ovate and with rounded or cordate base, then margin often toothed, surface not turning glossy brown in fruit; stigma linear or clavate 11
11. Corolla tube cylindrical, barely widened towards mouth, stamens attached in distal half of tube; limb in 2+3 arrangement or subregular, abaxial lobe not offset from remaining lobes. 12
    – Corolla tube campanulate to funnel-shaped above attachment point of stamens in proximal half of tube; limb in 4+1 arrangement or weakly so, abaxial lobe offset by 2–9.5 mm from remaining lobes 13
12. Leaves subsessile, attenuate in basal portion but base abruptly cordate and often somewhat amplexicaul when young . . . . **2.** *whytei*
    – Leaves petiolate, base cuneate or obliquely so . . . . **20.** *fissimuroides*
13. Leaves elliptic with long and narrow attenuate or cuneate base, sessile or with short petiole to 6 mm; bracteole pairs markedly unequal, larger of each pair ovate or elliptic, (3)6–12 mm wide . . . . **1.** *holstii*
    – Leaves ovate(-elliptic), base rounded, truncate or shortly and broadly attenuate, petiole conspicuous, 6–27 mm long; bracteole pairs subequal, linear or narrowly elliptic-lanceolate, 0.5–2.5 mm wide . . . . **33.** *gueinzii* *
14. Bracteoles large, pairs often unequal, larger of each pair lanceolate to broadly ovate, 4–20 mm wide at widest point (excluding lateral spines); low spreading subshrub or suffrutex, stems often trailing or weakly decumbent, sometimes radiating from a woody base; stigma subcapitate to broadly capitate 15
    – Bracteole pairs subequal to unequal, linear, lanceolate or reduced to paired spines, 0.3–5 mm wide (excluding lateral spines or teeth); subshrub or perennial herb, stems more erect or straggling, if spreading or weakly decumbent then bracteoles very different to above; stigma variously linear, clavate or subcapitate 17

15. Corolla white, tube extremely long and slender, 80–130 mm long; bracteole
    margins with numerous slender flexuose spines . . . .**29.** *capitata*
–   Corolla limb pale blue to purple or rarely white, tube 18–31 mm long; bracteole
    margins entire or minutely toothed 16
16. Bracteoles lanceolate, 4–7.5 mm wide, not imbricate; outer calyx lobes with
    laciniate margins; leaves 2.8–5 cm long, lacking a thickened pale margin; capsule
    20–22 mm long . . . . **28.** *sunzuana*
–   Bracteoles broadly ovate, (7)10–20 mm wide, imbricate and largely enclosing
    calyx; outer calyx lobes with densely hairy margins, swollen hair bases sometimes
    forming minute teeth but otherwise entire; leaves 0.8–3 cm long, with thickened
    pale margin; capsule 14.5–17 mm long. . . . . **30.** *macrostegia*
17. Bracteoles and axillary spines (if present) simple, margin entire. 18
–   Bracteoles and axillary spines (if present) with 1 to several lateral spines or teeth,
    pinnately arranged or along one margin only 25
18. Corolla 14–24 mm long including a 7.5–14.5 mm tube; flowers solitary at each
    axil; stems with short white retrorse hairs, sometimes restricted to 2 opposite
    lines; anthers less than 2 mm long 27
–   Corolla larger, often markedly so, or if down to 22–25 mm with a 13.5–15 mm
    tube, then flowers held in contracted unilateral cymes of 2 or more flowers and/
    or stems buff- or yellowish-strigose; anthers 2–7.5 mm long 19
19. Adaxial pair of corolla lobes 3.5–6 mm long, less than half length of other lobes,
    limb highly zygomorphic; calyx densely clothed in silky mixed glandular and
    eglandular hairs over entire surface . . . . **19.** *bremekampii*
–   Adaxial pair of corolla lobes 5.5–18 mm long, subequal to or more than half the
    length of other lobes; calyx not densely hairy, if glandular hairs numerous then
    lacking silky eglandular hairs, calyx surface clearly visible 20
20. Stems with numerous fine white retrorse or spreading hairs dense throughout,
    rather shaggy in appearance; plant very spiny with numerous sterile spinose bracteoles
    pairs between contracted internodes; anthers 2–2.5 mm long . . . .**5.** *eylesii*
–   Stems with numerous appressed to spreading coarse yellow or buff-coloured
    hairs and/or densely glandular-hairy, finer spreading eglandular hairs sparse
    and inconspicuous; internodes not so contracted and plants less spiny; anthers
    2.7–7.5 mm long 21
21. Stems with dense spreading glandular hairs; corolla tube cylindrical 22
–   Stems lacking glandular hairs; corolla tube funnel-shaped or campanulate above
    attachment point of stamens 23
22. Cymes single-flowered; corolla tube 35–65 mm long, corolla white fading to pale
    blue; capsule 20–25 mm long . . . .**17.** *molensis*
–   At least some cymes 2–9-flowered; corolla tube less than 26 mm long, corolla limb
    pale blue to mauve or purple, rarely white; capsule 11–16 mm long . .**25.** *spinulosa*
23. Stem indumentum ± spreading; stem leaves broadly ovate; stigma short, c.0.5 mm
    long . . . . **14.** *sp. A*
–   Stem indumentum appressed or strongly ascending; stem leaves oblong-elliptic,
    obovate or suborbicular; stigma 1–2 mm long 24
24. Corolla 22–38.5 mm long including tube to 23.5 mm; leaf apex obtuse or rounded
    and apiculate, or if more acute and spinulose then outer calyx lobes elliptic or
    obovate with attenuate or cuneate base (var. *ramulosoides*) . . . . **12.** *crassa*
–   Corolla 40–50 mm long including tube over 25 mm; most leaves with apex acute
    and spinulose; outer calyx lobes ovate(-elliptic) with truncate or cordate base. . .
    . . . . **13.** *nyasensis*

25. Leaves coriaceous with sinuate white-cartilaginous margin, often toothed or spinose. . . . . **7.** *rigida*
 − Leaves not coriaceous, margin not white-cartilaginous, entire or with bulbous-based hairs forming minute teeth only 26
26. Flowers solitary at each axil; corolla 14–24 mm long including 7.5–14.5 mm tube, campanulate above attachment point of stamens 27
 − Flowers in contracted unilateral cymes, each 2–14-flowered; corolla 22–42 mm long including 13.5–25 mm tube, cylindrical or only narrowly campanulate above attachment point of stamens 28
27. Leaves 3–6 cm long; bracteoles 12–21 mm long; calyx green at first with paler margin, outer lobes 3.5–5 mm wide, with 6–9 prominent stiff spines on each margin. . . . . **8.** *oxyphylla*
 − Leaves less than 3 cm long; bracteoles less than 12 mm long; calyx purplish at first, if green then without paler margin, outer lobes with 2–5 teeth or short spines on each margin or entire, rarely up to 6 spines then lobes less than 3 mm wide . . . .
 . . . . **9.** *virgula*
28. Stem with dense spreading glandular hairs . . . .**25.** *spinulosa*
 − Stem without spreading glandular hairs, or if present hairs sparse and inconspicuous. 29
29. Anterior and posterior calyx lobes broadly ovate, length to width ratio 1.05–1.65:1, margins entire or minutely toothed. . . . . **6.** *lanceolata*
 − Anterior and posterior calyx lobes elliptic, lanceolate or ovate, length to width ratio 1.9–4.4:1, margins with 2–9 lateral spines. 30
30. Corolla white, with only eglandular hairs externally; calyx lacking glandular hairs; leaves large, (3.7)5–10 cm long when mature, usually with 5–6 pairs of lateral veins; stem with white retrorse hairs restricted to 2 opposite lines . . . . . .**3.** *elegans*
 − Corolla pale pink, lilac or mauve (sometimes a few individuals white-flowered), with mixed glandular and eglandular hairs externally; calyx often with a few short spreading glandular hairs, particularly on marginal spines; leaves smaller, 0.9–3(4.7) cm long when mature, usually with 3–4 pairs of lateral veins; stem with white retrorse or spreading hairs restricted to 2 opposite lines or numerous throughout 31
31. Bracteoles with a narrow green (later brown-scarious) blade 1–3.5 mm wide excluding lateral spines, paired sterile bracteoles few or absent at lower nodes, not forming densely spiny stems; leaves 1.2–4.7 cm long; stems with white hairs less dense at maturity, sometimes restricted to 2 lines. . . . .**4.** *saxatilis*
 − Bracteoles spinose, usually without a green blade, to 1 mm wide excluding lateral spines, sterile bracteoles very numerous between short internodes, plants very spiny; leaves very small, to 1.4 cm long, numerous between short internodes; stems with dense short white hairs throughout . . . .**5.** *eylesii*
32. Bracteoles broadly elliptic to suborbicular, 7.5–17.5 mm wide; posterior calyx lobe markedly larger and broader than anterior lobe, often partially enveloping it, suborbicular, 17–25 mm wide . . . . **24.** *torrei*
 − Bracteoles linear-lanceolate, narrowly oblanceolate or narrowly elliptic, 0.5–5.5 mm wide; posterior calyx lobe equal to or slightly larger than anterior lobe, not partially enveloping it, variously shaped. 33
33. Leaves reniform, broadly ovate or deltate, length to width ratio 0.6–1.3:1; corolla tube more than 2.5 times longer than limb; stellate hairs on leaves and mature stems pale grey or pale buff coloured. 34
 − Leaves elliptic, ovate or lanceolate, length to width ratio 1.5–3:1, or if broadly ovate to suborbicular then corolla tube less than 2 times longer than limb; leaves and stems with rich golden-brown stellate hairs. 35

34. Outer calyx lobes lanceolate, narrowly oblong or rarely (obovate-)elliptic, base acute or attenuate, 2–4(8) mm wide, length to width ratio 1.6–2.9:1, lateral lobes and fruit usually clearly exposed; stems with dense glandular hairs; anthers 3–4.2 mm long . . . . **26.** *heterotricha*
- Outer calyx lobes broadly ovate, base cordate or truncate, (5)7–16 mm wide, length to width ratio 1–1.8:1, enclosing lateral lobes and fruit; stems with glandular hairs sparse or absent; anthers less than 3 mm long. . . . . **27.** *affinis*
35. Long-armed stellate hairs on stems and leaves rich golden or golden-brown, with a conspicuous fasciculate-stellate base (less conspicuous in *sp. B*); bracteoles held ± erect, often adpressed to calyx; outer calyx lobes usually tardily scarious, often with an involute margin 36
- Long-armed stellate hairs on stems and leaves pale grey, pale-buff, or if golden-yellow then with a rather inconspicuous and sometimes caducous stellate base; bracteoles more spreading or deflexed; outer calyx lobes papery and soon-scarious, margin (in our region) not involute 37
36. Outer calyx lobes entire or inconspicuously denticulate; corolla tube cylindrical throughout, barely widened towards mouth. . . . . **21.** *fulvostellata*
- Outer calyx lobes conspicuously dentate; corolla tube narrowly campanulate above attachment point of stamens. . . . . **22.** *sp. B*
37. Outer calyx lobes (in our region) broadly ovate or ovate-orbicular, margin subentire or denticulate; old stems never floccose; leaf buds lacking dense pale grey stellate hairs . . . . **23.** *mutabilis*
- If outer calyx lobes broadly ovate then margin coarsely spinose-toothed (margin rarely shortly toothed in subsp. *kirkii* but then outer calyx lobes obovate or narrowly oblong); old stems sometimes long white-floccose; leaf buds with dense pale grey stellate hairs but these ± falling early . . . . **25.** *spinulosa*

* Hybrids between *B. obtusa* and *B. gueinzii* may also key out here – see taxon 32a.

1. **Barleria holstii** Lindau in Bot. Jahrb. Syst. **20**: 19 (1894). —Clarke in F.T.A. **5**: 166 (1899). —Darbyshire in F.T.E.A., Acanthaceae 2: 332 (2010). Type: Tanzania, Lushoto Dist., Mashewa (Mascheua), fl. vii.1893, *Holst* 3516 (B† holotype, HBG, K, M, W, Z).

Perennial herb or subshrub, 50–200 cm tall, sometimes scrambling; stems sparsely to densely buff- or yellow-strigose, hair bases swollen with age and persisting on mature woody stems. Leaves elliptic, 6.5–15.5 × 2–5.5 cm, base long-attenuate or -cuneate, apex attenuate or acuminate, surfaces strigose, hairs most numerous on margin and veins beneath; lateral veins 5–6 pairs; petiole 0–6 mm long. Inflorescence axillary, contracted (1)2–8-flowered unilateral cymes to 2 cm long; bracteoles pale green or tinged purple, pairs unequal and offset, the larger ovate or elliptic, often asymmetric, 12–22 × (3)6–12 mm, the smaller proportionally narrower, more strongly asymmetric and declinate, each with denticulate margin, teeth narrow and an apical bristle. Calyx green, usually paler towards base and centre, with darker reticulate venation; anterior lobe ovate or elliptic, 12–20 × 7–13 mm, base obtuse or rounded, margin as bracteoles or subentire, apex emarginate or truncate, surface strigose; posterior lobe broadly ovate, 17–30 × 10–19 mm, base rounded, apex acute or subattenuate; lateral lobes lanceolate, 3.5–6 mm long. Corolla 30–43 mm long, white, pale-blue or pale-purple, pubescent externally with mixed eglandular and glandular hairs; tube 19–23 mm long, funnel-shaped above attachment point of stamens; limb in 4+1 arrangement; abaxial lobe 12–22 × 8–18 mm, offset by 3–5 mm; lateral and adaxial lobes 9–18 × 6–13 mm. Stamens attached 7–10 mm from corolla tube base; filaments 18–29 mm long; anthers 2.5–3.2 mm long; lateral staminodes 1–3 mm long, pubescent, antherodes 0.5–0.9 mm long. Ovary glabrous; stigma linear, 1–2 mm long. Capsule 13.5–15 mm long, glabrous. Seeds 5.5–6 mm long and wide.

Zambia. N: Samfya Dist., 2 km N of Samfya Mission, Lake Bangweulu, fr. 28.viii.1952, *White* 3132 (BR, FHO, K). **Zimbabwe**. E: Mutare Dist., Vumba, above Burma Valley, bud 26.vii.1976, *Müller* 3076 (SRGH). **Malawi**. C: Dowa Dist., Ntchisi, Chinthembwe (Cintembwe) Mission, fl. 9.v.1961, *Chapman* 1298 (K, LISC, SRGH).

Also in Tanzania. Mixed evergreen or semi-evergreen forest and degraded *Albizia* forest; in Tanzania also in lowland and riverine forest and thicket; 900–1300 m.

Conservation notes: Rare in the Flora area but more frequent in lowlands of Tanzania; assessed as Near Threatened in F.T.E.A.

Flowers from the isolated Zimbabwe population are needed to confirm its identity.

2. **Barleria whytei** S. Moore in J. Bot. **41**: 138 (1903). —Darbyshire in F.T.E.A., Acanthaceae **2**: 332 (2010). Type: Kenya, S of Mombasa to Shimoni, fl.& fr. 1902, *Whyte* s.n. (BM holotype, K).

Perennial herb or subshrub, 40–100 cm tall; stems buff-yellow strigose when young, hair bases somewhat swollen with age, persisting on mature stems. Axillary spines absent. Leaves subsessile, blade elliptic or subpandurate, 7.5–18 × 2.5–7 cm, base abruptly cordate, somewhat amplexicaul when young, attenuate above, apex acuminate, surfaces sparsely strigose at least on margin and veins beneath; lateral veins 4–6 pairs. Inflorescences often clustered towards stem apex, of contracted 1–8-flowered unilateral cymes, to 4.5 cm long; bracteoles paler than leaves at least towards base, pairs unequal and offset, the larger ovate, 10–17 × 4–11.5 mm, margin entire. Anterior calyx lobe broadly ovate, 14–20 × 8–15 mm, base rounded or subcordate, margin entire or obscurely denticulate, apex acute, obtuse or subattenuate, surface paler towards base, strigose particularly on margin and principal veins, hairs on margins sometimes with a swollen base; posterior lobe 16–28.5 × 10–20 mm; lateral lobes lanceolate, 4.5–8 mm long. Corolla 32–40.5 mm long, white or rarely mauve (see note), salver-shaped, pubescent externally with spreading eglandular and glandular hairs and occasional strigose hairs; tube narrowly cylindrical, 22–30 mm long, barely expanded towards mouth; limb subregular or in weak 2+3 arrangement, widely spreading; abaxial lobe 8.5–10.5 × 8–11 mm wide; lateral pair as abaxial lobe but 5–7.5 mm wide; adaxial pair 5–6 mm wide. Stamens attached in distal half of corolla tube; filaments 11.5–17 mm long, glabrous; anthers 2–2.5 mm long; lateral staminodes 1–2.5 mm long, pubescent towards base, antherodes c.0.75 mm long. Ovary glabrous; stigma linear, 0.6–1 mm long. Capsule 12–15.5 mm long, glabrous. Seeds 4–6 mm long and wide.

Mozambique. N: Palma Dist., Vamizi Is., road E of carpentry yard, fl. 20.vi.2012, *J. & S. Burrows* 12897 (BNRH, K).

Also in coastal Kenya and Tanzania. Coastal forest and bushland on coral rag; c.10 m.

Conservation notes: Assessed as Endangered under IUCN criterion B in F.T.E.A. Although discovery of the Mozambique population greatly extends its Extent of Occurrence, it still qualifies as Endangered based on Area of Occupancy and as the Kenyan site is threatened by development.

Specimen *J. & S. Burrows* 12896, also from Vamizi, is recorded as having mauve flowers; all other specimens seen of this species have white flowers. Only a scan of this specimen has been seen, but it looks a good match for *B. whytei*.

3. **Barleria elegans** C.B. Clarke in F.T.A. **5**: 154 (1899); in Fl. Cap. **5**: 49 (1901). — Obermeijer in Ann. Transv. Mus. **15**: 154 (1933). —Meyer in Merxmüller, Prodr. Fl. SW Afr. **130**: 16 (1968). —Compton, Fl. Swaziland: 552 (1976). —Bandeira *et al.*, Fl. Nat. Sul Moçamb.: 191 (2007). Type: Angola, Luanda Dist., fl.& fr. 1858, *Welwitsch* 5068 (BM lectotype, C, K, LISC, P), lectotypified here.

Subsp. **orientalis** I. Darbysh., subsp. nov. Differs from subsp. *elegans* in having white, not lilac or blue, flowers and in the external surface of the calyx having only a strigose indumentum, not both strigose and (sparsely) rather densely glandular-

and/or eglandular-puberulous. Type: Mozambique, Maputo, fl. 2.iv.1947, *Hornby* 2640 (K holotype, BM, ?LMA, PRE).

Barleria pungens L. var. *grandifolia* Nees in De Candolle, Prodr. **11**: 237 (1847). —Moore in J. Bot. **18**: 269 (1880) in part. Type: Mozambique, Delagoa Bay, fl. 1822, *Forbes* s.n. (K holotype).

Perennial herb or subshrub, often straggling or scandent, 40–150 cm tall; stems sparsely strigose, with two opposite lines of short white retrorse hairs, later glabrescent. Leaves (ovate-) elliptic to narrowly so, (3.7)5-10 × 1.5–3.5 cm, base (cuneate-)attenuate, apex attenuate to obtuse, apiculate or upper leaves mucronate, surfaces strigose, hairs most numerous on or restricted to midrib, margin and veins beneath; lateral veins (4)5–6 pairs; petiole to 12 mm long. Inflorescence axillary, a contracted unilateral cyme 1.5–4 cm long, 3–14-flowered; bracteoles green, (linear-)lanceolate, 12–23 × 2–5 mm long excluding the 3–7 lateral spines each 2–6.5 mm, apex spinose, triplinerved with pale prominent midrib, often with few broad sessile glands towards base; paired sterile bracteoles often present at lower nodes. Calyx green, sometimes with paler margin, later turning pale-scarious; anterior lobe elliptic or elliptic-lanceolate, 11–17.5 × 3.5–7 mm, base attenuate, margin with 4–9 slender lateral spines 3–6.5 mm long ending in a short bristle, apex spinose or bispinose, surface sparsely strigose; posterior lobe as anterior but 12.5–23 × 4–9 mm, marginal spines to 9 mm long, apex attenuate-spinose; lateral lobes linear-lanceolate, 6.5–10 mm long. Corolla 32–42 mm long, white, eglandular-pubescent externally; tube 17–25 mm long, cylindrical, narrowly campanulate towards mouth; limb in weak 4+1 arrangement or subregular; abaxial and lateral lobes 10–15 × 6–10 mm, abaxial offset by 2–4.5 mm; adaxial lobes somewhat narrower, 6–8 mm wide. Stamens attached 9–12 mm from base of corolla tube; filaments 18–21 mm long; anthers 2.5–3.5 mm long; lateral staminodes 3–5.5 mm long, pubescent, antherodes 0.7–1.1 mm long. Ovary with ring of minute crisped white hairs at apex; style glabrous; stigma linear, 0.4–0.9 mm long. Capsule 13–16.5 mm long, glabrous. Seeds 4.5–5 mm long and wide.

Zimbabwe. S: Mwenezi Dist., Bubi R., near Bubye Ranch homestead, fl. 8.v.1958, *Drummond* 5688 (BR, K, LISC, LMA, SRGH). **Mozambique**. Z: Quelimane, fr. 1908, *Sim* 20776a (PRE). MS: Machaze Dist., Massangena, margin of R. Save, fl.& fr. 11.vi.1942, *Torre* 4303 (BR, LISC, WAG). GI: Mabalane Dist., Mabalane, fl. 4.vi.1959, *Barbosa & Lemos* 8613 (COI, K, LISC, LMA). M: Boane Dist., between Matola and Umbeluzi, fr. 29.iv.1947, *Pedro & Pedrógão* 892 (LMA, LMU).

Also in Swaziland and South Africa (KwaZulu-Natal, Limpopo, Mpumalanga). Drier types of riverine forest, woodland and *Acacia–Combretum* thicket, coastal bushland, sometimes between exposed rocks, can persist in disturbed bushland; 0–550 m.

Conservation notes: Although losses of riparian habitat are likely to have led to some population declines, it remains locally common in suitable habitat; Least Concern.

Subsp. *elegans* is recorded from Angola and N Namibia. It is quite variable and some Angolan populations are very similar to subsp. *orientalis* except for differences listed in the diagnosis, whilst others have bracteoles and calyces with only short lateral spines, and tend to have smaller flowers (< 30 mm long) with the ovary having a few long ascending hairs towards the apex. Variation across the Angolan populations, however, appears clinal. The material of subsp. *orientalis* is much more uniform.

The flowers on *Müller* 2338 from Zimbabwe are recorded as bluish, but in all other material from our region the flowers are pure white; they appear blue in the dry state, which may be the reason.

4. **Barleria saxatilis** Oberm. in Bothalia **7**: 445 (1961). —Compton, Fl. Swaziland: 554 (1976). —Fabian & Germishuizen, Wild Fl. Nthn. S. Afr.: 398, pl.191 in part (1997). —Pooley, Field Guide Wild Fl. KwaZulu-Natal: 490 (1998). —Makholela *et al.* in S. Afr. J. Bot. **70**: 515 (2004). Type: South Africa, Zoutpansberg Dist., along road to Dandy farm, leading off to N at Sana R bridge on Louis Trichardt–Mara–Vivo road, fl. 3.iv.1957, *Meeuse* 10203 (PRE holotype, K).

Spiny compact shrublet to sprawling or subscandent perennial herb, 30–90 cm tall; stems sparsely strigose, with numerous fine white retrorse or spreading hairs evenly distributed or restricted to two opposite sides, nodes strigose. Leaves oblong, elliptic or somewhat ovate to obovate, 1.2–3(4.7) × 0.6–1.5(2) cm, base cuneate, attenuate or obtuse, apex acute to rounded with ± prominent mucro, surfaces strigose, hairs most numerous on margin and veins beneath; lateral veins 3–4(5) pairs; petiole to 5 mm. Inflorescence axillary, a contracted unilateral cyme 1–3.5 cm long, each (1)3–10-flowered; bracteoles green, linear-lanceolate or lanceolate, pairs unequal, the longer 6–17 × 1–3.5 mm excluding the (1)2–6 lateral spines to 1–5 mm, apex spinose, midrib pale and prominent, sterile pairs sometimes persisting at lower nodes. Calyx at first pale green or pinkish with dark green or purple venation, later turning pale brown-scarious; anterior lobe elliptic or ovate, 8–17(20) × 3.5–8 mm, base shortly attenuate or rarely obtuse, margin with 4–8 slender lateral spines 1.5–5 mm long ending in a short bristle, apex spinose, surface strigose along veins, often with short spreading glandular hairs on spines; posterior lobe ovate or lanceolate, 10–22 × 4.5–10.5 mm, marginal spines to 6 mm long, apex attenuate into a long spine; lateral lobes linear-lanceolate, 3.5–10 mm long. Corolla (chasmogamous flowers) 27–39 mm long, pale pink, lilac or (blue-)mauve, sometimes white-flowered plants within a population, eglandular-pubescent and glandular-pilose externally; tube 16.5–25 mm long, cylindrical, narrowly campanulate towards mouth; limb in weak 2+3 arrangement or subregular; abaxial lobe 11–14 × 5.5–7.5 mm; lateral lobes 10–13 × 4.5–8.5 mm; adaxial lobes 9–12.5 × 3–5 mm; cleistogamous flowers sometimes present, appearing as cylindrical buds. Stamens attached 10–11 mm from base of corolla tube; filaments 15–18 mm long; anthers (1.7)2–2.4 mm long; lateral staminodes 1.5–2.5 mm long, pubescent, antherodes c.0.7 mm long. Ovary with ring of few crisped white hairs at apex; style glabrous; stigma linear, 0.35–0.7 mm long. Capsule 10–13.5 mm long, glabrous. Seeds 3.5–4.5 mm long and wide.

Mozambique. GI: Mabalane Dist., near Combomune, margin of R. Limpopo, fl. 12.v.1948, *Torre* 7779 (LISC, WAG).

Also in Swaziland and South Africa (KwaZulu-Natal, Limpopo, Mpumalanga). Dry rocky hills or sandy plains, dry bushland in full sun or shade, also riverine bushland, roadside slopes and cliff ledges; c.100 m.

Conservation notes: Known only from a single collection in the Flora area, but locally abundant in N South Africa where it can be the dominant undershrub in open bushland – *Meeuse* 10203 states it is "very frequent and aspect-forming for miles"; Least Concern.

This is the dry-country vicariant of *Barleria elegans*; more vigorous plants of *B. saxatilis* are similar to that species but separable by the smaller leaves and fruits, the presence of glandular hairs on the corolla and, in our region, the different flower colour. The white-flowered variant of *B. saxatilis* illustrated in Fabian & Germishuizen (1997) may well depict *B. elegans* subsp. *orientalis*.

Makholela *et al.* (2004, p.519) note the regular occurrence of cleistogamous as well as chasmogamous flowers and conclude that "cleistogamy is responsible for the widespread distribution of *B. saxatilis*" as well as accounting for the low within-population genetic diversity recorded.

5. **Barleria eylesii** S. Moore in J. Bot. **43**: 50 (1905). Type: Zimbabwe, Matobo Dist., Matopos Hills, fl. ii.1903, *Eyles* 160 (BM holotype, SRGH).

Spiny shrublet c.40 cm tall, much-branched; stems densely pubescent with white retrorse or spreading hairs, nodes strigose; internodes short. Leaves elliptic to lanceolate, 0.9–1.4 × 0.3–0.6 cm, base attenuate to obtuse, apex narrowed into a short spine, strigose mainly on margin, midrib and veins beneath; lateral veins 3–4 pairs; petiole to 2.5 mm. Inflorescence a contracted unilateral cyme 1–2.5 cm long in upper axils, each 2–8-flowered, sometimes only one flower developed at a time; bracteoles spinose, pairs unequal in length, the longer 9–17 × 0.5–1 mm, margin entire or with 1–3 short lateral spines to 1 mm long, numerous sterile bracteoles present at lower nodes as axillary spines, often two pairs per axil. Calyx at first mauve, paler towards base and with darker venation, later turning pale-scarious; anterior lobe lanceolate, 8.5–14 × 2.5–4 mm, base shortly attenuate, margin with 2–5 short lateral spines 0.5–1.5(2.5) mm long ending in a short bristle, apex spinose or bifidly so, surface strigulose along veins, often with scattered patent glandular hairs; posterior lobe as anterior but 11.5–20 × 3.5–5 mm, marginal spines 1.5–3(6) mm long, apex attenuate into a long spine; lateral lobes linear-lanceolate, 7–9.5 mm long. Corolla (22)25–37 mm long, lilac or blue-mauve, pubescent externally with mixed glandular and eglandular hairs; tube (13.5)15–24 mm long, cylindrical, narrowly campanulate towards mouth; limb in weak 2+3 arrangement or subregular; abaxial lobe (8)10.5–14 × (6)7–10.5 mm; lateral lobes (8)9.5–13 × (5)6–10 mm; adaxial lobes (7.5)9.5–12 × 4–7.5 mm. Stamens attached ± midway along corolla tube; filaments (12)19–22 mm long; anthers 2–2.5 mm long; lateral staminodes (1)2–3 mm long, pubescent towards base, antherodes to 0.6 mm long or barely developed. Ovary with a ring of minute crisped white hairs at apex; style glabrous; stigma linear, 0.4–0.6 mm long. Capsule 9–12.5 mm long, glabrous. Mature seeds not seen.

Zimbabwe. W: Matobo Dist., Matopos Research Station, Hazelside, fl. 13.iii.1949, *West* 2859 (K, SRGH). E: Mutare Dist., Dora Farm, fl.& fr. 12.vi.1948, *Chase* 776 (BM, K, SRGH). S: Chivi Dist., Msumbiti kopje, fr. xii.1955, *Davies* 1811 (BR, K, SRGH).

Not known elsewhere. Granite hillslopes and inselbergs, usually amongst large boulders in partial shade or in open woodland; 600–1550 m.

Conservation notes: Endemic to the central watershed of Zimbabwe, a very local species known from few collections but its favoured habitats are not threatened; probably Least Concern.

The few specimens seen from S and E Zimbabwe differ from the Matobo populations in having spreading, not retrorse, stem hairs and in tending to have more flowers per cyme (though this latter character varies considerably even within a population, see *Chase* 776).

A further vicariant species within the *Barleria elegans* group, this species is closely allied to *B. saxatilis*. The two are usually easily separated but *Wright* T198 (SRGH) from Mushandike Nat. Park in Zimbabwe S, although closer to *B. eylesii* in most characters, tends towards *B. saxatilis* in having broader bracteoles with a distinct green blade.

6. **Barleria lanceolata** (Schinz) Oberm. in Ann. Transv. Mus. **15**: 155 (1933). — Meyer in Merxmüller, Prodr. Fl. SW Afr. **130**: 14 (1968). Type: Namibia, "Gross-Namaland", Gamochab, fl. 24.iv.1885, *Schinz* 61 (Z lectotype), lectotypified here, see note.

> *Barleria acanthoides* Vahl forma *lanceolata* Schinz in Verh. Bot. Vereins Prov. Brandenburg **31**: 199 (1890).
> *Barleria acanthoides* sensu Clarke in Fl. Cap. **5**: 49 (1901), non Vahl.

Spiny shrublet 15–100 cm tall; stems with short white retrorse hairs restricted to or concentrated in two opposite lines or bands, young stems sparsely strigose, sometimes with scattered minute spreading glandular hairs. Leaves oblanceolate, spathulate or narrowly oblong-elliptic, 1.2–4 × 0.4–0.8 cm, base cuneate, apex acute to rounded, mucronate, strigose mainly on margin and veins beneath, young leaves often with short white retrorse hairs at least on midrib above; lateral veins 3–4 pairs; petiole to 5 mm. Inflorescence a contracted unilateral cyme 1–3 cm long in upper axils, each 2–7-flowered; bracteoles linear-lanceolate, pairs unequal in length, the longer 7–20 × 0.8–1.8 mm, usually with a narrow green blade and a prominent pale midrib, apex spinose, margin with 1–11 lateral teeth or spines to 1–3.5 mm long, sterile bracteoles sometimes present at lower nodes as paired axillary spines. Calyx at first bluish, maturing green with darker, often reddish venation, soon turning pale-scarious; anterior lobe broadly ovate, 13–17 × 8.5–11.5(16) mm, base abruptly narrowed and rounded or subcordate, margin entire or with minute bristle-tipped teeth, apex shortly attenuate to obtuse and with a short mucro, rarely notched, surface strigose along veins, with short white retrorse hairs mainly towards base, with or without minute patent glandular hairs; posterior lobe as anterior but 14–19 mm long, apex with a more prominent mucro; lateral lobes lanceolate, 5–9 mm long. Corolla 25–42 mm long, lilac, pink, pale blue or rarely white, eglandular-pubescent externally with interspersed glandular hairs on lateral lobes; tube 15–25 mm long, cylindrical, barely widened or narrowly campanulate towards mouth; limb subregular; abaxial lobe 9–15.5 × 5–13.5 mm; lateral lobes 9.5–16.5 × 4–10 mm; adaxial lobes as laterals but 3–7.5 mm wide. Stamens attached 9–13.5 mm from base of corolla tube; filaments 13–27 mm long; anthers 2–2.6(2.9) mm long; lateral staminodes 1.3–4 mm long, pubescent towards base, antherodes 0.7–1 mm long. Ovary with ring of minute crisped white hairs at apex; style glabrous; stigma clavate, 0.5–0.8 mm long. Capsule 10.5–12 mm long, glabrous or with few hairs towards apex. Seeds 3.5–4 × 3.4–3.7 mm.

Botswana. SW: Ghanzi Dist., Mamuno, border gate between Namibia and Botswana, bud 13.ii.1970, *Brown* Mamuno 4 (SRGH).

Also in Namibia. Thin clay over quartzite and rocky ground with wooded and shrubby grassland; in Namibia in open bushland and sparse grassland on sandy soils or rocky ground, usually in full sun; 1050 m.

Conservation notes: Although known only from a single collection in the Flora area, it is fairly common and widespread in Namibia; Least Concern.

Variation within this species is currently being studied in preparation for an account of *Barleria* in Namibia; the description above covers only the form from C and S Namibia and Botswana which is *B. lanceolata* sensu stricto.

When promoting this taxon to species rank, Obermeijer (1933) listed *Dinter* 2094 as the type. However, the original material was that of Schinz from Gamochab in Namaqualand, for which there is one sheet in Zurich, chosen here as the lectotype, and one sheet in Kew. The latter has a different number (*Schinz* 2) and date (20. iv.1885) to the lectotype and so is not considered an isolectotype.

7. **Barleria rigida** Nees in De Candolle, Prodr. **11**: 242 (1847). —Obermeijer in Ann. Transv. Mus. **15**: 153 (1933). —Meyer in Merxmüller, Prodr. Fl. SW Afr. **130**: 16 (1968). Types: "Africa australi extratropica", n.d., *Hoffmannsegg* in Herb. Willdenow 1160 (B-W syntype); South Africa, Grootriviers Poort, n.d., *Lichtenstein* s.n. (B† syntype); South Africa, Grootriviers Poort, fl. n.d., *Burchell* 1991 (G-DC syntype, K).

Barleria ilicina T. Anderson in J. Linn. Soc., Bot. **7**: 28 (1863). Type: South Africa, Garip R., 19.ix.1830, *Drège* s.n. (S holotype).

Barleria schenckii Schinz in Verh. Bot. Vereins Prov. Brandenburg **31**: 198 (1890). Types: Namibia, "Gross-Namaland", Aus, n.d., *Schenck* 310 (?Z syntype, not located); Aus, "Gross-Namaland", fl. 1885, *Pohle* s.n. (Z syntype); Tiras, fl. 1885, *Schinz* 82 [number not listed in protologue; *Schinz* 3 on K sheet] (Z syntype, K).

Barleria irritans Nees var. *rigida* (Nees) C.B. Clarke in Fl. Cap. **5**: 48 (1901) in part.

Barleria rigida Nees var. *ilicina* (T. Anderson) Oberm. in Ann. Transv. Mus. **15**: 153 (1933).

Spiny shrublet 10–45 cm tall, much-branched and spreading; stems retrorsely white-puberulent mainly or exclusively on two opposite-decussate sides, nodes often strigose. Leaves rather coriaceous, narrowly oblong(-lanceolate), elliptic or rarely ovate, 1–3.2 × 0.3–1 cm, base cuneate, margin white-cartilaginous, sinuate, with or without marginal spines, apex acute and apiculate or spinose, surfaces sparsely yellow-strigose at least along midrib beneath, sometimes also minutely puberulent when young; lateral veins indistinct, midrib prominent beneath; petiole 0–5 mm. Inflorescence a contracted unilateral cyme 1–2.5 cm long in upper axils, each (1)2–6-flowered; bracteoles green with white margin or whitish throughout, (linear-)lanceolate, 12–25 × 1–4 mm, margin with 4–10 lateral spines per side 1.5–4.5 mm long, apex long-spinose. Calyx coriaceous, pale green at first with whitish margin and base, turning whitish throughout; anterior lobe elliptic or lanceolate, 10–16.5 × (3)4–5 mm, base attenuate, margin with 5–13 lateral spines 1.5–4.5 mm long ending in a stiff bristle, apex spinose or bifidly so, surface sparsely strigose and sparsely puberulent or glabrous, palmate-reticulate venation ± prominent; posterior lobe ovate or lanceolate, 12–23 × (3.2)4.5–6 mm, apex attenuate into a long spine; lateral lobes linear-lanceolate, 6–8.5 mm long. Corolla 25–33 mm long, white, pink, pale blue or violet, pubescent externally with mixed glandular and eglandular hairs; tube 15–19.5 mm long, cylindrical, very narrowly campanulate towards mouth; limb subregular; abaxial lobe 8–13 × 6.5–9.5 mm; lateral lobes as abaxial but 5.5–8.5 mm wide; adaxial lobes 8.5–15 × 4–7 mm. Stamens attached ± midway along corolla tube; filaments 16.5–22 mm long; anthers 1.5–2.5 mm long; lateral staminodes 1.2–2.7 mm long, pubescent at base, antherodes 0.5–0.9 mm long. Ovary with ring of minute crisped white hairs at apex and extending onto style base; stigma clavate or subcapitate, 0.4–0.65 mm long, apex bilobed. Capsule 9.5–12.5 mm long, glabrous; seeds c.4 mm long and wide.

Botswana. SW: Kgalagadi Dist., above Tshane pan, 256 km N of Tshabong, fl. 3.iii.1977, *Mott* 1117 (SRGH); Masetlheng Pan, fl.& fr. 21.iii.1978, *Skarpe* 289 (K, SRGH).

Also in Namibia and South Africa (Northern Cape, Free State). Sandy pans with dwarf shrubby vegetation, heavily grazed areas; elsewhere recorded from open sandy, gravelly or rocky terrain, e.g. dry riverbeds, dry hillslopes and sand flats; c.1100 m.

Conservation notes: Widespread and locally common, favouring dry areas with little vegetation; Least Concern.

Barleria rigida sensu lato is a variable species and, together with *B. irritans* Nees and *B. bechuanensis* C.B. Clarke, forms a difficult species complex. This group is currently under revision by Florence Nyirenda and Kevin Balkwill who will recognise three varieties within *B. rigida* (pers. comm.). Only the form recorded from Botswana, inland Namibia and just extending into South African Namaqualand is treated in the above description. Whilst this form will fall within their var. *rigida*, it nevertheless differs somewhat from plants elsewhere in South Africa which often have only 1- or 2-flowered inflorescences, and the outer calyx lobes have prominent green-reticulate patterning against a white background and shorter marginal spines. This form includes *Burchell* 1991 which Nyirenda & Balkwill will propose as the lectotype (pers. comm.).

8. **Barleria oxyphylla** Lindau in Bot Jahrb. Syst. **38**: 69 (1905). —Obermeijer in Ann. Transv. Mus. **15**: 158 (1933). Type: South Africa, Mpumalanga, Komatipoort, fl.& fr. 18.xii.1897, *Schlechter* 11813 (K lectotype, BM, GRA, HBG, WAG, Z), lectotypified here.

Spiny suffrutex 10–30 cm tall, 1 to several erect branches from a woody base; stems with fine white retrorse hairs mainly arranged in two opposite-decussate rows. Leaves linear or narrowly oblong, 3–6 × 0.5–1 cm, base cuneate, apex acute, mucronate, surfaces yellowish-hispid particularly on veins beneath and margin, hairs along margin with a ± bulbous base, with few to numerous fine white retrorse hairs when young; lateral veins 3–4 pairs; petiole 0–3 mm. Inflorescence axillary, single-flowered, sessile; bracteoles spinose, green with paler margin, 12–21 × 1–1.5 mm, margin entire or with 1–3 lateral teeth; pairs persisting at lower nodes as paired axillary spines, then whitish. Calyx green with paler margin, later scarious; anterior lobe ovate-elliptic or lanceolate, 12–18 × 3.5–5 mm, margin with 6–9 stiff spines 1–3 mm long ending in a bristle, apex spinose or bifidly so, surface with sparse to numerous fine crisped or straight white hairs particularly along margin, with sparse stiff yellowish hairs along main veins; posterior lobe as anterior but 15–21 mm long, apex long-spinose; lateral lobes linear-lanceolate, 6.5–9 mm long. Corolla 19–24 mm long, blue (according to Lindau), sparsely pubescent externally; tube 12–14.5 mm long, throat campanulate; limb subregular; abaxial lobe 6.5–8.5 × 5.5 mm; lateral lobes 7–10 × 4.5–5 mm; adaxial lobes as lateral pair but 3.5–4 mm wide. Stamens attached c.5 mm from base of corolla tube; filaments c.10 mm long; anthers c.1.7 mm long; lateral staminodes 1.3–1.7 mm long, pubescent, antherodes 0.7 mm long. Ovary with apical ring of minute crisped hairs; style glabrous; stigma subcapitate, 0.3–0.4 mm long. Capsule c.13 mm long, glabrous. Seeds c.4.5 × 4 mm.

Mozambique. M: Manhiça Dist., left bank of R. Incomati, c.30 km from Posto Administrativo de Sábiè, fr. 1.v.1953, *Myre & Balsinhas* 1647 (K, ?LMA).

Also in Swaziland and South Africa (Mpumalanga). Open grassy shrubland with sparse *Acacia* trees on compacted, rugged ground; c.300 m.

Conservation notes: Very rare, known from only three localities; recently (2012) confirmed to occur in Swaziland. Habitat data suggests it is tolerant of some disturbance; possibly threatened.

9. **Barleria virgula** C.B. Clarke in Fl. Cap. 5: 48 (1901). —Obermeijer in Ann. Transv. Mus. **15**: 158 (1933). Type: South Africa, Limpopo, Marico Dist., fl. n.d., *Holub* s.n. (K holotype).

 Barleria transvaalensis Oberm. in Ann. Transv. Mus. **15**: 156 (1933). —Obermeijer *et al.* in Bothalia **3**: 254 (1937). —Fabian & Germishuizen, Wild Fl. Nthn. S. Afr.: 390, pl.187 (1997). —Welman in Germishuizen *et al.*, Checklist S. Afr. Pl.: 80 (2006). Type: South Africa, Limpopo, Messina, *Rogers* s.n. (PRE-TRV holotype).

Slender spiny shrublet, 5–50 cm tall; stems with short white retrorse hairs evenly distributed or restricted to two opposite-decussate lines. Leaves often clustered on short lateral branches, oblanceolate, linear or narrowly oblong-elliptic, 0.8–2.7 × 0.15–0.6 cm, base cuneate, margin often involute when dry, apex acute to rounded, mucronulate, with ± appressed white to yellowish hairs mainly on midrib beneath, young leaves sometimes with short crisped white hairs; lateral veins inconspicuous; petiole 0–2 mm. Inflorescence axillary, single-flowered, sessile; bracteoles spinose, 2.5–11.5 × 0.3–1 mm, entire or with 1–3 lateral spines on each margin to 4.5 mm long, midrib prominent, pairs persisting at lower nodes as paired axillary spines to 16 mm long. Calyx purplish or green when young, later turning pale-scarious; anterior lobe oblong-lanceolate to narrowly elliptic, 6.5–12 × 1.3–4 mm, margin entire or usually with 2–5 flexuose spines or teeth to 2 mm long ending in a bristle, apex mucronate or bifidly so, surface sparsely strigulose; posterior lobe as anterior but 8.5–16 mm long, apex mucronate; lateral lobes linear-lanceolate, 5–9.5 mm long. Corolla 14–21.5 mm long, pale pink, mauve or blue, pubescent externally with retrorse eglandular hairs, limb also with spreading glandular hairs; tube 7.5–12 mm long, declinate towards base, throat campanulate; limb in weak 4+1 arrangement; abaxial lobe 5–8.5 × 4.5–7 mm, offset by 1.5–2 mm; lateral lobes as abaxial but 3.5–5.5 mm wide; adaxial lobes 2.5–5 mm wide. Stamens attached 4–5.5 mm from base of corolla tube; filaments 6–11.5 mm long; anthers 1–1.8 mm long; lateral staminodes 1–2.5 mm long, pubescent, antherodes 0.4–0.8 mm long. Ovary with apical ring of minute crisped hairs; style glabrous; stigma subcapitate, 0.2–0.4 mm long. Capsule 10–11 mm long, glabrous. Seeds c.3.5–4 × 3–3.5 mm.

Botswana. SE: Kgatleng Dist., Mochudi, fl.& fr. i.1915, *Harbor* in *TRV* 17014 (PRE); Kgatleng Dist., near Derde Poort on Botswana (Bechuanaland) side, fl. 30.xi.1954, *Codd* 8871 (K, PRE, SRGH). **Zimbabwe**. E: Mutare or Mutasa Dist., Odzi R., fl. xi.1931, *Myres* 620 (K). S: Beitbridge Dist., near Samilala R., c.14.5 km due W from Fort Tuli on old road to Maklautos, Tuli Circle, fr. 13.v.1959, *Drummond* 6128 (K, SRGH).

 Also in N South Africa (Limpopo, Mpumalanga, Gauteng). Mopane scrub, dry bushland and scrub on sandy flats; c.900 m.

 Conservation notes: Rather scarce and localised but its preferred habitats are unlikely to be threatened; probably Least Concern.

 Barleria transvaalensis was separated primarily on the stem hairs being evenly distributed throughout, not restricted to two lines. However, some specimens are intermediate in this character – hairs being most dense along but not restricted to two lines. Some populations of "*B. transvaalensis*" have slightly broader leaves and calyx, but this variation does not correlate with differences in stem indumentum.

 Obermeijer et al. 50 from the Soutpansberg in South Africa (K) was recorded as "to 6 ft high", much taller than otherwise known in this species; this may be a recording or transcription error.

 Barleria virgula appears to be allied to the *B. paolii* complex in NE Africa, particularly to *B. paolioides* I. Darbysh. with which it shares a short corolla tube. It differs from that species in the narrower leaves (obovate in *B. paolioides*), the bracteoles often having spines on both margins (always restricted to one margin in *B. paolioides*), and the shorter, pink or mauve corolla (white or cream, 19–28.5 mm long in *B. paolioides*).

10. **Barleria velutina** Champl. in Pl. Ecol. Evol. **144**: 83, fig.1 (2011). Type: D.R. Congo, Haut-Katanga, Shinkolobwe, fl. 15.v.1985, *Malaisse & Goetghebeur* 885 (BR holotype, B, C, CAS, K, LISC, MO, P, UPS, WAG).

Barleria grandicalyx sensu Phiri, Checklist Zamb. Vasc. Pl.: 18 (2005), non Lindau.

Much-branched perennial herb or suffrutex, erect to procumbent, 30–75 cm tall; stems with dense long appressed or ascending buff-coloured hairs, with shorter, finer spreading hairs on two opposite sides. Leaves (oblong-)elliptic to ovate, 1.7–5.2 × 0.8–2.8 cm, base cuneate to rounded or subcordate, margin revolute when young, apex acute or obtuse, margin and veins beneath with long appressed hairs, elsewhere with few to numerous fine spreading hairs; lateral veins 3–5 pairs, prominent beneath; petiole 0–5 mm long. Inflorescence usually congested into a short spike 3.5–11 cm long terminating main and short lateral branches, sometimes more widely spaced and axillary, each cyme single-flowered, sessile; bracts as leaves but reducing upwards, typically 10–20 × 3.5–10 mm in spike; bracteoles linear, lanceolate or narrowly elliptic, 7–18(25) × 1–2.5(4.5) mm, margin entire, apex acute, ± apiculate, hairs as on leaves, often with scattered capitate short-stalked glands. Calyx blue- or pink-tinged at first with darker venation, soon turning pale-scarious, accrescent; outer lobes subequal, broadly ovate(-elliptic), 15–30(34) × 10–21 mm, base rounded, cordate or cuneate, margin coarsely to shallowly dentate, teeth often antrorse with a flexuose apical bristle, apex acute- to obtuse-apiculate or anterior lobe with 2–3 apical teeth, surfaces with numerous long subappressed pale buff-coloured or silvery hairs most dense along margin, rarely with scattered short spreading glandular hairs; lateral lobes linear-lanceolate, 7–12(16) mm long. Corolla 26–37 mm long, pale blue, mauve or pink, limb glandular-pilose externally; tube cylindrical, 10.5–13.5 mm long, curved; limb in 0+3 arrangement; abaxial and lateral lobes 14–23 × 9–18 mm; adaxial lobes absent or at most 2 × 2 mm. Stamens attached 6–7 mm from base of corolla tube; filaments 12–16 mm long; anthers 3–5 mm long; lateral staminodes 1.7–2.5 mm long, pilose, antherodes not or barely developed. Ovary with apical ring of pale hairs; style with short glandular hairs towards apex, sometimes eglandular-hairy towards base; stigma capitate, 0.2–0.5 mm long. Capsule 15–19.5 mm long, glabrous or with sparse minute glandular hairs in distal half. Seeds 5.5–7 × 4–5 mm.

Zambia. B: Kabompo Dist., Kabompo, fl. 12.vi.1974, *Chisumpa* 168 (K, NDO). W: Mwinilunga Dist., Kabompo road, 27 km from Mwinilunga, fl. 6.vi.1963, *Edwards* 675 (K, LISC, SRGH).

Also in southern D.R. Congo (Shaba plateaux). Miombo woodland, grassland, rocky slopes over laterite and roadsides, in Congo sometimes recorded on mineral-rich deposits; c.1100 m.

Conservation notes: Although very localised it appears common within its range and suitable habitat remains widespread; Least Concern.

Barleria obtecta Champl. from southern D.R. Congo is a closely allied species, sharing the much reduced or absent adaxial corolla lobes, but differs in having a more obovate outer calyx lobes with an entire margin and larger fruits (to 27 mm long).

11. **Barleria sceptrum-katanganum** Champl. in Pl. Ecol. Evol. **144**: 87, fig.5 (2011). Type: D.R. Congo, Katanga, Dubié, fl. 24.vi.1957, *Duvigneaud* 3676 Ac3 (BRLU holotype).

Stout suffrutex with erect stems from a woody base, 25–50 cm tall; stems with dense appressed or ascending long buff to golden hairs, upper internodes also with short spreading eglandular and/or glandular hairs. Leaves somewhat coriaceous, narrowly oblong-elliptic to lanceolate, 8.5–12.5 × 2–3 cm, base cuneate to rounded, apex acute, surfaces with long buff-coloured hairs most numerous on margin and veins beneath; lateral veins (5)6–8 pairs, prominent beneath; petiole 2–6 mm long. Inflorescences compounded into a dense cylindrical or conical spike 4–15 cm long, each cyme 1 to few-flowered, sessile; bracts imbricate and largely enclosing calyx, elliptic, 26–34 × 11–18 mm, apex obtuse or rounded, apiculate, indumentum as leaves with additional short spreading eglandular and minute glandular hairs; bracteoles narrowly elliptic or obovate,

22–33 × 5–10 mm, often triplinerved, base attenuate, margin entire, apex acute, subattenuate or obtuse, mucronulate. Calyx at first purplish but turning pale-scarious, somewhat accrescent; anterior lobe broadly ovate, 20–27 × 13–22 mm, base shortly and broadly attenuate then rounded or subcordate, margin coarsely and irregularly dentate, teeth with an apical bristle, apex acute-mucronate or bifidly so, surfaces with numerous long subappressed buff-coloured hairs most dense within and along margin, with minute spreading glandular hairs most numerous along margin; posterior lobe as anterior but 24–33 mm long, apex acute-mucronate; lateral lobes lanceolate, 10–16 mm long. Corolla 40–45 mm long, white, limb pubescent externally with mixed eglandular and glandular hairs; tube subcylindrical, 25–30 mm long; limb in 2+3 arrangement; abaxial and lateral lobes 14–17 × 9–12 mm; adaxial lobes 8–11 × 3.5–6.5 mm. Stamens attached ± midway along corolla tube; filaments c.12 mm long; anthers shortly exserted, 2.5–3 mm long; lateral staminodes c.3 mm long, pilose in proximal half, antherodes c.0.7 mm long. Ovary with apical ring of short pale hairs; style glabrous; stigma clavate, 0.6–0.8 mm long. Capsule c.20 mm long, glabrous. Seeds c.7.5 × 5.5 mm.

Zambia. W: Kitwe, fl.& fr. 6.iii.1964, *Mutimushi* 674 (K, NDO).

Also in southern D.R. Congo (Shaba plateaux). *Marquesia* woodland, but in Congo in *Brachystegia* and *Julbernardia* woodland; c.1250 m.

Conservation notes: Rare, currently known from only 7 collections with only one in Zambia; probably best considered Data Deficient, but may prove to be threatened.

From the few fruits seen, two of the seeds abort very early and the capsule at first appears 2-seeded, though there is clearly locular capacity for four seeds.

12. **Barleria crassa** C.B. Clarke in F.T.A. **5**: 151 (1899). —White, For. Fl. N. Rhod.: 381 (1962). —Tredgold & Biegel, Zimb. Wild Fl.: 52, pl.35 (1996). —Darbyshire in F.T.E.A., Acanthaceae **2**: 339 (2010). Type: Zimbabwe, "South African Goldfields", fl. 1870, *Baines* s.n. (K holotype). FIGURE 8.6.**38**.

    Barleria venosa Oberm. in Ann. Transv. Mus. **15**: 168 (1933). Type: Zimbabwe, no locality, n.d., *van Son* in Vernay-Lang Kalahari Expedition s.n. (PRE holotype).

    Barleria nyasensis sensu auct., non C.B. Clarke. —White, For. Fl. N. Rhod.: 382 (1962), in part for *Fanshawe* 3512. —Phiri, Checklist Zamb. Vasc. Pl.: 18 (2005), in part.

Spiny shrub or perennial herb 45–250 cm tall; stems buff- or yellowish-strigose with interspersed shorter spreading hairs when young. Leaves subsessile, rather fleshy, (oblong-) elliptic, obovate or suborbicular, 0.7–4.5 × 0.4–1.6 cm, base cuneate or subattenuate to obtuse, margin revolute when young, apex obtuse, rounded or rarely acute, with a short spine-tip or apiculum, upper surface with coarse subspreading to subappressed hairs, lower surface strigulose on veins with numerous spreading whitish hairs elsewhere, at least when young; lateral veins 3–6 pairs, prominent beneath. Inflorescence axillary and/or subterminal, 1–3-flowered, subsessile; bracteoles spinose, 5–18 × 0.5–2.5 mm, straight or curved, conduplicate or triangular in cross-section, margin entire, pairs sometimes persisting at lower nodes as paired axillary spines. Calyx often purplish towards apex and on veins when young, turning pale-scarious; outer lobes subequal, broadly ovate to elliptic or obovate, 10.5–23 × 5–16 mm, base rounded or cordate to cuneate or attenuate, margin denticulate to spinulose-dentate, rarely subentire, apex obtuse to attenuate in outline, ± spine-tipped, anterior lobe rarely emarginate, surfaces sparsely strigulose and with scattered minute stalked glands; lateral lobes lanceolate-attenuate, 3–8.5 mm long. Corolla (22.5)25–38.5 mm long, blue, mauve or white, glandular-pilose externally or largely glabrous; tube 15–23.5 mm long, funnel-shaped above attachment point of stamens; limb in 2+3 arrangement; abaxial and lateral lobes 9–18 × 8–16.5 mm; adaxial lobes 6–15 × 4–10 mm. Stamens attached 6–12 mm from base of corolla tube; filaments 12–24 mm long; anthers 3–5 mm long; lateral staminodes 0.7–3.7 mm long, with minute stalked glands and often pilose, antherodes to 1.4 mm or barely developed. Ovary glabrous; style with declinate straight hairs at base; stigma linear, 1–2 mm long. Capsule (11.5)14.5–17 mm long, glabrous. Only immature seeds seen.

Fig. 8.6.**38**. BARLERIA CRASSA subsp. CRASSA. 1, habit (× ²/₃); 2, detail of young stem indumentum (× 4); 3, leaf, abaxial surface (× 2); 4, calyx and bracteoles (× 2); 5, dissected corolla with androecium and gynoecium (× 1½); 6, capsule and immature seeds (× 3). BARLERIA CRASSA subsp. RAMULOSOIDES. 7, leaf, abaxial surface (× 2); 8, calyx and bracteoles (× 2). BARLERIA CRASSA subsp. MBALENSIS. 9, calyx and bracteoles (× 1½); 10, corolla, side view (× 1). 1–4 from *Hopkins* 10274, 5 from *Carter & Coates-Palgrave* 2222, 6 from *Gilliland* 619, 7 & 8 from *Greenway* 5468, 9 from *Robinson* 3781, 10 from *Hutchinson & Gillett* 3962. Drawn by Juliet Williamson.

A variable species with three subspecies:

1. Outer calyx lobes elliptic or obovate with attenuate or cuneate base; young leaves with conspicuous apical spine 1–2 mm long. . . . . b) subsp. *ramulosoides*
   - Outer calyx lobes ovate with runded, truncate or cordate base (sometimes above a short attenuate portion); young leaves with minute apical spine to 0.5 mm long 2
2. Corolla blue or mauve, rarely pale blue, tube clearly longer than limb; outer calyx lobes with denticulate or subentire margin, rarely spinulose-dentate, base (above short attenuate portion) rounded, truncate or subcordate . . . . . a) subsp. *crassa*
   - Corolla white or pinkish, tube only marginal longer than limb; outer calyx lobes with spinulose-dentate margin, base (sub)cordate. . . . . c) subsp. *mbalensis*

a) Subsp. **crassa**. Darbyshire in F.T.E.A., Acanthaceae **2**: 340 (2010) in notes.

Leaves with spine tip minute to 0.5 mm long, or reduced to an apiculum. Young calyx often purplish; outer lobes ± broadly ovate, 11.5–22 × (7)9–13(15) mm, shortly attenuate into a rounded, truncate or subcordate base, margin denticulate with apiculate teeth or subentire, rarely spinulose-dentate; lateral lobes 4.5–8 mm long. Corolla (pale)blue, mauve or purple, pilose to sparsely so externally; tube (17.5)19.5–23.5 mm long; abaxial and lateral lobes 10–16 mm long; adaxial lobes 7.5–13.5 mm long. Lateral staminodes 0.7–2.5 mm long, pilose at least towards base; antherodes usually well-developed, 0.7–1.4 mm long, rarely absent.

Zambia. B: Kalabo Dist., no locality, old fl. 16.x.1963, *Fanshawe* 8076 (K, NDO). N: Kawambwa Dist., fl.& fr. 22.viii.1957, *Fanshawe* 3512 (K, NDO). W: Mwinilunga Dist., S of Mwinilunga on Kabombo road, fl. 3.vi.1963, *Edwards* 595 (K, SRGH). C: Mumbwa Dist., near Mumbwa, st. 1911, *Macauley* 782 (K). S: Mazabuka, no locality (escarpment miombo), fl.& fr. 20.v.1961, *Fanshawe* 6589 (K, NDO). **Zimbabwe**. N: Makonde Dist., 8 km N of Banket, vicinity of Mvurwi (Umvukwe) Mts, fl. 23.iv.1948, *Rodin* 4408 (K, SRGH). C: Makoni Dist., Chiduku communal land (Reserve), fl. 20.vi.1967, *Plowes* 2860 (K, SRGH). E: Mutare Dist., S of Mutare, Dora Farm, fl. 26.vi.1948, *Chase* 801 (BM, K, LMA, SRGH). S: Masvingo Dist., S side of Lake Kyle, just E of dam wall, fl. 9.viii.1988, *Carter & Coates-Palgrave* 2222 (K, SRGH). **Malawi**. C: Lilongwe Dist., Dzalanyama Forest Reserve, few km S of ranch house, st. 26.iii.1977, *Brummitt et al.* 14937 (K, SRGH, see note).

Also in E Angola; though not yet recorded, likely to occur in C Mozambique and southern D.R. Congo. Open or wooded rocky slopes, *Cryptosepalum* and *Brachystegia* woodland on sand, often on granite outcrops but also on serpentine and ironstone, wooded gulleys and roadsides; 900–1650 m.

Conservation notes: Widespread and often common in rocky or sandy habitats; Least Concern.

The few specimens seen from *Cryptosepalum–Brachystegia* woodlands on Kalahari sands in W Zambia and E Angola tend to have larger and less fleshy leaves than specimens from rocky habitats further east. *Mutimushi* 3438 from miombo woodland at Mwinilunga is close to *Barleria crassa* sensu lato but has broader elliptic leaves (to 2.4 cm wide), a dense velutinous indumentum throughout and narrower subentire outer calyx lobes. The material is insufficient to draw further conclusions, and more material is required from here to fully delimit the taxa.

The single specimen seen from Malawi is sterile and only provisionally placed in this subspecies.

b) Subsp. **ramulosoides** I. Darbysh., subsp. nov. Differs from subsp. *crassa* and subsp. *mbalensis* in the outer calyx lobes being elliptic or obovate with base cuneate, not ovate with base rounded, truncate or cordate, and in having prominent apical spines on the young leaves. Type: Zambia, Mpika Dist., Koloswe, fl.& fr. 24.vii.1930, *Hutchinson & Gillett* 4065 (K holotype, BM, SRGH).

Leaves with spine tip conspicuous at least when young, typically 1–2 mm long. Young calyx pale green to mauve(-brown), outer lobes broadly elliptic or obovate, 10.5–15 × 5–7.5(9) mm, base attenuate or cuneate, margin denticulate to spinulose-dentate; lateral lobes 4–7.5 mm long. Corolla blue or purple, sparsely pilose externally; tube 15–22 mm long; abaxial and lateral lobes 9–14 mm long; adaxial lobes 6–9.5 mm long. Lateral staminodes 1.2–2.8 mm long, with pilose hairs throughout or at base only; antherodes minute, 0.2–0.4 mm long.

Zambia. N: Chinsali Dist., Shiwa Ngandu, fl.& fr. 22.vii.1938, *Greenway* 5468 (EA, K, PRE). C: Serenje Dist., Kundalila Falls, fl.& fr. 2.vi.1965, *Robinson* 6681 (K).

Not known elsewhere. Rocky slopes and hill summits, including in miombo woodland; 1200–1650 m.

Conservation notes: Known from only 7 collections in EC and NE Zambia, it is clearly scarce but is found in a largely unthreatened habitat; not threatened.

In addition to the diagnostic characters above, this subspecies typically has short stem internodes with crowded leaves and numerous persistent spinose bracteoles at the lower nodes, reminiscent of the E African *Barleria ramulosa* C.B. Clarke, hence the epithet. *Pole-Evans* 2898 matches the habit and leaf form of this subspecies exactly but the calyx is broader (to 12 mm wide) and less clearly narrowed at the base, so tending towards subsp. *crassa*.

c) Subsp. **mbalensis** I. Darbysh. in F.T.E.A., Acanthaceae 2: 340 (2010). Type: Zambia, Kalambo Falls, fl.& fr. 16.vii.1950, *Bullock* 3000 (K holotype, BR, EA, K).

Barleria nyasensis sensu auct., non C.B. Clarke. —White, For. Fl. N. Rhod.: 382 (1962) in part for *Bullock* 3000. —Champluvier in Pl. Ecol. Evol. **144**: 85 (2011) in part for *Richards* 10132.

Leaves as in subsp. *crassa*. Calyx pale green(-brown) outside with darker reticulate venation, outer lobes broadly ovate, 15–22 × 11–16 mm, base (sub)cordate, margin spinulose-dentate; lateral lobes 3–5.5 mm long. Corolla white or pinkish, sparsely pilose or largely glabrous externally; tube 16.5–20 mm long, broadly campanulate above attachment point of stamens; abaxial and lateral lobes 14–18 mm long; adaxial lobes 10–15 mm long. Lateral staminodes 3–3.7 mm long, largely lacking pilose hairs; antherodes barely developed or to 0.5 mm long.

Zambia. N: Mbala Dist., 13 km S of Mpulungu, fl.& fr. 20.vii.1930, *Hutchinson & Gillett* 3962 (BM, BR, K, LISC, SRGH); Mbala Dist., Kalambo Falls, fl. 13.vii.1962, *Tyrer* 1 (BM, LISC, SRGH).

Also in SW Tanzania. In *Brachystegia* and *Pterocarpus* woodland on sandy soils, dry rocky hillslopes and cliffs, riverbanks; 1000–1250 m.

Conservation notes: With a highly restricted range but habitats not under particular threat; assessed as Data Deficient in F.T.E.A. but probably not threatened.

The pale corollas with a short, broadly funnel-shaped upper portion of the tube and large lobes readily separate this from other subspecies and it is tempting to separate it out at the species rank. However, specimens from Ntumbachushi Falls, Kawambwa in N Zambia (*Richards* 9316, *Lawton* 862) are largely intermediate between subsp. *mbalensis* and subsp. *crassa* sensu stricto, having a calyx close to the former but with a corolla form approaching the latter.

13. **Barleria nyasensis** C.B. Clarke in F.T.A. 5: 150 (1899). —Darbyshire in F.T.E.A., Acanthaceae 2: 340 (2010). —Champluvier in Pl. Ecol. Evol. **144**: 85 (2011), in part excl. *Richards* 10132. Type: Mozambique, mountains E of Lake Nyasa, fl. n.d., *Johnson* s.n. (K holotype).

Spiny perennial herb or subshrub 30–170 cm tall; stems buff- or yellowish-strigose with interspersed shorter spreading hairs when young. Leaves oblong-elliptic, 2–5 × 0.7–1.5 cm, base acute or subattenuate, margin revolute when young, apex ± acute, spine-tipped, upper surface with coarse spreading hairs, lower surface strigose on veins and with numerous spreading whitish hairs when young; lateral veins 4–6 pairs, ± prominent beneath; petiole 0–4 mm long. Inflorescence axillary, 1(3)-flowered, subsessile; bracteoles linear-lanceolate, 10–23 × 1–3 mm, often curved, conduplicate in cross-section, margin entire, apex spinose. Calyx at first (brown-) green with darker reticulate venation, turning pale-scarious; outer lobes subequal, broadly ovate(-elliptic), 14.5–26 × 10.5–17.5 mm, base truncate or cordate, margin spinulose-dentate, apex obtuse to attenuate in outline with apical spine, surface strigose, often with scattered minute stalked glands; lateral lobes lanceolate-attenuate, 7.5–13 mm long. Corolla 40–50 mm long, white, pale blue or pale mauve, glandular-pilose externally; tube (25)28–34 mm long, gradually funnel-shaped above attachment point of stamens; limb in 2+3 arrangement; abaxial and lateral lobes 12–15.5 × 11.5–15 mm; adaxial lobes 11.5–14.5 × 7–9.5 mm. Stamens attached 10–16 mm from base of corolla tube; filaments 19–22 mm long; anthers 4–5 mm long; lateral staminodes 2–2.8 mm long, pilose and with minute stalked glands, antherodes barely developed. Ovary glabrous; style with ring of declinate hairs at base or glabrous; stigma linear, 1–1.5 mm long. Capsule and seeds not seen.

Mozambique. N: Mountains E of Lake Nyasa, fl. n.d., *Johnson* s.n. (K).

Also in S Tanzania. In Tanzania found in miombo woodland and fringes of riverine woodland, often on dry rocky slopes.; c.1000 m.

Conservation notes: A highly localised species, with no further collections in the Flora area since the type gathering; assessed as Least Concern in F.T.E.A.

Close to *Barleria crassa*, of which it could be treated as a further allopatric subspecies. However, although it shares the spinulose calyx margin and pale corollas with subsp. *mbalensis*, the corolla form is very different, particularly the length and shape of the corolla tube, the throat being much longer and more gradually and narrowly funnel-shaped in *B. nyasensis*.

14. **Barleria sp. A** (= *Fanshawe* 9283)

Barleria nyasensis sensu Phiri, Checklist Zamb. Vasc. Pl.: 18 (2005) in part?, non C.B. Clarke.

Spiny perennial herb or subshrub to 60 cm tall; stems with ± numerous spreading buff or yellow hairs and interspersed paler and shorter spreading or retrorse hairs. Leaves subsessile, blade rather fleshy, broadly ovate or uppermost pairs elliptic, 1.5–3 × 0.9–1.8 cm, base rounded or obtuse, margin revolute when young, apex obtuse with a short deflexed spine, surfaces with numerous coarse hairs, subspreading above, paler and more appressed on veins beneath, finer pale spreading hairs between veins when young; lateral veins 3–4 pairs, prominent beneath. Inflorescences clustered in upper axils of main and short lateral branches, 1–3(5)-flowered, subsessile; bracteoles spinose, 9–15 × 1 mm, margin entire. Calyx green or mauve at first with darker reticulate venation, soon turning pale-scarious; outer lobes subequal, ovate, 14–20 × 7.5–13.5 mm, base shortly attenuate then rounded or truncate, margin spinulose-dentate, apex spinose, surface with pale ascending hairs mainly along veins, with short spreading glandular hairs most numerous along margin; lateral lobes linear-lanceolate, attenuate, 9–11 mm long. Corolla 35–40 mm long, mauve or pale purple with brownish throat, rather sparsely glandular-pilose externally; tube 20–22 mm long, funnel-shaped above attachment point of stamens; limb in 2+3 arrangement; abaxial lobe 13–15 × 12 mm; lateral lobes 14–20 × 10–12 mm; adaxial lobes 14–17 × 6–8.5 mm. Stamens attached c.6.5 mm from base of corolla tube; filaments 25–28 mm long; anthers 3.3–3.8 mm long; lateral staminodes c.2 mm long with minute stalked glands, antherodes barely developed. Ovary glabrous; style with declinate ring of hairs at base; stigma linear, c.0.5 mm long. Immature capsule 15 mm long, glabrous. Seeds not seen.

Zambia. N: Mpika Dist., Mbwingimfumu, fl. 17.ix.1958, *Lawton* 483 (K, NDO); Mpika, fl.& fr. 13.viii.1965, *Fanshawe* 9283 (K, NDO, SRGH).

Not known elsewhere. Rocky slopes within miombo woodland or with *Xerophyta–Myrothamnus–Brachystegia* community; c.1450 m.

Conservation notes: Only known from 4 specimens from NC Zambia; Data Deficient. Differs from both *Barleria crassa* and *B. nyasensis* in the spreading, not appressed, stem indumentum, the more ovate mature leaves, the relatively shorter cylindrical base to the corolla tube and the shorter stigma. *Fanshawe* 9283 is additionally striking in its elongate, acuminate corolla lobes; however, these are shorter and more rounded at the apex in *Lawton* 483.

15. **Barleria aromatica** Oberm. in Ann. Transv. Mus. **15**: 169 (1933). —Tredgold & Biegel, Zimb. Wild Fl.: 52, pl.35 (1996). Type: Zimbabwe, Mutare Dist., Ngambi R., fl.& fr. vi.1919, *Eyles* 1664 (SAM lectotype, BM, PRE, SRGH), lectotypified here.

Bushy perennial herb 60–90(150) cm tall, aromatic; stems with mixed short white spreading or retrorse hairs and long coarse pale-buff to golden-yellow hairs, appressed or ascending at first but spreading on older stems; uppermost internodes sometimes glandular-pubescent. Leaves often immature at flowering, elliptic, ovate or oblong, 1–4.2 × 0.5–2.2 cm, base obtuse or rounded, margin revolute when young, apex acute to rounded, apiculate, surfaces with buff to golden-yellow coarse appressed or ascending hairs, longest and most dense on veins beneath, often with finer and more spreading hairs between veins; lateral veins 3–4 pairs, prominent beneath; petiole 0–3 mm long. Inflorescence axillary, 1(2)-flowered, on patent or spreading-ascending peduncles 13–42 mm long, hairs as on stems, short spreading glandular hairs usually present; bracteoles variously ovate or elliptic to linear, 2–6.5(10) × 1–2(5) mm, margin entire, apex often recurved, apiculate; pedicels 0–6 mm long. Calyx green or pinkish at first with darker reticulate venation, later turning scarious; outer lobes subequal, elliptic, oblong or lanceolate, 9.5–14.5(21) × 2.5–6.5 mm, margin entire, often involute, apex acute to attenuate, apiculate, anterior lobe sometimes notched, surfaces buff- to yellow-strigose, usually with scattered short spreading glandular hairs; lateral lobes lanceolate, 9–13.5(16) mm long. Corolla 31–48 mm long, mauve or rarely white to pale pink, sometimes with yellow throat, pubescent externally with mixed eglandular and glandular hairs; tube cylindrical, 17–28.5 mm long; limb in 2+3 arrangement; abaxial lobe 14–23 × 11–17(21) mm; lateral lobes 12–21 × 11–15(20) mm; adaxial lobes as laterals but 7–11(14) mm wide. Stamens attached 10–16 mm from base of corolla tube; filaments 20–29 mm long; anthers exserted, 3.5–5 mm long; lateral staminodes 4–8 mm long, sparsely hairy at base, apex clavate, antherodes undeveloped. Ovary with tuft of deflexed hairs at apex and extending onto style base, distal portion of style with minute spreading glandular hairs; stigma subcapitate or clavate, 0.4–0.8 mm long. Capsule 17–21 mm long, glabrous except apical tuft of hairs. Seeds (?immature) c.6 × 5 mm.

Zimbabwe. N: Bindura Dist., Kingston Hill, fl. 7.v.1969, *Wild* 7756 (K, LISC, SRGH). W: Gwayi Dist., no locality, fl. n.d., *Eyles* 7568 (SRGH). E: Mutare Dist., Odzi, Ardwell Mine, fl. 28.iii.1966, *Wild* 7559 (BR, K, LISC, SRGH). S: Masvingo Dist., Mushandike Nat. Park, dam wall, fl. vii.1971, *Wright* T314 (K, SRGH).

Not known elsewhere. Open rocky hillslopes, often over serpentine deposits, but also recorded from granite outcrops; 1000–1600 m.

Conservation notes: Endemic to Zimbabwe and known from at least 10 localities; mining is likely to have resulted in loss of some populations although mine-spoil may create favourable conditions; Near Threatened.

Whellan 915 from the Mavuradonha Mts lacks the characteristic glandular hairs on the distal portion of the style, but is otherwise a close match for this species; further material from that site is desirable.

16. **Barleria glutinosa** Champl. in Pl. Ecol. Evol. **144**: 85 (2011). Type: D.R. Congo, Haut-Katanga, Mt Mumbwelume, fl. 8.iv.1987, *Kisimba & Muzinga* 46 (BR holotype, CAS, EA, K, LISC, MO).

Spreading perennial herb or shrublet 30–80 cm tall, branches decumbent or procumbent; stems with mixed short white ± retrorse hairs and long coarse ± spreading pale-buff or yellow hairs, often also viscid glandular-pubescent at least on upper internodes. Leaves sometimes immature at flowering, subsessile, ovate-elliptic to lanceolate, 1–3.5 × 0.5–1.5 cm, base obtuse or rounded, margin revolute when young, apex acute, mucronulate, surfaces with ± dense long pale-buff coarse spreading or ascending hairs, young leaves often viscid glandular-pubescent; lateral veins 2–4 pairs. Inflorescence axillary, 1(2)-flowered, densely viscid glandular-pubescent throughout; peduncles spreading-ascending, 9–26 mm long; bracteoles linear(oblanceolate), 5–13 × 1–1.5 mm, margin entire, apex spinulose; pedicels (1)4–16 mm long. Calyx pale green or brownish at first with darker reticulate venation, later turning scarious; outer lobes subequal (posterior lobe marginally longer), elliptic, oblong or lanceolate, 11–19 × 3.5–6 mm, margin entire, apex acute, attenuate or anterior lobe sometimes notched to 2.5 mm, mucronate, surfaces densely glandular-pubescent and sparsely strigose; lateral lobes linear-lanceolate, 9–15 mm long. Corolla 45–56 mm long, pale pink to mauve or white, limb sometimes with darker magenta stripes, pubescent externally with mixed eglandular and glandular hairs; tube cylindrical, 26–35 mm long; limb in 2+3 arrangement; abaxial lobe c.18 × 10–15 mm; lateral lobes 17–21 × 10–17 mm; adaxial lobes as laterals but 7–13 mm wide. Stamens attached 15–18 mm from base of corolla tube; filaments 21–25 mm long; anthers 3–5 mm long; lateral staminodes 2.5–6 mm long, sparsely hairy at base, apex clavate, antherodes undeveloped. Ovary with tuft of deflexed hairs at apex, proximal portion of style often with scattered longer hairs; stigma subcapitate or clavate, 0.3–0.5 mm long. Capsule 18–19 mm long, glabrous. Seeds not seen.

Zambia. C: Chongwe Dist., 80 km E of Lusaka, fl. 15.vi.1956, *King* 381 (EA, K, SRGH); Chongwe Dist., Rufunsa, fl.& fr. 6.vi.1958, *Fanshawe* 4542 (EA, K, NDO, SRGH). E: Petauke Dist., 16 km W of Kachololo Rest House, Great East Road, fl. 26.v.1952, *White* 2888 (FHO, K).

Also in southern D.R. Congo. Dry rocky or sandy hillslopes with open vegetation or stunted *Brachystegia* woodland; 1000–1250 m.

Conservation notes: Very localised and scarce, known from only 10 collections, mostly in the vicinity of the Great East Road in SE Zambia; no information available for the Congo population; probably Vulnerable under IUCN criterion B.

The type locality is quite disjunct from the Zambian sites; however, the morphological differences between the two are very slight – the Congo plant has shorter pedicels (1–2.5 mm vs. usually over 4 mm long) and somewhat smaller glandular heads to the inflorescence hairs.

17. **Barleria molensis** Wild in Kirkia **5**: 80 (1965). —Champluvier in Pl. Ecol. Evol. **144**: 86 (2011). Type: Zimbabwe, Chikomba Dist., Mhlaba Hills near Windsor crome mine, fl. 16.i.1962, *Wild* 5604 (SRGH holotype, BR, K, M, PRE).

Spreading spiny (sub)shrub 40–100 cm tall, often drying dark brown; stems viscid glandular-pubescent to pilose, with or without interspersed long spreading pale-buff hairs numerous at nodes. Leaves elliptic to obovate, 1.5–3.7 × 0.5–1.8 cm, base cuneate or attenuate, apex acute to rounded, mucronate, surfaces glandular-pubescent, with long pale-buff or greyish coarse spreading or ascending hairs; lateral veins 3–4 pairs; petiole 0–5 mm long. Inflorescence axillary, single-flowered, densely viscid glandular-pubescent to pilose throughout; peduncle 1–11 mm long; bracteoles spinose, 15–33 × 1–2.5 mm, margin entire, sometimes with a narrow blade either side of prominent midrib; pedicels 1–5 mm long. Calyx green-brown at first with darker reticulate venation, later turning scarious; anterior lobe oblong-elliptic to obovate, 17–24 × 4–7 mm, margin entire or with minute teeth formed by swollen hair bases towards apex, apex attenuate into apical spine, rarely bifidly so, surfaces densely glandular-pubescent, hairs often longest and most dense at base, sparsely strigose along veins and margin; posterior lobe as anterior or sometimes lanceolate, 20–26 × 4–9 mm; lateral lobes linear (-oblanceolate), 16–23 mm long. Corolla 51–85 mm long, white, fading to pale blue, drying dark blue to blue-black, externally pilose with eglandular and glandular hairs; tube cylindrical, 35–65 mm long, somewhat widened towards mouth; limb in 2+3 arrangement; abaxial and lateral lobes 15–21 × 8–13 mm; adaxial lobes 15–18 × 6–10 mm. Stamens attached 20–40 mm from base of corolla tube; filaments 27–36 mm long; anthers exserted, 5–7.5 mm long; lateral staminodes 3.5–7 mm long, sparsely to densely hairy at base, antherodes undeveloped or obsolete. Ovary with dense tuft of white crisped-spreading hairs at apex; style glabrous; stigma clavate or linear, 0.8–2.3 mm long. Capsule 20–25 mm long, glabrous except apical tuft of hairs. Only immature seeds seen.

Zimbabwe. C: Kadoma Dist., Great Dyke near Ngezi (Umgesi Poort), fl. 16.iii.1964, *West* 4774 (K, SRGH); Chirumanzu Dist., Mvuma–Kwekwe road, c.1 km E of Lalapansi turn-off on Great Dyke, fl. 12.v.1976, *Biegel* 5297 (K, SRGH).

Not known elsewhere. Open bushland, grassland or *Brachystegia* woodland on dry hillslopes over serpentine rocks; 900–1400 m.

Conservation notes: Endemic to a northern section of the Great Dyke of Zimbabwe. Golding (Sthn. Afr. Pl. Red Data Lists: 171, 2002) assessed it as of Least Concern, citing its favoured habitat of open rocky slopes. Mining may have resulted in the loss of some populations, although spoil resulting from small-scale mining can create favourable conditions. In view of the inferred threat, probably best considered Vulnerable under IUCN criterion B.

Flower size is rather variable in this species, particularly the corolla tube length. Specimens with a short tube such as *Biegel* 5312 appear quite different to those with a long-tube, as in the type. Intermediates do occur and it is possible that the corolla tube continues to lengthen rapidly even as the corolla matures to anthesis. Field observations are required to see how much population-level variation there is in flower size.

Barleria molensis appears most closely allied to two Ethiopian species: *B. longissima* Lindau and *B. sp. 12* of Fl. Ethiopia & Eritrea (= *Ensermu & Zerihun* 612).

18. **Barleria hydeana** I. Darbysh., sp. nov. Differs from *B. bremekampii* in the more lax inflorescences which are always single-flowered (mature inflorescences 2 to several flowered in *B. bremekampii*), the short glandular-viscid inflorescence indumentum with only sparse interspersed eglandular hairs (silky mixed eglandular and glandular hairs on calyx which cover the surface in *B. bremekampii*) and the proportionally narrower leaves (length to width ratio 2.4–3:1 vs. 1.3–1.65:1 in *B. bremekampii*). Type: Zimbabwe, Kwekwe Dist., Mazuri Ranch, end of dam wall, fl. 29.iv.1995, *Hyde* 95/119/1 (K holotype).

Spiny (?)shrub; stems with ± dense short spreading pale eglandular hairs and long spreading or ascending buff-coloured hairs, upper internodes densely short viscid glandular-pubescent. Leaves subsessile, (oblong)elliptic, 1.2–2 × 0.5–0.75 cm, base obtuse or acute, apex acute, mucronate, upper surface, margin and veins beneath with numerous long pale-buff ascending or subappressed hairs, lower surface with numerous fine short spreading hairs between veins, uppermost leaves also glandular-pubescent; lateral veins 3–4 pairs. Inflorescence axillary, single-flowered, densely viscid glandular-pubescent throughout; peduncle 10–16 mm long; bracteoles spinose, 13–19 × 1 mm, triangular in cross-section, margin entire, pairs persisting at lower nodes and on old leafless stems as stalked paired axillary spines; pedicels 1.5–8 mm long. Calyx dark brown in dry state, accrescent; anterior lobe oblong-obovate or obovate-elliptic, 10–12 × 4–4.5 mm in flower to 15–16 × 7–7.5 mm in fruit, margin at first subentire but 2–6 prominent spinulose teeth sometimes developing in distal half in fruit, apex acuminate into a prominent spine, surfaces densely short glandular-pubescent and sparsely strigose along veins and margin; posterior lobe as anterior but 12.5–15 × 4–5.2 mm in flower to 15.5–18 × 7.5–8.5 mm in fruit; lateral lobes linear-lanceolate, 6.5–8.5 mm long in flower to 9.5–12 mm in fruit. Corolla c.35 mm long (one flower measured), colour in life unknown, drying blue-black, eglandular-pubescent and glandular-pilose externally; tube c.18 mm long, cylindrical in proximal half, campanulate above; limb in 2+3 arrangement; abaxial lobe c.15.5 × 12 mm; lateral lobes c.17 × 13.5 mm; adaxial lobes much reduced, 3.8–5 × 3.5–3.7 mm. Stamens attached c.8 mm from base of corolla tube; filaments c.17 mm long; anthers exserted, c.5.5 mm long; lateral staminodes 2.5–3.5 mm long, densely pilose, antherodes undeveloped. Ovary glabrous except for small tuft of white hairs at apex; style glabrous; stigma clavate, 0.6 mm long. Capsule 18–19 mm long, glabrous except apical tuft of hairs. Only immature seeds seen.

Zimbabwe. C: Harare Dist., fl.& fr. 1.ix.1943, *Hopkins* in SRGH 10389 (K, SRGH). Not known elsewhere. Grazed grassland with shrubs; 1250–1500 m.

Conservation notes: Endemic to C Zimbabwe where it is known from only two collections, probably scarce and highly localised; possibly a serpentine specialist. Currently assessed as Data Deficient but extreme range restriction may make it Vulnerable.

This striking species shares with *Barleria bremekampii* the much-reduced adaxial corolla lobes less than one-third the length of the lateral lobes. It is easily separared from that species by the differences listed in the diagnosis. In fruit, *B. hydeana* could easily be mistaken for the Great Dyke endemic *B. molensis* but differs in its more lax inflorescence and broader, sometimes conspicuously toothed outer calyx lobes (length to width ratio of anterior lobe 2–2.7:1 vs. usually 3–4.5:1 in *B. molensis*).

The specific epithet honours Mark Hyde, collector of the type specimen and eminent field botanist in Zimbabwe.

19. **Barleria bremekampii** Oberm. in Ann. Transv. Mus. **15**: 162 (1933). —Phillips in Fl. Pl. Afr. **23**: t.893 (1943). —Fabian & Germishuizen, Wild Fl. Nthn. S. Afr.: 390, pl.187 (1997). Type: Zimbabwe, Lupane, fl.& fr. 16.vii.1929, *Bremekamp* in TRV 27520 (PRE lectotype, PRU), lectotypified here (see note).

Spiny shrub, erect or straggling, 90–150 cm tall; stems densely hairy with mixed long ascending buff-coloured or greyish hairs and shorter finer retrorse or spreading hairs, upper internodes usually with interspersed spreading glandular hairs. Leaves sometimes absent at fruiting, broadly ovate or elliptic, 1.2–3 × 0.8–2.2 cm, base rounded to cuneate, apex acute or obtuse, mucronate, surfaces with numerous ascending pale buff or greyish hairs, most dense on veins beneath; lateral veins 3–4 pairs, prominent beneath; petiole 1–7.5 mm long. Inflorescence axillary, numerous in upper portion of branches, each a contracted (1)2–6-flowered unilateral cyme; primary peduncle 2–7(9) mm long; bracteoles stoutly spinose, first pair (10)17–31 × 1–2 mm, often disproportionately longer than others in cyme, soon turning straw-coloured, triangular in cross-section, margin entire, pairs persisting at lower nodes as long paired axillary spines. Calyx green or purplish at first with darker venation but soon turning straw-coloured; anterior lobe narrowly oblong, lanceolate or somewhat oblanceolate, 7.5–15 × 2.5–5 mm, base cuneate, margin entire or with minute teeth near apex, apex usually with (1)2–3 mucronate teeth, surface densely silky-hairy with mixed ascending eglandular and spreading glandular hairs, usually obscuring surface; posterior lobe oblong-lanceolate, 8.5–16.5 mm long, apex acute, stiffly mucronate; lateral lobes linear-lanceolate, 7–14.5 mm long. Corolla (23)26–31(36) mm long, blue or mauve, with numerous pale retrorse eglandular hairs and few spreading glandular hairs externally; tube 14.5–21 mm long, cylindrical below attachment point of stamens, widening gradually towards mouth; limb in 2+3 arrangement; abaxial and lateral lobes 8.5–14 × 7.5–13.5 mm; adaxial lobes 3.5–6 × 2.5–4.5 mm. Stamens attached 7–10 mm from base of corolla tube; filaments 11–21 mm long; anthers 3.5–5.5 mm long; lateral staminodes 1.5–2.5 mm long, minutely glandular towards base, antherodes undeveloped or rarely to 0.8 mm long. Ovary with few to numerous pale straight hairs towards apex, extending onto style base; stigma linear, 0.8–1.3 mm long. Capsule 12.5–16.5 mm long, with few pale hairs at apex. Seeds c.5 × 5 mm.

Zimbabwe. W: Lupane Dist., Lupane, fl.& fr. 16.vii.1929, *Bremekamp* in TRV 27520 (PRE).

Also in South Africa (North West Province, Limpopo). Open bushland on dry rocky slopes and outcrops; c.1100 m.

Conservation notes: Although highly restricted in its range, it is locally common in NE South Africa and was plentiful in the Nylstroom Mts (Obermeijer 1933). Its current status in Zimbabwe is not known, though it is clearly rare having not been recollected since 1929. The main habitats are not threatened.

The adaxial corolla lobes of *Barleria bremekampii* are usually ≤ half the length of other lobes, as in the type specimen, but in small-flowered specimens from South

Africa such as *Germishuizen* 368 they can be up to 70% the length of other lobes. It is odd that the species has not been collected in Zimbabwe since 1929, particularly as it is so conspicuous and, in South Africa at least, can be locally common. The type locality is some considerable distance from the remaining range, but there is no reason to doubt that the locality details are correct.

The Transvaal Museum sheet of Bremekamp's collection, since transferred to PRE, has been selected as the lectotype. According to Obermeijer (1933) there is a duplicate of this specimen at PRU under the herbarium number 1614.

20. **Barleria fissimuroides** I. Darbysh., sp. nov. Resembles *B. ventricosa* and allies in sect. *Fissimura* but differs in having a 4-seeded (not 2-seeded) capsule, a linear (not clavate or subcapitate) stigma and a corolla with a cylindrical tube and a limb in the 2+3 arrangement (not funnel-shaped with limb in a 4+1 arrangement). Type: Mozambique, Manica Dist., Quinta da Fronteira, fl.& fr. 19.iii.1961, *Chase* 7438 (K holotype, LISC, LMA, M, SRGH).

Shrub to 100 cm tall, drying blackish throughout; stems with long spreading and ascending pale buff-coloured hairs with a ± conspicuously bulbous base, opposite furrows with two lines of short appressed multicellular hairs with dark cell walls. Leaves ovate, 8–11 × 3–4.7 cm, base cuneate or obliquely so, apex attenuate into a narrow tip, surfaces with long pale hairs evenly distributed above, mainly on veins beneath; lateral veins 6–7 pairs; petiole 12–32 mm long. Inflorescences axillary but clustered in upper portion of branches to form a loose terminal thyrse, each cyme 2 to several-flowered, partially dichasial; bracts foliaceous, gradually reducing upwards; primary peduncle to 2.5 mm long; bracteole pairs somewhat unequal, narrowly elliptic to oblanceolate, 9–18 × 2–4 mm, base attenuate from a curved stalk, margin entire, apex acute; pedicels to 2 mm long. Calyx somewhat accrescent; anterior lobe elliptic(-obovate), 11–12.5 × 6–8 mm in flower, to 17.5 mm long in fruit, base acute, margin entire, apex truncate to rounded in outline with 2 apiculate teeth, surface with long pale ascending hairs, those on margin with swollen base in fruit, also with minute crisped hairs towards base; posterior lobe ovate-elliptic, 13–16.5 × 6.5–8 mm in flower to 21 mm long in fruit, apex acute-apiculate; lateral lobes lanceolate, 6–7 mm long in flower. Corolla 32–37 mm long, blue, pubescent externally with mixed eglandular and glandular hairs; tube subcylindrical, c.23 mm long, curved towards base; limb in 2+3 arrangement; abaxial lobe c.13 × 11 mm; lateral lobes 14–15 × 11–12 mm; adaxial lobes c.12 × 8.5 mm. Stamens attached 13 mm from base of corolla tube; filaments c.18 mm long; anthers 2.4–2.7 mm long; lateral staminodes 2.5–3.2 mm long, pubescent, antherodes 0.9–1.3 mm long. Ovary with few long pale hairs in 2 opposite lines; style glabrous; stigma linear, 0.9–1.5 mm long. Immature capsule 18 mm long, glabrous. Only immature seeds seen.

Zimbabwe. E: Mutare Dist., Vumba, immediately behind Leopard Rock Hotel, fl. 7.iii.1950, *Chase* s.n. (E). **Mozambique**. MS: Manica Dist., Quinta da Fronteira (Border Farm), fl.& fr. 19.iii.1961, *Chase* 7438 (K, LISC, LMA, M, SRGH).

Not known elsewhere. Mountain slopes with grass and small trees; 1100–1550 m.

Conservation notes: Endemic to the Vumba area along the Zimbabwe/Mozambique border and known from only two collections. Clearly rare and highly localised, its scarcity suggests it is likely to be threatened by even small-scale disturbance; probably Vulnerable.

The specific epithet "fissimuroides" refers to this species' resemblance to taxa within sect. *Fissimura*: non-spiny plants drying blackish, entire calyx lobes which are not scarious and which lack conspicuous reticulate venation. Indeed, the type specimen was filed under *Barleria scindens* (= *B. ventricosa*) at Kew. However, the combination of 4-seeded fruit, linear stigma and corolla in the 2+3 arrangement clearly separate it from all members of that section and place it in sect. *Barleria*, although its affinities within that section are unclear. It is perhaps most likely to be confused with *B. gueinzii* or *B. holstii* but the corolla is very different to those species, lacking the expanded throat and offset abaxial lobe. The corolla form is more reminiscent of *B. spinulosa, B. fulvostellata* and allies but, amongst many differences to species in that group, it lacks stellate hairs. It is easily separated from the few specimens of *B. spinulosa* that lack stellate hairs in drying blackish, in the bracteoles being rather leafy, elliptic or oblanceolate and innocuous (not linear-lanceolate and spine-tipped), in the outer calyx lobes being entire (not toothed), and in drying black and lacking conspicuous venation (vs. papery and soon pale-scarious, with conspicuous reticulate venation).

21. **Barleria fulvostellata** C.B. Clarke in F.T.A. **5**: 163 (1899). —Darbyshire in Kew Bull. **64**: 675 (2010) in part. —Darbyshire in F.T.E.A., Acanthaceae **2**: 342 (2010) in part. Type: Malawi, Nyika Plateau, fl. ?vi.1896, *Whyte* 181 (K lectotype), lectotypified by Darbyshire (2010).

Perennial suffruticose herb or subshrub 30–150 cm tall; stems with dense golden(-brown) long-armed stellate hairs with a dense fasciculate-stellate (sometimes dendritic) base, often with spreading glandular hairs. Leaves ovate, elliptic or suborbicular, 1.8–9 × 1.3–6 cm, base truncate, subcordate, obtuse or cuneate, apex rounded to subattenuate, sometimes abruptly narrowed to a short acumen, surfaces densely long-armed stellate-pubescent, margin sometimes with spreading glandular hairs; lateral veins 3–6 pairs, these and reticulate tertiary venation prominent beneath; petiole 0–17 mm long. Inflorescences axillary or clustered towards stem apex to form an interrupted thyrse, each cyme unilateral, 1–4-flowered, sessile or peduncle to 5(9) mm long; bracteoles linear(-lanceolate), narrowly elliptic or oblanceolate, 7–23 × 0.5–5.5 mm, margin entire, apex ± mucronate; pedicels 0–4.5 mm long. Calyx yellow-green to green-brown with darker brown venation, tardily scarious; anterior lobe (oblong-)elliptic to somewhat obovate, 9–18 × 5–12.5 mm, margin entire or minutely denticulate, involute, apex rounded, obtuse or shallowly emarginate, indumentum as leaves, glandular hairs mainly on margins and apex; posterior lobe slightly larger, apex rounded to attenuate, ± mucronulate; lateral lobes lanceolate, ovate or oblong-elliptic, 6–14.5 mm long. Corolla 29–50 mm long, pale blue, pink or purple, pilose externally with mixed glandular and eglandular hairs; tube cylindrical, 14–29 mm long, ± curved below attachment point of stamens; limb in 2+3 arrangement; abaxial and lateral lobes 12.5–22 × 9–15 mm; adaxial lobes 11–19.5 × 6.5–11.5 mm. Stamens attached 8.5–16.5 mm from base of corolla; filaments 12–28 mm long; anthers 3.2–6.2 mm long; lateral staminodes 2–3.5 mm long, sparsely pubescent or pilose and with minute stalked glands, antherodes absent or poorly developed, to 0.6 mm long. Ovary pubescent towards apex or only with apical tuft of hairs; style pubescent towards base; stigma clavate to linear, 0.5–1 mm long. Capsule 13.5–17 mm long, sparsely pilose towards apex. Seeds 4–5 mm long and wide.

A variable species, with three recognised subspecies in the Flora area. Differences between these taxa are quite marked and they may prove to be closely related but discrete allopatric species.

1. Stamens inserted in lower half of corolla tube, filaments 18–28 mm long. . . . . c) subsp. *mangochiensis*
 – Stamens inserted in upper half of corolla tube, filaments 12–17 mm long 2
2. Lower leaves broadly ovate to suborbicular, largest ≥3.5 cm wide; bracteole pairs shorter than calyx; corolla tube 18.5–28.5 mm long . . . . a) subsp. *fulvostellata*
 – Lower leaves ovate-elliptic, largest <3.5 cm wide; bracteoles held ± upright and exceeding calyx; corolla tube 14–17 mm long . . . . b) subsp. *axillaris*

a) Subsp. **fulvostellata**. —Darbyshire in Kew Bull. **64**: 676 (2010); in F.T.E.A., Acanthaceae **2**: 343 (2010).

Stellate hairs on stems with a short arm 0.5–0.8 mm long; fertile stems with few to numerous spreading glandular hairs. Mature leaves broadly ovate to suborbicular, 4–9 × 3.5–6 cm, length to width ratio 1–1.7:1, base rounded, truncate or subcordate. Cymes usually clustered towards branch apex, forming a loose thyrse. Bracteoles linear-lanceolate to narrowly elliptic, 7–17 × 1–3 mm. Corolla tube 18–29 mm long; stamens attached in distal half of tube, filaments 15–17 mm long, anthers 5–6.2 mm long.

Malawi. N: Chitipa Dist., between Mpata and start of Tanganyika Plateau, fl. vii.1896, *Whyte* s.n. (K).

Also in S Tanzania. Habitat not recorded; in both *Acacia–Commiphora* and *Brachystegia* woodland in Tanzania; 600–2150 m.

Conservation notes: A scarce subspecies, endemic to the Malawi/Tanzania border area, that has not been recollected in Malawi since its original discovery; assessed as Endangered under IUCN criterion B in Darbyshire (2010).

b) Subsp. **axillaris** I. Darbysh. in Kew Bull. **64**: 677 (2010). Type: Malawi, Kasungu Dist., 24 km on Kasungu–Mzuzu road, fr. 31.iii.1991, *Bidgood & Vollesen* 2162 (K holotype, BR, C, CAS, MAL).

Stellate hairs on stems with a long arm 0.8–2.2 mm long; fertile stems lacking glandular hairs. Mature leaves ovate-elliptic, 3.3–5.8 × 1.5–3.3 cm, length to width ratio 1.3–2.3:1, base cuneate to obtuse or rounded. Cymes axillary, few per branch, sometimes restricted to uppermost axils but not forming a thyrse. Bracteoles narrowly elliptic or oblanceolate, 12–23 × 1.5–5.5 mm, held erect and often extending beyond calyx. Corolla tube 14–17 mm long; stamens attached in distal half of tube, filaments 12.5–15 mm long, anthers 3.5–4.2 mm long.

Zambia. N: Mpika, fl. 9.ii.1955, *Fanshawe* 2031 (K, NDO). **C**: Serenje Dist., Muchinga escarpment near Kapamba R., fl. 19.i.1966, *Astle* 4454 (K, SRGH). **E**: Chipata Dist., Sumbi Hills, fl. 3.i.1959, *Robson* 1028 (BM, K, LISC, PRE, SRGH). **Malawi. C**: Kasungu Dist., 24 km on Kasungu–Mzuzu road, fr. 31.iii.1991, *Bidgood & Vollesen* 2162 (BR, C, CAS, K, MAL).

Not known elsewhere. Miombo woodland often on hillslopes, sometimes in degraded woodland and roadsides; 600–1250 m.

Conservation notes: Endemic to central Zambia and Malawi but locally common in E Zambia; assessed as Least Concern in Darbyshire (2010).

c) Subsp. **mangochiensis** I. Darbysh. in Kew Bull. **64**: 677 (2010). Type: Malawi, Mangochi–Chiponde road, 4 km from Mangochi, fl. 8.vi.1978, *Iwarsson & Ryding* 748 (K holotype, C).

Stellate hairs on stems with an intermediate arm 0.7–1.5 mm long; fertile stems with few to dense glandular hairs. Leaves ?immature at flowering, elliptic or suborbicular, 1.8–3.2 × 1.3–1.8 cm, to 5.5 × 3 cm on sterile branches, length to width ratio 1–2:1, base obtuse. Cymes axillary but clustered towards stem apex and can form a short thyrse. Bracteoles linear or oblanceolate, 8.5–14 × 0.5–1.5 mm. Corolla tube 20–27 mm long; stamens attached in proximal half of tube, filaments 18–28 mm long, anthers 3.2–4.2 mm long.

Malawi. S: Mangochi Dist., near Mangochi (Fort Johnston), fl. 2.v.1960, *Leach* 9893 (K, LISC, PRE, SRGH); Mangochi Dist., Malindi turn-off, fl. 21.vii.1984, *Salubeni & Patel* 3798 (K, MAL). **Mozambique. N**: Mandimba Dist., Mt Mandimba, Amaramba, fl.& fr. 31.vii.1934, *Torre* 547 (COI, LISC).

Not known elsewhere. Bushland and open woodland on dry rocky hillslopes; 500–600 m.

Conservation notes: Endemic to a small area around the south end of Lake Malawi. The Mozambique specimen has marginally extended the known range of this subspecies (it was recorded as frequent there) although it is still very restricted; assessed as Endangered under IUCN criterion B in Darbyshire (2010).

22. **Barleria sp. B** (= *Robinson* 814).

Low spreading perennial herb or shrublet c.90 cm tall; stems with dense ± spreading golden long-armed stellate hairs, arm typically 2–2.5 mm long, often flexuose. Leaves subsessile, elliptic, 3.8–5 × 2–3 cm, base obtuse, apex obtuse or subacute, surfaces with numerous golden long-armed stellate hairs, stellate base sometimes falling early or absent, lateral veins 3–4 pairs, these and reticulate tertiary venation ± prominent beneath. Inflorescence of 1 to few-flowered congested unilateral cymes, clustered at apex of main and short lateral branches to form a short thyrse; bracts as leaves but much-reduced, typically 8–14 × 5–7 mm, more densely hairy, often with some spreading glandular hairs along margin; bracteoles held upright, linear-lanceolate, longer of each pair 10–12.5 × 1–1.5 mm, margin entire, apex mucronate. Calyx dark purple-brown at first with darker venation, turning pale-scarious; anterior lobe elliptic or somewhat obovate, 12.5–14 × 6–9 mm, base cuneate-attenuate, margin dentate, teeth slender with an apical bristle, apex rounded to acute, with several teeth, surface with dense mixed ascending golden long-armed stellate hairs and spreading glandular hairs; posterior lobe somewhat longer and narrower than anterior, apex obtuse to acute, mucronulate; lateral lobes linear-lanceolate, 9–10 mm long. Corolla 38–41 mm long, pale blue-violet, mixed glandular and eglandular pilose externally, mainly on lateral lobes; tube c.21.5 mm long, cylindrical below attachment point of stamens, narrowly campanulate above; limb in 2+3 arrangement; abaxial lobe 13–16 × c.13 mm; lateral lobes 14.5–18 × 10–13.5 mm; adaxial lobes 10–12.5 × 6.5–9 mm. Stamens attached in distal half of corolla tube; filaments c.15.5 mm long; anthers c.4 mm long; lateral staminodes 2–2.5 mm long, pubescent, antherodes absent. Ovary pale-pubescent towards apex; style pubescent towards base; stigma linear, c.1 mm long. Only immature capsule seen.

Zambia. E: Chama Dist., 97 km NW of Lundazi, East Luangwa escarpment, fl. 2.vi.1954, *Robinson* 814 (BR, K, SRGH).

Not known elsewhere. Dry stony ground; c.1000 m.

Conservation notes: An incompletely known taxon, apparently endemic to E Zambia; Data Deficient.

The single specimen seen is closely allied to the *Barleria fulvostellata* complex but quite different to the sympatric subsp. *axillaris*. It is possible that this is either (a) a further distinct and rare entity within that complex, or (b) a hybrid between *B. fulvostellata* and a second species. The dentate margin to the outer calyx lobes and the somewhat campanulate distal portion to the corolla tube (distinguishing it from all variants of *B. fulvostellata*) would suggest *B. crassa* or *B. sp. A* as possible candidates. Further material is required to draw firm conclusions.

23. **Barleria mutabilis** I. Darbysh., stat. & nom. nov. Type: Tanzania, Beho Beho, fl. 6.vi.1977, *Vollesen* in MRC 4617 (C holotype, DSM, K).

Barleria fulvostellata C.B. Clarke subsp. *scariosa* I. Darbysh. in Kew Bull. **64**: 678 (2010). —Darbyshire in F.T.E.A., Acanthaceae 2: 342 (2010). Type as for *B. mutabilis*.

Perennial herb or shrub 50–150 cm tall; stems with dense buff-yellow to golden long-armed stellate hairs, arm typically 1–2 mm long, stellate base sometimes falling early, young stems with few to numerous spreading glandular hairs or these absent. Leaves ovate or elliptic, 3.3–13.5 × 1.4–6 cm, base attenuate or cuneate to obtuse or rounded, apex acute, subattenuate or obtuse, surfaces with long-armed stellate hairs as on stems, stellate base sometimes falling early particularly above, margin of young leaves sometimes with spreading glandular hairs; lateral veins 4–5 pairs, prominent beneath; petiole 2–12 mm long. Inflorescences axillary towards stem apex, sometimes together forming a loose thyrse, each cyme contracted, (2)3–12-flowered, unilateral or often branching at maturity; bracteole pairs spreading and often deflexed, linear, narrowly elliptic or oblanceolate, 5–20 × 0.8–4.5 mm, margin entire or minutely toothed, apex mucronate; pedicels 0–2.5 mm long. Calyx papery, at first bluish or purple-tinged with darker reticulate venation, soon pale-scarious; anterior lobe obovate, elliptic or (in our region) ovate-orbicular, 8–21.5 × 4.5–19.5 mm, base cuneate to truncate or rounded, margin subentire or denticulate, apex rounded to attenuate or emarginate, indumentum as leaves, glandular hairs mainly on margins and apex or absent; posterior lobe slightly larger, apex obtuse to attenuate, mucronulate; lateral lobes (ovate)lanceolate, 6.5–12.5 mm long. Corolla 26.5–46.5 mm long, pale blue or lilac to purple, glandular-pilose externally; tube cylindrical, 17–23 mm long, curved below attachment point of stamens; limb in 2+3 arrangement; abaxial lobe 12–18.5 × 9–14 mm; lateral lobes 14–19.5 × 10–14.5 mm; adaxial lobes 8–14.5 × 7–11 mm. Stamens attached 10–16 mm from base of corolla tube; filaments 12–16 mm long; anthers 3–5.2 mm long; lateral staminodes 1.2–3.5 mm long, pilose and with minute stalked glands, antherodes barely developed or to 0.8 mm long. Ovary with an apical tuft of short hairs, sometimes with a few pale hairs towards apex; style pubescent towards base; stigma clavate, 0.7–1.5 mm long. Capsule 13–16.5 mm long, sparsely pilose towards apex or with an apical tuft of hairs only. Seeds 4–5.5 × 3.8–4.5 mm.

Mozambique. N: Mecula Dist., Serra Mecula, R. Kuthi, fl. 8.vi.2003, *Boane* 21 (K, LMU).

Also in E Tanzania. Riverine dry forest, in Tanzania recorded from low altitude woodland and wooded grassland; c.1000 m.

Conservation notes: Scarce and known from only two specimens within the Flora area; originally assessed as Least Concern by Darbyshire (2010) (under *B. fulvostellata* subsp. *scariosa*).

I have previously treated this taxon as a subspecies of *Barleria fulvostellata*, to which it is undoubtably closely allied. The less conspicuous stellate base and longer arm to the complex hairs, the thinner, more scarious calyx lobes, the thinner leaves with less prominent tertiary venation and the often more spreading bracteoles all separate this from subsp. *fulvostellata*. These characters tend more towards the variable *B. spinulosa* than to *B. fulvostellata* and in many ways *B. mutabilis* is intermediate between the two. It has proven easier to key out separately and so is elevated to full species status here.

The new epithet mutabilis ('changeable') refers to the considerable variation in calyx shape in this taxon from north to south. The plants from Niassa Reserve, N Mozambique have the largest and broadest calyx, which are ovate-orbicular. Those from east-central Tanzania have a much narrower calyx, obovate or elliptic in shape. However, from the limited material seen, the variation appears clinal and I prefer to keep it as a single variable species at present.

24. **Barleria torrei** I. Darbysh., sp. nov. Differs from *B. mutabilis* in the bracteoles being broadly elliptic to suborbicular, 7.5–17.5 mm wide (vs. linear, oblanceolate or narrowly elliptic, 0.5–4.5 mm wide), and in the posterior calyx lobe being markedly larger and broader than the anterior lobe and sometimes partially enveloping it (vs. outer calyx lobes subequal, the posterior lobe only marginally larger). Type: Mozambique, Lago Dist., Maniamba, fl.& fr. 24.viii.1934, *Torre* 549 (LISC holotype, COI).

Barleria sp. A sensu Darbyshire in Kew Bull. **64**: 678 (2010).

Shrub 100–200 cm tall; stems with dense golden long-armed stellate hairs, stellate base sometimes falling early, arm 0.5–1.3 mm long. Leaves broadly ovate, 2–8 × 2–4.8 cm, base subcordate or rounded, apex obtuse, acute or shortly attenuate, surfaces densely golden-hairy, hairs long-armed stellate or simple; lateral veins 5–6 pairs, these and reticulate tertiary venation prominent beneath; petiole 0–8 mm long. Inflorescences axillary but clustered towards stem apex, forming a terminal thyrse with reduced foliaceous bracts, each cyme contracted and strobilate, 3–8-flowered, unilateral or becoming branched at maturity, sessile or primary peduncle to 7.5 mm long; bracteoles broadly elliptic to suborbicular, 9.5–23 × 7.5–17.5 mm, apex obtuse to attenuate with a conspicuous mucro; flowers sessile. Calyx papery, green at first with a purple-brown apex, soon turning brown-purple scarious; anterior lobe suborbicular or broadly elliptic, 14–19 × 12–16 mm, base obtuse, rounded or cordate, margin irregularly dentate, somewhat undulate, apex obtuse or rounded, surface with golden long-armed stellate or mixed stellate and simple hairs, most numerous on veins; posterior lobe suborbicular 16–

25 × 17–25 mm, margin often involute and partially enveloping anterior lobe; lateral lobes lanceolate-attenuate, 9–15.5 mm long. Corolla 37–44 mm long, lilac or blue, pilose externally with mixed glandular and eglandular hairs; tube cylindrical, 21–22 mm long, curved below attachment point of stamens; limb in 2+3 arrangement; abaxial and lateral lobes 12–19 × 10–15 mm; adaxial lobes 7.5–10 × 5–8 mm. Stamens attached in distal half of corolla tube; filaments 16–20 mm long; anthers 4.5–5 mm long; lateral staminodes c.3.5 mm long, pubescent, antherodes obsolete or poorly developed, to 0.8 mm long. Ovary with apical tuft of hairs; style pubescent towards base; stigma clavate, 0.6–0.7 mm long. Capsule c.17.5 mm long, with few pale hairs towards apex. Only immature seeds seen.

Mozambique. N: Sanga Dist., Unango, fl. 27.v.1948, *Pedro & Pedrógão* 3966 (EA, LMA); Lago Dist., Maniamba, Malolo, Serra Géci, fl. 28.v.1948, *Pedro & Pedrógão* 4023 (EA, LMA).

Not known elsewhere. Open forest, wooded slopes above river, and abandoned cultivated land; 1100–1200 m.

Conservation notes: Endemic to Niassa Province in Mozambique and known from only the three cited specimens plus a photographic record taken by Meg Coates-Palgrave in May 2009 at Simanje R. near Lichinga, while Torre recorded it as widespread at Maniamba; best considered as Data Deficient.

The recent discovery of the Torre collection with flowers and fruits at LISC and COI confirms this to be a distinct species within the *Barleria fulvostellata* complex and so it is formally described here.

Label data on the EA and LMA sheets of *Pedro & Pedrógão* 3966 differ – on the former the locality is given as "Malolo, Jessi [Njesi] Mt" whilst on the latter it is recorded as "Unango".

25. **Barleria spinulosa** Klotzsch in Peters, Naturw. Reise Mossamb. **6**: 208 (1861). —Clarke in F.T.A. **5**: 152 (1899). —Obermeijer in Ann. Transv. Mus. **15**: 174 (1933). —Moriarty, Wild Fl. Malawi: 85, pl.43(3) (1975). —Darbyshire in F.T.E.A., Acanthaceae **2**: 343 (2010). Type: Mozambique, Quirimba Is. (Kerimba), fl. n.d. (1842–1846), *Peters* s.n. (B† holotype, PRE lectotype), lectotypified here. FIGURE 8.6.39.

Barleria consanguinea Klotzsch in Peters, Naturw. Reise Mossamb. **6**: 206 (1861). —Clarke in F.T.A. **5**: 153 (1899). Type: Mozambique, Rios de Sena, 1842–1846, *Peters* s.n. (B† holotype).

Barleria squarrosa Klotzsch in Peters, Naturw. Reise Mossamb. **6**: 207 (1861). Type: Mozambique, Tete, Rios de Sena, 1842–1846, *Peters* s.n. (B† holotype).

Barleria clivorum C.B. Clarke in F.T.A. **5**: 153 (1899). Types: Mozambique, Lower Zambezi, Chupanga, fl. viii.1858, *Kirk* s.n. (K syntype); Malawi, between Mpata and start of Tanganyika Plateau, fl. vii.1896, *Whyte* s.n. (K syntype).

Erect, scrambling or trailing (?short-lived) perennial herb or shrub 15–200(300) cm tall; stems with dense short spreading glandular hairs and sparse to numerous interspersed buff or yellow coarse ascending hairs, with or without a stellate base; older stems sometimes long white-floccose. Leaves sometimes immature at flowering, ovate, elliptic, lanceolate or sometimes obovate when immature, 3–13 × 1.5–6 cm, base attenuate, cuneate or rarely obtuse, apex acute to acuminate or obtuse to rounded when immature, leaf buds ± densely pale grey stellate-pubescent throughout but hairs often falling early, upper surface, margin and veins beneath with few to numerous coarse yellow or buff-coloured hairs with or without a stellate base, often with spreading glandular hairs at least along margin; lateral veins 4–6 pairs; petiole 3–32 mm long. Inflorescences axillary or crowded towards stem apex and forming a loose thyrse, each cyme (1)2–9-flowered, contracted or somewhat lax, unilateral or few-branched; primary peduncle 0.5–5 mm long, secondary peduncles 1–8.5 mm long; bracts foliaceous but often reduced and more obovate, then typically 6–21 × 3–14 mm at uppermost axils; bracteoles linear-lanceolate, pairs unequal, the larger (5.5)8.5–23(26) × 0.5–3(4) mm, margin with widely-spaced short teeth or subentire, apex spinose; flowers subsessile. Calyx papery, green to purple with paler base and conspicuous darker reticulate venation, later pale-scarious; anterior lobe ovate(orbicular) or lanceolate to obovate or oblanceolate, (6.5)8–20 × 2.5–14 mm, base truncate to cuneate or attenuate, margin with coarse spinose teeth throughout or in distal portion only, sometimes reduced to short teeth, apex acute to rounded or emarginate in outline, surfaces with sparse simple and/or long-armed stellate hairs, basal portion often with ± dense short crisped eglandular hairs, glandular hairs widespread or restricted to margin; posterior lobe (7)8.5–23.5 mm long, apex acute or attenuate, ± spinose; lateral lobes lanceolate-attenuate, 5–10.5 mm long. Corolla 23–41(46) mm long, limb pale blue, mauve or purple, tube or rarely whole flower white, glandular- and eglandular-pilose externally; tube cylindrical, 13.5–26 mm long, curved towards base; limb in 2+3 arrangement; abaxial lobe 7–17.5 × 5.5–12.5(14) mm; lateral lobes as abaxial but 4.5–10(12.5) mm wide; adaxial lobes divergent, 5.5–13.5 × 4–8.5(10) mm. Stamens attached 8–16 mm from base of corolla tube; filaments 8.5–21.5 mm long; anthers 2.7–5(6) mm long; lateral staminodes 0.8–3.5(4.7) mm long, pilose to subglabrous, antherodes to 1.4 mm long or absent. Ovary with apical ring of minute crisped hairs, with or without straight white hairs towards apex; style glabrous or sparsely pubescent towards base; stigma clavate or linear, 0.5–1.3 mm long. Capsule 11–16 mm long, glabrous or with sparse long white hairs towards apex. Seeds 3–5 × 2.7–4.5 mm.

Barleria kirkii T. Anderson is usually readily separated from *B. spinulosa* by the different calyx shape, often larger corolla lobes, reduced or absent antherodes, longer stamens and sparsely hairy (not glabrous) capsules. However, whilst this list of differences appears considerable, none of these characters are wholly consistent and there appears to be considerable intergrading in some areas where their ranges meet, as well as considerable variation within the two taxa. I therefore consider infraspecific rank to be more appropriate. The extent of range overlap is less than it may appear from the Flora geographic subdivisions, hence *kirkii* is treated as a subspecies. The following key will enable separation of most plants:

Anterior calyx lobes broadly ovate(-orbicular), elliptic or lanceolate, base truncate or
    shortly and broadly attenuate, margin with coarse spinulose teeth throughout;
    lateral staminodes with well-developed antherodes 0.6–1.4 mm long; staminal
    filaments 8.5–14.5(16.5) mm long . . . . .a) subsp. *spinulosa*
–   Anterior calyx lobes obovate, oblanceolate or narrowly oblong, base cuneate or
    more gradually attenuate, margin ± entire towards base, (spinulose) teeth mostly
    restricted to upper portion; lateral staminodes with antherodes absent or poorly
    developed, to 0.6(0.75) mm long; staminal filaments (12)15–21.5 mm long . . . .
    . . . . .b) subsp. *kirkii*

a) Subsp. **spinulosa**. FIGURE 8.6.**39**, 1–8.

    Barleria kirkii sensu da Silva *et al.*, Prelim. Checklist Vasc. Pl. Mozamb.: 18 (2004) in part.

Inflorescence unilateral or rarely branched, contracted, secondary peduncles 1–5.5(7.5) mm long. Anterior calyx lobe broadly ovate(-orbicular), elliptic or lanceolate, (4)6.5–14 mm wide, base truncate or shortly and broadly attenuate, margin with coarse spinulose teeth throughout. Corolla 23–35(38.5) mm long, lateral and abaxial lobes 7–13 mm long, adaxial lobes often only slightly smaller or subequal, ratio 0.75–1:1. Staminal filaments 8.5–14.5(16.5) mm long; staminodes with well-developed antherodes 0.6–1.4 mm long. Ovary and capsule glabrous or rarely with few pale hairs towards apex.

Botswana. SE: South East Dist., Mochudi, fl.& fr. i-iv.1914, *Harbor* in *Rogers* 6546 (J, K, KMG, PRE, SRGH). **Zambia**. N: Mpika Dist., North Luangwa Nat. Park, fl. 22.iv.1994, *P.P. Smith* 516 (K). W: Mpongwe Dist., Lake Kashiba, fl. 5.vi.1957, *Fanshawe* 3308 (BR, K, NDO). C: Kafue Dist., Mt Makulu Res. Station, fl.& fr. 13.v.1956, *Angus* 1285 (BR, FHO, K, PRE). E: Chipata Dist., Luangwa Valley, Mkania, fr. 28.vi.1968, *Astle* 5361 (K, SRGH). S: Mazabuka Dist., Mazabuka, fl. 20.v.1961, *Fanshawe* 6586 (K, LISC, NDO). **Zimbabwe**. N: Gokwe Dist., Ganderowe Falls, 43 km NW of Copper

Fig. 8.6.**39**. BARLERIA SPINULOSA subsp. SPINULOSA. 1, habit (× ²/₃); 2, mature leaf (× ²/₃); 3, long-armed stellate hair (from external surface of calyx) (× 40); 4, detail of long white floccose hairs on mature stems (× 1½); 5, calyx and bracteoles (× 2); 6, dissected corolla with androecium (× 2); 7, capsule including open valve showing retinacula (× 3); 8, seed (× 4). BARLERIA SPINULOSA subsp. KIRKII. 9, calyx and bracteoles (× 2). 1 & 6 from *Moriarty* 150, 2 & 4 from *Robinson* 2736, 3 from *Brummitt* 10280, 5 from *Chase* 334, 7 from *Borle* 185, 8 from *Hornby* 3320, 9 from *van Rensburg* 2227. Drawn by Juliet Williamson.

Queen, fl.& fr. 3.vii.1961, *Leach* 11161 (BR, K, SRGH). W: Matobo Dist., 3 km E of Antelope Mine, fl. 13.iv.1974, *Ngoni* 389 (K, SRGH). C: Goromonzi Dist., Marsala Farm off Shamva road at 48.6 km peg, fl. 20.iii.1983, *Best* 1960 (LMA, SRGH). E: Mutare Dist., near Mozambique border, fl. iii.1947, *Chase* 334 (K, SRGH). S: Bikita Dist., Save Valley, Umkondo Hill, fl.& fr. 9.ii.1966, *Wild* 7528 (K, SRGH). **Malawi**. N: Karonga Dist., by Sere R. below Kayelekera, 30 km WSW of Karonga, fl. 7.vi.1989, *Brummitt* 18411 (K, MAL). C: Salima Dist., Lifidzi Breeding Centre, fl. 19.v.1985, *Patel & Nachamba* 2147 (K, MAL). S: Chikwawa Dist., Kapichira (Livingstone) Falls, W bank of Shire R., fl. 21.iv.1970, *Brummitt* 10000 (BR, EA, K, MAL, SRGH). **Mozambique**. N: Murrupula Dist., near Murrupula, fl. 17.vii.1936, *Torre* 621 (COI, LISC). Z: Lugela Dist., Namagoa, 200 km inland from Quelimane, fl.& fr. vii.1945, *Faulkner* PRE 235 (COI, EA, K, LMA, PRE, SRGH). T: Changara Dist., 20 km from Tete towards Changara, 3 km from crossroads for Cabora Bassa, fl.& fr. 7.v.1971, *Torre & Correia* 18338 (LISC, LMU). MS: Chemba Dist., 29 km towards Tambara by slopes of Serra Lupata (Mt Nhamalongo), fl.& fr. 14.v.1971, *Torre & Correia* 18411 (LISC, LMU, MO, PRE). GI: Massinga Dist., road from Massinga to Vilanculos, fl. 7.vi.2012, *J.E.& S.M. Burrows* 12699 (BNRH, K).

Also in Tanzania and NE South Africa (Limpopo). In various types of dry forest, woodland, wooded grassland and bushland including *Brachystegia, Acacia, Combretum,* often associated with water courses, also along roadsides, more rarely in woodland on rocky hillslopes; 50–1300 m.

Conservation notes: The most common and widespread *Barleria* in the Flora region; Least Concern.

Calyx size and shape is quite variable, and plants from NE Mozambique and down into the lower Zambezi valley often have the smallest and narrowest calyx, so tending towards subsp. *kirkii.*

b) Subsp. **kirkii** (T. Anderson) I. Darbysh., comb. & stat. nov. Type: Mozambique, Cahora Bassa (Kaurabassa), 48 km above Tete, fr. xi.1858, *Kirk* s.n. (K holotype). FIGURE 8.6.**39**, 9.

> *Barleria kirkii* T. Anderson in J. Proc. Linn. Soc., Bot. 7: 30 (1863). —Clarke in F.T.A. 5: 152 (1899). —Obermeijer in Ann. Transv. Mus. **15**: 174 (1933). —White, For. Fl. N. Rhod.: 382 (1962). —da Silva *et al.,* Prelim. Checklist Vasc. Pl. Mozamb.: 18 (2004) in part. — Mapaura & Timberlake, Checklist Zimb. Vasc. Pl.: 13 (2004). —Phiri, Checklist Zamb. Vasc. Pl.: 18 (2005) in part. —Pickering, Wild Fl. Victoria Falls: 13 (2009).
>
> *Barleria variabilis* Oberm. in Ann. Transv. Mus. **15**: 174 (1933). Type: Zimbabwe, Victoria Falls, fl. v.1915, *Rogers* 13309 (BOL 138573 lectotype), lectotypified here.
>
> *Barleria lateralis* Oberm. in Ann. Transv. Mus. **15**: 176 (1933). Type: Zimbabwe, Kadoma Dist., Umsweswe, fl. 13.iv.1921, *Borle* 170 (PRE holotype, K, SRGH).

Inflorescence unilateral or often becoming branched at fruiting; contracted or more lax with secondary peduncles 2.5–8.5 mm long. Anterior calyx lobe obovate, oblanceolate or narrowly oblong, 2.5–7 mm wide, base cuneate(-attenuate), teeth largely restricted to distal portion, spinulose or more rarely short; posterior lobe as anterior or sometimes lanceolate. Corolla 28–41(46) mm long, abaxial and lateral lobes 9.5–17.5 mm long; adaxial lobes often clearly smaller, ratio 0.55–0.8(0.9): 1. Staminal filaments (12)15–21.5 mm long; staminodes lacking antherodes or these poorly developed, to 0.5(0.75) mm long. Ovary and capsule often with long white hairs towards apex, more rarely glabrous.

Zambia. C: Kafue Dist., 13 km N of Kafue, fl.& fr. 10.v.1957, *Angus* 1572 (BM, BR, FHO, K). S: Mazabuka Dist., Nega Nega, fl. 4.v.1964, *van Rensburg* 2919 (K, SRGH). **Zimbabwe**. N: Guruve Dist., Hunyani R., 9.6 km S of Dande Mission, fl.& fr. 15.v.1962, *Wild* 5746 (K, LISC, SRGH). W: Hwange Dist., near Dete (Dett), fl.& fr. 5.vii.1930, *Hutchinson & Gillett* 3410 (BM, K, LISC, SRGH). C: Kwekwe Dist., between Kadoma (Gatooma) and Kwekwe (Que Que), fl. 2.v.1958, *West* 3594 (BR, K, SRGH). E:

Chipinge Dist., Muzani R., Mutema communal land (Reserve), fl.& fr. 18.iv.1965, *Chase* 8282 (K, SRGH). S: Buhera Dist., Devuli Project, fl. v.1958, *Davies* 2456 (BR, K, SRGH). **Mozambique**. T: Changara Dist., 19 km W of Changara, fl.& fr. 30.iv.1960, *Leach & Brunton* 9877 (K, LISC, SRGH).

Not known elsewhere. Habitat as for subsp. *spinulosa;* 200–1200 m.

Conservation notes: A locally common subspecies associated with the Zambezi River and its larger tributaries, with a disjunct population along the Save River catchment; Least Concern.

Within subsp. *kirkii* there are two regional variants that are possibly worthy of recognition: (a) plants from the Victoria Falls region and W Zimbabwe (*Barleria variabilis*) tend to have a more persistent, dense stellate indumentum and contracted, few-flowered inflorescence units (usually 1–2 mature flowers or fruits present) with the outer calyx lobes only shortly toothed; they also have corollas towards the lowest end of the size range and with broad lobes (see photos in Pickering 2009); (b) plants from the Save River catchment (Zimbabwe E & S) have a markedly long corolla tube relative to the limb (ratio 1.8–2.3:1 vs. 1.2–1.75:1 in typical *kirkii*). This corolla form approaches that of subsp. *spinulosa* (ratio (1.4)1.7–2.2(2.5):1), and so could be considered a third, intermediate, subspecies. However, I have only seen a few collections of both these forms and would prefer to see more material to assess the consistency of these differences before describing further infraspecific taxa.

In the protologues of *B. consanguinea* and *B. squarrosa*, both described from the Sena River (an uncertain locality, possibly Tete), the outer calyx lobes are described as obovate. It is possible that they belong within subsp. *kirkii* but the descriptions are insufficient to be sure. Clarke (1899) considered *B. squarrosa* to be a form of *B. spinulosa* with a narrowly oblong calyx.

26. **Barleria heterotricha** Lindau in Bot. Jahrb. Syst. **38**: 69 (1905). Type: Mozambique, Moamba Dist., Ressano Garcia, fl. 18.xii.1897, *Schlechter* 11822 (B† holotype, GRA lectotype), lectotypified here (see note).

> *Barleria affinis* sensu auct., non C.B. Clarke. —Obermeijer in Ann. Transv. Mus. **15**: 174 (1933). —Fabian & Germishuizen, Wild Fl. Nthn. S. Afr.: 398, pl.191 (1997). —Mapaura & Timberlake, Checklist Zimb. Vasc. Pl.: 13 (2004) in part. —Setshogo, Prelim. Checklist Pl. Botswana: 17 (2005) in part. —Welman in Germishuizen *et al.*, Checklist S. Afr. Pl.: 79 (2006) in part.

Small shrub or straggling perennial herb 30–120 cm tall; stems densely spreading glandular-pubescent and with few to numerous pale buff or greyish long-armed stellate hairs, often with two opposite lines of short retrorse eglandular hairs. Leaves broadly ovate, deltate or reniform, 0.9–5.2 × 1–4 cm, base truncate, rounded or shallowly cordate, apex rounded to shortly attenuate, surfaces with dense pale greyish long-armed stellate hairs throughout, with spreading glandular hairs numerous at least along margins; lateral veins 2–4 pairs; petiole 5–30 mm long. Inflorescence axillary, flowers solitary or in contracted 2–3-flowered unilateral cymes; primary peduncle 0–5 mm long, secondary peduncles if present 2–5 mm long; bracteoles linear or narrowly oblanceolate, pairs unequal in length, the longer 2–10(15.5) × 0.3–2 mm, margin entire or minutely toothed, apex apiculate or spinulose; pedicels 0–2.5 mm long. Calyx dark blue, purple or green at first, with darker reticulate venation, soon turning pale brown- or whitish-scarious; outer lobes subequal, lanceolate, narrowly oblong or rarely (obovate-)elliptic, anterior lobe 5–10(15) × 2–4(8) mm, posterior lobe 6–12(17.5) mm long, base acute or attenuate, margin and apex irregularly denticulate, teeth sometimes elongate, apiculate with a long-armed stellate bristle at tip, more rarely subentire, apex obtuse, acute or truncate in outline, surfaces with rather sparse long-armed stellate hairs most numerous along veins and towards base and with scattered spreading glandular hairs; lateral lobes ovate or lanceolate, 3–6.5(8) mm long. Corolla 32–54 mm long, blue or mauve to almost white, mixed retrorse-pubescent and glandular-pubescent externally; tube 25–42 mm long, cylindrical;

limb in weak 2+3 arrangement; abaxial and lateral lobes 6.5–11 × 5.5–10 mm; adaxial lobes 5–7.5(9.5) × 4–6(7.5) mm. Stamens attached in distal third of corolla tube; filaments 10.5–15.5 mm long; anthers 3–4.2 mm long; lateral staminodes 1.2–2.3 mm long, pilose, antherodes to 0.6 mm long or obsolete. Ovary with a dense apical tuft of pale crisped hairs; style glabrous; stigma linear or clavate, 0.6–1 mm long. Capsule 11.5–16 mm long, glabrous. Seeds 3.8–5 × 3.8–4.5 mm.

a) Subsp. **heterotricha**

Outer calyx lobes lanceolate or narrowly oblong, anterior lobe 5–10 × 2–4 mm; lateral lobes usually ovate, 3–6.5 mm long.

Botswana. N: Ngamiland Dist., Moremi Gorge, fl.& fr. 23.iii.1999, *Turner* 100 (GAB, K). SE: Kweneng Dist., kloof 1.6 km S of Molepolole, fl. 18.xii.1983, *Woollard* 1238 (PRE). **Zimbabwe**. S: Beitbridge Dist., Tshiturupadzi Dip camp area, c.88 km E of Beitbridge, fl. 18.iii.1967, *Mavi* 238 (BR, K, LISC, SRGH). **Mozambique**. M: Moamba Dist., Ressano Garcia, fl. 18.xii.1897, *Schlechter* 11822 (GRA).

Also in NE South Africa (North West, Limpopo, Mpumalanga). Bushland and woodland on rocky hillslopes, often amongst large boulders or in crevices, also in woodland along riverbanks; 150–1100 m.

Conservation notes: Locally common with no obvious threats; Least Concern.

The type, *Schlechter* 11822, was widely distributed (COI, HBG, K, S, STU) as having been collected from Komati Poort in South Africa at 300 m altitude (1000 ft). However, in the protologue it is recorded as from Ressano Garcia in Mozambique on hills to 150 m – this is largely matched by the duplicate at GRA (where the altitude is recorded as c.100 ft). To add to the confusion, the Z sheet has both localities recorded, on two separate labels. It is possible that two separate collections were combined under a single number by Schlechter.

b) Subsp. **mutareensis** I. Darbysh., subsp. nov. Differs from the typical subspecies in the broader, elliptic or obovate (not lanceolate or narrowly oblong) outer calyx lobes and in the longer and more lanceolate lateral calyx lobes. Type: Mutare Dist., Zimunya communal land (Zimunya's Reserve), fl.& fr. 6.v.1956, *Chase* 6098 (K holotype, BM, BR, LISC, PRE, SRGH).

Outer calyx lobes obovate-elliptic (anterior) or elliptic (posterior), anterior lobe 12–15 × 5–8 mm; lateral lobes lanceolate, 6–8 mm long.

Zimbabwe. E: Mutare Dist., Marange communal land (Maranke Reserve), fl. 27.ii.1953, *Chase* 4803 (BM, K, LISC, SRGH). S: Bikita Dist., Moodie's pass, fl.& fr. 24.v.1959, *Noel* 2002 (SRGH).

Not known elsewhere. Amongst boulders and in rock crevices on hillslopes in light shade; 800–1100 m.

Conservation notes: Endemic to the Mutare–Save Valley area of E Zimbabwe and known from only four collections; clearly scarce and possibly threatened since the area is now heavily disturbed.

With its enlarged calyx, this subspecies could be confused with *Barleria affinis*, but the outer calyx lobes are still proportionately narrower and not ovate, and the indumentum, corolla and androecium clearly places it within *B. heterotricha*. Care should also be taken to separate this subspecies from *B. spinulosa* subsp. *kirkii*, particularly as sympatric populations of that species in E Zimbabwe have a rather long corolla tube relative to the limb. However, leaf shape clearly separates the two and the corolla tube is still considerably longer in *B. heterotricha*.

27. **Barleria affinis** C.B. Clarke in Fl. Cap. 5: 50 (1901). —Mapaura & Timberlake, Checklist Zimb. Vasc. Pl.: 13 (2004), excl. synonym *B. heterotricha.* —Setshogo, Prelim. Checklist Pl. Botswana: 17 (2005), excl. synonym *B. heterotricha.* Type: South Africa, Limpopo, Marico Dist., fl.& fr. n.d., *Holub* s.n. (K holotype).

Barleria cordata Oberm. in Ann. Transv. Mus. **15**: 172 (1933). Type: Zimbabwe, Bulawayo, on road to Khami R., fl. ii.1903, *Eyles & Johnson* 1199 (GRA lectotype, BM, SRGH), lectotypified here.

Subshrub 30–100 cm tall; stems with dense pale buff, greyish or yellow (when young) long-armed stellate hairs, upper internodes with or without few (rarely numerous, see note) interspersed spreading glandular hairs. Leaves reniform or broadly ovate, 0.6–2.3 × 0.8–2.8 cm, base truncate or shallowly cordate, apex rounded, often with a short attenuate tip, surfaces with ± dense pale long-armed stellate hairs, sometimes with a few glandular hairs along margin; lateral veins 2–4 pairs; petiole 2.5–14 mm long. Inflorescence axillary, 1(2)-flowered; peduncle 1–3 mm long, secondary peduncle if present 3–8 mm long; bracteoles linear or narrowly oblanceolate, 6.5–19 × 0.7–2(3.5) mm, straight or somewhat declinate, margin entire or minutely spinulose, apex spinulose. Calyx green or mauve- to blue-tinged at first with ± darker reticulate venation, soon turning pale-scarious, somewhat accrescent; outer lobes subequal, broadly ovate, (8)10.5–18.5 × (5)7–16 mm, base cordate or truncate and shortly narrowed, margin denticulate to spinulose-dentate, each tooth with an apical stellate-based bristle, apex acute, obtuse or subattenaute in outline, spinulose, anterior lobe sometimes 2–3-toothed, surface with long-armed stellate hairs most numerous along veins and towards base; lateral lobes lanceolate, 4–8 mm long. Corolla 29–48 mm long, pale blue or mauve with whitish tube and throat, retrorse-pubescent externally, rarely with a few interspersed glandular hairs; tube 21–37 mm long, narrowly cylindrical; limb in weak 2+3 arrangement; abaxial and lateral lobes 7–9.5 × 4–6.5 mm; adaxial lobes 5.5–7.5 × 2.5–5 mm. Stamens attached in distal third of corolla tube; filaments 9.5–11.5 mm long; anthers 1.8–2.8 mm long; lateral staminodes c.1.5 mm long, pilose, antherodes c.0.5 mm long. Ovary with dense apical tuft of pale crisped hairs; style glabrous; stigma linear or clavate, 0.5–0.9 mm long. Capsule 9.5–13 mm long, glabrous or with few long pale hairs on dorsal side of each valve. Seeds 3–4 mm long and wide.

a) Subsp. **affinis**

Corolla 29–32 mm long, including tube 21–25 mm long.

Botswana. SE: Central Dist., Shoshong, hill next to kgotla office, fl.& fr. 21.iv.2005, *Darbyshire* 452 (GAB, K). **Zimbabwe.** W: Bulawayo, Khami, fl.& fr. 25.iv.1946, *Wild* 1065 (K, SRGH); Matopos Nat. Park, fl. 24.ii.1981, *Philcox & Leppard* 8814 (K, SRGH). S: Mwenezi Dist., NW slopes of Mateke Hills, fl.& fr. 7.v.1958, *Boughey* 2741 (SRGH).

Also in NE South Africa (Limpopo). Amongst boulders and crevices on rocky slopes and kopjes; 1100–1450 m.

Conservation notes: Rather scarce and localised but with no obvious threats; Least Concern.

The shorter corolla tube is an additional character for separation of typical *Barleria affinis* from *B. heterotricha*, but see subsp. *producta* below. *Biegel & Pope* 3273 and *Boughey* 2741 from S Zimbabwe have the broadly ovate calyx of *B. affinis* but tend towards *B. heterotricha* in having dense glandular hairs throughout the vegetative parts; some of the corollas in *Biegel & Pope* 3273 are intermediate, others being closer to *B. affinis*.

b) Subsp. **producta** I. Darbysh., subsp. nov. Differs from subsp. *affinis* in the markedly longer corolla tube, 35–37 mm long, not 21–25 mm long. Type: Botswana, Selebi-Pikwe, xii.1977, *Kerfoot & Falconer* 45 (PRE holotype, J).

Corolla 42–48 mm long, including tube 35–37 mm long.

Botswana. SE: Central Dist., Selebi-Pikwe, fl. xii.1977, *Kerfoot & Falconer* 45 (J, PRE). Not known elsewhere. Habitat not recorded, probably as for subsp. *affinis;* 900 m.

Conservation notes: Apparently endemic to E Botswana and known only from the type; best considered Data Deficient.

The shape of the corolla in subsp. *producta* is similar to that of *Barleria heterotricha* but it is easily separated by the characters in the key.

28. **Barleria sunzuana** Brummitt & Seyani in Kew Bull. **32**: 721, fig.1 (1978). — Champluvier in Pl. Ecol. Evol. **144**: 83 (2011). Type: Zambia, Mbala Dist., Sunzu Mt, fl. 23.iv.1962, *Richards* 16369 (K holotype, LISC).

> *Barleria acanthoides* sensu auct., non Vahl. —Richards & Morony, Check List Fl. Mbala Dist.: 227 (1969). —Phiri, Checklist Zamb. Vasc. Pl.: 18 (2005).

Trailing or decumbent suffrutex, branches to 30 cm long radiating from a woody base and rootstock; stems with numerous long coarse spreading or ascending pale hairs, often interspersed with shorter finer spreading hairs. Leaves elliptic or narrowly oblong to somewhat obovate or oblanceolate, 2.8–5 × 1–2 cm, base cuneate or attenuate, apex acute to rounded, surfaces with pale coarse subappressed or ascending hairs, evenly distributed above, most dense on veins beneath; lateral veins (3)4–5 pairs; petiole to 5 mm long. Inflorescences clustered in uppermost axils, often forming dense heads, each cyme unilateral, contracted, 2 to several flowered; bracteoles lanceolate, 18–33 × 4–7.5 mm, margin entire or minutely toothed, apex spinose, sometimes falcate, triplinerved with prominent midrib, indumentum as leaves beneath but with finer more spreading hairs between veins and often with scattered glandular hairs. Calyx green or mauve-tinged at first with darker venation, later turning scarious, accrescent; anterior lobe broadly ovate, 23–31 × 17–21 mm, base cordate or rounded, margin laciniate with teeth to 4 mm long, bristle-tipped, apex usually bifidly spinose, indumentum as bracteoles, glandular hairs most conspicuous along laciniae; posterior lobe as anterior but 25–33 mm long, apex attenuate into a short spine; lateral lobes linear-lanceolate, 10–12 mm long. Corolla 43–48 mm long, pale blue or pale purple with white tube, pubescent externally, limb with mainly glandular hairs; tube 27–31 mm long, gradually and narrowly expanded above attachment point of stamens; limb in weak 2+3 arrangement; abaxial lobe 15–16 × 10–12 mm; lateral lobes 14–17.5 × 9–11 mm; adaxial lobes 13–14 × 6–8 mm. Stamens attached ± midway along corolla tube; filaments 14–16 mm long; anthers held at corolla mouth, 3–3.5 mm long; lateral staminodes 2–3 mm long, pilose, antherodes 1–1.5 mm long. Ovary with few white hairs at apex; style glabrous; stigma subcapitate, 0.5–0.7 mm long. Capsule 20–22 mm long, shortly beaked, glabrous. Seeds 7.5–8 × 6.5–7 mm.

Zambia. N: Mbala Dist., Sunzu Hill, fl.& fr. 28.iv.1936, *Burtt* 6101 (BM, BR, EA, K); Mbala Dist., Zambia Govt. Ranch, Saisi Valley, fl. 20.v.1968, *Richards* 23274 (BR, K).

Not known elsewhere. Open rocky slopes, rough rocky grassland, sandy flats amongst exposed laterite; 1500–2100 m.

Conservation notes: Endemic to the slopes of Mt Sunzu in NE Zambia and adjacent Saisi Valley flats. Although much suitable habitat remains on the mountain, satellite imagery indicates the adjacent valley is widely farmed; probably Vulnerable under IUCN criterion B.

29. **Barleria capitata** Klotzsch in Peters, Naturw. Reise Mossamb. **6**(1): 210 (1861). — Clarke in F.T.A. **5**: 153 (1899), in part excl. *Schweinfurth* 1071 from Eritrea. Type: Mozambique, Zambezi, Rios de Sena, imm.fl. 1842–1846, *Peters* s.n. (B† holotype, PRE).

> *Barleria megalosiphon* Mildbr. in Notizbl. Bot. Gart. Berlin-Dahlem **11**: 62 (1930). — Obermeijer in Ann. Transv. Mus. **15**: 158 (1933). —Setshogo, Prelim. Checklist Pl. Botswana: 17 (2005). Type: Zimbabwe, Victoria Falls, fl. i.1923, *Wilson* s.n. (AAH, B† syntypes).

Spiny subshrub with spreading, decumbent or subprostrate branches 20–70 cm long; stems with dense short white antrorse or mixed antrorse and retrorse hairs, interspersed with a few subappressed long pale hairs. Leaves often clustered on short lateral branches, elliptic or oblanceolate, 1.8–4 × 0.5–1.3 cm, base cuneate or attenuate, apex acute or obtuse, mucronate, surfaces with pale coarse (sub)appressed hairs most numerous on veins beneath, young leaves also with minute curled white hairs; lateral veins 3–4 pairs; petiole to 6 mm long. Inflorescences contracted unilateral cymes 3–7 cm long in upper axils, 3–10+-flowered, axis usually inrolled at maturity; bracteole pairs unequal, the larger held erect or horizontal, lanceolate, 18–30 × 5–11 mm, asymmetric and falcate, margin with numerous slender flexuose spines, apex spinose, midrib pale and prominent, smaller bracteoles declinate; reduced sterile pale-scarious bracteoles present at lower nodes, lanceolate or ovate, 6–15 × 1.5–5 excluding lateral spines. Calyx tardily scarious; anterior lobe broadly ovate, 28–32 × 15–16 mm, base rounded or cordate, margin with ± numerous spreading bristles thickened at maturity and forming slender teeth, apex acute, surface with sparse ascending or spreading hairs along veins and minute white curled hairs, at least when young; posterior lobe 32–42 × 21–23 mm long, apex shortly spinose, surface with conspicuous flexuose submarginal spines 3–5.5 mm long; lateral lobes subulate, 14–17 mm long, apex acute to rounded, surface with long subappressed hairs especially along margin and scattered short glandular hairs. Night flowering; corolla white (drying blue-black), glandular-pilose externally, limb with shorter eglandular hairs; tube narrowly cylindrical, (80)100–130 mm long; limb subregular, lobes 13–23 × 10–18 mm or adaxial lobes somewhat smaller. Stamens attached in distal third of corolla tube; filaments 31–35 mm long; anthers shortly exserted, 4–5 mm long; lateral staminodes 0.7–3.5 mm long, pilose, antherodes to 0.7 mm long or obsolete. Ovary shortly pubescent towards apex, with apical ring of minute crisped hairs; style puberulous towards base; stigma subcapitate, 0.5–0.8 mm long. Capsule c.21 mm long, antrorse-puberulous towards apex. Seeds c.6.5 × 6 mm.

Caprivi. Impilila Is., fl. 13.i.1959, *Killick & Leistner* 3354 (K, PRE). **Botswana**. N: Chobe Dist., Kasane, fl. 10.i.1965, *Henry* 37 (SRGH). **Zambia**. S: Livingstone Dist., 21 km W of Livingstone on Katombora road, fl. 6.i.1952, *Angus* 1113 (BM, BR, FHO, K). **Zimbabwe**. N: Gokwe Dist., Sengwa Research Station, fl. 22.i.1969, *Jacobsen* 458 (SRGH). W: Lupane Dist., near Gwayi R. 32 km N of Lupane, fl. ii.1965, *Walter* 6 (K, SRGH). **Malawi**. S: Chikwawa Dist., 5 km W of Ngabu by airfield, fl.& fr. 22.iv.1980, *Brummitt & Osborne* 15515 (BR, K, SRGH). **Mozambique**. T: Mutarara Dist., 6 km from Dôa, fr. 21.vi.1949, *Barbosa & Carvalho* 3204 (LMA). MS: Caia Dist., Rios de Sena, fr. 1842–1848, *Peters* s.n. (PRE).

Not known elsewhere. Woodland and wooded grassland with e.g. mopane, *Acacia* and *Combretum*, open grassland, often on rocky hillslopes, and roadsides; 100–1100 m.

Conservation notes: Endemic to the Flora area and widespread, but rather scarce and scattered with few collections despite the highly conspicuous flowers and inflorescence; probably Least Concern.

The extant isotype at PRE comprises only two leaves, but Clarke (1899), who saw the type, also named *Holub* s.n. from Leshumo valley, Zimbabwe as *Barleria capitata* and that collection clearly matches other material here assigned to this species. Klotzsch described the type as being erect, differing from material I have seen, but otherwise the description seems to match. As the type lacked flowers and fruits, Mildbraed probably missed the link between his *B. megalosiphon* from Victoria Falls and *B. capitata*. Despite the slight disjunction in distribution, plants from the Zambezi above Victoria Falls clearly match those from S Malawi and Tete with only minor differences, namely a slightly longer stigma (0.7–0.8 vs. 0.5 mm) and a somewhat shorter stem indumentum.

The long slender corolla tube and night-flowering suggest this species is moth-pollinated, one of a number of species in this section including *B. acanthoides* Vahl, *B. gracilispina* (Fiori) I. Darbysh. and *B. inclusa* I. Darbysh. from East Africa.

This is one of several species in sect. *Barleria* (including *B. macrostegia*) in which the mature inflorescences are several to many-flowered with the calyx adpressed and enclosed within or cupped by the large bracteoles; the axis often curves upwards at fruiting so that the inflorescence becomes inrolled and it is possible that the whole inflorescence becomes detached and dispersed by rolling along the ground.

30. **Barleria macrostegia** Nees in De Candolle, Prodr. **11**: 235 (1847). —Clarke in Fl. Cap. **5**: 50 (1901). —Obermeijer in Ann. Transv. Mus. **15**: 159 (1933). —Meyer in Mitt. Bot. Staatssamml. München **2**: 382 (1957); in Merxmüller, Prodr. Fl. SW Afr. **130**: 16 (1968). —Fabian & Germishuizen, Wild Fl. Nthn. S. Afr.: 386, pl.185 (1997). Type: South Africa, Vet River (Katrivier), bud n.d., *Burke* s.n. [cited in error as Burchell in protologue] (K holotype).

Barleria burchelliana Nees in De Candolle, Prodr. **11**: 235 (1847). Type: South Africa, Vet (Fat) R., fl. iii.1848, *Burke* 457 [cited in error as Burchell 457 in protologue] (K 000394589 lectotype, K 000394588), lectotypified here.

Barleria burkeana Sond. in Linnaea **23**: 92 (1850) as "= B. burchelliana et macrostegia N. ab Es. l.c. p. 235 (species non diversae)". —Anderson in J. Linn. Soc., Bot. **7**: 31 (1863).

Perennial herb with much-branched prostrate stems 15–50 cm long from a woody base; stems with short white retrorse hairs often in two opposite rows, and few to numerous ± spreading yellowish hispid hairs. Leaves often clustered on short lateral branches, narrowly oblong, (ovate-)elliptic or lanceolate, 0.8–3 × 0.3–0.7(1) cm, base cuneate, margin thickened and pale, apex acute, mucronate, yellowish-hispid on margin and midrib beneath, hairs along margin with ± bulbous base forming minute serrations; lateral veins indistinct, midrib prominent beneath; petiole 0–4 mm long. Inflorescence a dense cylindrical or subglobose unilateral cymes 2–7.5 cm long in the upper axils, 4–10+-flowered; axis sometimes inrolled at fruiting; bracteoles imbricate, largely enclosing calyx, green, sometimes with white margin, ± broadly ovate, (15)20–32 × (7)10–20 mm, margin entire or minutely toothed due to swollen hair bases, apex spinulose, surface tripli- or palmately nerved, hispid along main veins and margin; paired reduced sterile bracteoles sometimes present at lower nodes. Calyx green or mauve-tinged, later scarious, accrescent; anterior lobe suborbicular or somewhat ovate or elliptic, 15–21 × 10–17 mm in flower, up to 25 × 28 mm in fruit, base rounded or subcordate, margin entire or with minute teeth formed by swollen hair bases, apex obtuse, rounded or subattenuate and mucronate, rarely bi-mucronulate, surface yellowish-hairy with softer hairs than leaves, dense along margin; posterior lobe as anterior but broadly ovate(-orbicular), 17–26 × 11–18 mm in flower, to 33 × 29 mm in fruit, apex attenuate to obtuse, spinulose; lateral lobes lanceolate, 7–9 mm long in flower, to 11.5 mm in fruit. Corolla 31–42 mm long, pale blue to purple with whitish tube, rarely white throughout, glandular-pilose externally; tube cylindrical, 18–28 mm long, somewhat expanded towards mouth; limb subregular; lobes 9–15 mm long, abaxial and lateral lobes 5–9.5 mm wide, adaxial lobes 4–7.5 mm wide. Stamens attached 12.5–18 mm from base of corolla tube; filaments 8–12.5 mm long; anthers held at mouth or shortly exserted, 2–2.8 mm long; lateral staminodes 1–3 mm long, pubescent at base, antherodes 0.3–1.3 mm long or absent. Ovary with apical ring of minute crisped hairs extending onto style base; stigma broadly capitate, 0.6–1.2 mm wide. Capsule 14.5–17 mm long, glabrous. Seeds 5–5.5 × 5.5–6 mm.

Botswana. N: North East Dist., edge of Tshesebe village, fl. 29.iv.2005, *Darbyshire* 461 (GAB, K). SW: Kgalagadi Dist., 27 km N of Kang, fl. 18.ii.1960, *Wild* 5026 (BM, K, LMA, SRGH). SE: Kweneng Dist., 48 km S of Khutse near borehole, fr. 24.iv.1972, *Baker* 27 (K, LISC, SRGH). **Zimbabwe**. W: Bulilimamangwe Dist., Plumtree, Nata, fr. iv.1953, *Davies* 528 (K, SRGH). C: Marondera Dist., fl. vii.1954, *Bunzle* s.n. (K, LISC, SRGH).

Also in Namibia and South Africa (Northern Cape, Free State, North West, Gauteng, Limpopo, Mpumalanga). Bare ground, grassland, open wooded grassland on sandy soils, often in disturbed areas such as roadsides, pathways and fallow land; 900–1250 m.

Conservation notes: Widespread and fairly common; Least Concern.

The size of the inflorescence parts is rather variable, perhaps due in part to gradual accrescence with maturity. *Lugard* 295 (K) from Inkonane Pan, Botswana, has particularly small bracteoles and calyx but is otherwise a good match for this species.

Clarke (1901) claimed that Nees' *Barleria macrostegia* and *B. burchelliana* were founded on one collection. However, although collected at the same locality, they have different dates and the type of *B. macrostegia* has no collecting number; they should therefore be treated as two collections but are nevertheless conspecific. Nees erroneously attributed the two collections to Burchell; Sonder attempted to correct this mistake by renaming *B. burchelliana* (including *B. macrostegia*) as *B. burkeana*, but in doing so he created a superfluous name. Clarke chose the more sensible option of using *B. macrostegia* as the accepted name and synonymising both *B. burchelliana* and *B. burkeana*.

31. **Barleria repens** Nees in De Candolle, Prodr. **11**: 230 (1847) in part, excl. *Bojer* s.n. —Hooker in Bot. Mag. **113**: t.6954 (1887) in part, excl. *Wakefield* s.n. —Clarke in F.T.A. **5**: 166 (1899) in part, excl. *Bojer* s.n. —Clarke in Fl. Cap. **5**: 54 (1901). —Obermeijer in Ann. Transv. Mus. **15**: 168 (1933). —Bandeira *et al.*, Fl. Nat. Sul Moçamb.: 160, 192 (2007). —Darbyshire in F.T.E.A., Acanthaceae **2**: 354 (2010). Type: Mozambique, Moma Dist., Raza Is., fl.& fr. n.d., *Forbes* s.n. (K000394495 lectotype, K000394433), lectotypified by Darbyshire (2010).

Barleria querimbensis Klotzsch in Peters, Naturw. Reise Mossamb. **6**(1): 205 (1861). — Clarke in F.T.A. **5**: 166 (1899). Type: Mozambique, Ibo Dist., Quirimba (Kerimba) Is., fl.& fr. 1842–1848, *Peters* s.n. (B† holotype).

Barleria swynnertonii S. Moore in J. Linn. Soc., Bot. **40**: 160 (1911). Type: Mozambique, Beira, fl.& fr. 25.xii.1906, *Swynnerton* 1958 (BM holotype).

Trailing or scandent subshrub, branches 15–350 cm long; stems buff- or yellowish-strigulose, usually with shorter ± retrorse hairs when young. Leaves often somewhat anisophyllous, (ovate-) elliptic or somewhat obovate, 1.7–7.3 × 0.7–3 cm, base attenuate, apex subattenuate to obtuse or rounded, ± apiculate, surfaces strigose, hairs most numerous on margin and veins beneath; lateral veins 4–6 pairs; petiole 4–11 mm long. Inflorescence axillary, flowers solitary or in a 2–3-flowered contracted unilateral cyme; peduncle to 2.5 mm long; bracteoles linear or oblanceolate, 1.5–10(17) × 0.2–1.5(3) mm, not spiny, entire; pedicels 0.5–5 mm long. Calyx initially green(-brown), soon turning glossy brown, accrescent; anterior lobe broadly ovate, (8.5)15–23 × (5.5)9–14 mm in flower, up to 30 × 23 mm in fruit, base rounded or cordate, apex acute- or obtuse-apiculate, rarely emarginate, margin (sub)entire, veins and margin sparsely strigulose particularly towards base, hairs along margin sometimes with swollen base; posterior lobe as anterior but (10)17–29 mm long in flower, to 35 mm in fruit, apex acute-apiculate; lateral lobes lanceolate, 4–7 mm long in flower. Corolla (33)38–61 mm long, bright red to rose-pink, glandular- and eglandular-pilose or sparsely so externally; tube (22)26–38 mm long, campanulate above attachment point of stamens; limb in 4+1 arrangement; abaxial lobe (10)15–21 × 9–17 mm, offset by (3)4.5–9.5 mm; lateral lobes (9)12–17 × 8–14 mm; adaxial lobes (8)11–15.5 × 5.5–10.5 mm. Stamens attached ± midway along corolla tube; filaments (14)16–22 mm long; anthers 2.5–3.5 mm long; lateral staminodes 1–3 mm long, pilose, antherodes 0.5–1.2 mm long. Ovary glabrous; stigma subcapitate, 0.4–0.7 mm long. Capsule 15–19 mm long, glabrous. Seeds 3.8–5 mm long and wide.

Mozambique. N: Palma Dist., Palma–Pundanhar road, c.4.5 km W of junction with road to Nhica do Rovuma, fl. 9.xi.2009, *Goyder et al.* 6045 (K, LMA). Z: Pebane Dist., Pebane, fl. viii.1950, *Munch* 259 (K, SRGH). MS: Beira Dist., Macuti, fl.& fr. 23.iii.1960, *Wild & Leach* 5189 (K, LMA, SRGH). GI: Inharrime Dist., Ponta Zavora, fl.& fr. 16.x.1957, *Barbosa & Lemos* 8066 (COI, K, LISC, LMA). M: Marracuene Dist., Marracuene, fl.& fr. 9.x.1980, *Schäfer* 7247 (BR, K).

Also in coastal Kenya, Tanzania and South Africa (KwaZulu-Natal). Coastal scrub and woodland, understorey and margins of dry coastal forest, *Brachystegia* thicket, on sandy soils including stabilised sand dunes; 0–175 m.

Conservation notes: Common along the Mozambique coast; assessed as Least Concern in F.T.E.A.

Specimen *Boane 365* (LMU) from Lago Naucati, Chibuene (Mozambique GI) is unusual in having branched inflorescences with up to 9 flowers developing, but is otherwise a good match for this species.

The species is widely cultivated as an ornamental in the tropics, e.g. in Harare *(Biegel* 5201, K) and in Eduardo Mondlane University in Maputo (pers. obs.).

32. **Barleria obtusa** Nees in Linnaea **15**: 358 (1841). —Nees in De Candolle., Prodr.
11: 231 (1847), excl. var. b. —Anderson in J. Linn. Soc., Bot. **7**: 31 (1863), excl.
synonym *B. barbata.* —Clarke in Fl. Cap. **5**: 52 (1901). —Obermeijer in Ann.
Transv. Mus. **15**: 169 (1933). —Verdoorn in Fl. Pl. Afr. **25**: t.998 (1946). —
Compton, Fl. Swaziland: 554 (1976). —Fabian & Germishuizen, Wild Fl. Nthn. S.
Afr.: 390, pl.187 (1997). —Bandeira *et al.,* Fl. Nat. Sul Moçamb.: 160, 191 (2007).
Types: South Africa, Eastern Cape, Albany, Mt "Bothas", "der gross fischfluss", fl.
n.d., *Ecklon & Zeyher* s.n. (WU syntype); E Cape, Uitenhage, Zwartkoprivier, Pauli
Maré, n.d., *Ecklon & Zeyher* s.n. (?JE syntype).

   Barleria obtusa Nees var. *cymulosa* Hochst. in Flora **28**: 72 (1845). Type: South Africa,
Uitenhage, fl. iv.1839, *Krauss* s.n. (M lectotype), lectotypified here.

Perennial herb or subshrub 30–300 cm tall, often scrambling or scandent, sometimes a
dwarf shrublet; stems with mixed ± dense short white retrorse or spreading hairs and few to
numerous long coarse pale-buff spreading or ascending hairs; upper internodes sometimes also
glandular-pubescent. Leaves ovate or elliptic, 2–8 × 1–3.5(4.5) cm, base and apex both shortly
attenuate to rounded, surfaces with pale spreading to subappressed hairs, most numerous on
margin and veins beneath where sometimes with a swollen base, sometimes with short glandular
hairs along margin; lateral veins 3–5 pairs; petiole 5–17 mm long. Inflorescence axillary, flowers
solitary or usually in a lax 2–7-flowered unilateral or partially dichasial cyme 3–11 cm long,
on spreading and often looping peduncles; primary peduncle 8–45 mm long, indumentum
as stem but glandular hairs often more numerous; bracteoles linear or narrowly oblanceolate,
recurved, 2–16.5 × 0.5–1.8 mm, margin entire, apex acute; pedicels 0–3(12) mm long. Calyx
green with conspicuous darker parallel primary venation giving a striate appearance; outer lobes
subequal or posterior lobe somewhat longer, narrowly oblong-lanceolate to oblanceolate, 8–18
× 2–5 mm, base cuneate, margin entire or with minute teeth formed by swollen hair bases,
apex often recurved, acute- to rounded-apiculate, anterior lobe more rarely notched, surfaces
with numerous pale-buff or greyish ascending to subappressed hairs and spreading glandular
hairs, latter sometimes restricted to margin; lateral lobes pale, linear-lanceolate, 5.5–12 mm
long. Corolla 26–44 mm long, pale blue to mauve, with paler tube and with purple stripes in
throat, rarely white throughout, mixed glandular-pilose and retrorse eglandular-pubescent
externally; tube 17–26 mm long, narrowly campanulate above attachment point of stamens;
limb subregular; abaxial lobe 9.5–19 × 8–18 mm, offset by 2(5) mm; lateral lobes as abaxial but
7–15.5 mm wide; adaxial lobes 9–16.5 × 5.5–10.5 mm. Stamens attached 7.5–13 mm from base
of corolla tube; filaments 16–25 mm long; anthers 3–4 mm long; lateral staminodes 2.2–4 mm
long, pubescent, antherodes 0.6–0.9 mm long. Ovary glabrous or with few short ascending hairs
towards apex; style glabrous; stigma linear, 0.8–1.8 mm long. Capsule 13–16.5 mm long, glabrous
or largely so. Seeds 4–4.5 mm long and wide.

Mozambique. M: Namaacha Dist., Goba, fl.& fr. 6.vi.1920, *Borle* 1101 (EA, K, PRE);
Namaacha, Pedreira das Obras Publicas, fl.& fr. 29.v.1964, *Moura* 94 (COI, LMU).

Also in Swaziland and South Africa (North West, Gauteng, Limpopo, Mpumalanga,
KwaZulu-Natal). Bushland on rocky hillslopes and sandy soils, margins of coastal or
riverine forest, streamside thicket, roadsides; to 300 m.

Conservation notes: In the Flora area only known from the Lebombo Mts, but
widespread and locally common in South Africa and Swaziland, usually in areas of low
agricultural potential; Least Concern.

Some populations from South Africa (including 'var. a' in protologue of Nees and
illustrated plant in Verdoorn 1946) have contracted, single-flowered inflorescences.
But in all material seen from Mozambique and neighbouring parts of South Africa
and Swaziland the inflorescences are clearly pedunculate and at least some are 2 or
more flowered; only this form is treated here.

32a. **Barleria obtusa** × **gueinzii**

Perennial herb or subshrub; stems with mixed ± dense short white retrorse hairs mainly on two opposite sides and numerous long coarse pale ascending or spreading hairs. Leaves ovate, 4–5.5 × 1.8–3 cm, base shortly attenuate or cuneate, apex acute-apiculate, surfaces with coarse pale hairs, most numerous along margin and veins beneath; lateral veins 4–5 pairs; petiole 5–18 mm long. Inflorescences axillary towards stem apex, together forming an interrupted terminal thyrse, flowers in a lax to more congested (1)2–3-flowered unilateral cyme to 3 cm long, primary peduncle 1–13 mm long, indumentum as stem and with short spreading glandular hairs; bracteoles linear or narrowly elliptic-oblanceolate, 5.5–15 × 0.5–1(2) mm, margin entire, apex acute-apiculate, straight or somewhat recurved; pedicels 1–4 mm long. Calyx (purple-)green with darker reticulate venation; anterior lobe narrowly oblong-elliptic or oblanceolate, 11.5–17 × 5–6.5 mm, base cuneate, margin denticulate, teeth with an apical bristle, apex acute or notched up to 3 mm, apiculate, surfaces with numerous pale ascending hairs and short spreading glandular hairs mainly along margin; posterior lobe as anterior but slightly longer and narrower, apex acute-apiculate; lateral lobes linear-lanceolate, 5.5–8.5 mm long. Corolla 35–43.5 mm long, blue, mixed glandular-pilose and retrorse eglandular-pubescent externally; tube 19–27.5 mm long, campanulate above attachment point of stamens; limb in weak 4+1 arrangement; abaxial lobe to 16 × 12.5 mm, offset by 2–2.5 mm; lateral lobes to 14.5 × 11 mm; adaxial lobes to 11 × 6.5 mm. Stamens attached in proximal half of corolla tube; filaments to 28 mm long; anthers 3–3.6 mm long; lateral staminodes 3–3.5 mm long, sparsely pubescent, antherodes 0.5–0.7 mm long. Ovary glabrous or with few minute hairs towards apex; style glabrous; stigma linear or clavate, 0.8–1.4 mm long. Only immature capsule seen.

Mozambique. M: Boane Dist., Catembe, fl. 18.iv.1920, *Borle* 451 (PRE); Namaacha Dist., Goba Fronteira, Fonte do Passo, fl.& fr. 21.iv.1955, *Mendonça* 4524 (LISC, LMU, WAG).

Not known elsewhere (but hybridisation possible in Swaziland and E South Africa). Bushland and edges of woodland above waterfalls; 10–400 m.

Of the two specimens cited, *Mendonça* 4524 has a lax inflorescence like *Barleria obtusa* but the broader, more papery outer calyx lobes with a toothed margin suggest hybridisation with *B. gueinzii*. *Borle* 451 is very similar but has a more contracted inflorescence and so is close to form B of *B. gueinzii* (see below), but the narrower outer calyx lobes, the predominantly eglandular hairs on the corolla (vs. glandular-pilose in true *B. gueinzii*) and the more dense stem indumentum all suggest this is a hybrid with *B. obtusa*.

33. **Barleria gueinzii** Sond. in Linnaea **23**: 91 (1850). —Anderson in J. Linn. Soc., Bot. **7**: 30 (1863). —Clarke in Fl. Cap. **5**: 49 (1901). —Obermeijer in Ann. Transv. Mus. **15**: 168 (1933). —Compton, Fl. Swaziland: 553 (1976). —Fabian & Germishuizen, Wild Fl. Nthn. S. Afr.: 398, pl.191 (1997). —Bandeira *et al.*, Fl. Nat. Sul Moçamb.: 160, 190 (2007). Type: South Africa, Durban (Port Natal), fl. n.d., *Gueinzius* 383 (S holotype).

Barleria obtusa Nees var. *b* in De Candolle, Prodr. **11**: 231 (1847).

Barleria barbata C.B. Clarke in Fl. Cap. **5**: 49 (1901). Types: South Africa, Eastern Cape ("Pondoland"), between St. John's R. and Umtsikaba R., fl. 1837, *Drège* s.n. (K syntype); KwaZulu-Natal, no locality, fl. n.d., *Gerrard* 1973 (K syntype).

Perennial herb or shrub 50–250 cm tall, often scrambling or scandent; stems with pale appressed or ascending stiff or flexuose hairs, upper internodes often with shorter spreading or retrorse hairs on two opposite sides. Leaves ovate(-elliptic), 3–9.5 × 1.8–5 cm, base rounded, truncate or shortly attenuate, apex acute to attenuate, surfaces with pale ascending or subappressed hairs most numerous on margin and veins beneath where sometimes with a swollen base; lateral veins 4–6 pairs; petiole 6–27 mm long. Inflorescences axillary, sometimes clustered in upper leaf axils and forming a leafy thyrse, flowers solitary or often in 2–3-flowered contracted unilateral cymes, peduncle 0–4 mm long; bracteoles linear to narrowly elliptic-lanceolate, 4.5–16.5 × 0.5–2.5 mm, margin entire or with minute teeth formed by swollen hair bases, apex acute; pedicels 0.5–2.5 mm long. Calyx membranous, green or pale green with darker reticulate venation, tardily scarious; anterior lobe ovate to obovate, 11–17(21) × 7–13(15) mm, base shallowly cordate to cuneate or attenuate, margin with apiculate teeth, minute to more elongate, each with an apical bristle, apex obtuse, rounded or emarginate in outline, with 1–3 teeth, surface sparsely strigulose mainly on veins, sometimes with short spreading glandular hairs mainly along margin; posterior lobe ovate or elliptic, 13.5–23 mm long, apex acute- or attenuate-apiculate; lateral lobes pale, ovate or lanceolate, 2.5–7 mm long. Corolla 27–49 mm long, pale blue to mauve or purple, with paler tube and purple stripes in throat, glandular-pilose externally; tube 16–25 mm long, campanulate to broadly so above attachment point of stamens; limb in 4+1 arrangement or weakly so; abaxial lobe 12–21 × 11–16 mm, offset by 2–6 mm; lateral lobes (7.5)10–17 × 8–15 mm; adaxial lobes (8)9.5–15 × 5.5–10 mm. Stamens attached 6.5–11 mm from base of corolla tube; filaments 21–29 mm long; anthers 2.5–3.5 mm long; lateral staminodes 1.5–3.5 mm long, pilose and/or minutely glandular-pubescent, antherodes 0.5–0.8 mm long. Ovary glabrous or with few minute hairs towards apex; style glabrous; stigma clavate or linear, 0.6–1.2 mm long. Capsule 16.5–17.5 mm long, glabrous. Seeds 4–5 mm long and wide.

Two forms can be recognised within our region but further investigation of the South African material is required before any formal recognition of varieties.

Form **A** (typical)

Anterior calyx lobe ovate or elliptic, posterior lobe ovate, at least the latter with base obtuse to shallowly cordate, surfaces lacking glandular hairs (but see note below form B); lateral calyx lobes ovate, 2.5–4.5 mm long.

Mozambique. M: Namaacha Dist., Changalane, Estatuene, fl.& fr. 10.v.1969, *Balsinhas* 1480 (LISC, LMA); Namaacha Dist., Mt Ponduine, fl. 31.iii.1975, *Marques* 2674 (DSM, LMU).

Also in Swaziland and South Africa (KwaZulu-Natal, Limpopo, Mpumalanga). Dry riverine forest and bushland, particularly along margins and clearings, rocky slopes and roadsides; to 800 m.

Conservation notes: In the Flora area only known from the Lebombo Mts, but fairly widespread and often common in South Africa and Swaziland; Least Concern.

In our region the corolla throat is not as broadly campanulate as is often the case further south in typical *Barleria gueinzii*.

Form **B**

Anterior calyx lobe obovate or obovate-elliptic, posterior lobe elliptic, both with an attenuate or cuneate base, lobes and bracteoles often with short glandular hairs, most numerous along margin; lateral calyx lobes lanceolate, 4.5–7 mm long.

Mozambique. M: Maputo, Inhaca Is., fl.& fr. 13.vi.1970, *Correia & Marques* 1797 (LMU); Maputo, Ponta Mamoli, fl. 9.vii.1971, *Correia & Marques* 2114 (LMU).

Also in South Africa (Eastern Cape). Coastal scrub with *Diospyros* and *Mimusops;* secondary coastal forest with *Casuarina;* c.10 m.

Conservation notes: In the Flora area only known from Maputo Bay, but more common in South Africa; Least Concern.

Plants from coastal SE Mozambique closely match populations from the vicinity of East London in Eastern Cape (e.g. *Galpin* 3245, K) and appear rather distinct from typical *Barleria gueinzii*. These populations may result from back-crosses between true *B. gueinzii* and its hybrid with *B. obtusa* (see above). To complicate matters further, populations intermediate between the two forms occur in northern South Africa, for example *Schlechter* 4603 (K) from the Soutpansberg which has the ovate calyx lobes of typical *B. gueinzii* but with glandular hairs along the lobe margins. Mozambique plants of both forms tend to have less widely funnel-shaped corolla throats than most material from South Africa.

In the absence of fruits, *B. gueinzii* could easily be misplaced in Sect. *Fissimura*, appearing very close to some species in that section in floral morphology. It is therefore included in the key to that section.

Barleria L. sect. II. **Fissimura** M. Balkwill in J. Biogeogr. **25**: 110 (1998). —M. & K. Balkwill in Kew Bull. **52**: 569 (1997).

Barlerites Oerst. in Vidensk. Meddel. Dansk Naturhist. Foren. Kjobenhavn: 137 (1854) in part, lectotypified by M. & K. Balkwill (1997).

Barleria subgen. *Eu-Barleria* 'Villosae' sensu Clarke in F.T.A. **5**: 144 (1899), in part excl. *B. antunesi*, *B. holstii*, *B. limnogeton*, *B. querimbensis*, *B. repens* and *B. rotundisepala.*

Barleria sect. *Eu-Barleria* subsect. *Dispermae* sensu Obermeijer in Ann. Transv. Mus. **15**: 137 (1933).

Plants unarmed. Indumentum of simple hairs. Inflorescence of unilateral, dichasial or single-flowered cymes, axillary or compounded into a terminal synflorescence; bracts foliaceous, reduced in species with a terminal synflorescence. Anterior and posterior calyx lobes with palmate-reticulate venation but often inconspicuous, principal veins sometimes prominent in fruit; lateral lobes linear-lanceolate. Corolla white, blue, purple or rarely reddish, drying blue with darker venation or blue-black; tube cylindrical below attachment point of stamens, funnel-shaped or campanulate above; limb in 4+1 arrangement. Staminodes 3, lateral staminodes with antherodes well-developed or rarely absent. Stigma subcapitate or clavate, of 2 subconfluent lobes. Capsule drying black, laterally flattened, fusiform or somewhat obovate, without a prominent beak; lateral walls thin, often partially tearing from the thickened flanks at dehiscence; septum largely membranous. Seeds 2, discoid with dense (purplish-)brown, bronze or golden hygroscopic hairs.

A section of c.25 species, mainly African, in which taxon delimitation is very difficult with several species complexes and few wholly discrete taxa, although the species recorded in our region are fairly easy to separate from one another. In addition to the three species recorded in the Flora region, *Barleria pseudosomalia* I. Darbysh., recently described from C Tanzania, has an isolated population in NE Namibia just beyond the western boundary of the Caprivi Strip, and so may occur in our region (see Darbyshire *et al.* in Kew Bull. **67**: 764–765, 2012).

Key to Sect. *Fissimura* (including *B. gueinzii*)

1. Corolla with purple stripes in throat; S Mozambique . . . . **33.** *gueinzii**
 – Corolla lacking purple stripes in throat; not recorded from S Mozambique 2
2. Suffruticose perennial producing annual shoots from a woody rootstock; corolla (in our region) 37–45 mm long . . . . **35.** *boehmii*
 – Prostrate, scrambling or scandent perennial herb or subshrub, not suffruticose; corolla (in our region) 17.5–34.5 mm long 3
3. Corolla tube with basal cylindrical portion shorter than expanded throat; leaves beneath and calyx lobes with ± numerous fine spreading hairs between veins, as well as coarser more ascending hairs along veins; anterior and posterior calyx lobes usually with glandular hairs at least on margin. . . . . **34.** *ventricosa*
 – Corolla tube with basal cylindrical portion longer than expanded throat; leaves beneath and calyx lobes with only (sub)appressed coarse hairs, mainly along veins; anterior and posterior calyx lobes lacking glandular hairs . . . . **36.** *sp. C*

* *Barleria gueinzii* (sect. *Barleria*) is remarkably similar to *B. ventricosa* agg. in flower but is easily separated in fruit by the 4-seeded capsule – it is included here in case specimens with flowers only are keyed to sect. *Fissimura.*

34. **Barleria ventricosa** Nees in De Candolle, Prodr. **11**: 230 (1847). —Clarke in F.T.A. **5**: 164 (1899) in part, excl. *Fischer* 135. —Darbyshire in F.T.E.A., Acanthaceae **2**: 360 (2010). Types: Ethiopia, Tigray, Mt Soloda (Scholoda), 26.x.1837, *Schimper* I.42 (?GZU syntype, BR, GOET, HAL, HBG, K, M, MPU, TUB); Ethiopia, Mt Kubbi, 20.xi.1838, *Schimper* II.797 (GZU syntype, BR, TUB); Ethiopia, no locality, n.d. *Schimper* III.1903 (GZU syntype, BR, GOET, HAL, K, M, MPU, TUB); Ethiopia, no locality, n.d., *Quartin Dillon* s.n. (GZU, P syntypes). FIGURE 8.6.**40**.

Barleria stuhlmannii Lindau in Bot. Jahrb. Syst. **20**: 20 (1894). —Clarke in F.T.A. **5**: 167 (1899). —Champluvier in Fl. Rwanda **3**: 438, fig.137.4 (1985). Types: Tanzania, Mpwapwa, fl. 10.vii.1890, *Stuhlmann* 289 (B† holotype); Mpwapwa, fl.& fr. 24.iv.1929, *Hornby* 109 (K neotype), neotypified by Darbyshire (2010).

Barleria scindens Oberm. in Ann. Transv. Mus. **15**: 171, pl.I7 (1907). —Mapaura & Timberlake, Checklist Zimb. Vasc. Pl.: 13 (2004). Type: Zimbabwe, Bulawayo, fl. iii.1914, *Rogers* 13592 (PRE holotype, BOL).

Scrambling or scandent perennial herb 15–350 cm tall, with numerous erect or decumbent leafy branches from sprawling or trailing woody stems; stems 4-angular, with ascending or appressed pale buff to yellowish eglandular hairs and often few to numerous spreading glandular hairs, with or without a line of short retrorse hairs in opposite furrows. Leaves ovate, 1.8–10.5 × 1–5.3 cm, base cordate to obtuse, rarely acute to attenuate, apex acute to obtuse, surfaces pubescent, often rather densely so beneath, hairs more coarse on margin and veins beneath, finer and more spreading elsewhere; lateral veins 4–6 pairs; blade sessile or petiole to 20 mm. Inflorescence axillary, 1–2-flowered, subsessile; bracts foliaceous; bracteoles linear or oblanceolate, 2–7.5 × 0.2–0.5(1) mm, with dense pale ascending or appressed hairs. Anterior calyx lobe ovate to elliptic, rarely obovate, 7–13.5 × 4–10.5 mm, base shortly attenuate to rounded, margin with 4–7 minute to elongate teeth, each with an apical bristle, rarely subentire, apex notched for 1–2 mm, indumentum as leaves, usually also with a few glandular hairs at least along margin; posterior lobe ovate, 10.5–18.5 × 5.5–12.5 mm, apex acute-apiculate; lateral lobes 4–10 mm long. Corolla (17.5)23–31(34.5) mm long, white, pale blue or mauve, limb glandular-pilose externally; tube (10.5)13–18 mm long, funnel-shaped above attachment point of stamens; abaxial lobe (7)8.5–13.5 × 7.5–10(12.5) mm, offset by 2.5–4.5 mm; lateral lobes (5)7–8.5(10.5) × (4)5.5–7.5(10.5) mm; adaxial lobes as lateral pair but (2.5)3.5–5.5(7) mm wide. Stamens attached 4.5–7.5 mm from base of corolla; filaments (10.5)14.5–18(24) mm long; anthers 2–3 mm long; lateral staminodes 0.8–1.7 mm long, pubescent with or without shorter glandular hairs, antherodes 0.8–1.2 mm long. Ovary and style glabrous; stigma shortly clavate, 0.3–0.8 mm long. Capsule 9–11 mm long, glabrous. Seeds 4–5 × 3.5–4.5 mm.

Zimbabwe. N: Mutoko Dist., Nyamahere Hill, fl. 14.iii.1978, *Pope* 1652 (K, SRGH). W: Bulawayo Dist., Bulawayo–Esigodini (Essexvale) road, c.200 m N of Hope Fountain Secondary School by road to Waterford, fl. 13.ii.1974, *Norrgrann* 519 (SRGH). C: Shurugwi Dist., Ferny Creek Hills, fl. 2.iv.1967, *Biegel* 2037 (K, SRGH). E: Mutare Dist., NW base of Cross Hill, fl. 9.ii.1962, *Chase* 7664 (EA, K, SRGH). S: Chivi Dist., 0.8 km N of Lundi Bridge along Hippo Pool turnoff, fl. 16.iii.1967, *Mavi* 217 (K, SRGH). **Malawi**. S: Chiradzulu Dist., Nyungwe Hill, fl. 25.iii.1967, *Agnew* 529 (SRGH). **Mozambique**. MS: Manica Dist., Mt Chicamba, fl. 24.iv.1948, *Andrada* 1178 (LISC).

Also in Yemen, Eritrea, Ethiopia, Somalia, Sudan, South Sudan, Uganda, Kenya, Tanzania, Rwanda, D.R. Congo and South Africa (Limpopo). Open woodland on rocky hillslopes, usually over granite, riverine fringing woodland, termite mounds; 700–1700 m.

Conservation notes: Widespread and often abundant, locally common in Zimbabwe; assessed as Least Concern in F.T.E.A.

Barleria ventricosa is a widespread and highly polymorphic aggregate species; a full list of synonymy and extensive notes on the variation across its range are given in F.T.E.A. The description here covers only the forms found in our region previously separated as *B. scindens*. Plants of typical "*scindens*" are close to the form most common in Tanzania, "*B. stuhlmannii*", differing mainly in having a rounded or shallowly

Fig. 8.6.**40**. BARLERIA VENTRICOSA. 1, habit (× ²/₃); 2, detail of stem indumentum (× 2); 3, example of leaf variation (× ²/₃); 4, calyx and bracteoles (× 3); 5, dissected corolla with androecium (× 1½); 6, detail of androecium (× 2); 7, capsule valves with seed (× 4). 1 & 2 from *Norlindh & Weimarck* 4456, 3 from *Wild* 6355, 4 from *Davies* 2907, 5 & 6 from *Chase* 7664, 7 from *Teague* 105. Drawn by Juliet Williamson.

cordate leaf base. However, larger leaved variants from E Zimbabwe (5–10.5 cm vs. less than 5 cm long) are occasionally encountered, such as *Wild* 6355 from Wedza Mt (K, SRGH) and *Phipps* 1064 from Honde Valley, Nyanga (K, LISC, SRGH), in which the leaf base is acute or shortly attenuate. These plants are largely inseparable from other forms of *B. ventricosa* from East Africa; they are most probably a result of lush growing conditions. The variation with typical "*scindens*" appears clinal.

The absence of this species complex from most of Malawi and N Mozambique is perplexing since much suitable habitat is to be found there.

35. **Barleria boehmii** Lindau in Bot. Jahrb. Syst. **20**: 19 (1894). —Clarke in F.T.A. **5**: 167 (1899). —Darbyshire in F.T.E.A., Acanthaceae 2: 368 (2010). Types: Tanzania, Tabora Dist., Kakoma, fl. iii.1881, *Böhm* 25 (B† holotype); Tabora Dist., Kakoma, fl. 10.i.1936, *Lloyd* 10 (K neotype), neotypified by Darbyshire (2010).

Suffruticose herb with 1 to many ± erect stems 25–60(150) cm tall from a woody rootstock; stems with few to numerous long appressed or ascending coarse buff(-golden) hairs, in time of inconspicuous short crisped hairs in opposite furrows. Leaves elliptic to broadly ovate, 2–6 × 1–3.8 cm, base cuneate to rounded, apex acute to obtuse, surfaces with rather sparse ascending to spreading hairs, most numerous on margin and veins beneath; lateral veins 3–4 pairs; petiole to 4 mm long. Inflorescence of 1–3-flowered axillary cymes in upper half of stems; bracts foliaceous but gradually reduced upwards; bracteoles linear to oblanceolate or narrowly elliptic, 5–21 × 0.5–4.5 mm. Anterior calyx lobe ovate to elliptic, 12–20.5 × 7.5–10.5 mm, base attenuate to rounded, margin entire, apex emarginate or acute, indumentum as leaves; posterior lobe larger, 16–23.5 × 10–14.5 mm, apex acute to obtuse; lateral lobes 6.5–13 mm long. Corolla 37–45 mm long, pale blue to mauve or white, glandular-pubescent externally; tube broadly funnel-shaped above attachment point of stamens, 16–21 mm long; abaxial lobe 13–19 × 11–13.5 mm, offset by 9–12 mm; lateral lobes 10.5–13.5 × 9.5–11.5 mm; adaxial lobes 10–12 × 7–8.5 mm. Stamens attached 7–12 mm from base of corolla tube; filaments 17–22.5 mm long; anthers 2.5–3.3 mm long; lateral staminodes 3–7 mm long, (sparsely) pubescent towards base, antherodes to 1 mm long. Ovary glabrous; stigma subcapitate, 0.4–0.7 mm long. Capsule 9–13.5 mm long, glabrous. Seeds (?immature) c.5 mm long and wide.

Zambia. N: Isoka, fl. 21.xii.1962, *Fanshawe* 7196 (K, NDO, SRGH). **Malawi**. N: Chitipa Dist., 4.8 km W of Great North Road, near Kaseye junction, fl. 27.xii.1972, *Pawek* 6147 (CAH, K, MO, SRGH, UC).

Also in Burundi and W Tanzania. Miombo woodland and dambos on sandy soil; c.1200 m.

Conservation notes: Scarce in the Flora region with only three records known; assessed as Least Concern in F.T.E.A.

The above description covers only the variation recorded in our region and similar populations from SW Tanzania. This species is more variable in W and NW Tanzania (see F.T.E.A.) where forms with glandular-pilose inflorescences and populations with considerably smaller flowers are recorded.

36. **Barleria sp. C** (= *Richards* 10903)

Prostrate perennial herb or subshrub, much-branched; stems sub-4-angular, (pale)yellow-strigose, hairs appressed when young, sometimes less strictly so with age, with a line of short spreading to crisped hairs in opposite furrows; mature stems soon woody. Leaves ovate or elliptic, 3.5–5 × 1.2–2.4 cm, base cuneate or attenuate, apex acute to subattenuate, surfaces sparsely strigose above and on main veins beneath, midrib above with or without short crisped hairs; lateral veins 3–4 pairs; petiole to 3 mm. Inflorescences axillary at upper nodes of leafy branches, 1–3-flowered, subsessile; bracts foliaceous; bracteoles narrowly elliptic, lanceolate or oblanceolate, 3.5–11 × 0.5–3 mm, strigose. Calyx gradually accrescent and eventually scarious; anterior lobe broadly ovate to rounded, 11–12 × 8–10 mm in flower, to 15.5 × 14 mm in fruit, base rounded, margin denticulate, each tooth with an apical bristle, apex rounded to attenuate and apiculate or shallowly emarginate and 2-toothed, surface strigose mainly on palmate principal veins, sometimes also with minute ascending hairs between veins; posterior lobe ovate, 11.5–17.5 × 9.5–10.5 mm in flower, to 17.5 × 15 mm or more in fruit, apex acute to attenuate, apiculate; lateral lobes 4–5.5 mm long. Corolla 21.5–26.5 mm long, pale mauve or lilac, glandular-pilose mainly on lateral lobes externally; tube funnel-shaped, 12–13.5 mm long, basal cylindrical portion longer than expanded throat; abaxial lobe 6–13.5 × 6–11 mm, offset by 1.5–4 mm, apex emarginate; lateral lobes 6–10 × 6.5–9 mm; adaxial lobes 5.5–9.5 × 4.5–7 mm. Stamens attached 5.5–7 mm from base of corolla tube; filaments 14–15 mm long; anthers 2–2.8 mm long; lateral staminodes 1.7–2.2 mm long, pubescent, antherodes 0.5–1 mm long. Ovary glabrous; stigma shortly clavate or subcapitate, 0.3–0.6 mm long. Capsule c.8 mm long (perhaps immature or malformed?), glabrous. Only immature seeds seen.

Zambia. N: Mporokoso Dist., escarpment road to Sumbu, Lake Tanganyika, fl.& fr. 5.iv.1957, *Richards* 9034 (K); Mbala Dist., Kawa R. Gorge, fl. 14.ii.1959, *Richards* 10903 (K).

Not known elsewhere (but see note). Moist woodland, sometimes growing over rocks, *Bussea* dry forest thicket, sandstone outcrops and roadside banks; 950–1200 m.

Conservation notes: Apparently endemic to NE Zambia, rare and highly restricted, known from only four collections in dry forest; Data Deficient.

This incompletely known species is very close to the West African *Barleria ruellioides* T. Anderson, sharing a similar growth habit, leaf shape including the long cuneate or attenuate base, broadly ovate calyx lobes and long basal cylindrical portion to the corolla tube. The principal difference is in flower size: *B. ruellioides* has corollas typically 29–40 mm long with the tube 17–22 mm long, and anthers 3–4.5 mm long. The fruits also appear larger (10.5–13.5 mm), but it is possible that the fruits of sp. *C* seen are not mature. Even if they prove to be consistent, these differences seem rather trivial in view of the great variation seen within some other species in sect. *Fissimura*. Sp. *C* may be best considered a variant of *B. ruellioides* although the range disjunction is very pronounced and unexpected, *B. ruellioides* being recorded only as far east as W Cameroon. Also falling within this species group, and geographically much closer to sp. *C*, are *B. neurophylla* C.B. Clarke (W Tanzania, eastern D.R. Congo) and *B. pauciflora* Champl. (southern D.R. Congo). These two species are more easily separated from sp. *C* by the presence of paler "windows" with darker reticulate venation towards the base of the outer calyx lobes and in the cylindrical basal tube of the corolla being shorter than the expanded throat.

Barleria L. sect. III. **Stellatohirta** M. Balkwill in J. Biogeogr. **25**: 110 (1998). —M. & K. Balkwill in Kew Bull. **52**: 569 (1997). —Darbyshire in Kew Bull. **63**: 261–268 (2008).

> *Barleria* subgen. *Eu-Barleria* 'Stellato-hirtae' sensu Clarke in F.T.A. **5**: 143 (1899) in part, excl. *B. fulvostellata*.
> *Barleria* sect. *Eubarleria* subsect. *Thamnotrichae* sensu Obermeijer in Ann. Transv. Mus. **15**: 139 (1933).

Axillary spines absent. Indumentum stellate or dendritic. Inflorescences of single-flowered subsessile cymes compounded into dense terminal globose heads or spikes; bracts highly modified. Corolla white, blue or purple; limb in 4+1 arrangement. Staminodes 2(3), lateral pair with antherodes well-developed, rarely absent. Style base swollen; stigma linear, single-lobed. Capsule laterally flattened, fusiform, unbeaked or shortly beaked; lateral walls often thin, remaining attached to or tearing slightly from thickened flanks at dehiscence; septum largely membranous. Seeds 2, discoid with dense buff-coloured hygroscopic hairs.

A section of c.15 species confined to Africa and Arabia with centres of diversity in East Africa and Angola. Balkwill & Balkwill (1997) recorded the capsules of this section as differing from those of sect. *Fissimura* as the lateral walls do not split away from the thickened flanks at dehiscence. However, tearing is clearly visible in the capsules of *B. splendens* E.A. Bruce within sect. *Stellatohirta*, therefore this character is not used in the key to sections.

Key to sect. *Stellatohirta*

1. Bracts, bracteoles and outer calyx lobes with short dark spinulose tips; inflorescence shortly spiciform or conical; corolla 27–40 mm long, including 11–14 mm tube . . . . **37.** *taitensis* subsp. *rogersii*
   – Bracts, bracteoles and outer calyx lobes without spiny tips; inflorescence globose or capitate; corolla 45–79 mm long, including 21–41 mm tube 2
2. Much-branched shrub or shrublet; leaves ovate with base rounded above a short attenuate portion, petiole 8–23 mm long; corolla white; stems with dense slender white, cream or (on uppermost internodes) yellow-buff dendritic hairs lacking a long central arm . . . . **38.** *albostellata*
   – Suffruticose perennials with annual shoots from a woody base; leaves (ovate-) elliptic or obovate with base cuneate or attenuate, petiole short, 2–10 mm long; corolla blue or purple, occasional white-flowered plants within a population; stems with dense short golden(-bronze) or buff-coloured dendritic hairs, many with a long central arm 3
3. Corolla tube 21–25 mm long, divided ± equally into cylindrical base and funnel-shaped upper portion; hairs on upper leaf surface with stellate base often falling early or absent . . . . **39.** *aenea*
   – Corolla tube 32–41 mm long, basal ²/₃ cylindrical, upper third funnel-shaped or narrowly so; hairs on upper leaf surface (if present) with persistent stellate base. . . . . **40.** *purpureotincta*

37. **Barleria taitensis** S. Moore in J. Bot. **40**: 343 (1902). —Darbyshire in F.T.E.A., Acanthaceae **2**: 380 (2010). Type: Kenya, Kiumbi (Makindu) R., fl. 14.iv.1902, *Kässner* 600 (BM holotype, K, MO).

   Barleria stellato-tomentosa S. Moore var. *ukambensis* Lindau in Bot. Jahrb. Syst. **20**: 23 (1894). Types: Kenya, Ukambani, fl. vi.1877, *Hildebrandt* 2722 & 2722a (B† syntypes); Ndara (Taita), fl. ii.1877, *Hildebrandt* 2457 (B† syntype).
   Barleria salicifolia sensu Clarke in F.T.A. **5**: 162 (1899) in part as regards *Hildebrandt* 2722, 2722a & 2457 from Kenya & *Smith* s.n. from Tanzania, non S. Moore.

   Erect or scrambling subshrub 30–250 cm tall; young stems with dense golden dendritic hairs, many with a long ± antrorse arm. Leaves elliptic(-obovate) or lanceolate, 2.5–13 × 0.8–3.3 cm, base cuneate or attenuate, apex acute to rounded, apiculate, upper surface with pale long-armed stellate hairs, lower surface with dense whitish(-buff) stellate hairs, veins and margin with golden to pale-buff long-armed stellate hairs; lateral veins 4–6 pairs, impressed above, prominent beneath; petiole 0–5 mm long. Inflorescences terminating main and short lateral branches, spiciform, conical or subcapitate, 1.5–7.5 cm long; bracts narrowly elliptic, obovate, oblong or lanceolate, 9.5–21 × 2–7 mm, apex dark-spinulose or -apiculate, ± outcurved; bracteoles (linear-)lanceolate or oblanceolate, 7–16 × 1–3(4) mm, conduplicate in cross-section, apex dark-spinulose, ± outcurved. Anterior calyx lobe ovate, subrhombic or rounded, 11.5–22 × 6–16 mm, base obtuse to attenuate, margin entire, often involute, apex attenuate into two parallel to widely divergent dark-spinulose or apiculate tips, venation palmate, prominent, surface with ± dense fine pale grey(-buff) stellate hairs and golden to pale-buff appressed or ascending long-armed stellate hairs along veins and margin, those along margin sometimes more spreading and with a swollen base; posterior lobe ovate, 12.5–21 × 4–11 mm, margin sometimes involute, apex attenuate or acuminate with a spinulose tip; lateral lobes linear-lanceolate, 6.5–10 mm long. Corolla 26–59 mm long, white, pale to bright blue or purple, pubescent externally mainly on lateral lobes; tube cylindrical, 11–30 mm long, curved; each lobe 12–25 × 5–13 mm, abaxial lobe offset by 1–4.5 mm, adaxial lobes sometimes narrowest. Stamens attached ± midway along corolla tube; filaments 15–33 mm long; anthers 2.4–4 mm long; lateral staminodes 0.8–2.5 mm long, pubescent, antherodes 0.5–1.3 mm long. Ovary glabrous; style puberulous towards base; stigma 0.8–2.2 mm long. Capsule 10–16 mm long, shortly beaked, glabrous. Seeds 5.5–9.5 × 4–6.5 mm.

Subsp. **rogersii** (S. Moore) I. Darbysh., comb. & stat. nov. Type: Zimbabwe, Hwange, fl. v.1915, *Rogers* 13239 (BM holotype, BOL).

> *Barleria rogersii* S. Moore in J. Bot. **56**: 38 (1918). —Obermeijer in Ann. Transv. Mus. **15**: 177 (1933). —Welman in Germishuizen *et al.*, Checklist S. Afr. Pl.: 80 (2006) in part.
> *Barleria taitensis* sensu auct., non S. Moore sensu stricto. —White, For. Fl. N. Rhod.: 382 (1962). —Mapaura & Timberlake, Checklist Zimb. Vasc. Pl.: 13 (2004). —Phiri, Checklist Zamb. Vasc. Pl.: 18 (2005).

Bracts narrowly oblong or lanceolate, 9.5–15 × 2–4 mm; bracteoles linear-lanceolate, 7–12.5 × 1–2 mm. Anterior calyx lobe ovate or rounded, 12.5–18.5 × 6–13 mm, base obtuse or shortly attenuate, margin tightly involute, folded around posterior lobe, apex very shortly divided, two spinulose tips parallel or ± divergent; posterior lobe with margin involute. Corolla bright (rarely pale) blue or purple, 27–40 mm long; tube 11–14 mm long, shorter than lobes.

Zambia. S: Kalomo Dist., Katombora, Zambezi R., fl. 14.iv.1949, *West* 2912 (K, SRGH). **Zimbabwe.** N: Binga Dist., Mwenda Research Station, fl. 30.iii.1966, *Jarman* B2-4a (K, SRGH). W: Hwange Dist., near Zambezi/Deka junction, fl. 31.iii.1963, *Leach* 11624 (K, LISC, SRGH). S: Beitbridge Dist., Nulli Range, fl. 26.ii.1961, *Wild* 5430 (K, SRGH). **Mozambique.** GI: Massingir Dist., above Massingir Dam, entrance to Limpopo Nat. Park, fl. 12.v.2004, *Jacobsen* 2203 (PRE).

Also in South Africa (Limpopo). Open mopane, *Commiphora* or *Combretum* woodland and thicket, sandy or stony soils and rocky hillslopes; 600–1000 m.

Conservation notes: Locally common in dry woodlands of S Zambia/N Zimbabwe, but scarce elsewhere; Least Concern.

Close to var. *occidentalis* S. Moore (to be elevated to subspecies rank in a forthcoming paper) from Angola and Namibia, from which it differs in the ovate anterior calyx lobe with the spine-tips usually closely parallel or weakly divergent (not elliptic with widely divergent tips) and in the calyx indumentum being finer and denser giving a velvety grey appearance to the inflorescence, with less conspicuous long-armed hairs along the veins and margin. Subsp. *taitensis* is restricted to E Africa.

38. **Barleria albostellata** C.B. Clarke in F.T.A. **5**: 162 (1899). —Obermeijer in Ann. Transv. Mus. **15**: 177 (1933). —Codd in Fl. Pl. Afr. **29**: t.1138 (1952). —Fabian & Germishuizen, Wild Fl. Nthn. S. Afr.: 392, pl. 188 (1997). Type: Zimbabwe, Matebele country, Shashe (Shasha) R., fr. 27.iii.1876, *Holub* 1397 (K holotype). FIGURE 8.6.41.

Shrub or shrublet (45)90–300 cm tall; young stems with dense pale dendritic hairs, often yellow-buff on uppermost internodes, white or cream below; woody stems glabrescent. Leaves ovate to broadly so, 5–14 × 3–9 cm, base rounded above a short attenuate portion, apex acute or shortly attenuate, surfaces at first white-felty with dense dendritic hairs, sometimes yellow-buff beneath, hairs ± falling early, mature leaves sometimes glabrescent above, more persistent beneath particularly on veins; lateral veins 4–6 pairs, prominent beneath; petiole 8–23 mm long. Inflorescence subglobose, 2.7–5.5 cm in diameter, bracts, bracteoles and calyx green at first with purple margin, apex and blotching, turning brown-scarious except sometimes the outer bracts; outermost 2(4) pairs of bracts imbricate, largely enclosing rest of inflorescence, broadly ovate, 24–40(50) × 18–30(38) mm, surface with unevenly distributed white dendritic hairs most dense along midrib and towards base and margin, pinnate-reticulate venation conspicuous; bracts

Fig. 8.6.**41**. BARLERIA ALBOSTELLATA. 1, habit (× ²/₃); 2, stem indumentum (× 6) with detail of dendritic hairs (× 12); 3, calyx and bracteoles, posterior view (× 1½); 4, dissected calyx, external surface, posterior lobe to right (× 1½); 5, dissected corolla with androecium and upper portion of style (× 1); 6, detail of staminodes (× 4); 7, detail of stigma (× 6); 8, capsule valves (× 3); 9, seed (× 3). 1 from *Norlindh & Weimarck* 5124, 2, 8 & 9 from *Biltone et al.* in MSB 611, 3 & 4 from *Rogers* 5637, 5–7 from *Crook* 383. Drawn by Juliet Williamson.

rapidly reducing in size inwards, those towards outside of head 21–37 × 3.5–9 mm, spathulate or obovate, the innermost linear and down to 13 × 1.5 mm; bracteoles linear to oblong-oblanceolate or oblong-lanceolate, 17.5–30(38) × 1.5–5 mm, ascending, margin and midrib with numerous long antrorse bulbous-based hairs, sometimes with a stellate base. Anterior calyx lobe oblong-ovate to obovate, 17–27 × 5.5–11 mm, base cuneate, apex bifid with divergent triangular lobes 2.5–8.5 mm long, venation palmate or subparallel, prominent, surface with long bulbous-based antrorse hairs numerous along margin, sometimes with a stellate base; posterior lobe ovate to lanceolate, 15.5–23 × 5.5–9 mm, apex acuminate, hairs along margin as for anterior lobe, external surface elsewhere largely glabrous; lateral lobes linear-lanceolate, 9–11.5 mm long. Corolla 45–68 mm long, white, glabrous externally; tube cylindrical, 27–37 mm long, curved, barely widened at mouth; each lobe 14–25 mm long, abaxial lobe widest, 7.5–14.5 mm wide, offset by 3.5–7.5 mm, adaxial lobes narrower, 6.5–12 mm wide. Stamens attached ± midway along corolla tube; filaments 28–35 mm long; anthers 4–4.8 mm long; lateral staminodes 3–4 mm long, pubescent towards base, antherodes 0.6–0.9 mm long. Ovary glabrous; style puberulous at base; stigma 2.5–3.5 mm long. Capsule 12.5–16 mm long, shortly beaked, glabrous. Seeds 6.5–8 × 6–7.5 mm.

Botswana. SE: Central Dist., Lecheng village, fr. 23.vi.2009, *Bittone et al.* in MSB Botswana 611 (GAB, K). **Zimbabwe**. N: Mutoko Dist., Mutoko (Makate) ruins, fl. 15.ii.1962, *Wild* 5658 (BR, K, SRGH). W: Matobo Dist., eastern Matopo Hills, fl. 14.iii.1971, *Plowes* 3456 (K, SRGH). E: Mutare Dist., 80 km on Chimanimani (Melsetter) road, fl. 18.ii.1951, *Crook* 383 (K, SRGH). S: Zvishavane Dist., c.16 km from Zvishavane (Shabani) towards Gweru, fl. 25.ii.1931, *Norlindh & Weimarck* 5124 (BM, BR, K). **Mozambique**. T: Cahora Bassa Dist., between Songo and dam, fl. 21.ii.1972, *Macêdo* 4880 (LISC, LMA, LMU, LUBA).

Also in South Africa (Limpopo). Amongst boulders or on rocky or sandy soil, often in partial shade of dry woodland and riverine thickets; 400–1450 m.

Conservation notes: Rather local with scattered distribution, common around Matopos in W Zimbabwe (*Plowes* 1439, 1687 notes "a very common shrub round bases of drier kopjes in the Matopo hills"); Least Concern.

Phiri (Checklist Zambian Vasc. Pl.: 18, 2005) records this species from N Zambia, probably in error. I am unsure which species this refers to, possibly a white-flowered variant of *B. aenea*.

39. **Barleria aenea** I. Darbysh. in Kew Bull. **63**: 264, figs.3, 4 (2008). —Darbyshire in F.T.E.A., Acanthaceae 2: 384 (2010). Type: Tanzania, Nkansi Dist., 45 km on Namanyere–Karonga road, fl. 5.iii.1994, *Bidgood, Mbago & Vollesen* 2655 (K sheet 1 holotype, BR, C, CAS, DSM, EA, NHT).

Barleria subglobosa sensu Phiri, Checklist Zamb. Vasc. Pl.: 18 (2005) in part, non S. Moore.

Suffruticose perennial with numerous erect stems from a woody rootstock 45–75 cm tall; stems with dense golden-bronze dendritic hairs, many with a long antrorse arm, ± persistent. Leaves (ovate-)elliptic or ± obovate, 6.5–12.5 × 2–5.5 cm, base cuneate or attenuate, apex acute or subattenuate, upper surface with pale long hairs, stellate base often falling early or absent, lower surface pale stellate-pubescent, with buff to golden long-armed dendritic hairs on principal veins and margins; lateral veins 5–7 pairs, these and reticulate tertiary venation prominent beneath; petiole 2–10 mm long. Inflorescence subglobose, 2.5–4.5 cm in diameter; bracts (ovate-)elliptic, 19.5–34 × 5.5–18 mm, with dense golden-bronze long-armed dendritic hairs on margin and principal veins interspersed with paler stellate hairs; bracteoles ± narrowly (ovate-) elliptic, 21.5–27.5 × 6–10 mm. Anterior calyx lobe ovate-elliptic, 19–22 × 7.5–10.5 mm, apex bifid, segments obtusely deltate, 1.5–3 mm long, venation subparallel, inconspicuous, indumentum as bracts but sparse towards base; posterior lobe elliptic, 20–22.5 × 7.5–9.5 mm, apex acute or subattenuate, long-armed stellate hairs restricted to distal third, largely glabrous towards base except for simple hairs on margin; lateral lobes linear-lanceolate, 9.5–11.5 mm long. Corolla 55–75 mm long, blue or purple with whitish throat and tube, rarely white (occasional plants in a population), glabrous externally or sparsely pubescent towards apex of lateral lobes; tube 21–25 mm long, cylindrical in proximal half, funnel-shaped above; lobes each 23–36 × 19–30 mm, abaxial lobe offset by 7–12 mm. Stamens attached ± midway along corolla tube; filaments (19)26–32 mm long; anthers (4)4.5–5.5 mm long; lateral staminodes 5.5–7 mm long, antherodes to 0.5 mm long. Ovary glabrous; style puberulous towards base; stigma 3.2–4 mm long. Capsule c.16 mm long, apex attenuate, glabrous. Seeds 10–10.5 × c.7 mm.

Zambia. N: Kawambwa Dist., Kapweshi, fl. 4.i.1963, *Bands* 752 (K, NDO); Kaputa Dist., 16 km N of Nsama, fl.& fr. 16.iv.1989, *Radcliffe-Smith, Pope & Goyder* 5703 (K).

Also in SW Tanzania, likely to occur in southeastern D.R. Congo. Miombo and chipya woodland including dambo fringes; c.1000 m.

Conservation notes: Clearly scarce, known from only four collections despite being highly conspicuous; potentially threatened.

Descamps 39 (BR) from the Buleji Valley in southern D.R. Congo looks very close to this species but the immature corolla has a long and slender tube c.28 mm long and the lobes look like they may be quite small. More material with mature flowers is required before any firm conclusion can be reached.

40. **Barleria purpureotincta** I. Darbysh. in Kew Bull. **63**: 267, fig.3 (2008). Type: Zambia, Machile, fl. 8.ii.1961, *Fanshawe* 6211 (K holotype, NDO).

   Barleria subglobosa sensu Phiri, Checklist Zamb. Vasc. Pl.: 18 (2005) in part, non S. Moore.

Suffruticose perennial 20–40 cm tall with numerous erect stems from a woody rootstock; stems with dense buff to golden dendritic hairs, many with a long antrorse arm, ± persistent. Leaves narrowly elliptic or somewhat obovate, 5–9 × 2–3 cm, base cuneate, apex acute, subattenuate or obtuse, upper surface with numerous pale long-armed stellate hairs, eventually glabrescent, lower surface with ± dense pale-buff stellate hairs and buff to golden long-armed dendritic hairs on principal veins and margins; lateral veins 4–6 pairs, these and reticulate tertiary venation prominent beneath; petiole 2–5 mm long. Inflorescence subglobose or capitate, 2–3.5 cm in diameter; bracts, bracteoles and outer calyx lobes tinged purple in distal half along margin and principal veins; bracts (ovate-)elliptic to obovate, 18.5–23.5(30) × 7.5–9(18) mm, dense buff to golden long-armed dendritic hairs particularly on margin and principal veins; bracteoles narrowly elliptic or oblanceolate, 14.5–22.5 × 1.7–3.5 mm. Anterior calyx lobe ovate-elliptic, 17.5–24 × 8–10 mm, apex bifid with deltate segments, 1.7–4.5 mm long, venation subparallel, indumentum as for bracts but largely glabrous in proximal third except for pale marginal hairs; posterior lobe elliptic, 18.5–25 × 6–10 mm, apex acute or obtuse, long-armed stellate hairs restricted to distal half or third, largely glabrous towards base except for simple hairs along margin; lateral lobes linear-lanceolate, 8.5–12.5 mm long. Corolla 62–79 mm long, blue or purple, glabrous externally; tube 32–41 mm long, cylindrical in proximal two-thirds, funnel-shaped or narrowly so above; lobes each 22–39 × 14–20 mm, abaxial lobe offset by 5.5–7 mm. Stamens attached ± midway along corolla tube or in distal half; filaments 23–30 mm long; anthers 3.4–4.3 mm long; lateral staminodes 2.5–4.5 mm long, pilose, antherodes 0.4–0.7 mm long. Ovary glabrous; style puberulous at base; stigma c.3.5 mm long. Capsule and seeds not seen.

   Zambia. B: Kaoma Dist., 80 km S from Luampa Mission along Kaoma–Mulobezi road, fl. 4.iii.1996, *Zimba, Harder & Luwiika* 757 (K, MO). S: Kalomo Dist., Mulobesi stream near Kembe Forest, Nkoya Native Reserve, fl. 8.ii.1965, *Mitchell* 26/04 (K, SRGH).

   Also in E Angola (see below). Grassland on sandy soils, miombo woodland and in dambos; 1100–1250 m.

   Conservation notes: Restricted distribution and only known from four collections; Data Deficient.

   The species has recently been found to occur in Angola, with a single collection (?Kassassa, 9.v.1925, *Pocock* 214 (BOL)). The exact location of this population has not been pinpointed but two days after this collection Mrs Pocock was on the Zambian side of the Angola border at 15° S, close to the known Zambian sites of *B. purpureotincta*. Although this expands the range somewhat, this species is still highly restricted.

Barleria L. sect. IV. **Cavirostrata** M. Balkwill in J. Biogeogr. **25**: 110 (1998). —M. & K. Balkwill in Kew Bull. **52**: 566 (1997). —Darbyshire in Kew Bull. **63**: 601–611 (2009). *Barlerianthus* Oerst. in Vidensk. Meddel. Dansk Naturhist. Foren. Kjobenhavn: 136 (1854).

Axillary spines absent. Indumentum simple or stellate. Inflorescence of 1–2-flowered subsessile cymules compounded into a verticillate or strobilate terminal thyrse; bracts foliaceous or reduced. Calyx lobe margins entire. Corolla white, blue or mauve; tube subcylindrical, somewhat widened towards base and apex; limb 5-lobed, sinus between adaxial pair of lobes markedly wider than other sinuses. Staminodes 3, antherodes absent. Stigma linear, single-lobed. Capsule barely compressed laterally, narrowed into a prominent apical beak, hollow towards base, lateral walls remaining attached to flanks at dehiscence; septum with a shallow membranous portion above upper retinacula, elsewhere woody. Seeds 4 (or 2 by abortion), subglobose or subellipsoid, partially or barely flattened, with minute blunt hairs unevely distributed on surface, glabrescent.

A section of nine species, five restricted to the Indian subcontinent and four to tropical Africa. In addition to the two species recorded from our region, *Barleria grandipetala* De Wild. has been recorded from SW Tanzania and may well occur in the miombo woodlands of NE Zambia, although it appears to be replaced there by the closely related *B. descampsii* from which it differs primarily in having considerably larger bracts which enclose the calyx. These three species form a discrete group, to which the above description applies (see Darbyshire 2009).

Key to sect. Cavirostrata

Plants with numerous long-armed stellate hairs on stem, leaves, bracts and calyx; thyrses usually congested throughout . . . . 41. *descampsii*
– Plants with simple hairs only; thyrses on main branches more lax, at least lower cymules clearly spaced . . . . .42. *richardsiae*

41. **Barleria descampsii** Lindau in Bot. Jahrb. Syst. **24**: 318 (1897). —Clarke in F.T.A. **5**: 169 (1899). —Phiri, Checklist Zamb. Vasc. Pl.: 18 (2005) as *deschampsii*. — Darbyshire in Kew Bull. **63**: 610 (2009). Type: D.R. Congo, Mt Pueto, fl. iii.1896, *Descamps* s.n. (BR lectotype, K photo), lectotypified by Darbyshire (2009).

Barleria briartii De Wild. & T. Durand in Compt. Rend. Soc. Bot. Belg. **38**: 212 (1900). —Burkill & Clarke in F.T.A. **5**: 513 (1900). Type: D.R. Congo, Mts of Nzilo Gorge, fl. n.d., *Briart* s.n. (BR holotype, K photo).

Barleria paludosa S. Moore in J. Bot. **48**: 251 (1910). Type: D.R. Congo, Kundelungu, fl. 13.iii.1908, *Kässner* 2619 (BM holotype, K).

Barleria argenteo-calycina De Wild. in Bull. Jard. Bot. État **5**: 9 (1915). Type: D.R. Congo, R. Buehi, fl. v.1891, *Descamps* s.n. (BR holotype, K photo).

Suffruticose perennial 40–100 cm tall; stems with dense ascending or appressed golden to silvery long-armed stellate hairs. Leaves ovate, elliptic or lanceolate, 5.5–12 × 2–5 cm, base cuneate to obtuse or attenuate, margin somewhat revolute, apex acute or attenuate, sometimes rounded in lower leaves, upper surface with few to numerous long-armed stellate hairs, stellate base sometimes falling early, lower surface with ± dense fine whitish long-armed stellate hairs and coarser, buff or golden long-armed stellate hairs on nerves and margins; lateral nerves 4–6 pairs; petiole 0–10 mm long. Inflorescence congested, elongate to subcapitate, rarely more lax and thyrsoid, 1.5–10 cm long, each cymule single-flowered; bracts foliaceous, much-reduced upwards, those in upper axils elliptic to obovate, 11–18 × 3–9 mm; bracteoles oblanceolate, elliptic or linear, (7.5)11–18 × 1.5–7 mm. Anterior calyx lobe oblong to obovate or elliptic, (11.5)16.5–20.5(25) × (4)8.5–13.5 mm, base cuneate, apex rounded, truncate or notched up to 1 mm, surface densely covered in ascending golden to silvery long-armed stellate hairs, venation (sub)parallel, obscured by indumentum; posterior lobe (13)18.5–25 × (4)7–10 mm, apex obtuse or rounded; lateral lobes lanceolate, (7)8.5–15 mm long. Corolla 40–53 mm long, mauve or lilac with whitish tube, whole flower rarely white, shortly glandular-pubescent externally; tube cylindrical, 20–24.5 mm long; lobes 17–28 mm long, abaxial lobe 10–11.5 mm wide, lateral lobes to 10.5–13.5 mm, adaxial lobes to 7–10.5 mm. Stamens attached c.8 mm from base of corolla tube; filaments c.23 mm long, with few minute short-stalked glands at base; anthers 6.5–8.5 mm long; lateral staminodes 1.2–3.5 mm long. Ovary densely pubescent in distal half; style pubescent towards base; stigma 0.5–1 mm long. Capsule 15–18 mm long including beak 4–5.5 mm long, pubescent in distal half. Seeds black, subellipsoid to subglobose, 3–4 × 3–3.2 mm.

Zambia. N: Mbala Dist., Chitimbwa road, fl. 18.iii.1969, *Sanane* 532 (BR, EA, K, LISC). W: Solwezi Dist., Kabompo Gorge road, fr. 14.v.1969, *Mutimushi* 3339 (K, NDO).

Also in southern D.R. Congo. Miombo woodland, open wooded grassland, often on rocky hillslopes and by roadsides; 1200–1850 m.

Conservation notes: Assessed as Least Concern by Darbyshire (2009), and rather common across its range.

42. **Barleria richardsiae** I. Darbysh. in Kew Bull. **63**: 608, fig.2 (2009). —Darbyshire in F.T.E.A., Acanthaceae 2: 387, fig.53 (2010). Type: Tanzania, Ufipa Dist., 33 km on Namanyere–Karonga road, fl. 5.iii.1994, *Bidgood, Mbago & Vollesen* 2649 (K sheet 1 holotype, BR, C, CAS, DSM, EA, NHT, P). FIGURE 8.6.**42**.

Suffruticose perennial 50–100 cm tall, several erect branches from a woody rootstock; stems densely hairy with shorter, whitish or buff-coloured, spreading or retrorse hairs and longer, buff or yellow, ascending or spreading hairs. Leaves often in whorls of 3–4; ovate, narrowly elliptic or lanceolate, 6–10.5 × 1–3.5 cm, base cuneate to obtuse, apex acute or subattenuate, surfaces with coarse ascending hairs, most dense on margin and veins beneath, midrib with additional shorter spreading hairs; lateral veins 3–5 pairs; petiole 0–5 mm long. Inflorescence a terminal thyrse, (3)10–28 cm long, cymules congested towards apex, more widely spaced below, each with flowers solitary or paired; bracts foliaceous, reducing upwards, those in uppermost axils 14–23 mm long; bracteoles oblanceolate, narrowly obovate-elliptic or linear, 12–31 × 1.5–7 mm. Anterior calyx lobe oblong or obovate, 13–22.5 × 5–12 mm, base cuneate or acute, apex shallowly notched up to 3 mm, surface with numerous coarse ascending hairs particularly on veins and margins, hair bases ± swollen, with few interspersed short spreading hairs, venation parallel or narrowly palmate, prominent but sometimes obscured by indumentum; posterior lobe 16–24.5 mm long, apex acute to rounded; lateral lobes (ovate-)lanceolate, 6–14 mm long. Corolla 40–54 mm long, mauve or lilac with whitish tube; shortly glandular-pubescent externally; tube 18–24 mm long; lobes each 20–28.5 mm long, abaxial lobe 11.5–17.5 mm wide, lateral lobes to 12.5–19.5 mm, adaxial lobes to 9–15.5 mm. Stamens attached 6–8.5 mm from base of corolla tube; filaments 21–25 mm long, with numerous short-stalked glands in distal half, sparse below; anthers 5.5–7.5 mm long; lateral staminodes 1.5–3.7 mm long. Ovary appressed-pubescent towards apex; style pubescent in proximal half or throughout; stigma 1–1.6 mm long. Capsule 16.5–19 mm long including beak 4.5–6 mm long, appressed- or ascending-pubescent in distal half. Seeds black, subellipsoid, 3.8–4.3 × 3–3.8 mm.

Zambia. N: Mbala Dist., new road to Iyendwe Valley from Kambole, fl. 31.i.1959, *Richards* 10804 (K); Mbala Dist., Kalambo Falls, fl. 18.ii.1967, *Richards* 22096 (K).

Also in SW Tanzania. Miombo woodland on sandy soils and rocky hillslopes; 900–1750 m.

Conservation notes: Very local species; assessed as Near Threatened by Darbyshire (2009).

Barleria L. sect. V. **Somalia** (Oliv.) Lindau in Engler & Prantl, Nat. Pflanzenfam. **IV**(3b): 315 (1895). —Clarke in F.T.A. **5**: 142 (1899). —Obermeijer in Ann. Transv. Mus. **15**: 130 (1933). —M. & K. Balkwill in Kew Bull. **52**: 563 (1997). — Darbyshire & Ndangalasi in J. E. Afr. Nat. Hist. Soc. **97**: 123–134 (2009).

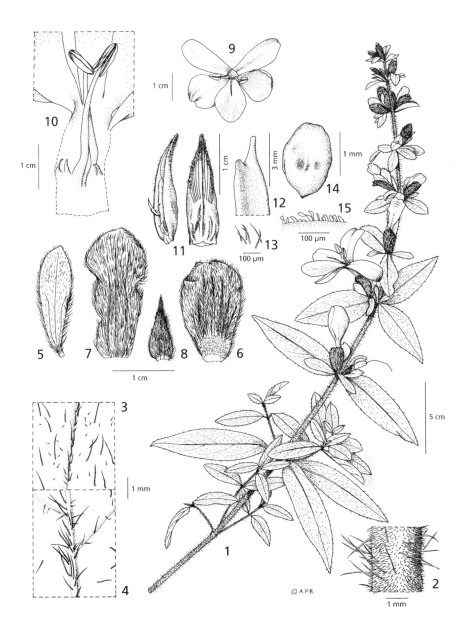

Fig. 8.6.**42**. BARLERIA RICHARDSIAE. 1, habit; 2, stem indumentum; 3, adaxial leaf indumentum at midrib; 4, abaxial leaf indumentum at midrib; 5, bracteole, inner surface; 6, anterior calyx lobe, outer surface; 7, lateral calyx lobe, outer surface; 8, posterior calyx lobe, outer surface; 9, face view of corolla; 10, dissected corolla tube with attached stamens and staminodes; 11, capsule, lateral and exterior view of single valve; 12, distal section of upper retinaculum; 13, capsule indumentum; 14, seed; 15, seed indumentum in profile. 1–8 from *Richards* 10804, 9 from photograph of *Bidgood et al.* 2649, 10 from *Richards* 19017; 11–15 from *Bidgood et al.* 3678. Drawn by Andrew Brown. Reproduced from Kew Bulletin (2009).

Somalia Oliv. in Hooker's Ic. Pl. **16**: t.1528 (1886).
Barleria subgen. *Eu-Barleria* "Glabratae" sensu Clarke in F.T.A. **5**: 143 (1899) in part, excl.
  B. grandis and *B. marginata.*

Axillary spines absent. Indumentum simple and/or of biramous hairs, these medifixed or 'anvil-shaped' (with one long arm and one short arm, the latter sometimes reduced to a swelling) or glabrous. Axillary buds often densely pale-hairy. Inflorescences single-flowered or dichasial cymes, either axillary, then bracts foliaceous, or compounded into terminal synflorescences, then bracts ± modified. Calyx lobes with margins entire. Corolla white, blue, mauve or purple, rarely yellow, sometimes with darker guidelines; limb variously in 4+1, 2+3, 1+3 or subregular arrangement. Staminodes 2–3, antherodes absent. Stigma linear, often curved, apex entire or minutely bilobed. Capsule drying (pale)brown, laterally compressed, fertile portion orbicular or obovate, ± abruptly narrowed into a prominent solid apical beak; lateral walls remaining attached to flanks at dehiscence, septum with a membranous portion above retinacula, elsewhere woody. Seeds 2, discoid with dense cream or buff-coloured, woolly (wavy) hygroscopic hairs.

A section of c.50 species, most diverse in eastern Africa and Angola; also in Madagascar, Arabia and the Indian subcontinent.

Key to sect. Somalia

1. Corolla 4-lobed, adaxial lobe emarginate 2
 – Corolla 5-lobed, adaxial pair of lobes sometimes partially fused but free for 4.5
   mm or more 3
2. Corolla 16–21 mm long, lateral lobes pubescent externally; flowers sessile. . . . . . .
   . . . . **53.** *lugardii*
 – Corolla 32–44 mm long, glabrous; flowers on short peduncles to 5 mm . . . .
   . . . . **54.** *quadriloba*
3 Mature stems minutely and densely white-velutinous, giving a pruinose appearance,
   if more sparse then corolla tube funnel-shaped with a short cylindrical basal tube
   and conspicuously widened throat (Fig. **43**); ovary and capsule puberulous with
   mixed glandular and eglandular hairs 4
 – Mature stems not minutely white-velutinous, often glabrescent; corolla tube
   cylindrical throughout or campanulate towards mouth; ovary and capsule
   glabrous or with only eglandular hairs, rarely a few glandular hairs at apex 5
4. Stems, bracteoles and external surface of outer calyx lobes lacking glandular
   hairs; outer calyx lobes broadly ovate with (sub)cordate base, strongly accrescent,
   20–35 × 18–25 mm in fruit. . . . . **47.** *mackenii*
 – Upper portion of stems, bracteoles and calyx lobes ± densely glandular-pubescent;
   outer calyx lobes elliptic or ovate with rounded to acute base, weakly accrescent,
   12.5–20 × 9.5–13.5 mm in fruit . . . . **48.** *lancifolia*
5. Outer calyx lobes broadly ovate, 10.5–30 mm wide, base cordate or rounded . . 6
 – Outer calyx lobes less than 10 mm wide, or if wider then elliptic, subrhombic or
   oblong with base acute, obtuse or attenuate. 8
6. Mature leaves, bracteoles and calyx coarsely hairy. . . . . **45.** *vollesenii*
 – Mature leaves, bracteoles and outer calyx lobes largely glabrous except sometimes
   for a few hairs along margins. 7
7. Suffrutex, branching mainly or exclusively from a woody base; corolla yellow or
   white, tube 25 mm long or more, lateral and adaxial lobes 22–29 mm long . . . . .
   . . . . **44.** *calophylloides*
 – Much-branched subshrub or perennial herb; corolla pale blue, mauve or white,
   tube to 15 mm long, lateral and adaxial lobes less than 20 mm long . . **46.** *galpinii*
8. Flowers in dense terminal heads or spikes, bracts clearly modified from leaves;
   corolla subregular or in 2+3 arrangement; suffruticose herb of miombo
   woodland. 9

- Flowers axillary but often crowded towards stem apex, bracts not clearly modified from leaves, gradually reducing in size upwards; corolla in 4+1 arrangement; habit and habitat various 11
9. Mature leaves glabrous; inflorescence lacking glandular hairs; outer calyx lobes large, elliptic or subrhombic, anterior lobe 7.5–17 mm wide; corolla limb subregular, adaxial lobes obovate-elliptic, 13.5–24.5 × 5.5–10.5 mm . . . . **43.** *lactiflora*
- Mature leaves hairy at least along margin and midrib below; inflorescence glandular-hairy; outer calyx lobes narrowly oblong-elliptic or ± obovate to subulate, anterior lobe 1.5–6.5 mm wide; corolla limb in 2+3 arrangement, adaxial lobes ± clearly fused towards base, free portions subulate or oblanceolate, 4.5–12 × 1.5–4 mm 10
10. Bracts* oblanceolate or narrowly elliptic, 1.5–4.5 mm wide; inflorescence densely glandular-pilose; leaves with prominent reticulate tertiary venation . . **55.** *phaylopsis*
- Bracts broadly obovate or orbicular, (8)10–20.5 mm wide; inflorescence glandular-puberulous, often sparsely so; leaves with inconspicuous reticulate tertiary venation . . . . **56.** *kaessneri*
11. Corolla tube and offset of abaxial lobe ± equal in length, tube 6.5–7 mm long; plant with sparse appressed medifixed hairs; dwarf suffrutex of coastal bushland . . . . **52.** *laceratiflora*
- Corolla tube clearly longer than offset of abaxial lobe, tube over 10 mm long; plant either without medifixed hairs, or if present then dense on stems; not in coastal bushland 12
12. Ovary and capsule densely appressed-hairy; stems with many short stiff appressed medifixed hairs, uppermost internodes often glandular-hairy. . . . **51.** *pretoriensis*
- Ovary and capsule glabrous or with few short spreading hairs towards apex; stem indumentum not as above 13
13. Leaves usually ovate or ovate-elliptic, rarely lanceolate, length to width ratio 1.5–4(5.7):1; anterior calyx lobe (oblong-)elliptic, length to width ratio 1.4–2.7:1, apex usually clearly bilobed; eglandular hairs on young leaves, bracteoles and calyx fine and soft, antrorse or crisped. . . . . **49.** *matopensis*
- Leaves lanceolate or linear-lanceolate, length to width ratio 4–9.5:1; anterior calyx lobe oblong-ovate to narrowly lanceolate, length to width ratio (2.5)3–5.5:1, apex entire or shallowly notched; eglandular hairs on leaves, bracteoles and calyx rather coarse and subappressed, some hairs 'anvil-shaped' . . . . **50.** *hirta*

* excluding enlarged pairs at base of inflorescence in both species

43. **Barleria lactiflora** Brummitt & Seyani in Kew Bull. **32**: 723 (1978). —Darbyshire in F.T.E.A., Acanthaceae 2: 392 (2010). Type: Malawi, Mzimba Dist., 8 km past Lunyangwa R., 9.6 km SW of M14, fl. 7.iii.1976, *Pawek* 10899 (K holotype, LISC, MAL, MO, PRE, WAG).

Suffruticose perennial 30–60(100) cm tall, with several to many ± erect stems from a woody rootstock; stems with two opposite lines of ascending hairs, later glabrescent. Leaves subsessile, somewhat coriaceous, narrowly oblong-elliptic to more broadly ovate, 8.5–16 × 2–6 cm, base cuneate to obtuse or rounded, apex acute to rounded, basal pairs often smaller and somewhat obovate, glabrous when mature; lateral veins 5–7(8) pairs. Inflorescence a dense terminal cylindrical or capitate head, 3.5–9.5 cm long, sometimes with additional solitary flowers in uppermost leafy axils; bracts ovate, elliptic or somewhat obovate, 15–33 × 3–14 mm, margin long-ciliate at least in proximal half and sometimes along veins, hairs usually with swollen bases; bracteoles as bracts but ovate-lanceolate to oblanceolate, 1.5–7.5 mm wide, hairs on margin more numerous; flowers subsessile. Anterior calyx lobe elliptic or subrhombic, 16–28.5 × 7.5–12(17) mm, base obtuse or acute, apex with two deltate lobes 4.5–8.5(11) mm long, indumentum as bracteoles, venation palmate, ± prominent; posterior lobe ovate, 16–31 × 5.5–11(14.5) mm, apex acute or obtuse; lateral lobes lanceolate, 7–13.5 mm long. Corolla 27–42 mm long, white, blue or mauve, limb pubescent externally, hairs mainly eglandular; tube cylindrical, 11.5–16.5 mm long, somewhat expanded towards mouth; limb subregular; lobes 13.5–24.5 mm long, abaxial lobe 9–14 mm wide, lateral lobes 7.5–13.5 mm, adaxial lobes 5.5–10.5 mm. Stamens attached ± midway along corolla tube; filaments 13–22.5 mm long; anthers 3–5 mm long; lateral staminodes 1–3(5.5) mm long. Ovary densely pubescent distally; style pubescent at base; stigma 2–3.3 mm long. Capsule 17.5–22 mm long, pubescent particularly on beak. Seeds c.8 × 6–7.5 mm.

Malawi. N: Chitipa Dist., 1.6 km S of Chendo, fl.& fr. 20.iv.1976, *Pawek* 11172 (K, MAL, MO, SRGH); Karonga Dist., escarpment just W of North Rukuru R. on Chitipa to Karonga road, fl. 4.iii.1982, *Brummitt, Polhill & Banda* 16325 (BR, K, LISC, MAL).

Also in S Tanzania. Miombo woodland and wooded grassland on sandy soils and rocky hillsides; 650–1450(2150) m.

Conservation notes: Locally common in N Malawi; assessed as Least Concern in F.T.E.A.

44. **Barleria calophylloides** Lindau in Bot. Jahrb. Syst. **20**: 17 (1894). —Clarke in F.T.A. **5**: 159 (1899). —Darbyshire in F.T.E.A., Acanthaceae **2**: 395 (2010). Types: Tanzania, Biharamulo Dist., Bukome, fl. 28.ii.1892, *Stuhlmann* 3431 (B† syntype); Tabora Dist., Igonda (Gonda), fl. 2.iii.1882, *Böhm* 164 (B† syntype, Z lectotype), lectotypified here.

Suffrutex to 50 cm long with several to many trailing to erect branches from a woody rootstock; stems with two opposite lines of short pubescence when young and/or crisped-pilose, later glabrescent. Leaves somewhat coriaceous, ovate or (oblong-)elliptic, 4–12.5 × 2–6.5 cm, base rounded or shallowly cordate to cuneate or attenuate, apex acute, obtuse or rounded, apiculate, lower leaves often smaller and somewhat obovate, surfaces glabrous or margin sparsely pilose towards base; lateral veins 4–7 pairs; petiole 0–12 mm long. Inflorescences single-flowered, axillary but often crowded towards stem apex; bracts foliaceous; bracteoles often held patent to calyx, obovate, oblanceolate or lanceolate, 4–25 × 0.5–10.5 mm, glabrous or with pilose margin; pedicels 2–4.5 mm long. Calyx accrescent; anterior lobe broadly ovate, 19–35 × 14.5–29 mm, base cordate, apex emarginate or rounded, surfaces glabrous or with few appressed marginal hairs at apex, rarely with scattered crisped pilose hairs on margin, venation palmate, prominent; posterior lobe 22–41 × 15.5–30 mm, apex obtuse or rounded; lateral lobes lanceolate, 7–15 mm long, ragged and hyaline towards base, ciliate. Corolla 48–62 mm long, yellow or white, glabrous; tube cylindrical, 25–33 mm long, barely expanded towards mouth; limb subregular; abaxial lobe 18–26.5 × 9.5–17.5 mm, offset by c.2 mm; lateral lobes as abaxial but 22–29 mm long; adaxial lobes as abaxial but 7–14 mm wide. Stamens attached c.10 mm from corolla base; filaments 20–28 mm long; anthers 3–5 mm long; lateral staminodes 0.3–2 mm long. Ovary glabrous; stigma 2–3.3 mm long. Capsule 16–23 mm long, glabrous. Seeds 9–9.5 × 8–8.5 mm.

a) Subsp. **calophylloides**. —Darbyshire in F.T.E.A., Acanthaceae **2**: 395 (2010).

Branches typically erect or decumbent. Stems with two lines of short pubescence; leaves and bracteoles glabrous. Upper stem leaves ovate or (oblong-)elliptic, largest leaf (5)6.5–12.5 cm long. Corolla pale yellow to sulphur yellow, rarely white.

Zambia. N: Mbala Dist., Katuka, fl. 12.iii.1950, *Bullock* 2623 (K); Mbala (Abercorn) Pans, fl. 22.ii.1959, *Richards* 10956 (K, LISC).

Also in Tanzania. Miombo woodland and wooded grassland on sandy soils, including roadsides, and in dense ravine thicket; 1050–1650 m.

Conservation notes: Locally common in NE Zambia; assessed as Least Concern in F.T.E.A.

Boaler 456 from Tanzania was selected as a neotype in F.T.E.A. (2010) as I had thought all the original material destroyed. However, the Z sheet of *Böhm* 164, an original syntype, has since come to light and so is selected as lectotype here; the Boaler neotype becomes superfluous.

b) Subsp. **pilosa** I. Darbysh. in F.T.E.A., Acanthaceae **2**: 396 (2010). Type: Tanzania, Njombe/Mbeya Dist., 54 km on Chimala–Iringa road, fl. 3.iv.2006, *Bidgood et al.* 5292 (K holotype, BR, CAS, DSM, NHT).

> *Barleria polyneura* sensu Brummitt & Seyani in Kew Bull. **32**: 726 (1978) in part for Malawi specimens, non S. Moore.

Branches trailing. Stems with crisped pilose hairs, multicellular with conspicuous cell-walls, these also present towards base of leaf and bracteole margins, rarely on outer calyx lobe margins. Upper stem leaves broadly ovate or elliptic, largest leaf 4–6.7 cm long. Corolla white.

Malawi. N: Mzimba Dist., 4.8 km SW of Lunyangwa R., 6.4 km SW of M14, fl. 7.iii.1976, *Pawek* 10904 (BR, K, LISC, MAL, MO, SRGH, UC). C: Kasungu Dist., 42 km N of Kasungu on M1, fr. 27.iii.1978, *Pawek* 14137 (K, MAL, MO). **Mozambique**. N: Cuamba Dist., Amaramba, 42 km from Cuamba (Nova Freixo) towards Mandimba, fl. 24.ii.1964, *Torre & Paiva* 10737 (LISC, UPS, WAG).

Also in S Tanzania. Miombo woodland on sandy soils, including dambo margins; 700–1250 m.

Conservation notes: Assessed as Vulnerable under IUCN criterion B in F.T.E.A. The discovery of the Torre & Paiva collection from Mozambique significantly extends the range; it is possible it is more widespread along the eastern side of Lake Malawi.

Salubeni 1009 from Lilongwe Dist., Malawi (K, SRGH) is recorded as having orange flowers, though the Kew sheet is in fruit; it is probable that either the label data are erroneous or they refer to wilted flowers which can stain an orange-brown colour.

45. **Barleria vollesenii** I. Darbysh. in F.T.E.A., Acanthaceae **2**: 397 (2010). Type: Tanzania, Masasi Dist., Chivirikiti village to Mbangala Forest Reserve, fl. 14.iii.1991, *Bidgood, Abdallah & Vollesen* 1988 (K sheet 1 holotype, BR, C, CAS, EA, LISC, MO, NHT, P, UPS, WAG).

Suffruticose herb, 20–60 cm tall, several ascending or erect unbranched stems from a woody rootstock; stems with long pale coarse hairs, spreading and subappressed, the latter ± concentrated in two opposite lines. Leaves oblong-elliptic or somewhat lanceolate, 7–11 × 1.8–2.8 cm, base cuneate, apex acute or obtuse, apiculate, indumentum as stem, hairs most dense on margin and veins beneath; lateral veins 5–6 pairs; petiole 0–4 mm long. Inflorescences single-flowered, axillary but crowded towards stem apex; bracts foliaceous; bracteoles narrowly oblong-elliptic or somewhat obovate, 14–40 × 2–10 mm, indumentum as leaves; pedicels 1–3 mm long. Anterior calyx lobe broadly ovate, 25–35 × 11–15 mm, base rounded or shallowly cordate, apex bifid with deltate-apiculate lobes 1.5–4 mm long, indumentum as leaves, hairs rather dense on veins and margin, hair bases often swollen, venation palmate, ± prominent; posterior lobe 31–41 × 14–17 mm, apex acute-apiculate; lateral lobes linear-lanceolate, 14–15 mm long. Corolla 62–67 mm long, white, limb with inconspicuous short glandular and eglandular hairs externally; tube cylindrical, 26–29 mm long, somewhat expanded towards mouth; limb in 4+1 arrangement; abaxial lobe c.30 × 17 mm, offset by 10–11.5 mm; lateral lobes c.28 × 13.5 mm; adaxial lobes c.24 × 13 mm. Stamens attached c.9 mm from corolla base; filaments 27–31 mm long; anthers 5–5.5 mm long; lateral staminodes 2–3 mm long. Ovary glabrous; stigma curved, c.2.5 mm long. Capsule and seeds not seen.

Mozambique. N: Marrupa Dist., 40 km on road Marrupa–Mukwajaja, near phenological observation field of Mademo, fl. 21.ii.1982, *Jansen & Boane* 7977 (C, K, LISC, LMA, LMU, SRGH).

Also in SE Tanzania. Miombo woodland with rock outcrops; c.700 m.

Conservation notes: Assessed as Data Deficient in F.T.E.A., but very likely to be globally threatened as only known from two localities.

46. **Barleria galpinii** C.B. Clarke in Fl. Cap. **5**: 54 (1901). —Obermeijer in Ann. Transv. Mus. **15**: 150 (1933). —Fabian & Germishuizen, Wild Fl. Nthn. S. Afr.: 394, pl.189 (1997). Type: South Africa, Mpumalanga, Sheba Battery, Kaap valley, Barberton, fl. iii.1891, *Galpin* 1331a (K holotype).

Subshrub or many-branched perennial herb, erect or decumbent, 15–70 cm tall; stems with two opposite lines of fine white antrorse hairs or with more numerous tomentellous hairs when young, soon glabrescent. Leaf buds densely white- or creamy-hairy. Leaves ovate or elliptic, 2.2–7 × 1.3–3.5 cm, base cuneate or attenuate, apex acute or obtuse, apiculate, soon glabrescent; lateral veins 4–6 pairs; petiole 0–5 mm long. Inflorescence axillary, 1-flowered, subsessile; bracts foliaceous; bracteoles held ± patent to calyx, (linear-)oblanceolate or elliptic, often narrowing up stem, 6.5–20 × 1.5–6.5 mm, glabrous; pedicels 1.5–4.5 mm long. Calyx somewhat accrescent; anterior lobe broadly ovate, 14–22.5 × 10.5–22.5 mm, base (sub)cordate, apex emarginate or notched up to 3.5 mm, rarely obtuse, exterior surface glabrous, interior surface puberulous towards centre, venation palmate, prominent; posterior lobe as anterior but 15–27 mm long, apex acute to rounded, apiculate; lateral lobes lanceolate, 4–7 mm long, shortly ciliate. Corolla 31–42.5 mm long, pale blue to mauve or white, glabrous or limb pubescent externally; tube cylindrical, 14–15 mm long, barely expanded towards mouth; limb in 4+1 arrangement; abaxial lobe 14–20 × 9–11 mm, offset by 4.5–8 mm; lateral lobes 12–19.5 × 8–10.5 mm; adaxial lobes as laterals but 5–9 mm wide. Stamens attached 5–6 mm from base of corolla; filaments 18–26.5 mm long; anthers 4–5 mm long; lateral staminodes 1.3–1.8 mm long. Ovary glabrous or with few pale hairs at apex; style glabrous; stigma 1.6–3.7 mm long. Capsule 15.5–18.5 mm long, glabrous. Seeds 7.5–8.5 × 6.5–7 mm.

Botswana. SE: Kanye Dist., Pharing, fr. ii.1950, *Miller* B/994 (PRE); Kgale Mt, 8 km S of Gaborone, fl. 12.ii.1976, *Mott* 878 (K, SRGH).

Also in South Africa (Limpopo, Mpumalanga). Open bushland and woodland on dry rocky hills; 1100–1250 m.

Conservation notes: Local but often common (although rare in our region) and can tolerate disturbance; Least Concern.

Galpin records the flowers as "yellow" on the type gathering but this is clearly erroneous as they are white through to mauve or blue on all subsequent material. In our region and adjacent areas of South Africa the corolla is glabrous, but some specimens from the central part of its range, including the type, have a hairy limb; two varieties could perhaps be recognised.

47. **Barleria mackenii** Hook. f. in Bot. Mag. **96**: t.5866 (1870). —Clarke in F.T.A. **5**: 160 (1899). —Obermeijer in Ann. Transv. Mus. **15**: 146 (1933). —Meyer in Mitt. Bot. Staatssamml. München **2**: 382 (1957); in Merxmüller, Prodr. Fl. SW Afr. **130**: 16 (1968). —A. & R. Heath, Field Guide Pl. N. Botswana: 22 (2009). Type: South Africa, 'Latin' Goldfields ('Tati Goldfields' in F.T.A.), fl. in Natal Bot. Gard. 1869–1870, *McKen* s.n. (K 000951846 lectotype, K 000951847), lectotypified here. FIGURE 8.6.**43**.

Barleria jucunda Benoist in Bol. Soc. Brot., sér.2 **24**: 18 (1950), illegitimate name, non *B. jucunda* Lindau. Type: Angola, Mossamedes, Hoque, fl. 15.iv.1938, *Abreu* 70 (COI holotype, BM, P).

Barleria jucunda var. *pilosula* Benoist in Bol. Soc. Brot., sér.2 **24**: 18 (1950), illegitimate name. Type: Angola, Benguela, Lengue, fl. 17.xii.1932, *Gossweiler* 9717 (COI holotype, BM, K).

Barleria cunenensis Benoist in Bol. Soc. Brot., sér.2 **24**: 19 (1950). Type: Angola, Huila, Ruacaná, R. Cunene, fl. 7.vi.1937, *Gossweiler* 10883 (COI holotype, BM).

Barleria gibsonii sensu Shendage & Yadav in Rheedea **20**: 119 (2010) in part, non Dalzell (see note).

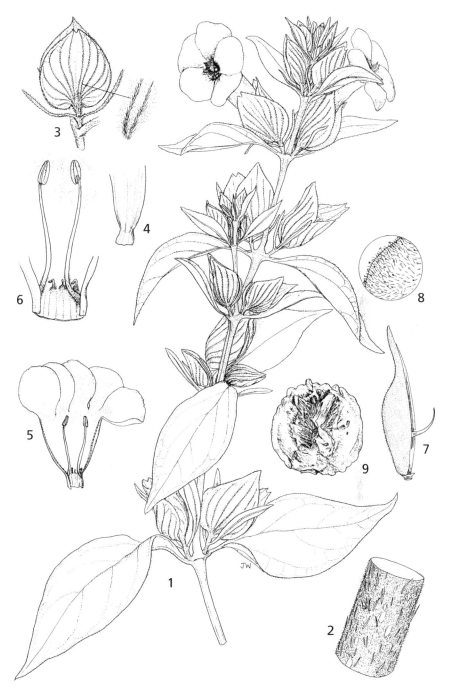

Fig. 8.6.**43**. BARLERIA MACKENII. 1, habit (× ²/₃); 2, detail of lower stem indumentum (× 8); 3, calyx and bracteoles, anterior view (× 1); 4, side view of corolla tube (× ²/₃); 5, dissected corolla with androecium (× ²/₃); 6, detail of androecium (× 1½); 7, capsule valve (× 2); 8, detail of capsule indumentum; 9, seed (× 4). 1, 4, 5 & 6 from *Brummitt et al.* 14245, 2 & 3 from *Fanshawe* 6417, 7–9 from *Smith* 818. Drawn by Juliet Williamson.

Subshrub or suffrutex 20–120 cm long, branches erect, straggling or trailing; stems with upper internodes strigulose, lower portion densely (rarely sparsely) and minutely white-velutinous. Leaf buds white-strigulose and often white-velutinous. Leaves ovate to narrowly so or elliptic, 3–12 × 1.2–4.2 cm, base attenuate, apex attenuate or acute, rarely obtuse, apiculate, margin and veins beneath strigulose, upper surface often sparsely strigose, later glabrescent; lateral veins 4–6 pairs; petiole 0–12 mm long. Inflorescence axillary in upper portion of branches, 1–3-flowered; peduncle 1–8 mm long; bracts foliaceous; bracteoles ± ascending, those at distal axils linear to narrowly oblanceolate, 7–20 × 0.5–2.5 mm, those at proximal fertile axils often larger, narrowly elliptic or oblanceolate, 20–37 × 2.5–9.5 mm, margin and midrib strigose; pedicels 1.5–5 mm long. Calyx accrescent; anterior lobe broadly ovate, 13.5–23 × 11–21 mm in flower, 20–32 × 18–25 mm in fruit, base (sub)cordate, apex notched to 4 mm or acute to obtuse, external surface strigose on main veins and margin, internal surface puberulous with mixed eglandular and glandular hairs, venation palmate, prominent; posterior lobe as anterior, 15.5–26 mm long in flower, 23–35 mm long in fruit, apex acute or obtuse, margin often somewhat involute; lateral lobes linear-lanceolate, 7.5–11.5 mm long in flower, 11–14.5 mm in fruit, eglandular- and glandular-pubescent. Corolla 31.5–51.5 mm long, blue, mauve or violet with a darker throat, rarely white throughout, upper tube and limb pubescent externally; tube 14.5–24 mm long, basal 3.5–6 mm cylindrical, 4–6 mm wide, then abruptly funnel-shaped to 8–12.5 mm wide in throat; limb subregular or in weak 4+1 arrangement, all lobes usually strongly imbricate even at anthesis; abaxial lobe 13.5–21 × 12–18 mm, offset by 1.5–4.5 mm; lateral lobes as abaxial but 12.5–21.5 mm wide; adaxial lobes 9–15 mm wide. Stamens attached 4–6 mm from base of corolla; filaments 12.5–22.5 mm long; anthers 3–4.2 mm long; lateral staminodes 0.5–2.2 mm long. Ovary densely eglandular- and glandular-puberulous; style glabrous; stigma 2–3 mm long. Capsule 18.5–21.5 mm long, eglandular- and glandular-puberulous. Seeds 5.5–7.5 × 5–7 mm.

Caprivi. Linyanti area, fl. 26.xii.1958, *Killick & Leistner* 3128 (K, PRE). **Botswana**. N: Okavango swamps, Matsaudi on road to Shorobe (Chorobe), fl. 28.ii.1967, *Lambrecht* 77 (K, SRGH). SW: Ghanzi Dist., road to Xabo, fl.& fr. 28.iii.2006, *Farrington et al.* in MSB 323 (K, MAH). SE: Mahalapye Dist., Expt. Station, fl.& fr. 16.ii.1961, *Yalala* 124 (K, SRGH). **Zambia**. B: Sesheke Dist., Masese Forest Station, fl.& fr. 2.ii.1975, *Brummitt, Chisumpa & Polhill* 14245 (K). C: Lusaka Dist., 30 km W of Lusaka, Mzimbili Ranch, fl.& fr. 10.iv.1998, *Bingham* 11647 (K). S: Mazabuka Dist., Nega-Nega, fl. 27.ii.1963, *van Rensburg* 1476 (K). **Zimbabwe**. W: Hwange Dist., Hwange Nat. Park, c.11 km beyond Mandavu Dam towards Shumba Camp, 27 km SW of Sinamatella Camp, fl. 25.ii.1967, *Rushworth* 233 (BR, SRGH). C: Gweru Dist., Gweru (Gwelo)–Shurugwe (Selukwe) road, side of Gweru kopje, fl.& fr. ii.1957, *Evans* 5 (K, SRGH). S: Gwanda Dist., between Tuli Breeding Station and Shashi R., fl. 6.i.1961, *Wild* 5309 (K, SRGH).

Also in Angola, Namibia and South Africa (Limpopo, Mpumalanga). Various types of dry woodland (mopane, *Acacia, Commiphora, Combretum*), often on margins, dambo margins, secondary grassland and open bushland on rocky hillslopes; 800–1500 m.

Conservation notes: A widespread species from a range of habitats, common in N Botswana; Least Concern.

Most material from W Angola (e.g. *Bamps et al.* 4454, *Gossweiler* 9717) differs from typical *Barleria mackenii* in having a glabrous or sparsely glandular-hairy ovary and capsule, the latter 15.5–19 mm long; *Pearson* 2519 from S Angola appears intermediate. In my opinion, the W Angolan plants should be considered a distinct subspecies; *B. jucunda* Benoist (an illegitimate later homonym) and its variety fall within this form. The type of *B. cunenensis* has more hairy outer calyx lobes but is otherwise a good match for typical *B. mackenii* (including a hairy ovary) and is synonymised here.

Shendage & Yadav (2010) synonymised this species within the Indian *B. gibsonii*, but that species belongs to sect. *Cavirostrata* and has 4 subglabrous seeds; the resemblance to *B. mackenii* is superficial.

A specimen at SRGH labelled as *Richards* 15077a from Sunzu Mt, N Zambia, is significantly outside the geographic and ecological range and thought to be mislabeled. *Richards* 15077 is a specimen of *B. sunzuana*.

48. **Barleria lancifolia** T. Anderson in J. Linn. Soc., Bot. **7**: 28 (1863). —Moore in J. Bot. **40**: 407 (1902). —Obermeijer in Ann. Transv. Mus. **15**: 147 (1933), excl. *B. alata* S. Moore. —Meyer in Mitt. Bot. Staatssamml. München **2**: 381 (1957); in Merxmüller, Prodr. Fl. SW Afr. **130**: 15 (1968). —Compton, Fl. Swaziland: 553 (1976). —Fabian & Germishuizen, Wild Fl. Nthn. S. Afr.: 390, pl.187 (1997). Type: Namibia, Damaraland, *Herb. coll. Trin. Dublin* s.n. (TCD holotype, missing).

Barleria hereroensis Engl. in Bot. Jahrb. Syst. **10**: 261 (1889). —Clarke in F.T.A. **5**: 147 (1899). Type: Namibia, Hereroland, near Daviep, fl.& fr. vi.1886, *Marloth* 1460 (B holotype, K photo, PRE).

Barleria latiloba Engl. in Bot. Jahrb. Syst. **10**: 261 (1889). Type: Namibia, Hereroland, Otjimbingue, fl.& fr. v.1886, *Marloth* 1316 (B† holotype, K photo, GRA).

Barleria cinereicaulis C.B. Clarke in Fl. Cap. **5**: 53 (1901). Types: South Africa, KwaZulu-Natal, no locality, fl. n.d., *Gerrard* 1266 (K syntype); Mpumalanga, Avoca, near Barberton, hillside above Sheba water-race, fl. 3.iii.1894, *Galpin* 1331 (PRE syntype).

Barleria gossweileri S. Moore in J. Bot. **49**: 305 (1911). —Makholela in Figueiredo & Smith, Pl. Angola, Strelitzia **22**: 21 (2008). Type: Angola, Benguella, fl.& fr., viii.1910, *Gossweiler* 4948 (BM holotype, K).

Barleria rautanenii Schinz in Vierteljahrsschr. Naturf. Ges. Zürich **61**: 437 (1916). Type: Namibia, Hereroland, Outjo, fl. vii.1898, *Rautanen* 779 (Z holotype).

Barleria exellii Benoist in Bol. Soc. Brot., sér.2 **24**: 19 (1950). —Makholela in Figueiredo & Smith, Pl. Angola, Strelitzia **22**: 21 (2008). Types: Angola, Mossamedes, Montemor, 27 km from Mossamedes, fl. 20.v.1937, *Exell & Mendonça* 2127 (BM, COI syntypes, P).

Perennial herb or subshrub 30–70 cm tall, woody at base; lower stems densely and minutely white-velutinous, uppermost internodes with ± numerous spreading glandular hairs, sparsely to densely puberulent, sometimes with scattered appressed-strigulose hairs. Leaf buds densely white-strigulose and/or white-velutinous. Leaves lanceolate, narrowly ovate-elliptic or ovate, 2.8–10.5 × 1–3 cm, base attenuate or cuneate, apex acute or rarely obtuse, apiculate, surfaces soon glabrescent except for margin and midrib beneath sparsely strigulose; lateral veins 4–5 pairs; petiole to 15 mm. Inflorescence axillary in upper portion of branches, 1–3-flowered; peduncle 0–8(25) mm long; bracts foliaceous, often much-reduced at upper nodes where typically linear or oblanceolate, 11–20 × 2–5 mm, often with scattered glandular hairs; bracteoles linear, oblanceolate or narrowly elliptic-lanceolate, 6–20.5 × 0.5–3.5 mm, spreading glandular-pubescent, veins strigulose; pedicels 0–4.5 mm long. Calyx somewhat accrescent; outer lobes equal, elliptic to ovate, 7–15.5 × 4.5–9.5 mm in flower, 12.5–20 × 9.5–13.5 mm in fruit, base rounded to acute, apex rounded to subattenuate, that of anterior lobe often minutely notched, external surface ± densely spreading glandular-pubescent, main veins also strigulose, internal surface puberulous, venation palmate or subparallel, prominent; lateral lobes lanceolate, 5–9.5 mm long in flower, 9–13 mm in fruit. Corolla 26.5–41.5 mm long, blue, mauve or lilac with paler tube, whole flower sometimes whitish at anthesis, lateral lobes eglandular- and glandular-pubescent externally, elsewhere glabrous; tube funnel-shaped or narrowly so, 13–18.5 mm long, basal 3.5–7.5 mm being 3.5–4.5 mm wide, rapidly to gradually widening to 5.5–10 mm at throat; limb in 4+1 arrangement; abaxial lobe 12–18.5 × 7.5–15 mm, offset by 3.5–7 mm; lateral lobes 10–16 × 7.5–14 mm; adaxial lobes as laterals but 6–10 mm wide. Stamens attached 4–5 mm from corolla base; filaments 16.5–25 mm long; anthers 3–4.5 mm; lateral staminodes 0.7–1.3 mm. Ovary densely puberulous in distal half; style glabrous; stigma 1.5–2.5 mm long. Capsule 14–17.5 mm long, puberulous mainly on beak. Seeds c.6.5 × 5 mm.

Botswana. N: Ngamiland Dist., Gcwihaba Hills, Cavern Hill, fl. 16.iii.1987, *Long & Rae* 303 (K). SW: Ghanzi Dist., Farm 48, fl.& fr. 28.iii.1969, *de Hoogh* 195 (K, SRGH). SE: Central Dist., Orapa, pan to old airport road, fl. 26.i.1975, *Allen* 247 (K, PRE). **Zimbabwe**. E: Chipinge Dist., Sabi–Tanganda Halt, Save Valley, fl.& fr. 28.iii.1956, *Whellan* 1024 (K, SRGH). S: Beitbridge Dist., 8 km S of Bubye R. on Masvingo (Fort Victoria)–Beitbridge road, fl. 17.iii.1967, *Mavi* 223 (K, SRGH). **Mozambique**. GI: Mabalane Dist., near Combomune, margin of R. Limpopo, fl. 12.v.1948, *Torre* 7778 (BR, LISC).

Also in Angola, Namibia, Swaziland and South Africa (Limpopo, North West,

Gauteng, Mpumalanga, KwaZulu-Natal, Northern Cape). Various types of dry woodland (*Acacia, mopane*), thicket and open wooded grassland, scrub on open rocky hillslopes, sandy or clay soils; 100–1200 m.

Conservation notes: Widespread and often common; Least Concern.

A rather variable species; only the variation seen within the Flora region is described above. Our material is referable to subsp. *lancifolia*; the highly disjunct subsp. *charlesii* (Benoist) J.-P.Lebrun & Monod (including *Barleria bonifacei* Benoist) is restricted to Mauritania in NW Africa and differs in lacking glandular hairs on the calyx and having a glabrous ovary and capsule – it should probably be treated as a distinct species. However, specimens largely lacking glandular calyx hairs also occur in S Namibia (e.g. *Pearson* 7993), where the fruits are hairy.

Barleria gossweileri S. Moore (including *B. exellii* Benoist) from coastal SW Angola is synonymised here. Whilst clearly falling within *B. lancifolia* sensu lato it differs from other southern African material in many of the appressed hairs on the stems, leaves and calyx being biramous; it could perhaps be considered a subspecies.

Although usually easily separated by the key characters, there is no doubt that *B. mackenii* and *B. lancifolia* are closely allied – they share similar habitats and an overlapping range, although *B. lancifolia* is generally more southerly and there are no records of them growing together. In addition to the key characters, *B. lancifolia* has a less hairy corolla with the upper portion of the tube glabrous and a more clearly offset abaxial corolla lobe. A few intermediate specimens are seen across the area of range-overlap including *West* 3874 (K) from N Botswana.

49. **Barleria matopensis** S. Moore in J. Bot. **45**: 91 (1907). —Obermeijer in Ann. Transv. Mus. **15**: 149 (1933). —White, Forest Fl. N. Rhod.: 382 (1962). —Pickering & Roe, Wild Fl. Victoria Falls: 13 (2009). Type: Zimbabwe, Matopo Hills, fl.& fr. ii.1903, *Eyles* 1165 (BM holotype, GRA).

> *Barleria albi-pilosa* Suess. & Merxm. in Mitt. Bot. Staatssamml. München **2**: 67 (1955). —Meyer in Mitt. Bot. Staatssamml. München **2**: 380 (1957); in Merxmüller, Prodr. Fl. SW Afr. **130**: 10 (1968). —M. & K. Balkwill in Kew Bull. **52**: figs.6s, 12j (1997). —Welman in Germishuizen & Meyer, Pl. Sthn. Afr., Strelitzia **14**: 93 (2003). —Setshogo, Prelim. Checklist Pl. Botswana: 17 (2005). Type: Namibia, Etosha Pan, Onguma, "Gelbholzdüne", fl. 8.xii.1952, *Walter* 412 (M holotype, B, PRE).

Shrub or woody perennial herb 10–300 cm tall, erect or scrambling; stems with fine white antrorse-appressed or tomentellous hairs, often dense when young, sometimes with numerous short spreading glandular hairs, later glabrescent. Leaf buds densely cream- to silvery-hairy. Leaves ovate, ovate-elliptic or lanceolate, 2–15 × 0.8–5.6 cm, base attenuate, often with a wing-like extension along petiole, apex acute to rounded, surfaces with numerous fine white antrorse or crisped hairs at first, soon (rarely tardily) glabrescent except along margin and midrib; lateral veins 4–6 pairs; petiole 0–26 mm. Inflorescence axillary but sometimes crowded towards branch apex, 1–3-flowered, sessile or peduncle to 2.5 mm long; bracts foliaceous, reducing upwards; bracteoles oblanceolate or obovate(-elliptic), 8–28 × 2–9.5 mm, indumentum as leaves, often with short glandular hairs along margin or more widespread above and/or with long-pilose hairs along margin of proximal half; flowers sessile. Calyx somewhat accrescent; anterior lobe (oblong-)elliptic, 6–13.5 × 2.5–9 mm, base acute or attenuate, apex with two acute lobes (1)2–6 mm long, surface with fine white antrorse or tomentellous hairs throughout or restricted to margin and two principal veins, often with short glandular hairs along margin, venation palmate, prominent in fruit; posterior lobe oblong-elliptic or oblong-lanceolate, 8.5–18.5 × 3–12 mm, apex acute to rounded; lateral lobes lanceolate, 4–7.5(9) mm long. Corolla 27–43 mm long, blue to mauve with paler throat or whitish throughout, pubescent mainly on lateral lobes externally with mixed glandular and eglandular hairs, rarely only the latter; tube 9.5–17 mm long, cylindrical towards base, campanulate towards mouth; limb in 4+1 arrangement; abaxial lobe (10)14–23 × 8–16 mm, offset by (2)4.5–8(10.5) mm; lateral lobes (10)12.5–21 ×

7–17 mm; adaxial lobes as laterals but 5.5–12.5 mm wide. Stamens attached 2.5–6 mm from base of corolla; filaments 19.5–28.5 mm; anthers 3.5–5.5 mm; lateral staminodes 0.5–2 mm. Ovary glabrous; stigma 1.8–4 mm long. Capsule 14–17 mm long, glabrous. Seeds 6–7.5 × 4.5–6.5 mm.

Form **A**

Foliage ± soon-glabrescent, only young leaf buds (rarely also bracts, bracteoles and calyx) densely white-hairy. Stems lacking glandular hairs. Leaves up to 15 cm long, length to width ratio 2.2–4.5:1, leaf apex usually acute or bluntly so. Anterior calyx lobe 6–12 × 2.5–5 mm, length to width ratio 2–2.7:1; posterior lobe 8.5–17.5 × 3–5.5 mm, length to width ratio 2.6–5.8:1.

Caprivi. Impalila Is., fl.& fr. 16.iii.1976, *du Preez* 9 (PRE, WIND). **Zambia**. S: Choma Dist., Mapanza, fl.& fr. 2.iii.1958, *Robinson* 2775 (BR, K, SRGH). **Zimbabwe**. N: Gokwe Dist., Marikiriki R. system near Gokwe R. fly gate, fl. 14.ii.1964, *Bingham* 1125 (K, LMU, SRGH). W: Matobo Dist., Matopos Nat. Park, fl. 24.ii.1981, *Philcox & Leppard* 8809 (K, SRGH). C: Chegetu Dist., 4.8 km W of Chegetu (Hartley)–Norton slip road, above banks of Serui R., fl. 24.i.1969, *Biegel* 2852 (K, LISC, SRGH). E: Chipinge Dist., near Rupisi, fl. 11.ii.1960, *Goodier* 879 (K, SRGH). **Mozambique**. MS: Gondola Dist., Mazare, Madanda, fl. 9.ii.1907, *Johnson* 109 (K).

Also in N Namibia and likely to occur in N Botswana. Commonly associated with dense riverine woodland and thicket or dry riverine forest, but also in drier woodland or among crevices on rocky hillslopes; 750–1550 m.

Conservation notes: Widespread and locally common, a dominant understorey shrub in the Victoria Falls area; Least Concern.

Form **B**

Foliage more tardily glabrescent, upper cauline leaves, bracts, bracteoles and calyx ± densely white- or silver-hairy, whole fertile portion of plants with a pale grey appearance. Young stems often with numerous short spreading glandular hairs. Leaves up to 6.5 cm long, length to width ratio 1.5–2(2.6):1; leaf apex often rounded or obtuse. Anterior calyx lobe 9.5–13.5 × 5.5–9 mm, length to width ratio 1.4–2:1; posterior lobe 15–18.5 × 6–12 mm, length to width ratio 1.4–2.4:1.

Zimbabwe. E: Chipinge Dist., Save–Runde R. junction, Chivirira (Chiribira) Falls, fl.& fr. 8.vi.1950, *Wild* 3450 (K, LISC, SRGH). S: Gwanda Dist., Marangudzi, fl.& fr. 10.v.1958, *Drummond* 5752 (K, LISC, SRGH). **Mozambique**. GI: "Merinqua Dist., Sabi R.", st. 25.vi.1950, *Chase* 2591 (BM, K, SRGH); Massingir Dist., near Mavodze village, c.30 km N of Massingir, Limpopo Nat. Park, fl.& fr. 13.v.2004, *Jacobsen* 6193 (PRE).

Also in NE South Africa (Limpopo). On dry rocky hillslopes and amongst large boulders, in open bushland or beneath *Androstachys* and *Brachystegia tamarindoides* woodland; 150–750 m.

Conservation notes: Moderately widespread across the broader Limpopo valley; Least Concern.

This is a distinctive lowland form of *Barleria matopensis* and should probably be treated as a subspecies. However, some material from N Namibia (e.g. *Bodenstein* 615, *Germishuizen* 7630) shares the large calyx of this form but has foliage closer to typical *B. matopensis*. The smaller leaves and often more persistent hairiness of this form are probably ecological adaptations to growing in drier, more open habitats.

50. **Barleria hirta** Oberm. in Ann. Transv. Mus. **15**: 148 (1933). —Plowes & Drummond, Wild. Fl. Zimbabwe, rev. ed.: pl.166 (1990). —Bandeira *et al.*, Fl. Nat. Sul Moçamb.: 160, 190 (2007). —Darbyshire in F.T.E.A., Acanthaceae **2**: 397 (2010). Type: Zimbabwe, Umtali, Odzani R. Valley, fl. 1914, *Teague* 8 (BOL holotype, GRA, K).

Perennial herb or subshrub 20–70 cm tall, several erect or ascending stems from a woody rootstock; stems wiry, white-pilose or -puberulous, usually with interspersed coarser subappressed anvil-shaped hairs 0.7–1.5 mm long, upper stems sometimes also glandular-pubescent. Axillary buds densely white-pilose. Leaves (linear-)lanceolate, 3.5–12 × 0.5–2.8 cm, base obtuse to shortly attenuate, apex acute, apiculate, lower surface with subappressed often anvil-shaped hairs, particularly on principal veins and margin, hairs more scattered or absent above; lateral veins 3–5 pairs; petiole 0–7 mm. Inflorescence single-flowered, axillary but often crowded towards stem apex; bracts foliaceous, much-reduced upwards, with scattered glandular hairs; bracteoles linear or oblanceolate, 6–21 × 0.5–2.5 mm, hairs as leaves with interspersed spreading glandular hairs; flowers sessile. Anterior calyx lobe oblong-ovate to narrowly lanceolate, 8.5–18 × 2.5–5 mm, apex entire or notched for 0.5–2.5 mm, indumentum as bracteoles, dense, venation parallel; posterior lobe 9.5–18.5 × 2–4.5 mm, apex acute or attenuate; lateral lobes lanceolate, 7.5–11 mm long, widened hyaline margin towards base. Corolla 25–45 mm long, pale to rich blue-purple with paler throat, rarely white throughout, lateral lobes pubescent externally, hairs mainly glandular; tube cylindrical, 11.5–21.5 mm long, somewhat expanded towards mouth; limb in 4+1 arrangement; abaxial lobe offset by 2.5–7 mm; all lobes 12–23 mm long, abaxial lobe 7.5–12.5 mm wide, lateral pair 6–11.5 mm, adaxial pair 4.5–8.5 mm wide, shortly fused at base for 1–3.5 mm. Stamens attached 3–6 mm from corolla base; filaments 16–25 mm long; anthers 2.3–3.7 mm long; lateral staminodes 0.5–1.3 mm long. Ovary glabrous or with few apical hairs; style glabrous; stigma 1.2–2.7 mm long. Only immature capsule seen, c.16 mm long, glabrous or with short hairs at apex. Only immature seeds seen.

Zimbabwe. W: Insiza Dist., Fort Rixon, fl. 8.ii.1974, *Mavi* 1506 (SRGH). C: Makoni Dist., Rusape, Lesapi Peninsula, fl. 8.iii.1964, *Masterson* 361 (K, SRGH). E: Mutare Dist., Mutare, Mt Chipondesmeve (Chipondinmwe), fl. 11.iii.1951, *Chase* 3659 (BM, BR, K, LISC, SRGH). S: Mberengwa Dist., Belingwe communal land (Reserve), fl. v.1958, *Davies* 2470 (K, SRGH). **Malawi**. S: Blantyre Dist., Michiru Mt, near CDC Estate, fl. 6.iii.1979, *Blackmore & Brummitt* 614 (BM, K, MAL). **Mozambique**. MS: Manica Dist., Vandúzi, fl. 23.iii.1948, *Garcia* 701 (BR, LISC, LMU, WAG). M: Namaacha Dist., Goba, near Fonte dos Libombos, fl. 13.xii.1961, *Lemos & Balsinhas* 302 (BM, COI, K, LISC, LMA, PRE, SRGH).

Also in Tanzania. Miombo woodland or more open scrub and grassland, bushland, usually amongst boulders on rocky hillslopes, roadsides; 900–1600 m.

Conservation notes: Locally common in E Zimbabwe but scarce and scattered elsewhere; assessed as Least Concern in F.T.E.A.

Significant regional variation was noted in F.T.E.A. for this species. A thorough examination of plants from Zimbabwe and Malawi reveals that, whilst the former are generally larger-flowered and with shorter styles, the overlap is considerable. A stronger candidate for subspecific separation is the variant from the Lebombo Mts of S Mozambique, separable from other material by the large corolla lobes (21–26 vs. 12–19.5 mm long) and clearly longer (not subequal) tube with the abaxial lobe more clearly offset; it also differs in having a more clearly diplotrichous stem indumentum (declinate-puberulous with long appressed or ascending anvil-shaped hairs) and in having very narrow linear-lanceolate and subsessile leaves. However, only two specimens have been seen and I would prefer to see more material before formally recognising it.

The related *Barleria meyeriana* Nees from E South Africa and Swaziland is also likely to occur in SE Mozambique, but has not yet been collected (see Balkwill, S. Afr. J. Bot. **58**: 290, 1992 for distribution map). It shares a glabrous ovary and similar corolla with *B. hirta* but differs in being less hairy throughout, having proportionately broader, ovate or lanceolate leaves and sparse to numerous conspicuously bulbous-based hairs along the bract, bracteole and/or calyx margins.

51. **Barleria pretoriensis** C.B. Clarke in Fl. Cap. **5**: 54 (1901). —Obermeijer in Ann. Transv. Mus. **15**: 149 (1933). —Killick in Flow. Pl. Afr. **41**: t.1623 (1971). —Fabian & Germishuizen, Wild Fl. Nthn. S. Afr.: 390, pl.187 (1997). Type: South Africa, Gauteng, Aapies Poort, Pretoria, fl. 1875–1880, *Rehmann* 4103 (K holotype, Z).

Much-branched subshrub or perennial herb 30–80 cm tall, erect, straggling or decumbent; stems with many white stiffly appressed medifixed hairs 0.2–0.5 mm long; upper internodes often with short spreading glandular hairs. Leaves lanceolate or narrowly oblong, 2–5 × 0.3–1 cm, base rounded to cuneate, apex acute, apiculate, surfaces with sparse biramous (medifixed and/or anvil-shaped) hairs 0.4–0.8 mm long, often restricted to margin and midrib beneath; lateral veins 3–4 pairs; petiole 1–5 mm. Inflorescences single-flowered in upper axils and subterminal, often one flower per node, widely spaced, sessile or peduncle to 4 mm long; bracts foliaceous, reduced upwards, typically 10–18 × 1.5–3.5 mm at uppermost nodes; bracteoles linear-lanceolate, narrowly elliptic or narrowly spathulate, (3)5–12(24) × 0.4–2(6) mm, hairs as on leaves, usually with short spreading glandular hairs; pedicels 0–3 mm long. Outer calyx lobes subequal, narrowly lanceolate, anterior lobe 8–17.5 × 2.5–4 mm, apex notched for 0.7(2) mm, surface with dense white appressed, simple and biramous (often anvil-shaped) hairs and sparse to numerous spreading glandular hairs, prominently 2-veined; posterior lobe with apex acute, surface with 1 or 3 prominent veins; lateral lobes linear-lanceolate, 7.5–13.5 mm long. Corolla 38–60 mm long, white at first, turning pale mauve, pink or violet; pubescent mainly on lateral lobes externally, hairs mainly glandular; tube cylindrical, 13.5–25 mm long, barely expanded towards mouth; limb in 4+1 arrangement; abaxial lobe 19.5–27.5 × 9–12.5 mm, offset by 3.5–7.5 mm; lateral lobes 19–26 × 7–13 mm; adaxial lobes 20–30 × 5–9.5 mm. Stamens attached 7–9 mm from corolla base; filaments 21–33 mm long; anthers 3.2–4.5(5) mm long; lateral staminodes 0.8–2.5 mm long. Ovary with dense appressed buff-coloured hairs throughout; style glabrous; stigma 2–3 mm long. Capsule 13–18 mm long, appressed-hairy including some anvil-shaped hairs. Seeds c.5 × 4 mm.

Botswana. SE: Kgatleng Dist., Mochudi, fl. i.1955, *Reyneke* 217 (PRE); 5 km S of Gaborone, fl. 24.i.1984, *Plowes* 7077 (PRE).

Also in northern South Africa (Limpopo, Mpumalanga, Gauteng, North West). Open rocky hillslopes, rock crevices or in sparse bushland and wooded grassland; 1000–1350 m.

Conservation notes: Local but apparently common in N South Africa; Least Concern.

Corolla size and shape is somewhat variable in this species. Most specimens seen have narrowly oblong-elliptic lobes, but in *Balkwill & Sebola* 7996 from the Waterberg Range, South Africa, the lobes are broadly elliptic.

52. **Barleria laceratiflora** Lindau in Bot. Jahrb. Syst. **38**: 68 (1905). —Darbyshire in F.T.E.A., Acanthaceae **2**: 406 (2010). Type: Tanzania, Lindi Dist., Ras Rungi, fl. 4.v.1903, *Busse* 2367 (B† holotype, EA).

Dwarf suffrutex 15–30 cm tall; stems with few stiff appressed medifixed and simple hairs at and below nodes and scattered short spreading glandular hairs when young, elsewhere glabrous. Leaves elliptic, 4.3–7 × 1.4–2 cm, base cuneate-attenuate, apex acute, apiculate; midrib, lateral veins beneath and margin with few stiff medifixed hairs, elsewhere glabrous; lateral veins 4 pairs; petiole to 6 mm. Inflorescence axillary, 1–3-flowered, subsessile or peduncle to 2.5 mm long; bracts foliaceous, often reduced and proportionally narrower than leaves; bracteoles oblanceolate, 6–16.5 × 0.5–3.5 mm; flowers subsessile. Anterior calyx lobe elliptic, 8.5–10 × 4–5.5 mm, apex bifid for 2.5–5 mm, surface with appressed medifixed hairs mainly on margin and veins, usually a few short glandular hairs towards base and along margin, two principal veins prominent; posterior lobe 12–14.5 × 5.5–6.5 mm, apex rounded or obtuse, apiculate, 3(5)-veined from base; lateral lobes lanceolate, 5–6 mm long. Corolla 19.5–24 mm long, pale blue; glandular-pubescent externally on limb; tube cylindrical, 6.5–7 mm long; limb in 4+1 arrangement; abaxial lobe 10–13 × 5.5 mm, offset by 6–7 mm; lateral lobes 8–10 × 4–4.5 mm; adaxial lobes 5.5–8 × 2.5–3.5 mm. Stamens attached 2.5–3.5 mm from corolla base; filaments 12.5–14 mm long; anthers 3.2–3.6 mm; lateral staminodes 0.5–0.8 mm. Ovary puberulous; style glabrous; stigma 1–1.6 mm long. Capsule and seeds not seen.

Mozambique. N: Ilha de Moçambique Dist., Goa Is., fl. 5.v.1947, *Gomes e Sousa* 3503 (K).

Also in SE Tanzania. Coastal bushland on sandy and coral-derived soils; sea level. Conservation notes: Known from only two localities; recorded as common on Goa Is. in 1947 but the site is now threatened with habitat degradation; probably Vulnerable, although assessed as Data Deficient in F.T.E.A.

This species falls within a distinctive subgroup in sect. *Somalia* with the combination of medifixed hairs on the vegetative parts, a strongly offset abaxial corolla lobe (4+1 arrangement) and pink to purple guidelines in the corolla throat. The group is otherwise largely restricted to the Somalia–Masai region of NE Africa and Arabia (*Barleria argentea* Balf. f. and allies).

53. **Barleria lugardii** C.B. Clarke in F.T.A. **5:** 161 (1899). —Meyer in Mitt. Bot. Staatssamml. München **2:** 382 (1957); in Merxmüller, Prodr. Fl. SW Afr. **130:** 15 (1968). —Welman in Germishuizen *et al.*, Checklist S. Afr. Pl.: 79 (2006) in part. —Darbyshire in F.T.E.A., Acanthaceae 2: 402 (2010). Type: Botswana, Khwebe, fl.& fr. i.1897, *Lugard* 128 (K holotype).

    Barleria breyeri Oberm. in Ann. Transv. Mus. **15:** 151 (1933). Type: Namibia, Klein Namutoni, n.d., *Breyer* in TVL 20642 (PRE holotype).

Perennial herb or subshrub 10–60 cm tall; stems strigose with white anvil-shaped hairs, young stems often spreading glandular-pubescent; mature stems with pale greyish bark, glabrescent. Leaf buds densely white-strigose. Leaves ovate-elliptic, 2.5–6 × 1–2.5 cm, base attenuate, apex acute or obtuse, margin and midrib beneath sparsely strigose, sometimes with short glandular hairs towards base; lateral veins 3–5 pairs; petiole 0–5 mm. Inflorescence axillary but in densely-flowered plants forming a terminal thyrse, each 1–3-flowered; bracts foliaceous, often narrowing upwards; bracteoles linear(-oblanceolate) to narrowly elliptic, 6–22 × 0.5–3.5 mm, indumentum as leaves but with glandular hairs often more widespread; flowers (sub)sessile. Calyx somewhat accrescent; anterior lobe ovate, 5.5–14 × 2–6.5 mm, apex attenuate, entire or bifid with linear lobes 0.5–4.5 mm long, strigose on margin and principal veins, often also with scattered glandular hairs, venation palmate to subparallel; posterior lobe 6.5–16.5 × 2.5–7.5 mm, margin often involute and partially enveloping anterior lobe, apex attenuate; lateral lobes lanceolate, 3.5–6.5 mm long. Corolla 16–21 mm long, white, pubescent on lateral lobes externally with mixed glandular and eglandular hairs; tube cylindrical, 7–10 mm long, expanded somewhat towards mouth; limb in 1+3 arrangement; abaxial lobe 9–11.5 × 5–7 mm, offset by 0.8–2 mm; lateral lobes 7.5–10.5 × 4.5–6.5 mm; adaxial lobe as laterals but 3.5–5 mm wide, apex emarginate. Stamens attached 2–3.5 mm from base of corolla; filaments 13–20 mm long; anthers 2–3 mm long; lateral staminodes 0.5–0.8 mm long. Ovary densely pubescent in distal half; style pubescent at base; stigma 0.8–1.2 mm long. Capsule 11–15.5 mm long, shortly pubescent. Seeds 5.5–7 × 4.5–6 mm.

Caprivi. Impalila Is., fl. 13.i.1959, *Killick & Leistner* 3355 (K, PRE). **Botswana. N:** Ngamiland Dist., 19 km SW of Maun towards Toteng, fl.& fr. 18.ii.1967, *Lambrecht* 73 (K, SRGH). **Zimbabwe. N:** Binga Dist., Mwenda, fl.& fr. 27.xii.1964, *Jarman* 102 (SRGH). **W:** Hwange Dist., 2.4 km from Mandavu Dam towards Shumba Camp, 18.5 km SW of Sinamatella Camp, Hwange Nat. Park, fl.& fr. 25.ii.1967, *Rushworth* 219 (BR, EA, K, SRGH). **Mozambique. T:** Cahora Bassa Dist., between Chicoa and Mágoè, 17 km from crossroads, 4 km along R-hand track to Manjericão, fl.& fr. 14.ii.1970, *Torre & Correia* 17975 (LISC, LMU, UPS, WAG).

Also in Tanzania and Namibia. Various dry woodland and thicket types, often over stony or sandy soils, sometimes open rocky hillslopes; 350–1150 m.

Conservation notes: Locally common in N Botswana, W Zimbabwe, C & S Tanzania; assessed as Least Concern in F.T.E.A.

The above description of *Barleria lugardii* covers only the small-flowered 'typical' form which occurs over most of its range; a variant with larger flowers (corolla 24–29 mm long with purple guidelines) is restricted to Tanzania. Having seen both forms in the field I suggest the Tanzanian taxon be formally recognised, at least at an infraspecific rank.

54. **Barleria quadriloba** Oberm. in Ann. Transv. Mus. **15**: 150 (1933). Type: South Africa, Limpopo Prov., Messina, fl. ii.1919, *Rogers* 22548 (PRE-TRV holotype, GRA, Z).

> *Barleria lugardii* sensu auct., non C.B. Clarke. —Setshogo, Prelim. Checklist Pl. Botswana: 17 (2005) in part. —Welman in Germishuizen *et al.*, Checklist S. Afr. Pl.: 79 (2006).

Perennial herb or subshrub 30–40 cm tall; stems with anvil-shaped strigose hairs, young stems with short spreading glandular hairs; mature stems with pale greyish bark, glabrescent. Leaf buds densely pale-strigose. Leaves ovate or elliptic, 3–5 × 1.2–2 cm, base cuneate or attenuate, apex acute, soon glabrescent except sparsely strigose on margin and midrib beneath; lateral veins 3–4 pairs; petiole 0–4 mm. Inflorescences axillary towards stem apex, single-flowered; bracts foliaceous; bracteoles linear or oblanceolate, 6–15 × 1–2.5 mm, strigose and with short spreading glandular hairs; peduncle 1.5–5 mm long; pedicel 0–1.5 mm long. Calyx somewhat accrescent; anterior lobe ovate, 10–13.5 × 5–6.5 mm, apex attenuate, entire or bifid with linear lobes 0.8–3 mm long, strigose especially along margin and principal veins, with short spreading glandular hairs, venation subparallel; posterior lobe 13–20 × 7.5–9 mm, margin involute, partially enveloping anterior lobe, apex attenuate; lateral lobes lanceolate, 6–9 mm long. Corolla 32–44 mm long, white, with or without purple guidelines, glabrous externally; tube cylindrical, 15–19 mm long, expanded somewhat towards mouth; limb in 1+3 arrangement; abaxial lobe 13–17 × 8–10.5 mm, offset by 2–5 mm, lateral lobes 15–21 × 7.5–9.5 mm, adaxial lobe 14.5–19 × 5–9 mm, apex emarginate or lobed up to 3 mm. Stamens attached 6–7 mm from corolla base; filaments 20–26 mm long; anthers 3–4 mm; lateral staminodes to 0.5 mm. Ovary densely pubescent; style glabrous or with few hairs at base; stigma 1.4–2 mm long. Capsule 13.5–16 mm long, shortly pubescent. Seeds 6–8 × 4.3–5 mm.

Zimbabwe. S: Beitbridge Dist., Chikwarakwara, fl.& fr. 23.ii.1961, *Wild* 5351 (K, SRGH).

Also in South Africa (Limpopo). Dry woodland (especially mopane) and open scrub on rocky hillslopes; c.350 m.

Conservation notes: Very local and known from few collections; currently considered Data Deficient.

This species has previously been included within *Barleria lugardii*, from which it is easily separated in the Flora region (see key). In Tanzania, a larger-flowered variant of *B. lugardii* occurs (see discussion in F.T.E.A.); although similar this form is readily separable from *B. quadriloba* in having subsessile (not short-stalked) flowers that are smaller than *B. quadriloba* (24–29 mm long, including tube 9.5–12.5 mm), stamens attached closer to base of corolla tube (2.5–3 mm), smaller anthers (2–3 mm long) and a hairy corolla limb (essentially glabrous in *B. quadriloba*).

55. **Barleria phaylopsis** Milne-Redh. in Bull. Misc. Inform., Kew **1940**: 65 (1940). Type: Zambia, Solwezi Dist., Mbulungu stream, W of Mutanda bridge, fl.& fr. 9.vii.1930, *Milne-Redhead* 690 (K sheet 1 lectotype, BR, PRE), lectotypified here.

> *Barleria affinis* De Wild. in Ann. Mus. Congo Belge, Bot. sér.4 **1**: 140 (1903), illegitimate name, non *B. affinis* C.B. Clarke. Type: D.R. Congo, Katanga, Lukafu, fl. vii.1900, *Verdick* 561 (BR holotype).

Suffrutex with several decumbent or creeping stems 15–30 cm long from a woody rootstock; stems hispid, hairs (buff-)golden, ascending or spreading. Leaves subsessile, oblong-elliptic or oblanceolate, 4–9.5 × 1.2–3 cm, base cuneate, apex acute or lower leaves obtuse, with numerous (buff-)golden hispid hairs on margin and midrib only or evenly distributed on both surfaces; lateral veins 4–8 pairs, these and reticulate tertiary veins prominent. Inflorescence terminal, sometimes also in uppermost leafy axils, densely spiciform or subcapitate, 2–5.5 cm long, each axil single-flowered or lowermost axils 2–3-flowered; bracts oblanceolate or narrowly elliptic, sometimes flushed purple distally or throughout, 9.5–15.5 × 1.5–4.5 mm (basal pairs to 23 × 7.5 mm), base cuneate, apex acute, ± densely glandular-pilose, with subappressed stiff hairs at least on margin and midrib, those on margin sometimes with swollen base; bracteoles as bracts but (elliptic-)oblanceolate to linear, 9–14.5 × 1–2 mm, glandular hairs more numerous; flowers sessile. Calyx often flushed purple at least towards apex, pale at base; anterior lobe narrowly oblong or subulate, 10–15.5 × 1.5–2.5 mm, apex bifid for 2.5–8 mm, indumentum as bracteoles, two parallel principal veins prominent; posterior lobe narrowly lanceolate or subulate, 10–16.5 × 2–2.7 mm, apex acute, midrib prominent; lateral lobes linear, 9–13.5 mm long. Corolla 17–22 mm long, white, pink, lilac or bright blue, with or without purple guidelines, lateral lobes sparsely pubescent externally; tube cylindrical, 9.5–11.5 mm long; limb in 2+3 arrangement, abaxial lobe 6–10.5 × 4.5–7 mm; lateral lobes 7–10.5 × 3–6 mm; adaxial pair fused for 0.7–5 mm, free portions oblanceolate to subulate, 4.5–10 × 1.5–4 mm. Stamens attached c.3 mm from corolla base; filaments 9.5–15.5 mm long; anthers 2.5–3 mm; lateral staminodes 0.25–0.8 mm. Ovary densely pubescent at least distally, sometimes short glandular hairs at apex; style with few hairs towards base; stigma 1.2–1.9 mm long. Capsule c.14.5 mm long, puberulous. Only immature seeds seen.

Zambia. N: Mbala Dist., Mbala (Abercorn), fl. 27.vi.1961, *Lawton* 732 (K, NDO). W: Mufumbwe Dist., Mufumbwe (Chizera), fl.& fr. 11.vi.1953, *Fanshawe* 68 (BR, EA, K, NDO).

Also in southern D.R. Congo. Miombo woodland and open wooded grassland; 1100–1500 m.

Conservation notes: Rather scarce and known from only seven collections; certainly rare in the Mbala region of Zambia with only a single specimen seen; probably threatened.

Lawton 732 from Mbala differs from the W Zambian material in having leaves hairy throughout, not restricted to the margin and midrib, and in the abaxial corolla lobes being fused for ± half their length and 1.5–2 mm wide vs. fused for only 0.7–1.5 mm at base and 3–3.7 mm wide. Two subspecies may be involved but more material is required to make any firm conclusions (see note on *B. affinis* below).

The type material of *Barleria affinis*, an illegitimate later homonym, is smaller than Zambian specimens of *B. phaylopsis*, with elliptic leaves less than 4 cm long (length to width ratio c.2.2:1 vs. 2.7–5.5:1 in Zambia) and few-flowered inflorescences. In addition, the anterior calyx lobe is only notched for 0.5 mm in *B. affinis*, whilst in Zambian *B. phaylopsis* this is conspicuously bifid. The leaf indumentum of this specimen agrees with that from Mbala and the adaxial corolla lobes are similarily narrow (2–2.5 mm wide) but are less clearly fused. For the present it is considered a depauperate form of *B. phaylopsis*.

Barleria griseoviridis I. Darbysh. from W Tanzania is closely allied to *B. phaylopsis* but is readily separated by the broader, obovate bracts (5–9 mm wide at spike midpoint) and inconspicuous reticulate leaf venation.

56. **Barleria kaessneri** S. Moore in J. Bot. **47**: 295 (1909). —Darbyshire & Ndangalasi in J. E. Afr. Nat. Hist. **97**: 131 (2009). —Champluvier in Pl. Ecol. Evol. **144**: 94, fig.3d (2011). Type: D.R. Congo, Katanga, Niembe R., fl. 27.v.1908, *Kässner* 3010 (BM holotype).

Suffrutex with 1 to several erect, ascending or procumbent stems 10–45 cm long from a woody rootstock; stems hispid, hairs (buff-)golden, antrorse; young stems with numerous shorter, finer hairs on two opposite sides; uppermost internodes sometimes with short spreading glandular hairs. Leaves pale glaucous- or yellow-green when dry, oblanceolate to narrowly elliptic or uppermost leaves ovate-elliptic, 6.3–12.3 × 1.8–3.3 cm, base cuneate or uppermost leaves obtuse, apex acute or lower leaves obtuse to rounded, glabrous except for sparse (buff-)golden hispid hairs on margin, midrib and sometimes lateral veins beneath; lateral veins 4–6 pairs; petiole 0–5 mm. Inflorescence a strobilate spike 3–7 cm long, terminating main and short lateral branches, each axil single-flowered; bracts broadly obovate to orbicular, (12)17–28 × (8)10–20.5 mm, base attenuate or rounded, apex broadly obtuse or subattenuate, external surface sparsely to more densely glandular-puberulous, with subappressed stiff hairs along margin and principal veins; bracteoles oblanceolate to obovate, (8)11–19 × (2)3–9 mm; flowers sessile. Calyx pale green, often flushed mauve towards apex; anterior lobe oblong-elliptic or somewhat obovate, (9.5)11.5–16.5 × (3.5)4.5–6.5 mm, apex acute or notched to 3.5 mm, glandular-puberulous towards margin and apex, short ascending eglandular hairs sparse to numerous, two principal veins prominent; posterior lobe as anterior but sometimes more oblong-lanceolate, (11)13.5–18 × 4–7 mm, apex acute-mucronulate; lateral lobes linear-lanceolate, 9.5–15 mm long. Corolla 23.5–30 mm long, white, pale blue or mauve, sometimes with darker guidelines, lateral lobes pubescent externally; tube cylindrical, 12.5–14 mm long; limb in 2+3 arrangement; abaxial lobe 8.5–14 × 7.5–11 mm; lateral lobes 10–16 × 6.5–7.5 mm; adaxial pair fused for 2.5–4.5 mm, free portions subulate or oblanceolate, 5–12 × 1.5–3.5 mm. Stamens attached c.3.5 mm from corolla base; filaments 15.5–19 mm long; anthers 2.2–3.3 mm; staminodes c.0.8 mm. Ovary densely pubescent; style sparsely hairy towards base; stigma c.1.5 mm long. Capsule 14–16 mm long, puberulous. Seeds 6–7 × 6 mm.

Zambia. W: Kitwe, fr. 13.vii.1967, *Mutimushi* 1975 (K, NDO); Masaiti Dist., Walamba, fr. 23.v.1954, *Fanshawe* 1255 (BR, EA, K, NDO).

Also in southern D.R. Congo. Miombo and chipya woodland including degraded areas; 1100–1350 m.

Conservation notes: Rather local and known from c.10 sites in Congo and 3 in Zambia, but probably under-recorded; likely to be Least Concern.

Barleria L. sect. VI. **Prionitis** Nees in De Candolle, Prodr. **11**: 237 (1847) in part. — Obermeijer in Ann. Transv. Mus. **15**: 129 (1933). —M. & K. Balkwill in Kew Bull. **52**: 561 (1997).

Prionitis (Nees) Oerst. in Vidensk. Meddel. Dansk. Naturhist. Foren. Kjøbenhavn: 137 (1854).
Barleria subgen. *Prionitis* sensu Clarke in F.T.A. 5: 141 (1899) in part, excl. *B. hereroensis* and *B. triacantha*.

Plants armed with 2–4(8)-rayed axillary spines, or unarmed. Stems ± quadrangular when young, later ± terete; usually strigose along interpetiolar line. Leaves often yellowish-green or glaucous, lower surface with minute sunken glands throughout, differing from broad sessile glands usually present towards base beneath; leaf base decurrent into petiole; indumentum of simple hairs or glabrous. Inflorescences single-flowered or dischasial cymes, either axillary (bracts foliaceous) or compounded into terminal synflorescences (bracts modified). Calyx lobes often spine-tipped, margin ± entire; lateral lobes linear-lanceolate or lanceolate-acuminate. Corolla (in our region) yellow, orange, apricot or red; limb subregular or usually in 4+1 arrangement. Stamens attached midway along or in upper half of corolla tube. Staminodes 2–3, lateral pair with antherodes present. Stigma linear or clavate, apex entire or minutely bilobed. Capsule drying (pale) brown, laterally compressed, with pronounced solid apical beak; lateral walls remaining attached to flanks at dehiscence, septum woody throughout. Seeds 2, ovate-discoid, with dense straight buff to golden hygroscopic hairs.

A section of c.45 species, most diverse in NE and E Africa and on Madagascar. In addition to the 11 species treated here, *Barleria marginata* Oliv. is likely to occur in the miombo woodlands of northern Mozambique.

Key to sect. Prionitis

1. Corolla limb zygomorphic, in 4+1 arrangement with abaxial lobe clearly offset from other lobes, if only shortly so (2–3.5 mm in *B. setosa*) then stem largely glabrous and external surface of calyx strigulose 2
－   Corolla salver-shaped, limb subregular; stem and external surface of calyx with numerous white spreading or retrorse hairs. 11
2. Corolla limb externally pubescent or pilose. 3
－   Corolla limb glabrous. 9
3. Bracts and calyx densely spreading-pubescent with mixed glandular and eglandular hairs; inflorescence a cylindrical (rarely subcapitate) spike with imbricate ovate, elliptic or obovate bracts 4
－   Without the above combination of characters 5
4. Capsule glabrous; most bracts in inflorescence elliptic or obovate with an obtuse or rounded mucronulate apex; corolla 45–62 mm long; axillary spines always absent . . . . **57.** *rhynchocarpa*
－   Capsule puberulous; most bracts in inflorescence ovate with an acute or shortly attenuate, mucronate apex; corolla 23.5–37 mm; often with minute 4-rayed axillary spines . . . . **58.** *crossandriformis*
5. Abaxial corolla lobe offset by 2–3.5 mm; outer calyx lobes with bulbous-based hairs along margin; axillary spines (if present) with longest rays 7–13.5 mm . . . .
. . . .**64.** *setosa*
－   Abaxial corolla lobe offset by 4–10.5 mm; outer calyx lobes lacking bulbous-based marginal hairs; axillary spines (if present) with longest rays 1–7.5 mm 6
6. Bracts lacking glandular hairs; calyx lobes clearly hairy, strigose, sometimes with finer more spreading hairs, sometimes also with few short glandular hairs towards apex 7
－   Bracts with numerous short (rarely subsessile) glandular hairs in distal half; calyx lobes glabrous except for few glandular hairs and minute eglandular hairs towards apex 8
7. Capsule densely puberulous; abaxial corolla lobe subequal to others, usually longer than tube, ratio 1–1.55:1; outer calyx lobes often with minute spreading hairs as well as ± flexuose appressed ones. . . . .**60.** *delagoensis*
－   Capsule glabrous or sparsely and minutely hairy; abaxial corolla lobe shorter than others, usually shorter than tube, ratio 0.65–0.8(1.1):1; outer calyx lobes with stiff appressed hairs only . . . . **61.** *ameliae*
8. Bracts obovate or broadly spathulate, 3.5–12.5 mm wide, length to width ratio 1–2.7:1, margin of proximal half usually with ± prominent bulbous-based spreading hairs . . . . **62.** *senensis*
－   Bracts narrowly spathulate, 1.5–4 mm wide, length to width ratio 4–8:1, margin strigose, hairs lacking bulbous bases . . . . **63.** *rhodesiaca*
9. Leaves and bracts with a prominent white cartilaginous marginal rim . . **66.** *randii*
－   Leaves and bracts lacking thickened white rim 10
10. Bracts soon reducing in size and modified from leaves up stem; leaves and bracts with no or few broad sessile glands beneath towards base; abaxial corolla lobe offset by (5.5)7.5–13.5 mm; anthers less than 3 mm long . . . . **59.** *tanzaniana*
－   Leaves and bracts subequal, with numerous broad sessile glands beneath towards base; abaxial corolla lobe offset by 4–6 mm; anthers 3.5–4.7 mm long. . **65.** *coriacea*
11. Leaves broadly obovate, to 3 cm long, length to width ratio 1.2–2:1; much-branched shrub forming dense rounded clumps; cymes single-flowered . . . .
. . . .**67.** *holubii*
－   Leaves obovate-elliptic, 2.6–4.3 cm long, length to width ratio (1.75)2–2.4:1; suffrutex branching mainly towards base; cymes sometimes 2-flowered. . **68.** *sp. D*

57. **Barleria rhynchocarpa** Klotzsch in Peters, Naturw. Reise Mossamb. **6**: 204 (1861). —Darbyshire in F.T.E.A., Acanthaceae **2**: 418 (2010). Types: Mozambique, Quirimba Is., fl.& fr. 1842–1846, *Peters* s.n. (B† holotype); Mozambique, Ibo Dist., Ibo Is., fl.& fr. 5.ix.1948, *Pedro & Pedrógão* 5046 (EA neotype, LMA), neotypified by Darbyshire (2010).

Barleria sacleuxii Benoist in Notul. Syst. (Paris) **2**: 17 (1911). —Brummitt in Bot. Mag. **182**: t.773 (1979). —Vollesen in Opera Bot. **59**: 79 (1980). —M. & K. Balkwill in Kew Bull. **52**: 561 (1997). Type: Zanzibar, fl. i.1888, *Sacleux* 545 (P holotype, BR, EA & K photos).

Crossandra rhynchocarpa (Klotzsch) Cuf. in Bull. Jard. Bot. Brux. État **34** (suppl.): 955 (1964) in part for the type.

Crossandra nilotica var. acuminata sensu Clarke in F.T.A. **5**: 115 (1899) in part for *Peters* s.n., non Lindau.

Barleria ukamensis sensu da Silva et al., Prelim. Checklist Vasc. Pl. Mozamb.: 18 (2004), non Lindau.

Perennial herb or subshrub 15–120 cm tall, erect or decumbent; stems minutely pubescent on two opposite sides, more widespread when young, soon glabrescent. Axillary spines absent. Leaves ovate or elliptic, 6–11.5 × 3–5.8 cm, base attenuate, apex acute or obtuse, apiculate, glabrous on margin and veins, sparsely strigose beneath, sometimes shortly spreading-pubescent when young; petiole 5–30 mm. Inflorescence a ± cylindrical strobilate terminal spike 3–10(15) cm long, each cymule 1(2)-flowered, subsessile; bracts imbricate, (oblong-)elliptic to obovate or lower pairs ovate, 12.5–24.5 × 5–18 mm, apex obtuse or rounded, mucronulate, surface densely pubescent with mixed shorter eglandular and longer glandular hairs, margin often glandular-pilose; bracteoles (elliptic-)lanceolate, 8.5–20 × 1–3.5 mm. Anterior calyx lobe oblong-elliptic or -ovate, 12.5–20.5 × 4.5–8 mm, apex acute or subattenuate, often notched, indumentum as bracts, venation subparallel; posterior lobe more ovate, apex often attenuate; lateral lobes 10.5–15.5 mm long. Corolla 45–62 mm long, yellow, orange or apricot, pilose externally, some hairs glandular; tube 20–28 mm long; limb in 4+1 arrangement; abaxial lobe 17–23 × 10–13.5 mm, offset by 9–14 mm; lateral lobes 13.5–18 × 11.5–15 mm; adaxial lobes as laterals but 8.5–12.5 mm wide. Stamens with filaments 22–26.5 mm long, pubescent in proximal two-thirds; anthers 3.2–4.3 mm; lateral staminodes 1.5–3.5 mm, pilose, antherodes 1–1.3 mm. Ovary glabrous; stigma 0.7–1 mm long. Capsule 16–20 mm long, glabrous. Seeds 9–11 × 6.5–8.5 mm.

Mozambique. N: Ibo Dist., Ibo Is., fr. 5.ix.1948, *Pedro & Pedrógão* 5046 (EA, LMA). Also in E Tanzania. Coastal bushland and grassland, foreshores, lowland riverine forest and thicket in Tanzania; 0–20 m.

Conservation notes: Known from only a single specimen in the Flora area; coastal bushlands on the NE Mozambique islands are now largely degraded. Assessed as Near Threatened in F.T.E.A. but should be considered nationally threatened.

In the protologue Klotzsch describes the upper lobe as being twice as short as the other lobes, but it is clear he mis-interpreted the position of the abaxial lobe – this species has the typical 4+1 arrangement. Otherwise the protologue decription is a good match for the species described here.

58. **Barleria crossandriformis** C.B. Clarke in Fl. Cap. **5**: 51 (1901). —Obermeijer in Ann. Transv. Mus. **15**: 144 (1933). —Compton, Fl. Swaziland: 552 (1976). — Fabian & Germishuizen, Wild Fl. Nthn. S. Afr.: 396, pl.190 (1997). —Bandeira et al., Fl. Nat. Sul Moçamb.: 160, 191 (2007). Type: South Africa, Mpumalanga, Avoca, Barberton, fl.& fr. iv.1890, *Galpin* 887 (K holotype, BOL, GRA, SAM, Z).

Shrub or perennial herb 60–200 cm tall; stems glabrous or uppermost internodes minutely pubescent on two opposite sides. Axillary spines absent or minute, 4-rayed, stalk to 0.8 mm long, longest ray 1–4.5 mm long. Leaves ovate or elliptic, 4.3–13 × 2.3–6 cm, base attenuate, apex acute, obtuse or attenuate, apiculate, glabrous or margin and veins, sparsely strigose beneath; petiole 7–23 mm. Inflorescence a cylindrical or subcapitate strobilate terminal spike 3–12 cm long, each cymule 1-flowered, sessile; bracts ovate or upper pairs elliptic, 14–23 × 7–14 mm, apex acute or shortly attenuate, mucronate, surface densely pubescent with mixed short eglandular and glandular hairs, some longer ascending hairs especially along veins; bracteoles elliptic to lanceolate, 7–13.5 × 2–3.5(5) mm, base attenuate, apex acuminate to cuspidate. Outer calyx lobes subequal, (ovate-)elliptic, anterior lobe 10.5–13.5 × 8–10.5 mm (flattened); posterior lobe 12–14.5 mm long, both with a strongly revolute margin, apex shortly attenuate, indumentum as bracts but glandular hairs very dense, venation subparallel; lateral lobes 11.5–13.5 mm long. Corolla (23.5)27.5–37 mm long, yellow, orange or salmon-coloured, pubescent externally; tube (10.5)13–16 mm long; limb in 4+1 arrangement; abaxial lobe (8.5)10–16 × (6)8–11.5 mm, offset by 3.5–7 mm; lateral lobes as abaxial but 5–11 mm wide, adaxial lobes 4.5–10.5 mm wide. Stamens with filaments (9)14–20 mm long, pubescent in proximal two-thirds; anthers 2.4–3.8 mm long; lateral staminodes 0.7–1.5 mm, pilose, antherodes 0.9–1.3 mm. Ovary puberulous; style pubescent at base; stigma 0.8–1.3 mm long. Capsule 16.5–18 mm long, puberulous. Seeds c.8.5–9 × 7 mm.

Zimbabwe. S: Mwenezi Dist., Mwenezi (Nuanetsi) R., gorge upstream from Buffalo Bend, fl.& fr. 28.iv.1961, *Drummond & Rutherford-Smith* 7552 (BR, EA, K, LISC, PRE, SRGH). **Mozambique**. GI: Massingir Dist., rio dos Elefantes, 70 km to NW of Lagoa Nova, fl.& fr. 21.iii.1973, *Lousã & Rosa* 361 (LMA). M: Boane Dist., Matola, Boane–Changalane road, fl.& fr. 12.v.1968, *Balsinhas* 1261 (COI, LMA).

Also in NE South Africa and Swaziland. Bushland and woodland on rocky hillslopes and gullies, riverine bushland and secondary grassland; 20–650 m.

Conservation notes: Scarce in our region but locally common in N South Africa and habitat not threatened; Least Concern.

Some collections from South Africa (including the type and *Balkwill et al.* 11670) have notably small flowers with lobes often relatively narrower than in large-flowered plants. I have not seen enough flowering material from our region to gauge whether this variation occurs throughout its range.

59. **Barleria tanzaniana** (Brummitt & J.R.I. Wood) I. Darbysh. in F.T.E.A., Acanthaceae 2: 422 (2010). Type: Tanzania, 2.4 km NE of Pangani, fl. 13.vii.1953, *Drummond & Hemsley* 3320 (K holotype, B, BR, EA, LISC).

    Barleria prionitis L. subsp. *tanzaniana* Brummitt & J.R.I. Wood in Kew Bull. **38**: 439 (1983), excl. *Harris* 285.

    Barleria diacantha sensu Clarke in F.T.A. **5**: 145 (1899) in part for *Holst* 3213 & *Stuhlmann* 8530, non Nees.

    Barleria prionitis sensu auct., non L. —Vollesen in Opera Bot. **59**: 79 (1980). —da Silva *et al.*, Prelim. Checklist Vasc. Pl. Mozamb.: 18 (2004) in part.

    Barleria delagoensis sensu da Silva *et al.*, Prelim. Checklist Vasc. Pl. Mozamb.: 18 (2004), non Oberm. (based on geography).

Subshrub 40–500 cm tall, scandent or spreading; stems glabrous or upper internodes with crisped, often retrorse hairs on two opposite sides. Axillary spines ± sparse, 2(4)-rayed, stalk to 1 mm long, longest ray (3.5)7–17 mm long. Leaves elliptic(-obovate), 4.8–13.5 × 1.7–5.2 cm, base cuneate-attenuate, apex acute or attenuate, mucronate, glabrous or margin, midrib and veins sparsely strigose beneath; petiole to 12 mm. Inflorescence a loose terminal spike 2–15.5 cm long, initially contracted but soon lax, each cymule 1-flowered, sessile or peduncle to 3 mm long, pedicels to 1.5 mm; bracts ± spreading, lower pairs foliaceous, soon reducing upwards where oblanceolate, 12.5–32(42) × 2–11(14) mm, apex acute or shortly attenuate, mucronate, hairs as leaves beneath, few broad sessile glands towards base; bracteoles spinose, 4–14 × 0.4–1.8 mm. Outer calyx lobes subequal, ovate or lanceolate, 8–12.5 × 3.5–6.5 mm, apex acute or attenuate, mucronulate, anterior lobe often notched, glabrous or with few minute subsessile glands towards apex, venation parallel, often inconspicuous; lateral lobes 6.5–10.5 mm long, usually with minute subsessile glands and/or eglandular hairs towards apex. Corolla (28)33–43.5 mm long, yellow, orange or apricot, glabrous externally; tube 9–12 mm long; limb in 4+1 arrangement; abaxial lobe (13)17–22.5 × 7–12 mm, offset by (5.5)7.5–13.5 mm; lateral lobes as abaxial but (11.5)15.5–20.5 mm long; adaxial lobes as laterals but 4.5–8.5 mm wide. Stamens with filaments

21.5–29.5 mm long, minutely pubescent in proximal half; anthers 2.3–3 mm; lateral staminodes 1.5–3.5 mm, pubescent or pilose, antherodes 0.9–1.2 mm. Ovary glabrous; stigma 0.6–1 mm long. Capsule 14–18.5 mm long, glabrous. Seeds 6.5–8 × 5–7 mm.

Mozambique. N: Pemba/Quissanga Dist., between Metuge and Mahate, fl.& fr. 3.ix.1948, *Pedro & Pedrógão* 5041 (EA, LMA); Pemba Dist., between Nagororo (Nangororo) and R. Tari, fl. 30.v.1959, *Gomes e Sousa* 4474 (COI, K, LISC).

Also in E Tanzania. Riverine forest and thicket, open lowland forest and margins; 50–300 m.

Conservation notes: Common in lowlands of E Tanzania but only just extends into Mozambique; assessed as Least Concern in F.T.E.A.

Brummitt & Wood (1983) treated this and the following two taxa as subspecies of a broadly circumscribed *Barleria prionitis*. Following this approach, many more African taxa would have to be considered as subspecies despite obvious and consistent morphological and ecological differences. I prefer to treat the African 'subspecies' of *B. prionitis* as separate species (see F.T.E.A. Acanthaceae **2**: 415 for further discussion).

60. **Barleria delagoensis** Oberm. in Ann. Transv. Mus. **15**: 143 (1933). Type: Mozambique, Masieni, mouth of Incomáti (Komatie) R., fl. ix.1924, *van Dam* in TVL 25321 (PRE holotype).

   Barleria prionitis L. subsp. *delagoensis* (Oberm.) Brummitt & J.R.I. Wood in Kew Bull. **38**: 440 (1983). —Bandeira *et al.*, Fl. Nat. Sul Moçamb.: 160, 192 (2007).

   Barleria prionitis sensu da Silva *et al.*, Prelim. Checklist Vasc. Pl. Mozamb.: 18 (2004) in part, non L.

Perennial herb or subshrub 40–150 cm tall; stems glabrous or minutely puberulous on two opposite sides, uppermost internodes sometimes strigulose; woody stems with sandy-coloured bark. Axillary spines small, 4(6)-rayed, stalk 0.3–1.5 mm long, longest ray 1.5–7.5 mm long, rarely absent. Leaves elliptic or ovate-elliptic, 3.2–10 × 1.6–4.2 cm, base attenuate, apex acute, obtuse or shortly attenuate, mucronulate, margin, midrib and veins strigose beneath, glabrescent; petiole to 18 mm. Inflorescence a terminal spike 2.5–8(22) cm long, contracted or lower cymules more widely spaced, each cymule 1- or 3-flowered, sessile; bracts ± spreading, lowermost pairs foliaceous, soon reducing upwards where oblanceolate, 8.5–26 × 1.5–7 mm, apex attenuate, mucronate, veins and margin strigose; bracteoles white-green, lanceolate-spinose, 6–12.5 × 0.9–1.5 mm. Outer calyx lobes subequal, ovate, anterior lobe (7.5)10.5–16 × 4–8 mm, posterior lobe (9.5)12–17.5 mm long, margin often involute, apex attenuate into apical spine, anterior lobe sometimes notched, surface with numerous ascending or appressed hairs, usually with minute spreading hairs with a few short glandular hairs towards apex, venation parallel, prominent only in fruit; lateral lobes (8)11.5–15 mm long. Corolla 29–39 mm long, (pale)yellow, orange or brick red, pilose externally; tube 11–14 mm long; limb in 4+1 arrangement; adaxial lobe 11.5–18 × 5–7 mm, offset by 5.5–8 mm; lateral lobes 10–17 × 5.5–8 mm; adaxial lobes as laterals but 4.5–7 mm wide. Stamens with filaments 15.5–29 mm long, sparsely pubescent towards base; anthers 2–3.3 mm; lateral staminodes 0.7–1.7 mm, pubescent, antherodes 0.7–1 mm. Ovary minutely hairy or glabrous; style glabrous; stigma linear, c.0.7 mm long. Capsule 15–18 mm long, densely puberulous. Seeds c.7.5 × 5 mm.

Mozambique. MS: Marromeu Dist., 15 km towards Lacerdónia, Inhaminga road, fl.& fr. 11.v.1942, *Torre* 4127 (BR, LISC, LMU, O, PRE, WAG). GI: Funhalouro Dist., route Mavume–Aluise R., near 60 km from Mavume, fl. ii.1939, *Gomes e Sousa* 2213 (COI, EA, K, LMA). M: Maputo city, Polana coast, road to Caracol, fl. 20.i.1960, *Lemos & Balsinhas* 12 (BM, COI, K, LISC, PRE, SRGH).

Also in South Africa (KwaZulu-Natal, ?Mpumalanga) and ?Swaziland. Littoral scrub including under coconut plantations, coastal bushland and dry coastal forest, lowland riverine woodland, usually on sandy soils; 0–150 m.

Conservation notes: Locally common in a range of coastal habitats, some of which are threatened, and extending along the lower Zambezi; Least Concern.

61. **Barleria ameliae** A. Meeuse in Bothalia **7**: 443 (1961). Type: Namibia, Caprivi Strip, Impalila Is., fl. 15.i.1959, *Killick & Leistner* 3391 (PRE holotype, K, L, M, SRGH, WIND).

Barleria priointis L. subsp. *ameliae* (A. Meeuse) Brummitt & J.R.I. Wood in Kew Bull. **38**: 441 (1983). —Mapaura & Timberlake, Checklist Zimb. Vasc. Pl.: 13 (2004).

Barleria prionitis sensu auct., non L. —Clarke in F.T.A. **5**: 145 (1899) in part for *Buchanan* s.n. from Malawi. —Phiri, Checklist Zamb. Vasc. Pl.: 18 (2005).

Barleria eranthemoides C.B. Clarke in F.T.A. **5**: 147 (1899) in part for *Buchanan* s.n. & *Meller* s.n. from Malawi, non lectotype chosen by Darbyshire in F.T.E.A. (2010).

Perennial herb or subshrub 15–120 cm tall, erect or decumbent, often rooting at lower nodes; stems glabrous or minutely puberulous; woody stems with pale grey or sandy bark. Axillary spines small, 4-rayed, stalk to 1 mm long, longest ray 2–7 mm long. Leaves elliptic(-obovate), 4.8–10.5 × 1.6–4.7 cm, base cuneate-attenuate, long-decurrent, apex acute or attenuate, mucronulate, margin, midrib and veins strigose beneath, glabrescent; petiole to 13 mm. Inflorescence axillary or a ± weakly defined terminal spike, 2.5–22 cm long, lax or rarely congested, each cymule 1- or 3-flowered, sessile; bracts ± spreading, lower pairs foliaceous, reducing upwards where oblanceolate, 11–38(48) × 2.5–15 mm, apex acute or shortly attenuate, mucronate, hairs as leaves beneath or more widespread; bracteoles white-green, lanceolate-spinose, 2.5–10.5 × 0.8–1.2 mm. Outer calyx lobes subequal, ovate or lanceolate, anterior lobe 10.5–14(16.5) × 3–5.5 mm, posterior lobe 11–17(20) mm long, apex acute or attenuate, spinose, anterior lobe sometimes notched, surface strigose, rarely with a few minute glandular hairs towards apex, venation parallel, prominent in fruit; lateral lobes 10–14 mm long. Corolla 27–40 mm long, yellow, orange or apricot, pubescent externally; tube 10–18 mm long; limb in 4+1 arrangement; abaxial lobe 9–14 × 5–7.5 mm, offset by 4.5–8 mm; lateral and adaxial lobes subequal, 11–16 × 6.5–8.5 mm or adaxial lobes narrower, minimum 4.5 mm wide. Stamens with filaments 11.5–21 mm long, pubescent in proximal two-thirds; anthers 2.5–4 mm; lateral staminodes 1.3–2 mm, pubescent, antherodes c.1 mm. Ovary glabrous or minutely hairy; style glabrous; stigma linear, 0.7–1.1 mm long. Capsule 13–16 mm long, glabrous or sparsely and minutely hairy. Seeds 7–8 × 4.5–6 mm.

Caprivi. Impalila Is., E portion, fl. 15.i.1959, *Killick & Leistner* 3391 (K, L, M, PRE, SRGH, WIND). **Zambia**. N: Mpika Dist., North Luangwa Nat. Park, fl. 19.ii.1994, *P.P. Smith* 337 (K). E: Chipata Dist., Mkania, 32 km SE of Mfuwe, fl.& fr. 28.ii.1969, *Astle* 5551 (K, SRGH). S: Kalomo Dist., Machile, fr. 16.iv.1961, *Fanshawe* 6499 (BR, K, NDO). **Zimbabwe**. N: Gokwe Dist., Sengwa Research Station, fl. 2.i.1968, *Jacobsen* 79 (SRGH). W: Hwange Dist., Hwange Nat. Park, Mandavu Dam area, c.18 km SW of Sinamatella Camp, fl.& fr. 25.ii.1967, *Rushworth* 217 (K, PRE, SRGH). C: Kadoma Dist., Sanyati communal land, fl.& fr. 3.iii.1995, *Hyde* 95/062/5 (K). E: Chipinge Dist., near Rupisi, fl.& fr. 11.ii.1960, *Goodier* 878 (K, PRE, SRGH). S: Mwenezi Dist., Runde R., Fishans, fl.& fr. 28.iv.1962, *Drummond* 7792 (BR, K, SRGH). **Malawi**. S: Machinga Dist., Press Ranching, fl. 13.vi.1986, *Nankhuni & Balaka* 31 (K, MAL). **Mozambique**. Z: Mopeia Dist., Mopeia, 4 km on road to Campo, fl. 29.xii.1967, *Torre & Correia* 16808 (LISC). M: Magude Dist., 1 km from Mapulanguene to Macaene, fr. 24.vi.1969, *Correia & Marques* 796 (LMU).

Not known elsewhere. Mainly in mopane and *Acacia* woodland on sandy soils, and dry open riverine forest and thicket; 100–1100 m.

Conservation notes: Widespread but patchily distributed, can be locally common or abundant in dry woodland; Least Concern.

This species is clearly closely related to *Barleria delagoensis*, sharing the small axillary spines and similar habit, foliage and inflorescence form. However, in addition to the difference in fruit indumentum, they differ in *B. ameliae* having (a) calyx with strictly appressed stiff hairs only (vs. more flexuose and ascending longer hairs usually intermixed with minute spreading hairs), (b) abaxial corolla lobe shorter than the other lobes and with the ratio of tube:abaxial lobe (0.9)1.2–1.45:1 (vs. abaxial corolla lobe subequal in length to other lobes and usually longer than the tube, ratio 0.65–

1:1), and (c) often a more lax synflorescence (though with considerable overlap). Meeuse (1961) also points to the ecological differences.

62. **Barleria senensis** Klotzsch in Peters, Naturw. Reise Mossambique **6**(1): 209 (1861). —Clarke in F.T.A. **5**: 146 (1899). —Meyer in Merxmüller, Prodr. Fl. SW Afr. **130**: 17 (1968). —Moriarty, Wild Fl. Malawi: 85, pl.43 (1975). —Fabian & Germishuizen, Wild Fl. Nthn. S. Afr.: 396, pl.190e, 191f (1997). Type: Mozambique, Rios de Sena, fl. 1842–1846, *Peters* s.n. (B sheet B10-0190529 lectotype, sheets B10-0190528, B10-0190530), lectotypified here. FIGURE 8.6.**44**.

　　Barleria eenii S. Moore in J. Bot. **45**: 229 (1907). —Obermeijer in Ann. Transv. Mus. **15**: 178 (1933). —Meyer in Mitt. Bot. Staatssamml. München **2**: 380 (1957). Type: Namibia, Damaraland, fl. 1879, *Een* s.n. (BM holotype).

　　Barleria petrophila Lindau in Bot. Jahrb. Syst. **43**: 353 (1909). —Obermeijer in Ann. Transv. Mus. **15**: 178 (1933). Type: Namibia, Omburo, fl. ii.1900, *Dinter* 1412 (B† holotype).

　　Barleria spathulata N.E. Br. in Bull. Misc. Inform., Kew **1909**: 128 (1909). —Obermeijer in Ann. Transv. Mus. **15**: 144 (1933). Type: Botswana, Kwebe, fl. 18.iv.1898, *Lugard* 5a (K holotype, GRA).

　　Barleria albida Lindau in Bot. Jahrb. Syst. **57**: 22 (1920). —Obermeijer in Ann. Transv. Mus. **15**: 178 (1933). Type: Namibia, Hereroland, Fransfontain–Outjo, fl.& fr. 9.vi.1912, *Dinter* 2646 (B† holotype).

Perennial herb or shrub 30–150(300) cm tall, erect or straggling; stems glabrous or uppermost internodes minutely puberulous on opposite sides. Axillary spines small, often sparse, rarely absent, 4(6)-rayed, stalk to 1.5 mm long, longest ray 1.5–7.5 mm long. Leaves sometimes immature at flowering; mature blade elliptic(-lanceolate) or ± obovate, 4–14 × 1–6 cm, base cuneate or attenuate, apex acute or attenuate, apiculate, leaf buds densely pale-strigose but soon glabrescent except along margin and midrib beneath; petiole 0–10 mm. Inflorescence a congested head or spike 1.5–8 cm long terminating main and short lateral branches, sometimes with additional flowers in upper leaf axils, each cymule 1(3)-flowered; bracts usually white in proximal half, green above and along main veins, rarely green throughout, obovate to broadly spathulate, 7.5–16 × 3.5–12.5 mm, apex rounded, truncate or emarginate then shortly attenuate into a mucro, main veins and margin of distal half strigose, margin of proximal half usually with spreading, bulbous-based hairs, surface with short (rarely subsessile) glandular hairs in distal half, sometimes with interspersed finer short eglandular hairs, few broad sessile glands usually present towards base; bracteoles white with green midrib, rarely green with pale margin, linear-lanceolate or lanceolate, (5.5)7.5–14 × 1–2 mm, mucronate. Calyx lobes green with paler margins, later turning whitish-green, outer lobes lanceolate, 7–12.5 × 2.5–4 mm, apex acute or attenuate, mucronate, anterior lobe sometimes notched, with a few short eglandular and glandular hairs towards apex, sometimes longer ascending hairs along margin, elsewhere glabrous, venation inconspicuous; lateral lobes somewhat narrower. Corolla 23–36 mm long, yellow, orange, apricot or ochre-coloured, pubescent externally, limb with mixed eglandular and glandular hairs; tube 9–15.5 mm long; limb in 4+1 arrangement; abaxial lobe 8.5–13.5 × 5.5–8.5 mm, offset by 4–9 mm; lateral lobes 10–15 × 6–9 mm; adaxial lobes 9.5–14 × 5.5–8 mm. Stamens with filaments 12–20 mm long, shortly pubescent in proximal half; anthers 2.5–3.5 mm; lateral staminodes 0.5–1.5 mm, pilose, antherodes 0.6–1.1 mm. Ovary glabrous; stigma linear, 0.6–1 mm long. Capsule 12–17 mm long, glabrous. Seeds 7.5–9 × 4.5–6 mm.

Botswana. N: Central Dist., near Tsoe–Rakops road, fl.& fr. 16.ii.1980, *P.A. Smith* 3043 (K, SRGH). SW: Central Kalahari Game Reserve, Deception Pan, Deception Valley, fl. iv.1975, *Owens* 27 (K, SRGH). SE: Central Dist., Boteti Delta area, NE of Mopipi, fl. 21.iv.1973, *Glanville* 7 (K, SRGH). **Zambia**. C: Kafue Dist., Kapongo, Old Lusaka Yacht Club, fl.& fr. 5.vi.1996, *Bingham et al.* 11039 (K). E: Petauke Dist., 16 km E of Luangwa R. bridge, escarpment road, fl. 30.v.1961, *Leach & Rutherford-Smith* 11091 (K, LISC, SRGH). S: Gwembe Dist., near crossing of Sinazongwe– Mwamba road with Zongwe R., Gwembe valley, fl. 2.vii.1961, *Angus* 2963 (FHO, K, SRGH). **Zimbabwe**. N: Kariba Dist., Kariba, fl.& fr. vi.1960, *Goldsmith* 71/60 (BM, K, SRGH). W: Hwange

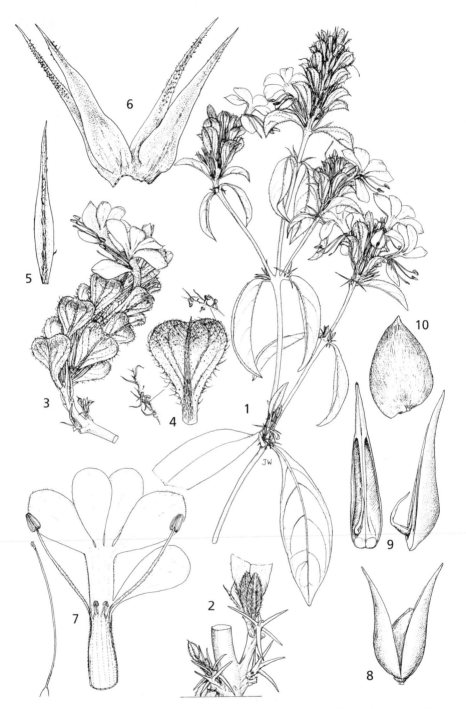

Fig. 8.6.**44**. BARLERIA SENENSIS. 1, habit (× ²/₃); 2, detail of rayed axillary spines and leaf buds (× 2); 3, inflorescence (× 1); 4, bract, external surface (× 2) with detail of surface and margin hairs; 5, bracteole, external surface (× 4); 6, dissected calyx, external surface, posterior lobe to right (× 4); 7, dissected corolla with androecium and pistil (× 2); 8, capsule with seeds (× 2); 9, capsule valves showing retinacula and solid septum (× 3); 10, seed (× 3). 1, 2 & 4–6 from *Drummond* 5977, 3 & 8 from *Goldsmith* 71/60, 7, 9 & 10 from *Smith et al.* 38. Drawn by Juliet Williamson.

Dist., c.8 km from Hwange on road to Sinamatella camp, fl.& fr. 30.vi.1974, *Raymond* 267 (BM, K, SRGH). **E**: Chipinge Dist., Save–Tanganda Halt, Save valley, fl. 28.iii.1956, *Whellan* 1023 (K, LISC, SRGH). **S**: Beitbridge Dist., Sentinel Ranch, c.6.4 km NE of Pazhi–Limpopo confluence, fl. 24.iii.1959, *Drummond* 5977 (BR, K, LISC, LMA, PRE, SRGH). **Malawi. S**: Chikwawa Dist., Livingstone (Kapachira) Falls, W bank of Shire R., fl. 21.iv.1970, *Brummitt* 10010 (BR, EA, K, MAL, SRGH). **Mozambique. T**: Changara Dist., Tomo–Mazoe road, fl.& fr. 21.ix.1948, *Wild* 2560 (K, LISC, PRE, SRGH). **MS**: Caia Dist., Rios de Sena, fl. n.d., *Peters* s.n. (B).

Also in SE Angola, Namibia and NE South Africa (Limpopo, Mpumalanga). Various types of dry woodland (*Acacia*, mopane), thicket and fallow, typically on sand, common on shaded riverbanks; 100–1000 m.

Conservation notes: Widespread and common; Least Concern.

Plants from the southwest of the Flora area and adjacent Namibia (*Barleria eenii*) often lack the conspicuous white portion to the bracts and can have less prominently bulbous-based hairs along the bract margin. This form can also have more widely spaced cymules at the base of the synflorescence. However, the variation is clearly clinal and cannot be readily separated into distinct taxa. *Brown* 8731 from SW Botswana is distinctive in having well-spaced axillary flowers with larger, spathulate bracts over 20 mm long, but otherwise agrees with *B. senensis* and is considered an extreme form. It would be useful to see further material from that area.

63. **Barleria rhodesiaca** Oberm. in Ann. Transv. Mus. **15**: 144 (1933). Type: Zimbabwe, Hwange (Wankie), fl.& fr. vi.1920, *Rogers* 13401 (GRA lectotype, BOL, K, PRE, Z), lectotypified here.

Perennial herb or subshrub 60–120 cm tall, erect or straggling; stems glabrous, older stems with grey bark. Axillary spines minute, often sparse, 4(6)-rayed, stalk to 0.7 mm long, longest ray 1–3 mm long. Leaves (linear-)lanceolate, narrowly oblong-elliptic or oblanceolate, 3.5–11.7 × 0.5–3.5 cm, base cuneate(-attenuate), apex acute or rarely obtuse, apiculate, surfaces strigulose when young, soon glabrescent except along margin and midrib beneath, sometimes sparse broad sessile glands towards base beneath; petiole 0–5 mm. Inflorescence a congested head 1.5–4 cm long on main and short lateral branches, each cymule 1–7-flowered; bracts green with whitish margins towards base, narrowly spathulate, 10–22 × 1.5–4 mm, apiculate, strigose mainly along midrib and margin, often with finer short hairs interspersed and short-stalked glands in distal half; bracteoles white with green midrib, linear-lanceolate, (6.5)8–13 × 1–2 mm, apex mucronate. Calyx lobes green at first with paler margins, later whitish-green throughout, subequal, outer lobes lanceolate, 8.5–14.5 × 2.3–3.5(4.3) mm, (sub)attenuate into apiculum, anterior lobe sometimes notched, with short eglandular and glandular hairs towards apex, elsewhere glabrous, venation inconspicuous or 2 (anterior) or 1 (posterior) veins prominent; lateral lobes somewhat narrower. Corolla 29–43.5 mm long, yellow, pubescent externally, limb with mixed eglandular and glandular hairs; tube 11–15 mm long; limb in 4+1 arrangement; abaxial lobe 8.5–14 × 3.5–6 mm, offset by 7–10.5 mm; lateral lobes 11.5–17 × 7–10 mm; adaxial lobes 11.5–19.5 × 5–8.5 mm. Stamens with filaments 24–30 mm long, glabrous or with short hairs towards base; anthers c.3 mm; lateral staminodes c.0.5 mm, pilose, antherodes c.1 mm. Ovary glabrous; stigma linear, 0.6–0.75 mm long. Capsule 13.5–15.5 mm long, glabrous. Only immature seeds seen.

Botswana. N: Chobe Dist., banks of Chobe R. near Kasane, Commissioner's Kop, fl. 20.iv.1989, *P.A. Smith* 5165 (SRGH). **Zambia. S**: Siavonga Dist., Siavonga turnoff to Kafue 2 km, fl. 17.vii.1996, *Bingham* 11105 (K). **Zimbabwe. N**: Sebungwe area, Zambezi road, fl.& fr. vi.1956, *Davies* 2001 (K, LISC, SRGH). **W**: Hwange Dist., Lukosi R., fl.& fr. 28.iv.1959, *Noel* 1919 (K, PRE, SRGH).

Not known elsewhere. Dry rocky slopes in open mopane or *Commiphora* woodland or wooded grassland and riverbanks; 450–950 m.

Conservation notes: Endemic to the Zambezi valley. Local and uncommon but not considered threatened; Least Concern.

This species is closely allied to *Barleria senensis* and could be considered a regional variant. However, the differences in bract shape and indumentum are striking, particularly in the area of range overlap (mainly in Hwange Dist. in W Zimbabwe) where the bracts of *B. senensis* are always broadly obovate and with conspicuously bulbous-based cilia. In addition, the staminal filaments of *B. senensis* are more hairy and the abaxial corolla lobe in *B. senensis* is longer than, subequal to or only slightly shorter than the four 'upper' lobes, whilst in *B. rhodesiaca* it is ± noticably smaller.

Flower size is quite variable within *B. rhodesiaca* and, with few intact corollas available for measurement, it is quite possible that the full size range is not recorded here.

64. **Barleria setosa** (Klotzsch) I. Darbysh., comb. & stat. nov. Type: Mozambique, Ilha de Moçambique Dist., Goa Is. (Insel Goa), fl.& fr. 10.vii.1843, *Peters* 29 [no number in protologue] (B lectotype, K photo), lectotypified here.

   Barleria prionitis L. var. *setosa* Klotzsch in Peters, Naturw. Reise Mossambique **6**(1): 209 (1861).

   Barleria eranthemoides C.B. Clarke in F.T.A. **5**: 147 (1899) in part for *Peters* s.n., non lectotype of Darbyshire in F.T.E.A., Acanthaceae **2**: 436 (2010).

Perennial herb or shrub to 90 cm tall, branches erect or decumbent, then rooting at lower nodes; stems glabrous or with few hairs immediately below nodes. Axillary spines, if present, 4-rayed, stalk 1–2.5 mm long, longest ray 7–13.5 mm long. Leaves (obovate-)elliptic, 3.2–8.5 × 1.5–3.6 cm, base cuneate, apex acute, mucronulate, glabrous or margin and veins beneath sparsely strigulose; petiole 0–9 mm. Inflorescence a contracted spike or subcapitate head 3.5–6 cm long; bracts erect, oblong(-oblanceolate) or shallowly pandurate, 12–20 × 2.5–6 mm, mucronate, sometimes paler towards base, strigulose on main veins and margin of distal half, margin of proximal half with longer, more spreading, bulbous-based hairs; bracteoles (linear-)lanceolate, 10.5–13 × 0.8–2.5 mm. Calyx whitish-green towards base, green above; outer lobes lanceolate, 11–14 × 2.7–4 mm, apex mucronulate, anterior lobe usually notched, venation prominent, (sub)parallel, veins strigulose, margin with longer ascending bulbous-based hairs, apical portion with minute ascending-eglandular and spreading-glandular hairs; lateral lobes 10.5–12.5 mm long. Corolla 30–35 mm long, yellow or orange, shortly pubescent externally with mixed eglandular and glandular hairs; tube narrowly cylindrical, 19–20 mm long; limb in weak 4+1 arrangement; abaxial lobe 8.5–13 × 6.5–7.5 mm, offset by 2–3.5 mm; lateral lobes 10–11.5 × 6–6.5 mm; adaxial lobes as laterals but 5–6.5 mm wide. Stamens with filaments 13.5–14 mm long, shortly pubescent in proximal half; anthers 2–3.2 mm; lateral staminodes 1.5–2.5 mm, pilose, antherodes 0.6–1 mm. Ovary glabrous; stigma linear, c.1 mm long. Capsule c.12.5 mm, glabrous. Seeds c.5.5 × 4.5 mm.

Mozambique. N: Ilha de Moçambique Dist., Cabeceira Pequena, fl. 1884–1885, *Carvalho* s.n. (COI); Goa Is., fl. 5.v.1947, *Gomes e Sousa* 3506 (K); Goa Is., fl.& fr. 19.x.1954, *Gomes e Sousa* 4268 (PRE).

Not known elsewhere. Sometimes common in coastal bushland over coral and sand; 0–20 m.

Conservation notes: Endemic to Goa Is., Ilha de Moçambique and the adjacent Cabeceira Pequena peninsula in Nampula Province, areas that have seen much intensive human settlement – nearly all of Ilha de Moçambique is now built up while bushland on Goa Is. is now restricted to the northern fringe; would qualify as Endangered under IUCN criterion B.

Clarke (1899) included Peters specimens from Goa Is. within his *Barleria eranthemoides* but appears not to have studied them in detail. The similarity is, in fact, superficial; *B. setosa* shares with *B. eranthemoides* the congested inflorescence spike with conspicuous oblong or subpandurate bracts with bulbous-based hairs along the margin. They differ most notably in *B. setosa* having a 4+1, not subregular, corolla limb. In addition, *B. setosa* has a hairy corolla and acute outer calyx lobes with strigulose prominent veins, the margins having bulbous-based hairs (vs. corolla glabrous, outer calyx lobes acuminate, venation inconspicuous and surface glabrous except for minute, non bulbous-based, hairs along the margin in *B. eranthemoides*). The real affinity appears to be with *B. senensis*, but it clearly differs from that species in corolla shape and the presence of bulbous-based hairs along the calyx margin.

Axillary spines are absent on *Gomes e Sousa* 3506 and *Leach* 12350, but well-developed on the Peters specimens and on *Carvalho* s.n.

65. **Barleria coriacea** Oberm. in Ann. Transv. Mus. **15**: 139 (1933). —Compton, Fl. Swaziland: 552 (1976). Type: South Africa, Lydenburg, Inkumpi R. at Zebedele's Kraal, fl.& fr. n.d., *Nelson* 370 (PRE holotype, K).

Barleria prionitis sensu auct. —Clarke in Fl. Cap. **5**: 46 (1901) for *Nelson* 370 & 400. — Fabian & Germishuizen, Wild Fl. Nthn. S. Afr.: 386, pl.185 (1997). —Setshogo, Prelim. Checklist Pl. Botswana: 17 (2005), non L.

Barleria dinteri sensu Welman in Germishuizen *et al.*, Checklist S. Afr. Pl.: 79 (2006) in part, non Oberm.

Compact spiny subshrub 20–50 cm tall. Stems white-puberulous when young, hairs concentrated on two opposite sides or more widespread, later glabrescent; base of stems woody and often stout. Axillary spines 2–4(6)-rayed, stalk 0–2 mm long, longest ray (5.5)9–18.5 mm long. Leaves somewhat coriaceous, narrowly oblong or oblanceolate, 2.7–6.3 × 0.8–1.7 cm, base cuneate, apex acute or lower pairs more rounded, mucronate, margin and veins beneath strigulose when young, later glabrescent, broad sessile glands ± numerous towards base beneath; petiole 0-5 mm. Inflorescence axillary, flowers solitary, sometimes clustered towards stem apex, sessile; bracts foliaceous, often with more numerous and conspicuous sessile glands towards base; bracteoles whitish, spinose, 11.5–18.5 × 1.5–2.5 mm. Outer calyx lobes subequal, lanceolate, 10.5–15 × 3–5 mm, long-attenuate into apical spine, anterior lobe sometimes bifid, surface and margin sparsely strigulose or glabrescent, with scattered broad sessile glands, 2 (anterior) or 1 (posterior) veins prominent; lateral lobes 9–12 mm long. Corolla 29–34 mm long, yellow, glabrous externally or with sparse short hairs on tube; tube 10.5–13 mm long; limb in 4+1 arrangement; abaxial lobe offset by 4–6 mm, this and lateral lobes 12–16 × 7–12 mm; adaxial lobes 6–9.5 mm wide. Stamens with filaments 15–17.5 mm long, shortly pubescent; anthers 3.5–4.7 mm long; lateral staminodes 1–1.5 mm, pilose, antherodes 1–1.2 mm. Ovary glabrous; stigma linear, 1.5–2 mm long. Capsule 13–17 mm long, glabrous. Seeds 7–8 × 5.5–6.5 mm.

Botswana. SE: Kgatleng Dist., Mochudi, fl.& fr. i–iv.1914, *Harbor* in Herb. Rogers 6615 (BOL, K, PRE).

Also in N South Africa (Limpopo and Gauteng) and Swaziland. Elsewhere in dry *Acacia* bushland, on limestone ridges and roadsides; c.1000 m.

Conservation notes: Very localised and apparently scarce, known from few herbarium collections but appears tolerant of some disturbance. Galpin (13396) recorded it as "steadily spreading with overstocking and become a pest" near Naboomspruit, South Africa; probably Least Concern.

Barleria dinteri Oberm. from Namibia is closely allied and probably best considered a regional subspecies. It differs primarily in having a more dense indumentum when young, the uppermost leaf buds and stems often covered with short white hairs. In addition, it tends to have a more stunted habit with stouter branches with many short internodes, often longer bracteoles (typically 14–27 mm long), and more acute corolla lobes. However, none of these characters are entirely consistent. From the distribution recorded, Welman appears to have treated material of both *B. dinteri* and *B. coriacea* under the former name in the Checklist of South African Plants (*Germishuizen et al.* 2006), but does not list *B. coriacea* as a synonym.

66. **Barleria randii** S. Moore in J. Bot. **38**: 203 (1900). —Burkill & Clarke in F.T.A. **5**: 512 (1900). —Obermeijer in Ann. Transv. Mus. **15**: 142 (1933). —Tredgold & Biegel, Rhod. Wild Fl.: 52, pl.35 (1979). Type: Zimbabwe, Bulawayo, fl. xii.1897, *Rand* 115 (BM holotype, GRA).

Spiny perennial herb or subshrub 5–60 cm tall, from a woody rootstock; stems with upper internodes puberulous on two opposite sides, later glabrescent. Axillary spines (2)4-rayed, sessile or stalk to 1 mm long, longest ray 6–16 mm long. Leaves coriaceous with conspicuous white cartilaginous marginal rim, elliptic or somewhat obovate, 2.5–5.5 × 1–3.5 cm, base cuneate or attenuate, apex (sub)attenuate into stiff mucro, glabrous or midrib strigose beneath, broad sessile glands clustered towards base beneath; petiole 0–6 mm. Inflorescnce axillary but clustered towards stem apex, forming a ± defined short spike to 5.5 cm long; bracts foliaceous but often reduced and more broadly obovate, typically 10–26 × 6–24 mm, apex often rounded and abruptly narrowed into a prominent mucro; bracteoles whitish except for darker midrib, linear-lanceolate or pairs at spike base oblanceolate, 8–15 × 0.7–1.5(2.5) mm, midrib strigose. Calyx green(-brown) with paler margin; outer lobes subequal, ovate or lanceolate, anterior lobe 6–10 × 3–4 mm, posterior lobe 7–11.5 mm long, apex long attenuate into flexible spine, often notched on anterior lobe, venation inconspicuous, external surface glabrous except for occasional sunken glands, rarely strigulose along midrib; lateral lobes 6.5–9 mm long. Corolla 34.5–45.5 mm long, yellow or apricot-yellow, glabrous externally; tube 12.5–19 mm long; limb in 4+1 arrangement; abaxial lobe 8–14.5 × 4.5–7 mm, offset by 7–11 mm; lateral lobes 12–19.5 × 7–11 mm, adaxial lobes slightly narrower. Stamens with filaments 18.5–19.5 mm long, shortly pubescent in proximal half; anthers (3)3.5–4.2 mm; lateral staminodes 1–2.7 mm, pilose, antherodes 0.6–1 mm. Ovary glabrous; stigma linear, 1.2–1.9 mm long. Capsule 13.5–17.5 mm long, glabrous. Only immature seeds seen, c.5.5 mm wide.

Botswana. SE: Kgatleng Dist., 56 kms N of Sikwane, fl. 26.xi.1955, *Reyneke* 436 (BM, K, PRE). **Zimbabwe**. W: Matobo Dist., Matopos, fl. xi.1922, *Eyles* 3755 (BOL, K). S: Mberengwa Dist., 16 km S of Mberengwa (Belingwe), fl. 26.xii.1959, *Leach* 9714 (K, SRGH).

Also in South Africa (Limpopo). On bare or sparsely vegetated sandy soils and stony road verges in *Acacia* woodland; 850–1500 m.

Conservation notes: Known from few localities but bare ground is a significant habitat; Least Concern.

Whilst in several species of sect. *Prionitis* the leaf margin is somewhat thickened and yellowish-green, the very prominent white margin is apparently unique to this species.

67. **Barleria holubii** C.B. Clarke in Fl. Cap. **5**: 47 (1901). —Obermeijer in Ann. Transv. Mus. **15**: 146 (1933). —Darbyshire in F.T.E.A., Acanthaceae **2**: 440 (2010). Type: South Africa, North West Province, Marico Dist., fl. n.d., *Holub* s.n. (K holotype).

Very spiny shrub forming dense, rounded clumps, 15–150 cm tall; stems with short white spreading or retrorse hairs, evenly distributed or concentrated in two opposite lines. Axillary spines 2–4-rayed, subsessile, longest ray 10–23 mm long. Leaves broadly obovate, 0.8–3 × 0.6–1.7 cm, base attenuate or cuneate, apex rounded or abruptly narrowed, mucronate, shortly spreading-pubescent at least beneath when young, with longer ascending or appressed hairs on margin and veins beneath; petiole 0–3 mm. Inflorescence axillary towards stem apex, flowers solitary, subsessile; bracts foliaceous; bracteoles spinose, (8.5)12–21 mm long, bracteoles of old inflorescence persisting. Anterior calyx lobe lanceolate, 8.5–13.5 × 2–3 mm, apex spinose or notched-spinose, surface with white spreading or retrorse hairs, with or without coarser ascending hairs along margin and short spreading glandular hairs towards apex, venation parallel; posterior lobe 9–14 mm long; lateral lobes narrower. Corolla salver-shaped, 30–37 mm long, yellow or pale orange, densely pubescent externally, limb with or without occasional glandular hairs; tube narrowly cylindrical, 20–26 mm long; limb subregular, lobes widely spreading; abaxial lobe 6.5–11.5 × 5–6.5 mm; lateral lobes 7–12 × 5–6 mm, adaxial lobes as abaxial but 3.5–5.5 mm wide. Stamens with filaments 8–9.5 mm long, pubescent in proximal half; anthers wholly or only partially exserted, 1.9–2.5 mm; lateral staminodes 0.6–1.5 mm, pubescent, antherodes 0.6–1 mm. Ovary glabrous; stigma c.0.5 mm long. Capsule 10.5–14.5 mm long, glabrous. Seeds c.7 × 5 mm.

Subsp. **holubii**

Anterior calyx lobe 11–13.5 mm long, posterior lobe 12–14 mm long, indumentum of short white spreading to retrorse eglandular hairs throughout, coarser ascending hairs along margin and ± along main veins, glandular hairs absent or very sparse towards apex. Corolla lacking glandular hairs externally.

Botswana. SE: Central Dist., Shoshong, hill next to kgotla office at entrance to valley, fl.& fr. 21.iv.2005, *Darbyshire* 451 (GAB, K, MAH). **Zimbabwe**. W: Matobo Dist., Matopos Nat. Park, fl. 30.iv.1967, *Biegel* 2105 (K, SRGH).

Also in northern South Africa (Limpopo, Mpumalanga). Open or lightly wooded rocky slopes with large boulders; 1000–1500 m.

Conservation notes: Highly range-restricted and apparently scarce but can be locally common in rocky hills; Least Concern.

The highly disjunct subsp. *ugandensis* I. Darbysh., recorded from NE Uganda and NW Kenya, differs mainly in having a smaller calyx (posterior lobe 9–12 mm long) with numerous glandular hairs towards the apex; the corolla also has a few glandular hairs on the limb.

68. **Barleria sp. D** (= *Lousã & Rosa* 354)

Spiny suffrutex c.30 cm tall, with erect or decumbent branches arising mainly from a woody base; stems with dense white spreading hairs throughout. Axillary spines 2- or very unevenly 4-rayed, subsessile, longest ray 7.5–18 mm long. Leaves obovate-elliptic, 2.3–4.3 × 1.2–2.1 cm, base attenuate, apex rounded or obtuse, mucronate, surfaces spreading white-pubescent, coarser ascending hairs on margin and veins beneath; petiole 0–3 mm. Inflorescence axillary, 1–2-flowered, subsessile; bracts foliaceous, sometimes with scattered glandular hairs beneath; bracteoles spinose, 6–12 mm long. Anterior calyx lobe lanceolate-attenuate, 8.5–11 × 2–3 mm, apex spinose or bifidly so, surface spreading white-pilose, coarser ascending hairs along margin, sometimes with scattered spreading glandular hairs towards apex, surface with two parallel veins; posterior lobe 9.5–12.5 mm long, apex spinose; lateral lobes 8.5–10.5 mm long, linear-lanceolate. Corolla salver-shaped, c.31 mm long (single corolla measured), colour unknown (?yellow); densely pubescent externally; tube narrowly cylindrical, 21 mm long; limb subregular, abaxial lobe 9 × 5.5 mm; lateral lobes 10.5 × 6 mm; adaxial lobes 9.5 × 4 mm. Stamens and staminodes not seen (anthers apparently barely exserted beyond throat). Ovary glabrous; stigma c.0.5 mm long. Capsule c.13 mm long, glabrous. Seeds not seen.

Mozambique. GI: Massingir Dist., left margin of Rio dos Elefantes, 75 km to NW of Lagoa Nova, fl.& fr. 17.iii.1973, *Lousã & Rosa* 354 (K, LMA).

Also in N South Africa (Limpopo). Amongst rocks and on rocky soils; 50–200 m.

Conservation notes: Known from only two collections, the second in Kruger Nat. Park at Lebomboberg Gorge (*van der Schijff* 2337 (PRE), 24.ii.1953); Data Deficient.

Closely allied to *Barleria holubii* but differing in the larger more elliptic leaves and bracts with a less conspicuous mucro, the different habit and in the inflorescences being sometimes 2-flowered, not always single-flowered. Some of these characters are shared by *B. penelopeana* I. Darbysh. from NW Tanzania, which differs primarily in having larger fruits, usually 17–22 mm long. More material is required, including further observations on the habit of this species and that of *B. holubii*.

The African Plants Database records several localities for *B. holubii* from the NE border of South Africa with Mozambique, which are likely to refer to this species. However, I have not seen the collections on which these records are based.

Excluded species.

Barleria acanthophora (Roem. & Schult.) Nees in De Candolle, Prodr. **11**: 726 (1847). —Clarke in F.T.A. **5**: 169 (1899). Type: Mozambique, no locality, n.d., *Loureiro* s.n. (?P-LOUR).

Eranthemum spinosum Lour., Fl. Cochinch. **1**: 19 (1790). Type as above, "Habitat agreste in suburbiis Mozambicci in Africâ".

Eranthemum acanthophorum Roem. & Schult. in Schultes, Mant. **1**: 154 (1822).

Erect caespitose subshrub to 30 cm tall with short assurgent branches. Stipules and bracts spinose. Leaves opposite, ovate, small, entire, pilose. Flowers pale violet, solitary, lateral, pedunculate. Calyx of two 'leaves', erect, acuminate. Corolla tube long, filiform, curved at base, limb 5-lobed, spreading. Stamens 2, anthers oblong, incumbent. Style short, stigma rather thick. Capsule ovate, 2-locular, many-seeded.

The above description is adapted from the protologue of *Eranthemum spinosum* by Loureiro. I have not located the type material, possibly housed in the Loureiro Herbarium at Paris. Clarke did not see it either but noted that in the original description "the many-seeded capsule with long filiform corolla-tube and spinous stipules form a combination of characters difficult to find in Acanthaceae, throwing aside even the calyx" (p. 169). The description of the calyx does indeed sound like a *Barleria*, whilst the "spinous stipules" could easily be a misinterpretation of paired spinose bracteoles, which could also fit with this genus. However, the many-seeded capsule (if correct) clearly rules out *Barleria*.

Barleria casatiana Buscal. & Muschl. in Bot. Jahrb. Syst. **49**: 496 (1913). —White, For. Fl. N. Rhod.: 381 (1962). —Phiri, Checklist Zamb. Vasc. Pl.: 18 (2005) as *cesatiana*.

In the protologue this species was recorded as having been collected from Zambia (*de Aorta* 446), but *Barleria casatiana* is an Eritrean endemic. The likely provenance of the type specimen is from *Schweinfurth* 2042 (see Ensermu in Fl. Ethiopia **5**: 421, 2006).

28. ASYSTASIA Blume [6]

Asystasia Blume, Bijdr.: 796 (1826). —Nees in in De Candolle, Prod. **11**: 163 (1847). *Salpinctium* T.J. Edwards in S. Afr. J. Bot. **55**: 7 (1989).

Herbs or shrubs. Leaves opposite, entire (rarely undulate to serrate). Flowers in terminal or axillary spikes or racemes; bracts persistent, prominent or not; bracteoles usually minute, rarely large. Calyx deeply divided into 5 lobes, one lobe longer and slightly wider than others. Corolla indistinctly 2-lipped with 5 subequal lobes, lower mid-lobe largest and usually distinctly rugulate, or moderately to strongly 2-lipped, with a 3-lobed lower and 2-lobed upper lip; basal tube cylindric, widening upwards into a ± distinct throat. Stamens 4, inserted in throat, subequal to distinctly didynamous, included in tube, more rarely exserted; anthers bithecous, thecae slightly superposed, each with 1 (rarely 2 or 3) spurs at base or apiculate. Style filiform, pubescent in basal part or glabrous; stigma capitate with 2 equal or subequal flat ellipsoid lobes; ovary with 2 ovules per locule. Capsule 2–4-seeded, clavate with sterile solid basal stalk about half total length. Seeds compressed, rarely globose, ± circular in outline, with or without a thickened irregularly crenate margin, surfaces tuberculate, rugose, rugose-tuberculate or rarely smooth.

A genus of about 50 species, centred in tropical Africa with the highest diversity in eastern Africa to South Africa; also in India, SE Asia, and through Indonesia to New Guinea, Australia and the Pacific. One species introduced to the Americas.

---

[6] By Ensermu Kelbessa and Kaj Vollesen

1. Pedicel and capsule reflexed against inflorescence axis; inflorescence with 2–4 flowers; capsule glabrous . . . . **6.** *retrocarpa*
– Pedicel and capsule held erect; some or all inflorescences with 5 or more flowers; capsule hairy, sometimes also glandular 2
2. Scandent shrub; corolla orange to deep orange or dull red, throat distinctly curved; stamens subequal with anthers protruding from throat. . . . . **1.** *zambiana*
– Erect to decumbent or scrambling annual or perennial herb; corolla white to mauve, pink or purple; stamens clearly didynamous with anthers included in throat 3
3. Corolla pure white or with pale yellow throat or pale yellow patches in throat and on palate 4
– Corolla white to mauve, pink or pale blue, with mauve, violet or purple (rarely reddish) patches on palate and often lines in throat. 5
4. Annual herb; bracts and bracteoles 5–20 mm long, longer than calyx; corolla 1–1.5 cm long; capsule 1.5–2.2 cm long . . . . **7.** *mysorensis*
– Perennial herb; bracts and bracteoles 3–5 mm long, shorter than calyx; corolla 3.2–3.7 cm long; capsule 2.5–2.8 cm long. . . . . **4.** *albiflora*
5. All flowers opposite (both bracts at a node supporting a flower) 6
– Some or all flowers alternate (only one bract at a node supporting a flower). . . 7
6. Calyx 5–8(9) mm long; corolla 1.2–2.2 cm long; capsule 2–2.6 cm long; seed 4–5 × 3.5–4 mm . . . . **3.** *glandulosa*
– Calyx 11–20 mm long; corolla 2.7–3.8 cm long; capsule 3.3–4.2 cm long; seed 6–7 × c.5 mm . . . . **8.** *welwitschii*
7. Basal cylindric corolla tube 10–16 mm long, with orange to red capitate glands; seed margin not thickened . . . . **2.** *malawiana*
– Basal cylindric corolla tube 2–7 mm long, with colourless capitate glands; seed with distinctly thickened margin . . . . **5.** *gangetica*

1. **Asystasia zambiana** Brummitt & Chisumpa in Kew Bull. **32**: 702 (1978). Type: Zambia, Mufulira, 28.viii.1955, *Fanshawe* 2431 (K holotype, NDO).

Scandent shrub; stems woody, scrambling or scandent, to 2 m long, dark-purplish when dry, retrorsely sericeous-pubescent when young, glabrescent. Leaves dark purplish and sparsely pubescent along main veins; petiole 0.5–2.5 cm long, ill-defined; lamina ovate to elliptic, largest 7–11 × 2.7–4.5 cm; apex acute to acuminate with obtuse tip, base cuneate to attenuate. Raceme terminal, rarely with a single branch from lowermost node, 0.5–2(6) cm long, lower 1–5 nodes bearing only bracts and bracteoles, sericeous-pubescent or sparsely so; flowers alternate, rarely opposite; peduncle 1–5(10) mm long; bracts and bracteoles purplish, 1–4 mm long, triangular, puberulous (at lower nodes pilose with yellowish hairs); pedicels 0.7–1.5 cm long, finely puberulous with non-glandular hairs and short capitate glands. Calyx 5–9 mm long, lobes lanceolate to narrowly triangular, finely puberulous with non-glandular hairs and sub-sessile capitate glands, ciliate. Corolla orange to deep orange or dull red with yellow mottling in throat, 32–42 mm long, sparsely puberulous and with scattered capitate glands; cylindric tube 9–11 × 3–3.5 mm; throat distinctly curved, 17–25 mm long, 7–9 mm wide at oblique mouth; limb indistinctly 2-lipped, lobes reflexed, ovate with rounded apex, with crenate margins, lower mid-lobe 6–7 × 6 mm, lateral lobes 5–6 × 4–5 mm, lobes in upper lip c.4 × 5 mm. Stamens subequal, with anthers slightly protruding; filaments 10–12 mm long, glabrous; thecae c.3 mm long, apiculate, glabrous. Style 28–30 mm long, pubescent at very base. Capsule 25–30 mm long, sparsely puberulous with short capitate glands and short non-glandular hairs. Seeds 5.5–6 × 4.5–5 mm, without distinct margin; testa tuberculate.

Zambia. W: Mufulira Dist., fl. 13.viii.1954, *Fanshawe* 1468 (K). N: Mbala Dist., Inono Valley, 1225 m, fl. 27.iv.1952, *Richards* 1542 (K).

Also in D.R. Congo and Angola. In riverine forest and thicket, mostly in rocky places; 900–1500 m.

Conservation notes: Moderately widespread; not threatened.

The corolla shape of this species is unusual – the tube is downwardly curved and oblique at the opening, with the lower side shortened and the upper elongate. The orange to red corolla colour is also unusual.

2. **Asystasia malawiana** Brummitt & Chisumpa in Kew Bull. **32**: 703 (1978). —Vollesen in F.T.E.A., Acanthaceae **2**: 449 (2010). Type: Malawi, Ntchisi Dist., Ntchisi Forest Reserve, 25.iii.1970, *Brummitt* 9375 (K holotype).

Perennial herb with tuberous roots; stems erect or ascending, to 0.8(1.5) m high, glabrous to sparsely pubescent, mostly in two bands; lateral branches sometimes bending down, rooting where touching the ground and producing new plants. Leaves glabrous below, with sparse appressed broad many-celled hairs above; petiole to 6(8) cm long; lamina ovate to elliptic or broadly so, largest 6–16 × 3.5–7 cm; apex acuminate to cuspidate, base attenuate and decurrent on lower leaves, cuneate to subcordate upwards. Raceme 0.5–4(10 in fruit) cm long, terminal or axillary from upper nodes, rarely branched from base, puberulous with non-glandular hairs in two bands; flowers alternate, rarely some opposite; peduncle 0.2–6 cm long; lower bracts often foliaceous, to 2 × 1.5 cm, middle and upper bracts and bracteoles 2–4 mm long, linear-lanceolate to narrowly triangular, glabrous but for ciliate margins; pedicels 2–7(10 in fruit) mm long, puberulous or densely so with non-glandular hairs. Calyx 4–9 mm long; lobes linear-lanceolate, puberulous or densely so with sparse to dense short dark red to yellow capitate glands. Corolla white with reddish purple veins and markings in throat and lower lip, 2.5–3.5 cm long, sparsely puberulous with orange to red capitate glands and non-glandular hairs; basal cylindric tube 10–16 × 1–2 mm, throat 1–1.5 cm long, 0.7–1.1 cm wide at mouth; limb weakly 2-lipped with subequal sub-orbicular rounded lobes, 5–8 × 5–8 mm. Stamens didynamous, included in throat; filaments 3–4 and 5–6 mm long, glabrous; thecae c.2 mm long, dorsally with sub-sessile capitate glands, without spur. Style 16–20 mm long, puberulous near base. Capsule 2.3–2.8 cm long, puberulous with short-stalked capitate glands. Seeds 5–6 × 4–5 mm, mid-surface nearly smooth, tuberculate towards margins, with 5–6 narrow and smooth channels and 5–6 broad tuberculate ridges forming projections on margin.

Malawi. C: Ntchisi Dist., Ntchisi Forest Reserve, 1000 m, fl.& fr. 14.iv.1991, *Bidgood & Vollesen* 2181 (C, CAS, K, MAL). S: Zomba Dist., Zomba Plateau, Chingwe's Hole, 1825 m, fl.& fr. 1.iv.1990, *Brummitt et al.* 18583 (K, MAL). **Mozambique.** Z: Gurué Dist., Mt Namuli, Muretha Plateau, 1725 m, fl. 23.xi.2007, *Timberlake et al.* 5292 (K, LMA, LMU).

Also in Tanzania. In montane evergreen forest and riverine forest; 1000–1900 m. Conservation notes: Widespread; not threatened.

The closely related South African *Asystasia varia* N.E. Br. has more densely hairy stems, inflorescence and calyx, a lilac or mauve corolla and a capsule indumentum with both eglandular and glandular hairs.

3. **Asystasia glandulosa** Lindau in Bot. Jahrb. Syst. **33**: 189 (1902). —Vollesen in F.T.E.A., Acanthaceae **2**: 450 (2010). Type: Tanzania, Songea Dist., Milonji R., ii.1901, *Busse* 992 (B† holotype, EA).

Perennial herb with one to several stems from a creeping woody rootstock; stems to 60 cm long, erect to procumbent, subglabrous to densely pubescent. Leaves subglabrous to densely pubescent, densest on veins; petiole ill-defined, to 5 mm long; lamina narrowly ovate or narrowly elliptic, rarely ovate or elliptic, largest 5.5–10 × 1.5–3(3.5) cm; apex rounded to acuminate, base cuneate to attenuate and decurrent. Raceme 1.5–8(10 in fruit) cm long, terminal, never branched from base, puberulous to pubescent; all flowers opposite; peduncle 1.5–6.5(8) cm long; bracts and bracteoles 1–4 mm long (lowermost pair to 6 mm), subulate to narrowly triangular, puberulous or sparsely so; pedicels 1–3(4 in fruit) mm long, puberulous, with stalked capitate glands towards apex. Calyx 5–8(9 in fruit) mm long; lobes subulate to linear-lanceolate, puberulous, with capitate glands, usually also with long pilose hairs. Corolla white to pale mauve, pale pink or purple, with purple markings in throat and on rugula, 1.2–2.2 cm long, puberulous or sparsely so with capitate glands and non-glandular hairs; cylindric tube 3–6 × 1.5–2.5 mm; throat 5–9 mm long, 4–7 mm wide at mouth; limb 2-lipped, midlobe in lower lip 4–7 × 5–7 mm, ovate with rounded apex, lateral lobes 4–5 × 4–5 mm, upper lip 4–6 mm long with lobes c.2 × 3 mm. Stamens didynamous, included in throat; filaments 3–4 and 6–7 mm long; thecae black, c.2 mm long, with a minute spur or apiculate. Style 10–12 mm long, pubescent. Capsule 2–2.6 cm long, puberulous or densely so with stalked capitate glands and non-glandular hairs. Seeds 4–5 × 3.5–4.5 mm, with evenly distributed tubercles apart from a concentric band near edge, margin slightly double-ridged, not thickened or toothed, roughened with tubercles.

Zambia. B: Kabompo Dist., 85 km on Zambezi (Balovale)–Kabombo road, fl. 26.iii.1961, *Drummond & Rutherford-Smith* 7381 (K, SRGH). N: Mpika Dist., Mpika, 1225 m, fl. 30.xii.1958, *D.R.M. Stewart* 158 (K). W: Mwinilunga Dist., Matonchi Farm, fr. 31.xii.1937, *Milne-Redhead* 3527a (K). C: Chongwe Dist., 100–130 km E of Lusaka, Chakwenga headwaters, fl. 10.i.1964, *Robinson* 6184 (K). **Malawi**. N: Mzimba Dist., 11 km S of Eutini, 1250 m, fl.& fr. 31.i.1976, *Pawek* 10773 (K, MAL, MO, SRGH). C: Kasungu Dist., Kasungu to Bua, 1000 m, fl.& fr. 13.i.1959, *Robson & Jackson* 1125 (BM, K, LISC). S: Zomba Dist., Thondwe (Ntondwe), fl.& fr. 21.x.1905, *Cameron* 117 (K). **Mozambique**. N: Marrupa Dist., Marrupa, fl. 28.i.1981, *Nuvunga* 425 (K, LMU).

Also in D.R. Congo, Tanzania and Angola. *Brachystegia* and *Julbernardia* woodland, wooded grassland and bushland, usually on sandy soils but occasionally on rocky hills; 750–1700 m.

Conservation notes: Widespread; not threatened.

4. **Asystasia albiflora** Ensermu in Kew Bull. **53**: 930 (1998). —Vollesen in F.T.E.A., Acanthaceae 2: 455 (2010). Type: Tanzania, Chunya Dist., near Mbangala Village, 13.ii.1994, *Bidgood, Mbago & Vollesen* 2252 (K holotype, BR, C, CAS, DSM).

Erect, procumbent or scrambling perennial herb from an irregular woody rootstock, never stoloniferous; stems to 1 m long, sparsely pubescent to retrorsely sericeous. Leaves glabrous to sparsely sericeous along veins; petiole to 1 cm long; lamina ovate or narrowly hastate, largest 9–14.5 × 2.5–3.5 cm; apex acuminate, base truncate to subcordate, upper decurrent; margin entire. Raceme (3)5–20 cm long, terminal, simple, rarely branched from base, puberulous to retrorsely sericeous in two bands; flowers all opposite or alternate upwards, rarely all alternate, single; peduncle 2.5–8(13) cm long; bracts and bracteoles 3–5 mm long, triangular to linear-lanceolate, subglabrous, ciliate; pedicels 1–2(3 in fruit) mm long, glabrous to puberulous. Calyx 5–9(12 in fruit) mm long; lobes linear-subulate, with white margins, subglabrous, sometimes ciliate. Corolla pure white with pale yellow throat, 3.2–3.7 cm long, puberulous and with scattered capitate glands; basal cylindric tube 5–8 × c.2 mm, straight; throat 1.2–1.7 cm long, 6–8 mm wide at mouth; limb indistinctly 2-lipped; lobes in lower lip c.1 × 1 cm, in upper lip c.0.8 × 0.8 cm. Stamens didynamous, included in throat; filaments c.5 and 8 mm long, glabrous; thecae c.2 mm long, acute at base but not spurred. Style 1.7–2 cm long, pubescent at base and apex. Capsule 2.5–2.8 cm long, densely puberulous with capitate glands and non-glandular hairs. Seeds 5–5.5 × 4–4.5 mm, margin with dense to confluent tubercles forming a thick rim; surface with scattered tubercles near edges, middle part nearly smooth.

Zambia. N: Mbala Dist., Lunzua Valley, above Kafakula, 950 m, fl.& fr. 5.iii.1955, *Richards* 4787 (K). W: Ndola, fl.& fr. 29.i.1974, *Fanshawe* 12186 (K).

Also in Central African Republic, Sudan, D.R. Congo, Uganda and Tanzania. In chipya woodland and 'waste places'; 950–1300 m.

Conservation notes: Widespread; not threatened.

The material from W Zambia differs from that from E Africa (including NE Zambia) in the hastate leaves, occasionally approached in Tanzania; no other differences have been seen.

5. **Asystasia gangetica** (L.) T. Anderson in Thwaites, Enum. Pl. Zeyl.: 235 (1860). —
Bolnick, Guide Wild Fl. Zambia: 41 (2007). —Bandeira, Bolnick & Barbosa, Fl.
Nat. Sul Moçamb.: 188 (2007). Type: India, Ganges, *Herb. Linn.* 28.27 (LINN
lectotype, K photo), lectotypified by Ensermu (1994).

Justicia gangetica L., Amoen. Acad. **4**: 299 (1759).

Asystasia coromandeliana Nees in Wallich, Pl. Asiat. Rar. **3**: 89 (1832); in De Candolle,
Prodr. **11**: 165 (1847). —Clarke in F.T.A. **5**: 131 (1899); in Fl. Cap. **5**: 42 (1901). Type: India,
no locality, *Herb. Wallich* 2399 (K lectotype), lectotypified here.

Subsp. **micrantha** (Nees) Ensermu in Proc. 13th Meeting AETFAT **1**: 343 (1994). —
Vollesen in F.T.E.A., Acanthaceae **2**: 459 (2010). Type: Sudan, Sennar, *Acerbi* in
Herb. DC. 687 (G holotype). FIGURE 8.6.**45**.

Asystasia coromandeliana Nees var. *micrantha* Nees in De Candolle, Prodr. **11**: 165 (1847).

Asystasia podostachys Klotzsch in Peters, Naturw. Reise Mossamb. **6**: 199 (1861). Type:
Tanzania, Zanzibar, *Peters* s.n. (B† holotype).

Asystasia subhastata Klotzsch in Peters, Naturw. Reise Mossamb. **6**: 200 (1861). Type:
Mozambique, Mocuba Dist., Boror, *Peters* s.n. (B† holotype).

Asystasia floribunda Klotzsch in Peters, Naturw. Reise Mossamb. **6**: 200 (1861). Type:
Mozambique, Mocuba Dist., Boror, *Peters* s.n. (B† holotype).

Asystasia acuminata Klotzsch in Peters, Naturw. Reise Mossamb. **6**: 201 (1861). Type:
Mozambique, Pemba Dist., Querimba, *Peters* s.n. (B† holotype).

Asystasia pubescens Klotzsch in Peters, Naturw. Reise Mossamb. **6**: 202 (1861). Type:
Mozambique, Mossuril Dist., Anjoana Is. and Mozambique (Mossambique) Is., *Peters* s.n.
(B† holotype).

Asystasia scabrida Klotzsch in Peters, Naturw. Reise Mossamb. **6**: 202 (1861). Type:
Mozambique, Mossuril Dist., Mozambique (Mossambique) Is., *Peters* s.n. (B† holotype).

Asystasia multiflora Klotzsch in Peters, Naturw. Reise Mossamb. **6**: 203 (1861). Type:
Tanzania, Zanzibar, *Peters* s.n. (B† holotype).

Asystasia querimbensis Klotzsch in Peters, Naturw. Reise Mossamb. **6**: 204 (1861). Type:
Mozambique, Pemba Dist., Querimba Is., *Peters* s.n. (B† holotype).

Asystasia coromandeliana Nees var. *linearifolia* S. Moore in J. Linn. Soc. Bot. **40**: 160 (1911).
Type: Mozambique, Beira Dist., Beira, 25.xii.1906, *Swynnerton* 1949 (BM holotype).

Asystasia pinguifolia T. Edwards in S. Afr. J. Bot. **53**: 231 (1987). Type: Mozambique,
Jangamo Dist., "Inharrime", 15 km S of Inhambane, near Mutamba, 6.vii.1971, *Edwards &*
Vahrmeyer 4228 (PRE holotype).

Perennial or suffrutescent herb, rarely annual, with an irregular rootstock, occasionally
rooting from lower nodes; stems erect, creeping-ascending or straggling, to 1.5(2) m long,
usually much-branched, glabrous to densely pubescent or retrorsely sericeous. Leaves glabrous
to pubescent, densest on veins; petiole to 5 cm long; lamina ovate or narrowly so, more rarely
lanceolate to broadly elliptic, largest 2–11.5 × 1–5(6.5) cm; apex acuminate to rounded, base
cordate to truncate or lower attenuate to rounded, sometimes abruptly narrowed into a winged
petiole; margin entire. Raceme 1–13(28) cm long, terminal, on axillary branches or rarely
axillary, simple, rarely branched, subglabrous to retrorsely sericeous, rarely puberulous or
pubescent; all flowers alternate, rarely the lower opposite; peduncle 0.5–10(13) cm long; bracts
and bracteoles 1–2 mm long, triangular, distinctly white-edged, glabrous to puberulous and
distinctly ciliate; pedicels 0.5–3 mm long, subglabrous to pubescent, erect in fruiting stage. Calyx
4–9 mm long; lobes subulate to linear-lanceolate, acuminate to cuspidate, sparsely puberulous
to densely ciliate-pubescent, sometimes with capitate glands. Corolla white, rarely pale mauve or
pale blue, lower mid-lobe with large violet or purple patch, 1–2.5 cm long, sparsely puberulous
with non-glandular hairs and few to many capitate glands; basal cylindric tube 2–7 × 1.5–2.5 mm;
throat 4–11 mm long, 4–9 mm wide at apex; limb 2-lipped; lobes unequal, 3–8 mm long, midlobe
in lower lip largest. Stamens didynamous, included in tube; filaments 3–5 and 4–6 mm long;
thecae c.2 mm long, spurred. Style 7–15 mm long, puberulous for ⅓ to ½ its length. Capsule
17–28 × 4–5 mm, held erect, puberulous or densely so with capitate glands and non-glandular
hairs, rarely subglabrous. Seeds 4–5.5 × 4–4.5 mm, margin ± distinctly thickened from confluent
tubercles, edge straight; surface tuberculate, densest towards margin.

Caprivi. 5 km N of Katima Mulilo on Ngema road, 925 m, fl. 22.xii.1958, *Killick & Leistner* 3018 (K, PRE). **Botswana**. N: Ngamiland Dist., Kwando Area, fl. 22.i.1976, *Williamson* 71 (K, LISC, SRGH). **Zambia**. B: Senanga Dist., 76 km S of Senanga, W bank of Zambezi R., Sioma Falls, 1050 m, fl. 1.ii.1975, *Brummitt et al.* 14218 (K). N: Mbala Dist., Chilongwelo, 1475 m, fl. 14.ii.1952, *Richards* 703 (K). W: Ndola, fl. 2.ii.1954, *Fanshawe* 773 (EA, K). C: Lusaka Dist., 13 km SE of Lusaka, Protea Hill Farm, 1300 m, fl.& fr. 8.ii.1996, *Bingham* 10895 (K). E: Chipata Dist., Chikowa Mission, 1000 m, fl.& fr. 13.x.1958, *Robson & Angus* 74 (BM, K, LISC). S: Mazabuka Dist., Kafue Gorge, below dam, 1000 m, fl.& fr. 9.ix.1972, *Strid* 2084 (C, K). **Zimbabwe**. N: Murewa Dist., Shavanhohwe (Shanawe) R., fl.& fr. 14.i.1937, *Eyles* 8930 (K). W: Hwange Dist., Victoria Falls, 900 m, fl. 30.i.1934, *Saunders-Davies* s.n. (BM). C: Harare, fl. 15.i.1933, *Eyles* 7286 (K). E: Mutare Dist., E Vumba Mts, 1100 m, fl. 19.v.1957, *Chase* 6497 (K, SRGH). S: Chiredzi Dist., N bank of Rundi R., Chipinda Pools, 300 m, 13.i.1961, *Goodier* 61 (K, LISC, SRGH). **Malawi**. N: Chitipa Dist., Misuku Hills, Mughesse Forest, 1575 m, fl. 25.iv.1972, *Pawek* 5191 (K). C: Lilongwe Dist., 1 km SW of Malingunde, 1100 m, fl.& fr. 24.ii.1982, *Brummitt et al.* 16060 (K, MAL). S: Machinga Dist., 8 km on Ntaje road from Zomba–Liwonde road, Chikala Hills, 900 m, fl.& fr. 3.iv.1991, *Bidgood & Vollesen* 2169 (K, MAL). **Mozambique**. N: Ilha de Moçambique Dist., Mossuril Bay, Goa Is., 5 m, fl. 10.viii.1964, *Leach* 12349 (K, SRGH). Z: Lugela Dist., M'gulumi Mission, fl. i.1945, *Faulkner* K125 (K). MS: Muanza Dist., Cheringoma, Chinizuia R. estuary, fl. v.1973, *Tinley* 2888 (K, LISC, SRGH). GI: Bilene Dist., between Magul village and Macia, fl.& fr. 1.vi.1959, *Barbosa & Lemos* 8558 (K, LISC). M: Matutuine Dist., E of Zitundo, Ponta Mamoli, 10 m, fl. 26.xi.2001, *Goyder* 5003 (K).

Throughout tropical Africa and South Africa, also in Madagascar and the Mascarene Is., tropical Arabia, introduced into Malaysia. On forest margins and in clearings and a wide variety of woodland, wooded grassland and bushland, shorelines, roadsides, ditches, plantations, often in disturbed or secondary vegetation; 0–1600 m.

Conservation notes: Widespread; not threatened.

Subsp. *gangetica*, an erect shrubby herb with a purple corolla 2.5–4 cm long, is native to India, Sri Lanka, Thailand, Malaysia, Philippines and the Pacific. It has been introduced as an ornamental to Mauritius, Tropical Africa, Florida, Central and South America and the West Indies. In our area it has only been recorded from Harare in Zimbabwe (*Biegel* 5184).

This taxon is relatively uniform considering its wide distribution. Material from the Flora area is variable but, compared with similarly widespread taxa in other genera, not excessively so. Specimens from inland areas usually have a white corolla while specimens from the coast often have a mauve or pale blue corolla. Coastal forms tend to have smaller flowers.

There is a subglabrous form along the Mozambique coast (*Asystasia pinguifolia* T. Edwards) with broad, rounded fleshy leaves that we consider to be a form adapted to places subject to salt spray. Similar fleshy-leaved forms occur along the East African coast. The form often has some or most flowers opposite, a feature that also occurs sporadically in inland forms.

6. **Asystasia retrocarpa** T. Edwards in S. Afr. J. Bot. **57**: 305 (1991). Type: South Africa, KwaZulu-Natal, Ingwavuma, Ndumu Nature Reserve, Mkonjane, 22.xii.1971, *Pooley* 1601 (NU holotype, PRE).

Much branched annual or short-lived perennial herb; stems erect to decumbent, to 60 cm long, glabrous to pubescent with non-glandular hairs. Leaves glabrous to sparsely pubescent, mostly along veins; petiole indistinct, to 5 mm long; lamina linear-lanceolate to ovate, largest 3–9 × 0.3–2.5 cm; apex acute, base attenuate; margin entire. Racemes axillary and terminal, with 2–4 alternate flowers, glabrous to sparsely pubescent in two bands; peduncles 1–4 cm long; bracts and bracteoles

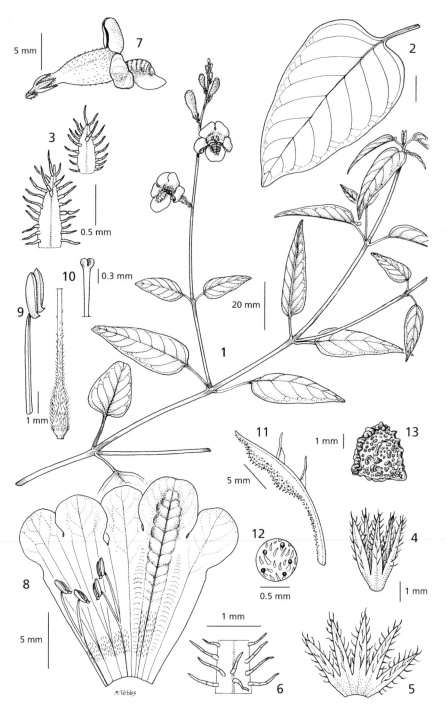

Fig. 8.6.**45**. ASYSTASIA GANGETICA subsp. MICRANTHA. 1, habit; 2, leaf; 3, bract (left) and bracteole (right); 4, calyx; 5, calyx opened up; 6, detail of calyx indumentum; 7, corolla; 8, corolla, opened up; 9, apical part of filament and anther; 10, ovary, basal part of style and stigma; 11, capsule; 12, detail of capsule indumentum; 13, seed. 1 & 3–10 from *Congdon* 380, 2 from *Mwangoka* 4366, 11–12 from *Peter* 60689, 13 from *Abeid* 119. Drawn by Margaret Tebbs. Reproduced from Flora of Tropical East Africa (2010).

1–3 mm long, linear-subulate, glabrous; pedicels 3–10 mm long, densely sericeous-pubescent with retrorse white hairs, sharply reflexed in fruiting stage. Calyx 5–7(10 in fruit) mm long; lobes subequal, linear-subulate, sparsely ciliate-pubescent, sometimes with scattered capitate glands. Corolla white with mauve markings on lower lip, 15–16 mm long, sparsely pubescent and with scattered glandular hairs; basal cylindric tube 4.5–5 × c.2 mm, throat c.6 mm long, c.4 mm wide at mouth; limb moderately 2-lipped; lips c.5 mm long with lobes 4–5 × 3–4 mm in lower lip and c.2 × 2 mm in upper. Stamens included or anthers just visible at corolla mouth; anthers muticous, thecae c.2 mm long. Style 6–7 mm long, glabrous. Capsule pendulous, 12–20 × 3–4 mm, glabrous. Seeds ellipsoid, 4.5–5 × 2.5–3 mm, margin and rim smooth; testa with indistinct ridges and tuberculate.

Zimbabwe. N: Makonde Dist., Trelawney, fl. 29.iii.1943, *Jack* 63 (K, SRGH). S: Chiredzi Dist., Chipinda Pools, fl.& fr. 20.i.1960, *Goodier* 854 (K, LISC, SRGH). **Mozambique**. M: Boane Dist., Maputo, 30 km on Boane–Moamba road, n.d., *Marques* 2697 (LMU, cited in Edwards 1991).

Also in N and E South Africa. In mopane woodland; 50–1200 m.

Conservation notes: Widespread; not threatened.

Corolla and stamen measurements have mostly been taken from Edwards (1991). Only one collection (*Jack* 63) has a corolla, but this is glued to the sheet.

7. **Asystasia mysorensis** (Roth) T. Anderson in J. Linn. Soc. Bot. **9**: 524 (1867) as *mysurensis*. —Vollesen in F.T.E.A., Acanthaceae **2**: 466 (2010). Type: India, Mysore (Maysure), 28.vii.1800, *Heyne* s.n. (B† holotype, K).

　　Ruellia mysorensis Roth in Nov. Pl. Sp.: 303 (1821).

　　Asystasia schimperi T. Anderson in J. Linn. Soc., Bot. **7**: 53 (1863). —Clarke in F.T.A. **5**: 135 (1899); in Fl. Cap. **5**: 43 (1901). —Setshogo, Checklist Pl. Botswana: 17 (2005). Type: Ethiopia, Goelleb, 24.viii.1854, *Schimper* in *Hohenacker* 2220 (K lectotype); Tigray, Dscheladscheranne, 31.vii.1840, *Schimper* III.1657 and *Schimper* III.1659 (K syntypes), lectotypified here.

　　Calophanes crenata Schinz in Bull. Herb. Boiss. **3**: 415 (1895). Type: South Africa, KwaZulu-Natal, Phoenix Station, 14.vii.1893, *Wood* 4967 in part (B† holotype, BM), see note below.

　　Glossochilus parviflorus Hutch. in Bull. Misc. Inform., Kew **1909**: 186 (1909). Type: South Africa, Limpopo Prov., Potgeiters Rust, 28.ix.1908, *Leendertz* 1265 (K holotype).

Annual herb; stems erect to decumbent, rarely straggling, to 0.5(1 in straggling plants) m long, sparsely to densely puberulous to pubescent when young. Leaves pubescent or sparsely so along veins; petiole to 2(4) cm long; lamina ovate to elliptic or narrowly so, largest 3–8.5(11) × 0.8–3(4) cm; apex acuminate to rounded, base attenuate to cuneate; margin entire, ciliate. Spikes 1–5(7) cm long, terminal, simple, with opposite flowers, rarely some alternate, pubescent or sparsely so; peduncle 0.5–3.5 cm long; bracts green, lower foliaceous, 8–20 mm long in middle of spike, narrowly ovate or narrowly elliptic to ovate or elliptic, acuminate to acute, sparsely puberulous and fimbriate-ciliate; bracteoles yellowish green, 5–12 mm long, lanceolate to narrowly elliptic or narrowly ovate, acuminate, subglabrous to sparsely puberulous and fimbriate-ciliate; pedicels 0–2 mm long. Calyx 4–7 mm long; lobes subulate to linear, puberulous with capitate glands only or intermixed with non-capitate glands. Corolla pure white, sometimes with greenish-yellow spots on lower midlobe and palate, 1–1.5 cm long, sparsely puberulous with capitate glands; basal cylindric tube c.2 × 1 mm; throat 4–6 mm long, 4–5 mm wide at mouth; limb indistinctly 2-lipped, lobes 4–7 × 3–6 mm, subequal, recurved, lower midlobe largest. Stamens didynamous, included; filaments 3–3.5 and 4.5–5.5 mm long, glabrous (short) or with a row of capitate glands; anthers c.1.5 mm long, thecae dorsally with capitate glands, apiculate or minutely spurred. Style 4.5–5.5 mm long, glabrous. Capsule 1.5–2.2 cm long, finely puberulous with capitate and non-capitate glands, with longer hairs at apex. Seeds 4–4.5 × 3–3.5 mm, margin toothed (lobate); testa rugose-tuberculate.

Botswana. N: Ngamiland Dist., Maun, fr. 5.iv.1967, *Lambrecht* 145 (K, SRGH). **Zambia**. W: Kitwe, fl.& fr. 8.ii.1967, *Mutimushi* 1804 (K, NDO). C: Lusaka Dist., 13 km SE of Lusaka, Protea Hill farm, 1300 m, fl.& fr. 8.ii.1996, *Bingham* 10894 (K). **Zimbabwe**. W: Matobo Dist., 12 km SE of Figtree on Matopos road, Figtree Trial Plot 2, fl.& fr. 29.ii.1974, *Mavi* 1490 (K, SRGH). E: Chimanimani Dist., Save Valley, Nyanyadzi, 550 m, fl.& fr. 3.ii.1948, *Wild* 2491 (K, SRGH). S: Gwanda Dist., Marangudzi, 600 m, fl.& fr. 10.v.1958, *Drummond* 5737 (K, LISC, SRGH). **Mozambique**. MS: Chibabava Dist., Chibabava, fl.& fr. 25.viii.1947, *Pimenta* 39 (K, SRGH). M: Maputo Dist., Matola, fl.& fr. 18.xi.1980, *Jansen* 7577 (K, LMA).

Also in Sudan, Ethiopia, D.R. Congo, Rwanda, Burundi, Uganda, Kenya, Tanzania, Namibia, South Africa, Yemen and India. In grassland and disturbed areas, sometimes a weed; 5–1300 m.

Conservation notes: Widespread; not threatened.

As already pointed out by Clarke (1901), *Wood* 4967 is a mixed collection; the other species involved is *Dyschoriste nagchana* (Nees) Bennet. The specimen at Kew is wholly the latter, while the one at BM is *Asystasia mysorensis* and thus an isotype.

The young leaves are edible (*Pimenta* 39).

8. **Asystasia welwitschii** S. Moore in J. Bot. **18**: 308 (1880). —Clarke in F.T.A. **5**: 134 (1899). Type: Angola, Huíla, Pungo Andongo, ii.1857, *Welwitsch* 5105 (BM lectotype, K), lectotypified here.

> *Asystasia welwitschii* var. *stenophylla* S. Moore in J. Bot. **49**: 298 (1911). Type: Angola, Malange Dist., 10.viii.1903, *Gossweiler* 1068 (BM lectotype, K), lectotypified here.

Perennial herb with several stems from a woody rootstock; stems usually pyrophytic and erect but becoming decumbent if not burnt, to 25 (60 in decumbent stems) cm long, sparsely to densely sericeous-pubescent. Leaves often immature in flowering state, glabrous to pubescent, densest along veins; petiole usually lacking, rarely to 2 cm long; lamina lanceolate to ovate, largest 3.5–8.5 × 0.8–3 cm; apex acute to acuminate, base rounded to attenuate; margin entire. Raceme 5–10(23) cm long, terminal, sometimes with pair of branches from lower axils, flowers opposite, sparsely pubescent, sometimes with sub-sessile glandular hairs; peduncle 0.5–3 cm long; bracts 5–11 × 1.5–4 mm, ovate to elliptic, acute, ciliate-pubescent or sparsely so, with dense small sub-sessile glands; bracteoles 3.5–9 × 1–3 mm, with similar indumentum; pedicels of lower flowers 0.5–3(5 in fruit) mm long, upper flowers subsessile. Calyx 11–20 mm long; lobes linear-lanceolate, cuspidate, densely finely puberulous, with dense sub-sessile glands, usually also with scattered long hairs. Corolla white with pale purple stripes or markings, 27–38 mm long, puberulous outside with glandular and non-glandular hairs; basal cylindric tube 7–11 × 2.5–4 mm, throat 14–18 mm long, 7–11 mm wide at mouth; limb indistinctly 2-lipped, lower mid-lobe 10–11(14) × 9–10 mm, lateral lobes 9–10(12) × 7–8 mm, upper lobes 5–6 × 4–5 mm. Stamens didynamous, filaments 7–8 and 8–10 mm long; anthers 3.5–4 mm long, thecae spurred. Style 15–17 mm long, pubescent at base or glabrous. Capsule 33–42 mm long, densely pubescent with glandular and non-glandular hairs. Seeds 6–7 × c.5 mm, elongate with 2 roundish projections, testa ± smooth.

Zambia. W: Mwinilunga Dist., Matonchi Farm, fl. 1930, *Milne-Redhead* 1021 (K); Solwezi Dist., Solwezi, fl. 26.vii.1964, *Fanshawe* 8847 (K).

Also in D.R. Congo, Angola and Namibia. In burnt dambo grassland on loamy soils, chipya woodland and bushland on limestone, and *Cryptosepalum* woodland; c.1400 m.

Conservation notes: Local in the Flora are but probably not threatened.

29. PSEUDERANTHEMUM Radlk.[7]

Pseuderanthemum Radlk. in Sitzungsber. Math.-Phys. Cl. Königl. Bayer. Akad. Wiss. München **13**: 282 (1883). —Milne-Redhead in Bull. Misc. Inform., Kew **1936**: 259 (1936). —Champluvier in Syst. Geogr. Pl. **72**: 33–53 (2002).

> *Pigafetta* Adans., Fam. **2**: 223 (1763), rejected name, non (Blume) Becc.
>
> *Eranthemum* sensu Clarke in F.T.A. **5**: 169 (1899) in part, non L.

---

[7] By Kaj Vollesen

Perennial herbs, subshrubs or shrubs; cystoliths conspicuous, linear; stems subquadrangular or rounded. Leaves opposite, entire to crenate. Flowers single or in 3–7-flowered cymules aggregated into long racemoid cymes. Bracteoles present, narrowly triangular. Calyx divided almost to base into 5 equal linear-lanceolate to narrowly triangular lobes. Corolla hairy and/ or glandular on outside; basal tube long and linear (rarely short), straight or slightly curved, widened into a short throat; lobes 5, wider in lower lip, spreading in lower lip, erect in upper. Stamens 2, inserted at base of throat and held dorsally, included to slightly exserted, glabrous, with 2 small staminodes; anthers bithecous, oblong, straight or curved, apiculate or rounded at both ends, glabrous, hairy or glandular on connective. Ovary with 2 ovules per locule; style glabrous; stigma of 2 equal, broadly ellipsoid erect lobes. Capsule (2)4-seeded, clavate with contracted solid, stalk-like basal part, glabrous, seed-bearing part ellipsoid, laterally flattened; retinacula strong, hooked. Seeds discoid to ellipsoid, rugose to reticulate on both sides or smooth on outer side.

A genus of 50–75 species widely distributed across all tropical regions. Particularly diverse in SE Asia and the Pacific Region where several species complexes are in severe need of monographic work.

Corolla orange-red to scarlet; seed deeply reticulate on inner surface, smooth or very
   slightly reticulate on outer surface, rim entire; branches pale yellow to orange-
   brown, smooth and glossy; plant drying black . . . . **2.** *hildebrandtii*
–   Corolla white to mauve; seed reticulate-rugose on both sides, rim jagged; branches
   not smooth and glossy; plant drying green. . . . .**1.** *subviscosum*

1. **Pseuderanthemum subviscosum** (C.B. Clarke) Stapf in Bot. Mag. **135**: t.8244 (1909). —Strugnell, Checklist Mt. Mulanje: 35 (2006). —Vollesen in F.T.E.A., Acanthaceae 2: 480 (2010). Type: Mozambique, Gurué Dist., 'Makua', Namuli Hills, 1887, *Last* s.n. (K lectotype, BM), lectotypified by Milne-Redhead (1936). FIGURE 8.6.**46**.

Eranthemum subviscosum C.B. Clarke in F.T.A. 5: 173 (1899).

Eranthemum lindaui C.B. Clarke in F.T.A. 5: 173 (1899). Type: Tanzania, Lushoto Dist., E Usambara Mts, Mashewa, vii.1893, *Holst* 3494 (K lectotype); E Usambara Mts, Nderema, n.d., *Holst* 2251 (B† syntype), lectotypified here.

Pseuderanthemum albo-coeruleum Champl. subsp. *robustum* Champl. in Syst. Geogr. Pl. **72**: 37 (2002). Type: Tanzania, Morogoro Dist., Mtibwa Forest Res., xi.1954, *Semsei* 1948 (K holotype, EA).

Pseuderanthemum tunicatum sensu Milne-Redhead in Bull. Misc. Inform., Kew **1936**: 264 (1936) in part. —White *et al.*, Evergr. For. Fl. Malawi: 118 (2001). —Phiri, Checklist Zamb. Vasc. Pl.: 19 (2005), non (Afz.) Milne-Redh.

Erect shrubby herb or shrub to 1.25(2?) m tall, plant drying green; young stems puberulous or sparsely so with broad curved or curly hairs with purple walls. Leaves glabrous or sparsely puberulous on midrib; petiole to 3(5.5) cm long; lamina elliptic to slightly obovate, largest 7.5–20 × 2.5–8 cm; apical part abruptly or gradually narrowing below an acute to acuminate apex, tip acute to obtuse; base attenuate to cuneate (rarely truncate). Cymes 2–21 cm long, unbranched or with lateral cymes to 7 cm long from lower nodes, sometimes with elongated cymes from upper leaf axils or with sessile cymules here; flowers all solitary or in 3(7)-flowered cymules towards base of cyme (rarely throughout); peduncle to 19 cm long, with 1–2(3) pairs of sterile bracts if cyme unbranched; axis puberulous with mixture of short straight and curved or curly hairs with purple walls; bracts linear to narrowly triangular, sparsely puberulous, 2–6 mm long; pedicels 1–4 mm long, puberulous with short straight stubby hairs; bracteoles as bracts, 2–4 mm long. Calyx puberulous with short straight stubby hairs, with or without stalked capitate glands, 3–6 mm long. Corolla white to mauve, subglabrous to sparsely puberulous outside with curly hairs with purple walls, and scattered capitate glands; tube straight or slightly curved, basal linear part 1.6–2.5 cm long, throat 2–3 mm long; lobes 8–11 × 5–7(9) mm in lower lip, elliptic, rounded or 3-toothed. Stamens included or slightly exserted; filaments c.2 mm long, glandular;

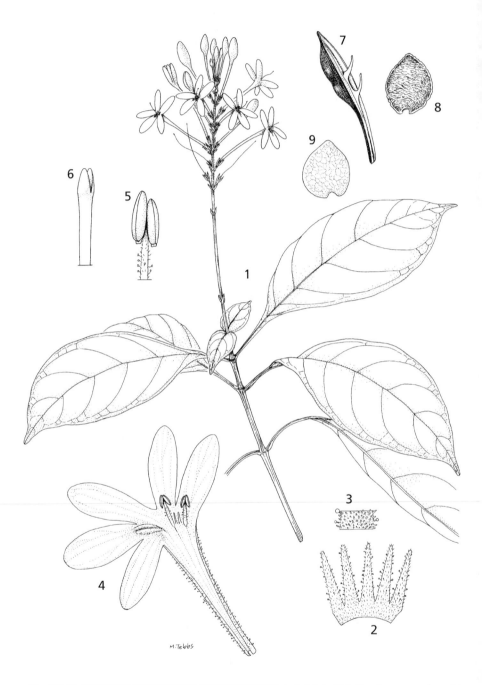

Fig. 8.6.**46**. PSEUDERANTHEMUM SUBVISCOSUM. 1, habit (× ²/₃); 2, calyx, opened up (× 6); 3, detail of calyx indumentum (× 15); 4, corolla, opened up (× 2); 5, apical part of filament and anther (× 10); 6, apical part of style and stigma (× 16); 7, capsule (× 2); 8 & 9, seed, inner (8) and outer (9) surface (× 4). 1, 6 from *Chapman* 6344, 2–5 from *Chapman* 7996, 7–9 from *Drummond* 5014. Drawn by Margaret Tebbs.

anthers pale pink to purple, 1–1.5 mm long, apiculate or rounded, hairy on connective. Capsule 1.8–2.3 cm long. Seeds c.4 × 4 mm, densely reticulate-verrucose on inside, smoother on outside, with jagged rim.

Zambia. N: Mbala Dist., Lunzua R. gorge, 900 m, fl.& fr. 11.vii.1960, *Richards* 12869 (EA, K). **Zimbabwe**. E: Chipinge Dist., Chirinda Forest, 1100 m, fl. ix.1962, *Goldsmith* 194/62 (K, LISC, SRGH). **Malawi**. N: Nkhata Bay Dist., Chombe Estate, fl.& fr. 6.ix.1955, *Jackson* 1754 (EA, K, MAL). S: Mt Mulanje, Pamba stream, fl. 25.viii.1986, *Chapman* 7996 (K, MO). **Mozambique**. Z: Lugela Dist., Mt Mabu, 1150 m, fl.& fr. 14.x.2008, *Mphamba et al.* 61 (K, LMA, LMU, MAL). MS: Sussundenga Dist., Dombe, Mt Tchianganhi, fr. 19.xi.1965, *Pereira & Marques* 780 (LMU). GI: Bilene Dist., Chirindzene, road to Chiconela, fr. 19.ix.1980, *Nuvunga et al.* 345 (BM).

Also in Cameroon, Gabon, Central African Republic, Congo (Brazzaville), D.R. Congo, Sudan, Ethiopia, Uganda, Kenya, Tanzania, Angola, South Africa, Madagascar and the Comoro Is. In lowland and montane forest, and riverine forest; 350–1500 m.

Conservation notes: Widespread; not threatened.

2. **Pseuderanthemum hildebrandtii** Lindau in Bot. Jahrb. Syst. **20**: 39 (1894). — Vollesen in F.T.E.A., Acanthaceae **2**: 482 (2010). Type: Zanzibar, Kidoti, xi.1873, *Hildebrandt* 981 (K lectotype, BM), lectotypified by Milne-Redhead (1936).

Eranthemum hildebrandtii (Lindau) C.B. Clarke in F.T.A. **5**: 172 (1899).

Erect shrubby herb or shrub to 2 m tall; plant drying black; young stems pale yellow to orange-brown, smooth and shiny, glabrous to puberulous. Leaves glabrous to sparsely pubescent, densest along veins; petiole to 2(4) cm long; lamina ovate, largest 2.5–16 × 1–7 cm; apex acuminate to obtuse, base attenuate and decurrent to cuneate, abruptly narrowed or not below middle. Cymes to 12 cm long, often much less, sometimes reduced to apparently solitary flowers in upper leaf axils, unbranched or branched, often with additional cymes from upper leaf axils; flowers solitary or in 3–7-flowered cymules, sometimes with 2 cymules per bract at lower nodes; axis glabrous to sparsely puberulous; lower pair(s) of bracts usually foliaceous (in reduced cymes looking as if flowers all solitary and axillary), others lanceolate to narrowly triangular, glabrous or sparsely puberulous-ciliate, 1–3 mm long; pedicels 1–3(4) mm long, glabrous; bracteoles as bracts, often with narrow scarious margin, 1–2 mm long. Calyx sparsely ciliate on outside, inside puberulous with thick hairs, 3–5 mm long. Corolla orange-red to scarlet, with darker centre, sparsely puberulous outside, rarely subglabrous, with a mixture of hairs and capitate glands, sometimes ± entirely glands; tube straight, basal linear part 2.2–3.2 cm long, throat 3–4 mm long; lobes 9–16 × 4–8 mm in lower lip, elliptic-obovate, rounded. Stamens exserted; filaments 2–3 mm long; anthers dark purple, 1–1.5 mm long, rounded to apiculate, glabrous on connective. Capsule 1.9–2.7 cm long. Seeds 4–6 × 3–5 mm, deeply reticulate on inner side, smooth to slightly reticulate on outer side, rim smooth.

Mozambique. Z: Serra do Sidi, fl. 10.ii.1905, *Le Testu* 689 (BM).

Also in Kenya, Tanzania and Zanzibar. Coastal and lowland forest and thicket; altitude unknown.

Conservation notes: Known from only one locality in the Flora area; probably Vulnerable here, but not threatened elsewhere.

This is the only collection of this species from the Flora area. It is otherwise widespread in lowland forests and thickets in Kenya and Tanzania. There is no reason to doubt the veracity of the specimen, and further collection in N Mozambique might well lead to its discovery here.

The red flowers and smooth glossy bark are abnormal in *Pseuderanthemum*, characters in which *P. hildebrandtii* tends towards *Ruspolia*. The pale yellow to reddish brown glossy bark and protruding leaf bases are also distinct.

30. RUSPOLIA Lindau [8]

Ruspolia Lindau in Engler & Prantl, Nat. Pflanzenfam. **IV**, 3b: 354 (1895). —Milne-Redhead in Bull. Misc. Inform., Kew **1936**: 269 (1936).
Eranthemum sensu Clarke in F.T.A. **5**: 169 (1899) in part, non L.

Shrubby herbs or shrubs; cystoliths conspicuous, linear; stems subquadrangular or rounded. Leaves opposite, entire or slightly crenate. Flowers single or in 3–7-flowered cymules aggregated into long racemoid cymes. Bracteoles present, narrowly triangular. Calyx deeply divided into 5 equal linear-lanceolate to filiform lobes. Corolla hairy and/or glandular on outside; basal tube long, linear, only slightly widened into a short throat; lobes 5, subequal or wider in lower lip, all spreading or deflexed. Stamens 2, inserted at base of throat and held dorsally there, usually slightly exserted, glabrous; anthers monothecous, straight or slightly curved, apiculate or rounded, glabrous. Ovary with 2 ovules per locule; style glabrous; stigma of 2 equal, broadly ellipsoid erect lobes. Capsule (2)4-seeded, clavate with contracted solid stalk-like basal part, glabrous, seed-bearing part ellipsoid, laterally flattened; retinacula strong, hooked. Seeds discoid, inner and outer surfaces different, with a strong raised marginal rim.

A genus of 5 species, 4 in tropical Africa with one extending to Madagascar, one endemic in Madagascar.

1. Inflorescence axis, bracts and calyx with stalked capitate glands; corolla outside with long curly hairs, without stalked capitate glands; capsule 2–3 cm long; seed 5–7 mm long, inner surface with irregular ridges; plant drying green . . . . **3**. *decurrens*
 – Inflorescence axis, bracts and calyx without stalked capitate glands; corolla outside with mixture of long curly hairs and capitate glands, or with capitate glands only; capsule 2.8–4.8 cm long; seed 9–10 mm long, inner surface with irregular ridges or smooth with broad marginal rim; plant drying green or black 2
2. Corolla outside with mixture of long curly hairs and stalked capitate glands; calyx glabrous or sparsely puberuous on lobes; capsule 2.8–4 cm long; inner surface of seed with irregular ridges; lower nodes in inflorescence 1.5–5 cm long; plant drying green . . . .**2**. *seticalyx*
 – Corolla outside with stalked capitate glands only; calyx puberulous-ciliate with curly hairs along whole length; capsule 3.8–4.8 cm long; inner surface of seed smooth; lower nodes in inflorescence to 1(2) cm long; plant drying black . . . . **1**. *australis*

1. **Ruspolia australis** (Milne-Redh.) Vollesen in F.T.E.A., Acanthaceae **2**: 486 (2010). Type: South Africa, Limpopo, Soutpansberg, Wyliespoort, iv.1934, *Schweikerdt & Verdoorn* 441 (PRE holotype, K).
 Ruspolia hypocrateriformis (Vahl) Milne-Redh. var. *australis* Milne-Redh. in Bull. Misc. Inform., Kew **1936**: 272 (1936). —Plowes & Drummond, Wild Fl. Zimbabwe: 131 (1990). —Mapura & Timberlake, Checklist Zimb. Vasc. Pl.: 14 (2004). —Setshogo, Checklist Pl. Botswana: 19 (2005).
 Ruspolia hypocrateriformis sensu Pooley, Wild Fl. KwaZulu-Natal: 78 (1998), non (Vahl) Milne-Redh.

Scrambling or decumbent (rarely erect?) shrub to 4 m tall; plant drying black; young stems glabrous or sparsely puberulous at nodes. Leaves glabrous to sparsely puberulous, densest on veins; petiole ill-defined, to 2 cm long; lamina ovate to elliptic, largest 7–14 × 3–7 cm; apex acute

---

[8] By Kaj Vollesen

to subacuminate with rounded tip, basal part below middle abruptly or gradually narrowed to a long decurrent base. Cymes 2–6 cm long, dense, with short lateral branches to 5 mm long in basal part (rarely unbranched), often also with lateral cymes from upper leaf axils; lower internodes to 1(2) cm long; flowers all in 3–7-flowered cymules; axis sparsely to densely sericeous-puberulous with curly hairs, no stalked capitate glands; lower 1–2 pairs of bracts sometimes foliaceous, others linear to narrowly triangular or ovate, ciliate with curly hairs, 6–12 mm long, persistent; pedicels 1–3 mm long, puberulous; bracteoles linear to narrowly triangular, ciliate with curly hairs, 6–10 mm long. Calyx 5–9 mm long, puberulous-ciliate with curly hairs. Corolla bright red, with darker spots towards centre, with scattered stalked capitate glands outside, no hairs; linear tube 2–2.8 cm long; throat 3–4 mm long; lobes 7–12 × 4–7(9) mm, elliptic, rounded. Filaments 3–4 mm long; anthers purple, c.1 mm long. Capsule 3.8–4.8 cm long. Seeds 9–10 × c.8 mm, smooth on outer surface, smooth with broad raised rim on inner surface.

Zimbabwe. W: Matobo Dist., Maleme Dam, fr. 30.iv.1967, *Biegel* 2099 (K, SRGH). E: Mutare Dist., commonage, Murahawa's (Murakwa) Hill, 1275 m, fl. 5.iii.1964, *Chase* 8133 (K, LISC, SRGH). S: Chivi Dist., Nyoni Mts, fl. 28.ii.1970, *Müller & Gordon* 1298 (K, SRGH). **Mozambique**. N: Mecúfi Dist., 5 km S of Pemba (Porto Amelia), fl. 21.iii.1960, *Gomes e Sousa* 4546 (K). T: Cahora Bassa Dist., Songo, bairro ZAMCO, fl. 18.iv.1972, *Macêdo* 5212 (LISC, LMU). M: Namaacha Dist., Namaacha, Cascata, fr. 22.iii.1983, *Zunguze & Boane* 457 (BM, LMU).

Also in Tanzania and N South Africa. Dry deciduous forest, coastal bushland, riverine scrub, rocky hills, termite mounds; near sea level–1000(1300) m.

Conservation notes: Widespread; not threatened.

2. **Ruspolia seticalyx** (C.B. Clarke) Milne-Redh. in Bull. Misc. Inform., Kew **1936**: 270 (1936). —Vollesen in F.T.E.A., Acanthaceae **2**: 486 (2010). Type: Malawi, Kondowa to Karonga, vii.1896, *Whyte* s.n. (K lectotype), lectotypified by Milne-Redhead (1936). FIGURE 8.6.**47**.

   Eranthemum seticalyx C.B. Clarke in F.T.A. **5**: 172 (1899).
   Pseuderanthemum seticalyx (C.B. Clarke) Stapf in Bot. Mag. **135**: t.8244 (1909).

Erect (rarely straggling) shrubby herb or shrub to 1.3 m tall; plant drying green; young stems glabrous to puberulous. Leaves glabrous below to puberulous along midrib and larger veins, glabrous above or with uniformly scattered hairs; petiole ill-defined, to 3 cm long; lamina ovate to elliptic or broadly so, largest 9–23 × 4.5–10.5 cm; apex acute to acuminate with acute to obtuse tip, basal part below middle abruptly narrowed to a long decurrent base. Cymes (3)5–21 cm long, unbranched, sometimes with lateral cymes from upper leaf axils; lower internodes 1.5–5 cm long, rarely some cymes condensed with almost absent internodes; flowers solitary or in 3–5(7)-flowered cymules towards base of cyme or throughout; axis puberulous or sparsely so, no stalked capitate glands; lower 1–2 bract pairs usually foliaceous, others subulate to narrowly triangular, 3–8 mm long, indumentum as axis; pedicels 0.5–2(4 in fruit) mm long; bracteoles subulate to narrowly triangular, 3–8 mm long. Calyx 5–8 mm long, glabrous or sparsely puberulous apically on lobes, without stalked capitate glands. Corolla orange-red to red or scarlet, with numerous darker spots towards centre, a mixture of long curly non-glandular hairs and stalked capitate glands outside; linear tube 1.7–3.2 cm long; throat 2–3 mm long; lobes 8–13 × 5–8 mm, elliptic, subacute to rounded. Filaments c.3 mm long; anthers purple, 1–1.5 mm long. Capsule 2.8–4 cm long. Seeds 9–10 × 7–8 mm, smooth on outer surface, with irregular ridges on inner surface.

Botswana. N: Ngamiland Dist., Dindinga Is., fl. 29.iii.1975, *P.A. Smith* 1329 (K, LISC, SRGH). **Zambia**. N: Mbala Dist., Mpulungu, 775 m, fl.& fr. 22.iii.1952, *Richards* 1196 (K). W: Kitwe Dist., Lake Kashiba, fr. 5.vi.1957, *Fanshawe* 3311 (EA, K). C: Lusaka Dist., Chilanga, fl. 4.iii.1966, *Lawton* 1365 (K). S: Kalomo Dist., 65 km on Kalomo–Kafue Nat. Park road, 1075 m, fl. 11.iii.1997, *Schmidt et al.* 2497 (K, MO). **Zimbabwe**. N: Kariba Dist., Chirundu to Zambezi R., 375 m, fl. 15.iii.1966, *Simon* 710 (EA, K, LISC, SRGH). **Malawi**. N: Rumphi Dist., Njakwa Gorge, 1075 m, fl. 4.iv.1969, *Pawek* 1917 (K). C: Dedza Dist., Mua Livulezi Forest, Sosola, fl.& fr. 15.v.1960, *Adlard* 380 (K, SRGH).

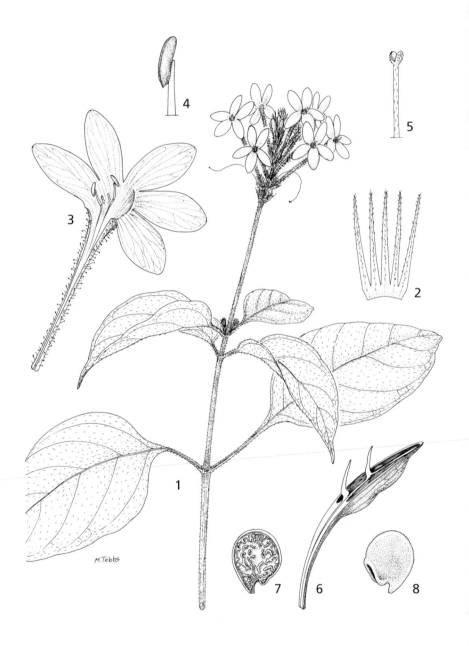

Fig. 8.6.**47**. RUSPOLIA SETICALYX. 1, habit (× ²/₃); 2, calyx, opened up (× 6); 3, corolla, opened up showing stamens and staminodes (× 2); 4, anther (× 10); 5, apical part of style and stigma (× 16); 6, capsule (× 2); 7, seed, inner lateral surface (× 2); 8, seed, outer lateral surface (× 2). 1–5 from *Welch* 576, 6–8 from *Mabberley* 1375. Drawn by Margaret Tebbs. Reproduced from Flora of Tropical East Africa (2010).

S: Mangochi Dist., Monkey Bay, Tumbi Is., 500 m, fl. 1.iii.1970, *Brummitt & Eccles* 8829 (EA, K). **Mozambique**. N: Lago Dist., Metangula, fl.& fr. 22.v.1948, *Pedro & Pedrógāo* 3842 (EA). MS: Cheringoma Dist., Gorongosa Nat. Park, 23 km from Inhaminga, Muzamba R. gorge, 225 m, fr. 9.vi.2012, *Bester* 11116 (K, PRE).
Also in D.R. Congo, Tanzania and Madagascar. Riverine forest and scrub, shady places in woodland, rocky hills, termite mounds; 200–1200(1500) m.
Conservation notes: Widespread; not threatened.

3. **Ruspolia decurrens** (Nees) Milne-Redh. in Bull. Misc. Inform., Kew **1936**: 269 (1936). —Bolnick, Guide Wild Fl. Zambia: 42 (2007). —Vollesen in F.T.E.A., Acanthaceae 2: 488 (2010). Type: Sudan, Kordofan, Milbes, Choor R., 29.ix.1839, *Kotschy* 276 (K lectotype); Ethiopia, Tigray, Tacazze Valley, 2.x.1838, *Schimper* II.773 (K syntype), lectotypified by Milne-Redhead (1936).

Eranthemum decurrens Nees in De Candolle, Prodr. **11**: 453 (1847). —Clarke in F.T.A. **5**: 170 (1899).

Eranthemum senense Klotzsch in Peters, Naturw. Reise Mossamb. **6**: 219 (1861). —Clarke in F.T.A. **5**: 171 (1899). Type: Mozambique, Lower Zambesi, Sena R., Boror, n.d., *Peters* s.n. (B† holotype).

Pseuderanthemum decurrens (Nees) Radlk. in Sitzungsber. Math.-Phys. Cl. Königl. Bayer. Akad. Wiss. München **13**: 286 (1883).

Pseuderanthemum senense (Klotzsch) Radlk. in Sitzungsber. Math.-Phys. Cl. Königl. Bayer. Akad. Wiss. München **13**: 286 (1883).

Shrubby herb to 1(1.5) m tall, erect, scrambling or decumbent with ascending flowering branches; plant drying green; young stems puberulous. Leaves glabrous or puberulous below along midrib and larger veins, glabrous or with uniformly scattered hairs above; petiole ill-defined, to 1.5 cm long; lamina ovate to elliptic, largest 5.5–17.5 × 2.5–7 cm; apex acute to acuminate with obtuse tip, basal part below middle abruptly (rarely gradually) narrowed to a long decurrent base. Cymes 5–40 cm long, unbranched, very rarely with lateral cymes from lowermost internodes, occasionally also with cymes from upper leaf axils, lower internodes (0.5)1–4(5) cm long (if short then cymes over 15 cm long); flowers solitary or in 3(5)-flowered cymules (or 1 flower with 2 non-developing buds) towards base of cyme; peduncle 0.5–1.5 cm long; axis puberulous and with stalked capitate glands; lower 1–3 pairs of bracts usually foliaceous, others subulate, 3–9(12) mm long, indumentum as axis; pedicels 0.5–2(4 in fruit) mm long; bracteoles subulate, 3–5(7) mm long. Calyx 5–9 mm long, puberulous with stalked capitate glands and non-glandular hairs. Corolla salmon-pink to orange red, with numerous darker spots towards centre, with long curly non-glandular hairs outside, without capitate glands; linear tube 1.8–2.8(3.2) cm long, throat 2–3 mm long, lobes 8–14 × 5–9 mm, elliptic, broadly rounded. Filaments 3–4 mm long; anthers purple, 1.5–2 mm long. Capsule 2–3 cm long. Seeds 5–7 × 5–7 mm, smooth on outer surface, with irregular ridges on inner surface.

Zambia. N: Mpika Dist., North Luangwa Nat. Park, 600 m, fl. 22.iv.1994, *P.P. Smith* 521 (K). W: Kitwe Dist., Lake Kashiba, Mpongwe, fl. 5.vi.1957, *Fanshawe* 3312 (K). C: Chibombo Dist., Chisamba, 1300 m, fl.& fr. 21.iv.1996, *Bingham* 10996 (K). S: Mazabuka Dist., Kafue Gorge, fl. 19.iv.1956, *Noak* 210 (K, SRGH). **Zimbabwe**. N: Guruve Dist., 10 km S of Dande Mission, Hunyani R., 450 m, fl. 15.v.1962, *Wild* 5745 (K, SRGH). W: Hwange Dist., 2 km N of Falls Road towards Inyantue R., 825 m, fl.& fr. 29.iii.1963, *Leach* 11610 (K, LISC, SRGH). E: Chimanimani Dist., Umvumvumvu R. Gorge, 700 m, fl.& fr. 21.iv.1963, *Chase* 7995 (K, LISC, SRGH). S: Mwenezi Dist., Lundi R., Fishans, fl. 28.iv.1962, *Drummond* 7794 (K, LISC, SRGH). **Malawi**. C: Salima Dist., Lifidzi, fl. 19.v.1985, *Patel & Nachamba* 2141 (K, MAL). S: Blantyre Dist., Shire R., Mpatamanga Gorge, 350 m, fl. 6.v.1989, *Goyder & Radcliffe-Smith* 3218 (K). **Mozambique**. N: Malema Dist., Mutuali, 600 m, fl.& fr. 22.vi.1947, *Hornby* 2766 (K). Z: Maganja da Costa Dist., from Namacurra (Inhamacurra) to Maganja da Costa, fl.& fr. 25.iii.1943, *Torre* 4980 (BR, COI, LISC, LMA, LMU, MO, WAG). T: Changara Dist., 20 km W of Changara,

500 m, fl. 30.iv.1960, *Leach* 9876 (K, SRGH). MS: Cheringoma Dist., Inhamitanga to Mupa, fl.& fr. 27.v.1948, *Mendonça* 4401 (BR, LISC, LMU).

Also in Central African Republic, Sudan, Ethiopia and Tanzania. In riverine forest and scrub, shady places in woodland and rocky hills, on termite mounds; 350–1300 m. Conservation notes: Widespread; not threatened.

The species is said to occur in Botswana (Germishuizen *et al.* in *Strelitzia* **14**: 105, 2003), but no collections are cited.

31. RUTTYA Harv. [9]

Ruttya Harv. in Hooker, London J. Bot. **1**: 27: (1842). —Nees in De Candolle, Prodr. **11**: 309 (1847).

Haplanthera Hochst. in Flora **26**: 71 (1843). —Nees in De Candolle, Prodr. **11**: 308 (1847).

Shrubs or shrubby herbs; cystoliths conspicuous, linear; stems rounded. Leaves opposite, entire. Flowers single, aggregated into short few-flowered to long many-flowered racemoid cymes. Bracteoles present, triangular. Calyx deeply divided into 5 equal subulate to narrowly triangular lobes. Corolla glabrous on outside; basal tube short and wide, widening slightly into a short throat; lobes 5, lower 3 strongly deflexed against tube, upper 2 erect, forming a hood. Stamens 2, with 2 minute to long staminodes; filaments attached near base of upper lip and held under it, glabrous, terete to flattened and winged; anthers monothecous, ellipsoid to oblong, straight or curved, with a short spur or mucronate to rounded at base, glabrous. Ovary with 2 ovules per locule; style glabrous; stigma of 2 equal, erect, broadly ellipsoid lobes. Capsule 4-seeded, clavate with contracted solid stalk-like basal part, glabrous, seed-bearing part ellipsoid, laterally flattened; retinacula strong, hooked. Seeds discoid, smooth on both sides, with a broad longitudinal ridge on outer side.

A genus of six species, one in N Ethiopia and Eritrea, one widespread in NE Africa and extending to Arabia, one in southern Africa and three on Madagascar.

Ruttya ovata Harv. in Hooker, London J. Bot. **1**: 27 (1842). —Clarke in F.T.A. **5**: 174 (1899); in Fl. Cap. **5**: 55 (1901). —Bandeira, Bolnick & Barbosa, Fl. Nat. Sul Moçamb.: 192 (2007). Type: South Africa, KwaZulu-Natal, Durban, Port Natal, n.d., *T. Williamson* s.n. (K holotype). FIGURE 8.6.**48**.

Duvernoia trichocalyx Lindau in Bot. Jahrb. Syst. **22**: 122 (1895). Type: South Africa, KwaZulu-Natal, Durban, iii.1894, *Kuntze* s.n. (B† holotype, K).

Erect or scrambling shrub to 3 m tall; young stems brownish, sparsely distichously puberulous, soon glabrescent; older branches greyish, with conspicuous cup-like scars. Leaves with a few curly hairs on midrib, otherwise glabrous; petiole to 3 mm long; lamina ovate, largest 4–7 × 2–3 cm; apex subacute with small mucro, base attenuate; margin slightly to distinctly wavy. Cymes 3–7 cm long, many-flowered, unbranched; flowers in 3-flowered cymules or single towards apex of cyme; axis puberulous; lower pair of bracts foliaceous, upper pairs linear to subulate, to 1 cm long, caducous; pedicels to 2 mm long, glabrous; bracteoles subulate, to 8 mm long. Calyx 1.2–1.5 cm long, divided to c.1 mm from base, glabrous; lobes subulate. Corolla white with small mauve dots on basal part of lower lip and extending into throat; tube c.5 mm long, c.3.5 mm wide at base, not widening upwards; throat c.3 mm long; lobes oblong, subacute to rounded, 8–10 × 3–4 mm. Filaments terete, glabrous, 6–8 mm long; anthers c.2 mm long, with small hooked spur at base. Capsule 2.8–4.8 cm long. Seeds broadly ellipsoid, 8–9 × 6.5–7.5 mm.

Mozambique. M: Maputo Dist., Maputo (Lourenço Marques), fl. 18.i.1948, *Faulkner* 188 (K); Maputo, Polana, road to Caracol, fl. 20.i.1960, *Lemos & Balsinhas* 11 (K, LISC).

---

[9] By Kaj Vollesen

Fig. 8.6.**48**. RUTTYA OVATA. 1, habit (× ²/₃); 2, calyx (× 4); 3, corolla, opened up (× 2); 4, apical part of filament and anther (× 4); 5, basal part of style and ovary (× 4); 6, apical part of style and stigma (× 16), 7, capsule (× 1); 8 & 9, seed, inner (8) and outer (9) surface (× 2). 1–6 from *Faulkner* 188, 7–9 from *Hobson* 2146. Drawn by Margaret Tebbs.

Also in E South Africa and Swaziland. Coastal bushland; 50 m.

Conservation notes: Possibly Near Threatened in the Flora area, but not threatened elsewhere.

32. BRACHYSTEPHANUS Nees[10]

Brachystephanus Nees in De Candolle, Prodr. **11**: 511 (1847). —Figueiredo in Kew Bull. **51**: 753–763 (1996). —Champluvier & Darbyshire in Syst. Geogr. Pl. **79**: 115–192 (2009).

Oreacanthus Benth. in Gen. Pl. **2**(2): 1104 (1876). —Champluvier & Figueiredo in Bull. Jard. Bot. Belg. **65**: 413–417 (1996).

Perennial herbs or subshrubs, sometimes monocarpic and cyclically mass-flowering (plietesial). Stems (sub-)4-angular, often sulcate between angles when young. Leaves evergreen or rarely deciduous, opposite-decussate, pairs equal to anisophyllous, blade margin crenulate to subentire; lower leaves petiolate, uppermost leaves often reduced and subsessile; cystoliths numerous, ± conspicuous. Inflorescences terminal and sometimes also in upper leaf axils; spiciform or paniculate thyrses, cymule branching dichasial or partially monochasial, or cymules reduced to a single flower; bract pairs equal, free; bracteoles paired, as bracts or much-reduced. Calyx tubular, divided almost to base into 5 linear(-lanceolate) lobes, subequal to unequal in length. Corolla bilabiate; tube cylindric to campanulate, long to very short; upper lip either subulate or more broadly ovate to elliptic, apex acute, obtuse, apiculate or emarginate; lower lip oblong to elliptic, apex minutely to conspicuously 3-lobed, rugula absent. Stamens 2; filaments attached at or just below mouth of corolla tube, ± long-exserted, glabrous; anthers monothecous, basally muticous. Staminodes absent. Disk annular or shallowly cupuliform. Ovary oblong-ellipsoid, glabrous or largely so, bi-locular, 2 ovules per locule; style filiform, ± long-exserted, glabrous; stigma either minutely clavate or conspicuously capitate, apex shallowly bilobed. Capsule stipitate, placental base inelastic. Seeds 4 per capsule, or 2 by abortion, held on retinacula, lenticular, tuberculate.

A genus of 22 species distributed in tropical Africa and Madagascar. Four species occur within the Flora area, one (*B. africanus*) within Sect. *Brachystephanus*, the remaining three within Sect. *Oreacanthus* (Benth.) Champl. For a discussion on the infrageneric classification see Champluvier & Darbyshire (2009).

1. Corolla tube over 20 mm long, much longer than limb; upper lip (ovate-)elliptic; lower lip with lobes minute, to 0.5(1.3) mm long; stigma conspicuously capitate. ·**4.** *africanus*
   – Corolla tube less than 10 mm long, shorter than limb; upper lip subulate; lower lip with lobes conspicuous, 3–7 mm long; stigma minute, barely wider than style ·2
2. Inflorescence paniculate; bracts inconspicuous, 0.8–4(5) × 0.5 mm . **3.** *montifuga*
   – Inflorescence spiciform; bracts conspicuous, 4.5–15 × 0.7–6.7 mm · · · · · · · · · · 3
3. Bracts obovate, obcordiform or broadly elliptic, with a prominent acumen, 2.8–6.7 mm wide; capsule at least sparsely puberulous. · · · · · · · · · · · · · · · **1.** *coeruleus*
   – Bracts linear-lanceolate, 0.7–1.8 mm wide; capsule glabrous · · · · · **2.** *calostachyus*

1. **Brachystephanus coeruleus** S. Moore in J. Bot. **45**: 332 (1907). —Darbyshire in F.T.E.A. Acanthaceae **2**: 471 (2010). Type: Uganda, Mpamba R., Lake Albert-Edward, fl. xii.1906, *Bagshawe* 1378 (BM holotype, US).

   Oreacanthus coeruleus (S. Moore) Champl. & Figueiredo in Bull. Jard. Bot. Belg. **65**: 416 (1996).

---

[10] By Iain Darbyshire

Erect, decumbent or scandent perennial herb or subshrub, 35–200 cm tall; stems pubescent in two opposite lines or on paired opposite ridges, hairs short and antrorse, rarely longer and both antrorse and retrorse. Leaves sometimes absent at flowering, ovate to elliptic, 8.5–17 × 4–9 cm, base (cuneate-)attenuate, apex acuminate, principal veins and margin shortly pubescent, upper surface sparsely pubescent or rarely pilose, lateral veins 7–11 pairs; petiole 10–80 mm long; . Inflorescence terminal, spiciform, 2–14.5 cm long, each cymule several-flowered; axis antrorse-to spreading-pubescent, sometimes additionally glandular-puberulous and/or glandular-pilose; bracts pale green with markedly darker acumen and midrib or dark green throughout, often purple-tinged, variously suborbicular, ovate, elliptic, obovate or obcordiform, 4.5–11.5(15) × 2.3–6.7 mm, apex attenuate to curved-caudate, margin and often midrib pubescent, surface sometimes glandular-puberulous and/or glandular-pilose; bracteoles as bracts but (linear-) lanceolate, 3.5–8(9.5) × 0.5–2.2 mm. Calyx lobes linear, 3.7–7.5 mm long, surface shortly pubescent and ± densely glandular-pilose. Corolla 10.5–17.5 mm long, tube white, limb pink, purple, blue or rarely white, limb pilose towards apex externally; tube subcylindrical, 4.5–7 mm long; upper lip subulate, 6–10.5 mm long, apex obtuse; lower lip 6.5–11 mm long, lobes oblong, 3–5.5 mm long. Stamens with filaments (6)8.5–13 mm long; anthers 1.5–2 mm long. Stigma minute, barely wider than style. Capsule 6–8 mm long, puberulous to sparsely so in upper half. Seeds 0.9–1.8 mm wide, rugulose-tuberculate.

Subsp. **apiculatus** Champl. in Syst. Geogr. Pl. **79**: 130, fig.1 (2009). —Darbyshire in F.T.E.A., Acanthaceae **2**: 472 (2010). Type: Tanzania, Kigoma Dist., Lugonezi R., Lubalisi, fl.& fr. 30.v.1975, *Kahurananga, Kibuwa & Mungai* 2720 (K holotype, BR, EA).

Inflorescence up to 9.5 cm long at maturity; bracts obovate, obcordiform or broadly elliptic, 6.5–11.5(15) mm long, including prominent curved caudate apex to 2–5(7) mm.

Zambia. N: Mbala Dist., Lunzua Gorge near upper fall, fl.& fr. 7.v.1936, *Burtt* 6261 (BM, BR, EA, K, P).
Also in W Tanzania. In forests and on shaded riversides; c.1400 m.
Conservation notes: Assessed as Vulnerable under criterion B by Darbyshire (2010).
Subsp. *coeruleus*, in which the bracts have an attenuate or shortly acuminate apex to only 2 mm long, is recorded from Uganda, eastern D.R. Congo and W Kenya.

2. **Brachystephanus calostachyus** Champl. in Syst. Geogr. Pl. **79**: 131, fig.1 (2009). Type: D.R. Congo, 15 km NW of Lubumbashi, fl., 5.iv.1955, *Schmitz* 5155 (BR holotype).

Perennial herb, 30–100 cm tall; stems shortly and sparsely antrorse-pubescent on paired opposite ridges. Leaves ovate(-elliptic), 5.5–10.5 × 2.5–4.2 cm, base cuneate to attenuate, apex subattenuate to acuminate, principal veins beneath, margin and midrib above shortly and sparsely pubescent, elsewhere glabrous, lateral veins 6–7 pairs; petiole 7–30 mm long. Inflorescence terminal, spiciform, 3–8 cm long, each cymule several-flowered; axis antrorse-pubescent, sometimes additionally glandular-pilose or -pubescent; bracts green, sometimes with paler margin, linear-lanceolate, 5.5–11 × 0.7–1.8 mm, straight to curved, margin and midrib shortly pubescent, surface sometimes glandular-pilose; bracteoles as bracts but 4–8.5 × 0.5–1 mm. Calyx lobes linear, 3.2–5.5 mm long, shortly pubescent and at least sparsely glandular-pilose. Corolla 16–19 mm long, pale mauve, limb pilose towards apex externally; tube subcylindrical, 5–7 mm long; upper lip subulate, 8.5–11 mm long, apex obtuse; lower lip 8.5–12 mm long, lobes oblong, 4–7 mm long. Stamens with filaments 9–12 mm long; anthers 1.4–1.7 mm long. Stigma minute, barely wider than style. Capsule 6.5–7 mm long, glabrous. Seeds rugulose.

Zambia. N: Mbala Dist., by Mwambeshi R., fl. 16.iv.1959, *Richards* 11246 (EA, K, SRGH).
Also in D.R. Congo and Angola. In riverine forest and woodland; 1500–1650 m.
Conservation notes: Although widespread in the northern Zambesian centre of endemism, this species is scarce, currently known from only 12 collections (only

two from Zambia). Its favoured riparian habitats are perhaps threatened by human encroachment, although site-specific threats are unknown.

The description of the fruits is taken from the protologue since I have not seen fruiting material. The remainder of the above description is derived from the Zambian material only.

This species is closely related to *Brachystephanus coeruleus* and could be considered only subspecifically distinct, but the difference in bract shape is conspicuous giving inflorescences of the two species a quite different appearance.

3. **Brachystephanus montifuga** (Milne-Redh.) Champl. in Syst. Geogr. Pl. **79**: 147 (2009). Type: Zambia, Solwezi Dist., just E of Kabompo R., fl.& fr. 31.vii.1930, *Milne-Redhead* 807 (K holotype).

    Oreacanthus montifuga Milne-Redh. in Hooker's Icon. Pl. **32**: t.3197 (1933). —Friis & Vollesen in Kew Bull. **37**: 467 (1982).

Decumbent perennial herb or shrub, 30–180 cm tall, rooting at lower nodes; stems shortly and sparsely antrorse-pubescent on paired opposite ridges or with scattered crisped-pilose hairs, soon glabrescent. Leaves ovate to elliptic, 4–17.5 × 1.8–11.5 cm, base cuneate to attenuate or uppermost leaves subcordate, apex acuminate, principal veins beneath and midrib above shortly and sparsely pubescent, upper surface sometimes pubescent, or blade glabrous except for short marginal hairs, lateral veins 6–10 pairs; petiole 4–90 mm long. Inflorescence terminal, laxly but often narrowly paniculate, 5–39 × 2–12 cm, cymules dichasial or partially monochasial; indumentum of axis variable, densely purple-pilose, most hairs with a minute gland tip, and/or pale-puberulous with interspersed pale glandular hairs, the gland tip more conspicuous; main axis bracts and bracteoles usually inconspicuous, narrowly triangular to linear, 0.8–4(5) × 0.5 mm. Calyx lobes linear, 2–4(6.5) mm long in flower, 3.5–5.5(10.5) mm long in fruit, puberulous and glandular-pilose or -pubescent. Corolla (11)15–18 mm long, white, lilac or mauve, glabrous or limb pilose externally; tube subcylindrical, 4.5–7 mm long; upper lip subulate, (6)8–9.5 mm long, apex obtuse; lower lip (6)8–12.5 mm long, lobes oblong, 3–4.5 mm long. Stamens with filaments (7.5)9–12.5 mm long; anthers 1.4–2.2 mm long. Stigma minute, barely wider than style. Capsule 7.5–12 mm long, glabrous. Seeds 1–2.4 mm wide, rugose-tuberculate.

Zambia. N: Kawambwa, fl.& fr. 26.viii.1957, *Fanshawe* 3651 (BR, K, NDO). W: Mwinilunga Dist., Kalene Hill?, fl.& fr. 18.v.1969, *Mutimushi* 3157 (K, NDO, SRGH).

Also in southern D.R. Congo. Shaded streamsides, wet riverine forest (mushitu) and dry evergreen forest; 1400–1600 m.

Conservation notes: Currently known from 10 localities but probably more widespread in under-collected regions of southern D.R. Congo and N Zambia. Its riverine and mid-altitude forest habitat is under threat from human encroachment; potentially threatened.

Inflorescence indumentum varies considerably in this species, from having long purple pilose hairs usually with minute gland-tips (e.g. the type) to being pale-puberulous with scattered longer glandular hairs with more conspicuous gland-tips (e.g. *Fanshawe* 5590 from Mbala). However, intermediate specimens are recorded in our region, for example *Mutimushi* 3157.

Fanshawe 3651 is unusual in having bracteoles and calyx lobes over twice the length of the other material seen. However, the inflorescence form and indumentum are a good match for the type and they are considered conspecific.

4. **Brachystephanus africanus** S. Moore in Trans. Linn. Soc., Bot. **4**: 31 (1894). —Clarke in F.T.A. **5**: 177 (1899). —Figueiredo & Jury in Kew Bull. **51**: 753 (1996). —White *et al.*, Evergr. For. Fl. Malawi: 113 (2001). —Darbyshire in F.T.E.A., Acanthaceae **2**: 474, fig.61 (2010). Type: Malawi, Mt Mulanje, fl. 1891, *Whyte* 56 (BM lectotype, K), lectotypified by Figueiredo & Jury (1996).

Perennial herb or subshrub, 30–150(250) cm tall, often decumbent and rooting at lower nodes; stems antrorse- to appressed-pubescent when young or largely glabrous. Leaves ovate, elliptic or oblong-elliptic, 5.5–24 × 1.8–10 cm, base often narrowly cuneate then abruptly attenuate, apex acuminate to caudate, lower surface sparsely to rather densely pubescent on veins beneath, upper surface glabrous or with scattered hairs, lateral veins 7–13 pairs; petiole 10–62 mm long. Inflorescence terminal, sometimes also in the upper leaf axils, narrowly spiciform, 5.5–25 cm long, bracts imbricate at least when young, sometimes more widely spaced at maturity; cymules single-flowered, sessile; axes antrorse-pubescent or glabrous; bracts dark green or tinged pink, purple or red-brown, ovate or lanceolate to oblong-elliptic or somewhat obovate to oblanceolate, (3.5)5–25 × 1.7–6.5 mm, apex attenuate to caudate, straight, incurved or recurved, margin narrowly hyaline, glabrous except for minute hairs on acumen or more rarely with scattered glandular hairs and/or eglandular-puberulous; bracteoles lanceolate, 2–6.5(8) × 0.5–1 mm, margin hyaline. Calyx lobes linear, ± unequal, longest (2.5)7.5–18 mm, usually with scattered glandular hairs in distal half. Corolla (blue-)pink to purple, rarely white, 29–51 mm long, tube and limb shortly pubescent externally, with scattered longer glandular hairs on limb; tube narrowly cylindrical, 22.5–38 mm long, straight to somewhat curved; upper lip (ovate-)elliptic, 6.5–13.5 × 5.5–8 mm, apex minutely apiculate; lower lip elliptic, 7–14.5 × 6–8.5 mm, lobes minute, to 0.5(1.3) mm long. Staminal filaments 13.5–36 mm long; anthers 1.7–3.5 mm long. Stigma capitate, 0.25–0.5 mm wide. Capsule 11.5–15 mm long, glabrous. Only immature seeds seen, tuberculate, tubercles elongating towards margin and with minute hair-like processes.

Var. **africanus**. —Champluvier in Syst. Geogr. Pl. **79**: 158 (2009). —Darbyshire in F.T.E.A., Acanthaceae **2**: 475 (2010). FIGURE 8.6.**49**.

Brachystephanus bequaertii De Wild. in Rev. Zool. Bot. Africaines **8**, suppl. Bot.: 36 (1920). Type: D.R. Congo, Masisi, fl.& fr. 24.xii.1914, *Bequaert* 6374 (BR lectotype), lectotypified by Figueiredo & Jury (1996).

Brachystephanus velutinus De Wild. in Rev. Zool. Bot. Africaines **8**, suppl. Bot.: 36 (1920). Type: D.R. Congo, Masisi–Walikale, fl. i.1915, *Bequaert* 6570 (BR holotype).

Brachystephanus africanus var. *velutinus* (De Wild.) Figueiredo in Kew Bull. **51**: 756 (1996). —Champluvier in Syst. Geogr. Pl. **79**: 163 (2009).

Brachystephanus africanus var. *longibracteatus* Champl. in Syst. Geogr. Pl. **79**: 161 (2009). Type: D.R. Congo, Mt Mbese-Mbese, fl. xii.1958, *Léonard* 1964 (BR holotype).

Bracts ovate, lanceolate, (oblong-)elliptic or at most somewhat oblanceolate, acumen straight or incurved.

Zimbabwe. E: Chimanimani Dist., Mutsangazi R., W of Mt Peni, fl. 13.xii.1972, *Müller & Goldsmith* 2066 (BM, K, SRGH). **Malawi**. N: Chitipa Dist., Misuku Hills, Mugesse Forest, fl. 28.ii.1983, *Dowsett-Lemaire* 651 (K). C: Ntchisi Dist., Ntchisi Forest Reserve, fl. 14.iv.1991, *Bidgood & Vollesen* 2180 (BR, K, MAL). S: Mt Mulanje, Litchenya Plateau, Nessa Path, fl. 15.xi.1986, *J. & E. Chapman* 8218 (K, MO, SRGH). **Mozambique**. Z: Gurué Dist., Namuli Mt, Muretha Plateau, fl. 18.xi.2007, *Mphamba* 18 (K, LMA, LMU, MAL).

Also in Nigeria, D.R. Congo, Rwanda, Burundi, Uganda, and Tanzania. In montane and submontane forest including streamsides and pathsides, often locally dominant in the understorey; 900–1900 m.

Conservation notes: Assessed as Least Concern by Darbyshire (2010).

In plants from Malawi and Mozambique, the calyx is longer than or subequal to the bracts and would fall within Champluvier's var. *africanus*. The specimens from Zimbabwe have bracts longer than the calyx lobes and would fall within Champluvier's var. *longibracteatus*. However, calyx and bract length appear to vary independently of one another in this species and the recognition of two varieties based upon these characters is not considered useful. From outside our region, var. *recurvatus* Champl. from eastern D.R. Congo and var. *madagascariensis* Figueiredo from Madagascar do appear to be good taxa with distinctive bract characteristics.

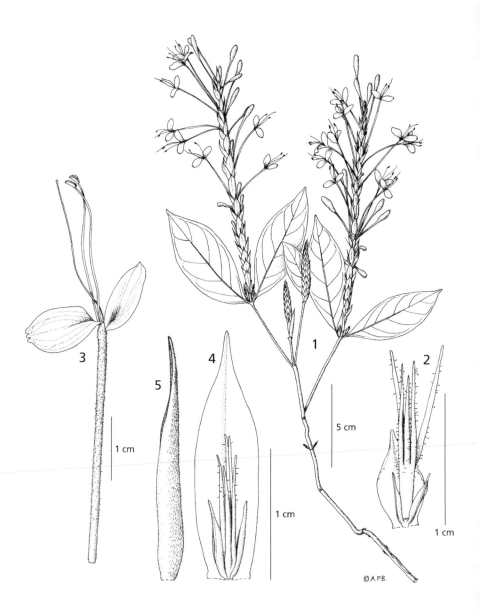

Fig. 8.6.**49**. BRACHYSTEPHANUS AFRICANUS var. AFRICANUS. 1, habit; 2, bract, bracteoles and calyx; 3, corolla with stamens and pistil; 4, bract, bracteoles and calyx, variant with short calyx lobes and long bracts ('var. *longibracteatus*'); 5, bract, lateral view. 1–3 from *Drummond & Hemsley* 1986, 4, 5 from *Bidgood & Vollesen* 2180. Drawn by Andrew Brown. Adapted from Systematics & Geography of Plants (2009), Botanic Garden, Meise, with permission.

33. ISOGLOSSA Oerst.[11]

Isoglossa Oerst. in Vidensk. Meddel. Dansk Naturhist. Foren. Kjobenhavn: 155 (1854), conserved name. —Brummitt in Kew Bull. **40**: 785–791 (1985). —Hansen in Nordic J. Bot. **5**: 5 (1985). —Kiel *et al.* in Taxon **55**: 683–694 (2006). —Darbyshire in Kew Bull. **64**: 401–427 (2009).

Rhytiglossa Lindl., Intr. Nat. Syst. Bot., ed.2: 444 (1836), rejected name, Brummitt in Taxon **23**: 440–441 (1974); McVaugh in Taxon **24**: 247–248 (1975).

Ecteinanthus T. Anderson in J. Linn. Soc., Bot. **7**: 45 (1864).

Homilacanthus S. Moore in J. Bot. **32**: 129 (1894).

Schliebenia Mildbr. in Notizbl. Bot. Gart. Berlin-Dahlem **12**: 99 (1934).

Annual or perennial herbs or shrubs, sometimes monocarpic and cyclically mass-flowering (plietesial); cystoliths present, linear, ± conspicuous. Stems (sub-)4- or 6-angular, often grooved between angles when young, at least in dry state. Leaves evergreen or deciduous, opposite-decussate, pairs equal to anisophyllous, blade margin entire, undulate or shallowly crenate, lower leaves petiolate or subsessile, uppermost leaves and those subtending inflorescence often sessile with rounded or cordate base. Inflorescences terminal and/or axillary in upper leaf axils, rarely on older leafless branches; spiciform or laxly paniculate thyrses, branching dichasial and/or monochasial, cymules 1 to several-flowered; bract pairs ± equal, free, shape and size variable, sometimes foliaceous, sometimes rapidly reducing in size along main inflorescence axis; bracteoles as bracts or much-reduced. Calyx usually divided almost to base into 5 subequal lobes, often lengthening in fruit. Corolla bilabiate; indumentum variable but always glabrous towards base of tube externally; tube cylindrical, campanulate or inflated; upper lip ± hooded, apex 2-lobed, lobes often reflexed; lower lip oblong, elliptic or obovate, apex 3-lobed, median lobe broader than lateral pair, palate and throat often raised upwards with a longitudinal central furrow and herring-bone-like transverse ridging. Stamens 2; filaments attached at various heights within corolla tube, exserted or held at mouth; anthers bithecous, thecae strongly offset or at subequal height, parallel to highly oblique, upper theca larger than lower or pair subequal in size, basally muticous; staminodes absent. Disk shallowly cupular. Ovary (oblong-)ovoid, largely glabrous, bilocular, 2 ovules per locule; style filiform; stigma shortly bilobed. Capsule stipitate, placental base inelastic. Seeds 4 per capsule or fewer by abortion, held on retinacula, lenticular, compressed-ellipsoid or subdiscoid, with a hilar excavation, variously tuberculate, rugose or smooth.

Genus of 50–70 species with a palaeotropical and subtropical distribution, most diverse in East Africa. Several species are plietesial, i.e. perennial but monocarpic, displaying population-level periodic mass-flowering on a cycle of several years, often becoming dominant in the forest understorey at maturity. Many species, including the majority in our region, have a highly restricted range and so are of high conservation concern.

1. Inflorescence a lax open panicle with clearly pedunculate dichasial and/or monochasial cymules, rarely reduced to a simple cyme. 2
– Inflorescence spiciform, if branched then cymules (sub)sessile, congested to widely spaced along axis, each cymule 1 to several-flowered. 4
2. Main axis bracts narrowly oblong to spathulate, 6.5–8 mm long; glandular hairs on inflorescence pilose, to 1.5 mm long. . . . . **4.** *sp.* A
– Main axis bracts triangular- to linear-lanceolate, if over 5 mm long then glandular hairs on inflorescence very shortly stalked or subsessile .3
3. Capsule puberulous, with short glandular hairs; seeds with elongate glochidiate tubercles; largest leaves with narrow cuneate wing along upper portion of petiole; calyx lobes to 3 mm long in flower; corolla 11–13 mm long*. . . . .**2.** *membranacea*
– Capsule glabrous; seeds rugose; leaves without wing along petiole; calyx lobes 3.5–6 mm long in flower; corolla 13–19.5 mm long. . . . .**3.** *strigosula*

[11] By Iain Darbyshire

4. Anther thecae superposed, either parallel to filament or one or both thecae oblique or spreading (Fig. **50**); corolla 7.5–24 mm long*................... 5
   - Anther thecae held at ± same height, parallel to filament (Fig. **51**); corolla 19–50 mm long 13
5. Corolla 20–24 mm long, tube cylindrical, c.2 mm wide in lower half then abruptly expanded laterally, 7–8 mm wide at mouth .... **12.** *cataractarum*
   - Corolla 7.5–17.5 mm long, tube cylindrical or subcampanulate throughout or strongly inflated from base 6
6. Corolla tube inflated from base; bracts minute and inconspicuous, triangular to subulate, to 1.5 mm long 7
   - Corolla tube cylindrical or subcampanulate, not inflated; bracts conspicuous, variously ovate, lanceolate, elliptic, obovate, flabellate or suborbicular, 4–11.5 mm long 8
7. Corolla glabrous outside; calyx lobes 3–6.5 mm long in flower; villose glandular hairs on inflorescence 2–3.5 mm long .... **5.** *multinervis*
   - Corolla pubescent outside, particularly dorsally; calyx lobes 2.5–3.5 mm long in flower; villose glandular hairs on inflorescence 0.5–1.5 mm long .. **6.** *eliasbandae*
8. Seeds with elongate glochidiate tubercles; anther thecae subparallel to filament; bracts largely glabrous except for minute cilia and rarely scattered glandular hairs; palate of lower corolla lip lacking prominent herring-bone pattern.....**1.** *gregorii*
   - Seeds rugose**; at least one theca of each anther held strongly oblique or patent to filament; bracts conspicuously ciliate and/or densely glandular-pubescent; palate of lower corolla lip with prominent herring-bone pattern............. 9
9. Bracts and bracteoles with a conspicuous pale-hyaline margin, densely ciliate; capsule 5–8 mm long, always with 1–2 seeds maturing; inflorescence fasciculate or shortly spiciform, 1–3 cm long .... **7.** *ciliata*
   - Bracts and bracteoles lacking a conspicuous hyaline margin, indumentum variable; capsule 10–15 mm long, at least some capsules with 4 seeds maturing; inflorescence spiciform or several-branched, 1.7–15 cm long 10
10. Cymule bracts with conspicuously toothed margin, 1–4 teeth per side, apex with a prominent linear acumen 4–4.5 mm long; lobes of lower corolla lip oblong, 3.5–4 mm long .... **11.** *namuliensis*
    - Cymule bracts with margin entire or at most minutely and inconspicuously toothed, apex acute, attenuate or shortly acuminate to 2.5 mm long (bracts subtending inflorescence branches, if present, can be larger and with longer acumen); lobes of lower lip rounded, 0.5–2.5 mm long 11
11. Largest leaves (sometimes absent at flowering/fruiting) to 6.5 cm long with 4–5 pairs of lateral veins; inflorescence densely glandular-pubescent, bracts with only minute marginal cilia; plant of lowland bush and thicket......**8.** *glandulosissima*
    - Largest leaves 8–22 cm long with 6–13 pairs of lateral veins; inflorescence indumentum variable, if bracts glandular then also conspicuously ciliate; plant of montane and submontane forest... 12
12. Mature bracts ± equal in size along inflorescence axis, broadly ovate to elliptic, 4–7 mm wide... **9.** *bracteosa*
    - Mature bracts reducing in width upwards, those in upper portion of axis narrowly elliptic to lanceolate, 1–2.5 mm wide .... **10.** *milanjiensis*
13. Leaf base narrowly attenuate forming a wing along midrib, then abruptly truncate, subsessile; calyx lobes imbricate at base; corolla 38–50 mm long......
    .... **13.** *pawekiae*
    - Leaf base not as above; calyx lobes not imbricate; corolla 19–33.5 mm long... 14

14. Inflorescence densely glandular-villose; corolla tube inflated immediately above base, not curved; upper lip of corolla bulbous at apex, not arcuate; bracts 10–19.5 mm long . . . .**17.** *eylesii*

– Inflorescence usually densely short glandular-pubescent; corolla tube not inflated, curved upwards below midpoint, gradually widened to mouth; upper lip arcuate, not bulbous; bracts 1.5–7.5 mm long 15

15. Seeds smooth; corolla limb glabrous externally; calyx lobes to 5 mm long in flower; at least main inflorescence axis with both bracts of each pair fertile . . . . .
. . . .**16.** *mbalensis*

– Seeds rugose; corolla limb pubescent externally, at least on upper lip; calyx lobes 5–9 mm long, if shorter then all partial inflorescences ± 1-sided 16

16. Partial inflorescence wholly or partially 1-sided with one bract of each pair usually sterile, axis often curved; inflorescence axis, bracts and/or calyx with few to numerous pale needle-like 3 mm long eglandular hairs . . . .**14.** *floribunda*

– Inflorescence with well-defined central axis on which both bracts of each pair are fertile, usually with several pairs of lateral branches which can be 1-sided when young; inflorescence lacking long pale needle-like hairs . . . . **15.** *grandiflora*

* Corolla measurements in key and descriptions are of the tube plus lower lip when straightened, except in *I. eylesii* where the upper lip is clearly longer than the lower; in this case the corolla is measured as tube plus upper lip.

** Seeds unknown in *I. namuliensis* but it is expected they will be rugose as with closely allied species.

1. **Isoglossa gregorii** (S. Moore) Lindau in Engler, Pflanzenw. Ost-Afr. **C**: 372 (1895). —Clarke in F.T.A. **5**: 232 (1900). —Mildbraed in Notizbl. Bot. Gart. Berlin-Dahlem **9**: 505 (1926). —Darbyshire in F.T.E.A., Acanthaceae **2**: 621 (2010). Type: Kenya, Mt Kenya, fl.& fr. 24.vi.1893, *Gregory* s.n. (BM holotype).

Homilacanthus gregorii S. Moore in J. Bot. **32**: 129, pl.343 (1894).
Isoglossa oerstediana Lindau in Bot. Jahrb. Syst. **20**: 56 (1894). —Clarke in F.T.A. **5**: 232 (1900), in part for *Scott Elliot* 6767, *Johnston* 11a, *Holst* 523 & *Volkens* 1852 (mixed collection with *I. punctata* from Tanzania). Type: Tanzania, Lushoto Dist., Usambara, fl.& fr. iii.1892, *Holst* 523 (B† holotype).
Justicia nyassae Lindau in Notizbl. Bot. Gart. Berlin-Dahlem **8**: 424 (1923). Type: Tanzania, Rungwe Dist., Kyimbila, N of Lake Nyasa, Rungwe, fl.& fr. 19.ix.1912, *Stolz* 1565 (B† holotype, BM, C, K).

Straggling or trailing decumbent perennial herb, 10–100(200) cm tall, rooting at lower nodes. Stems pubescent with mainly retrorse hairs in two opposite lines, later glabrescent. Leaves ovate, 1.5–12 × 0.8–7.5 cm, base attenuate to rounded, apex (sub)acuminate, margin and midrib pubescent, upper surface and veins beneath sparsely pilose or veins glabrous; lateral veins (4)5–7(8) pairs; petiole 4–60(90) mm long. Inflorescence terminal on main and short lateral branches, sometimes with reduced inflorescence in upper leaf axils, spiciform or with 1–2 branches, main axis 2–24 cm long, often curved, cymules subsessile along axes, clustered with bracts imbricate towards apex, ± distantly spaced below at maturity, ± 1-sided, one bract of each pair often sterile; axis retrorse-pubescent in 2 opposite lines or more widespread, sometimes also with scattered glandular and/or long crisped eglandular hairs; bracts often purple towards apex, elliptic to somewhat obovate or basal pairs ovate, 4.5–9 × 2.5–9 mm, apex attenuate to acuminate, glabrous except for inconspicuous short hairs along margin, occasionally with scattered short-stalked glands; bracteoles elliptic-lanceolate, 3–7 mm long. Calyx lobes tinged purple towards apex, subulate, 3.5–6.5 mm long in flower, extending to 5.5–11 mm in fruit, with short eglandular hairs particularly along margin, often also with scattered glandular hairs and/or long crisped eglandular hairs. Corolla 7.5–16.5 mm long, pink, mauve or rarely white, with pink to purple markings on palate and throat; largely glabrous or pubescent externally on limb and upper tube; tube subcylindrical, (4)5–9.5 mm long, 1–1.5 mm wide at base, 1–2.7 mm at mouth; upper lip hooded, 2.5–6 mm long, lobes reflexed, 0.5–1.5 mm long; lower lip 3–7 mm long, lobes 1.3–2.5 mm long, palate lacking prominent herring-bone patterning. Stamens attached in upper third of corolla tube; filaments free for 1.5–3.5 mm; anther thecae superposed and subparallel, each 0.65–1.1 mm long. Capsule 7.5–10 mm long, glabous or with minute eglandular hairs, occasionally with scattered glandular and/or long eglandular hairs. Seeds 1.5–2.5 × 1–2 mm, with elongate, glochidiate tubercles most dense towards rim.

Zimbabwe. E: Mutare Dist., Vumba Mts, NE of Castle Beacon, fl.& fr. 19.viii.1956, *Chase* 6180 (BR, K, LISC, ?SRGH). **Malawi**. C: Dedza Dist., Dedza Mt, W side of summit, fl.& fr. 25.vi.1970, *Brummitt & Salubeni* 11690 (K, MAL, SRGH). S: Zomba Dist., Zomba Plateau, fl.& fr. 11.x.1979, *Banda & Salubeni* 1556 (K).

Also in northeastern D.R. Congo, Uganda, Kenya and Tanzania. Moist montane and submontane forest, including open areas, sometimes by water; 1400–2200 m.

Conservation notes: Assessed as Least Concern in F.T.E.A. Uncommon in the Flora region and apparently absent from several seemingly suitable montane areas (e.g. Misuku Hills of Malawi), but widespread and common in the highlands of Kenya and N Tanzania.

Bract shape is less variable in our region than in East Africa; only dimensions for our plants are given in the above description.

2. **Isoglossa membranacea** C.B. Clarke in F.T.A. **5**: 230 (1900). —Darbyshire in Kew Bull. **64**: 410 (2009). —Darbyshire in F.T.E.A., Acanthaceae 2: 622 (2010). Type: Malawi, Misuku Hills (Masuku Plateau), fl. vii.1896, *Whyte* s.n. (K holotype).

Straggling or decumbent perennial herb or subshrub, 20–200 cm tall. Stems largely glabrous. Mature leaves (ovate-)elliptic to broadly so, 6–21.5 × 3–13 cm, base of largest leaves narrowly cuneate, often forming a wing along upper petiole, then abruptly to gradually attenuate, apex acuminate, midrib antrorse-pubescent above, margin with or without multicellular hairs shortening upwards, surfaces glabrous or with scattered hairs above; lateral veins 6–11 pairs; petiole 12–65 mm long. Inflorescence terminal, a large lax panicle, ± pyramidal to narrowly so, 10–36 cm long, much-branched, dichasial or largely so; axis puberulous and sparsely to densely glandular-pubescent, hairs spreading, 0.2–1 mm long, sometimes also with scattered long eglandular hairs to 1.5 mm; bracts and bracteoles minute, (triangular-)lanceolate, 0.4–2.5 mm long; flowers sessile or pedicels to 1 mm. Calyx lobes lanceolate, 1.7–3 mm long in flower, 2.5–3.5 mm in fruit, indumentum as axis. Corolla 7.5–13 mm long, white, pale pink or pale violet, often with pink to purple markings on palate; pubescent externally, glabrous within; tube 4–5.7 mm long, subcampanulate, floor declinate, 1–1.5 mm wide at base, 2.5–3.5 mm at mouth; upper lip broadly triangular, hooded, reflexed distally, 2.5–5.5 mm long, lobes 1–2.5 mm; lower lip protruding, 3.5–7 mm long, lobes 1–3 mm, palate with shallow herring-bone patterning. Stamens attached ± midway along corolla tube; filaments free for 2–2.5 mm; anther thecae superposed, slightly overlapping or separated by up to 0.7 mm, each 0.5–1.1 mm long. Capsule 8.5–11.5 mm long, indumentum variable (see subspecies). Seeds 2–2.5 × 1.5–2 mm, with elongate glochidiate tubercles most dense towards rim.

Subsp. **membranacea**. Darbyshire in Kew Bull. **64**: 410 (2009).

Petiole and lower portion of leaf margin with conspicuous multicellular hairs. Corolla 11–13 mm long, white and usually with pink to purple markings on palate. Anther thecae slightly overlapping, each 0.8–1.1 mm long. Capsule eglandular-puberulous and with short glandular hairs.

Malawi. N: Chitipa Dist., Mugesse, fl.& fr. 3.viii.1977, *Phillips* 2728 (K, LMU, SRGH); Misuku Hills, Mugesse, fl.& fr. 12.ix.1977, *Pawek* 12990 (BR, K, MAL, MO, SRGH, UC).

Not known elsewhere. Moist montane and submontane forest including pathsides; 1500–2200 m.

Conservation notes: Subspecies endemic to the Misuku Hills of Malawi and known from only five collections; assessed as Vulnerable under IUCN criterion D2 by Darbyshire (2009). Subsp. *septentrionalis* I. Darbysh. is recorded from South Sudan, northeast D.R. Congo, Uganda and W Kenya. It differs in having largely glabrous leaves and petioles, smaller corollas (7.5–10 mm long), smaller and fully superposed anther thecae (each 0.5–0.8 mm long) and in the capsule indumentum including long eglandular hairs.

3. **Isoglossa strigosula** C.B. Clarke in F.T.A. **5**: 231 (1900). —Milne-Redhead in Mem. New York Bot. Gard. **9**: 25 (1953) as *I. strigulosa.* —Burrows & Willis, Pl. Nyika Plateau: 51 (2005) as *I. strigulosa.* —Darbyshire in F.T.E.A., Acanthaceae **2**: 628, fig.80,5 (2010). Type: Malawi, Misuku Hills (Masuku Plateau), fl. vii.1896, *Whyte* 288 (K holotype).

   Isoglossa dichotoma sensu B. Hansen in Nordic J. Bot. **5**: 10 (1985) in part, non (Hassk.) B. Hansen.

Straggling perennial herb or subshrub, 20–300 cm tall, rooting at lower nodes. Stems retrorse-pubescent in two opposite lines, hairs with conspicuous cell walls, soon glabrescent. Leaves ovate or elliptic to narrowly so, 3–11.5 × 1.5–5.5 cm, base attenuate or cuneate, apex acuminate, margin, midrib above and main veins beneath pubescent, upper surface sometimes with scattered hairs; lateral veins 5–8(9) pairs; petiole 6–30(45) mm long. Inflorescence terminal and in upper leaf axils, laxly paniculate or reduced to a simple cyme, 2–12(19) cm long, branching dichasial or minor branching often 1-sided, cymules pedunculate with flowers terminating branches; axis eglandular-puberulous, sparse or glabrous towards base, more dense on minor branches, usually also with capitate short-stalked glands; bracts linear-lanceolate, 2.5–9 mm long; bracteoles 1.5–3 mm long; flowers subsessile or pedicels to 2 mm. Calyx lobes linear-lanceolate, 3.5–6 mm long in flower, 4.5–7 mm in fruit, eglandular-puberulous particularly along margin, with scattered capitate short-stalked or subsessile glands. Corolla 13–19.5 mm long, white or tinged pink, with red or purple markings on palate and throat; pubescent externally; tube 7–9.5 mm long, somewhat inflated, subcampanulate, 1.5–3 mm wide at base, 4.5–7 mm at mouth, floor pubescent within, hairs most dense or restricted to near base; upper lip broadly triangular, hooded, 4.5–7 mm long, lobes reflexed, 1–2.2 mm long; lower lip protruding, 6–11 mm long, lobes 2–3 mm long, palate sparsely pubescent or glabrous, upraised and with prominent herring-bone patterning. Stamens attached ± midway along corolla tube; filaments free for 2.5–4 mm; anther thecae superposed, subparallel, upper theca 1.6–2.1 mm long, lower theca 1.2–1.6 mm long. Capsule 14–18 mm long, glabrous. Seeds 1.75–2 mm wide, rugose, inconspicuously so at maturity.

Zambia. N: Isoka Dist., Mafinga, fl.& fr. 24.v.1973, *Chisumpa* 71 (K, NDO). E: Nyika Nat. Park, Chowo Forest, fl. 18.iv.1986, *Philcox, Pope & Chisumpa* 10000 (BR, K). **Malawi.** N: Chitipa Dist., Musitu Forest, 21 km SE of Chisenga, fl.& fr. 11.vii.1970, *Brummitt* 11992 (EA, K, LISC, MAL, SRGH). C: Kasungu Dist., Kantorongonda and Kasungu Mts, fl. x.1893, *Crawshay* s.n. (BM).

   Also in the Southern Highlands of Tanzania. Moist montane forest understorey, including clearings and margins, sometimes near streams or in swamp forest; 1900–2300 m.

   Conservation notes: Common in montane forests of S Tanzania and N Malawi, can become dominant in undergrowth of forest margins; assessed as Least Concern in F.T.E.A.

4. **Isoglossa sp. A** (= *de Koning* 7459).

   Erect herb. Uppermost stems with two opposite lines of retrorse hairs with conspicuous cell walls, soon glabrescent. Leaves ovate, c.8 × 4.5 cm, base cuneate-attenuate, apex acuminate, midrib shortly antrorse-pubescent above, elsewhere glabrous; lateral veins 10–12 pairs; petiole 40–50 mm long. Inflorescence terminal, a lax slender panicle, 26 cm long, one of each pair of cymules largely undeveloped, cymules dichasial or partially so; lower portion of main axis with indumentum as uppermost stem, upper portion and cymule axis puberulous and sparsely to densely glandular-pilose, the latter with hairs ± crisped, 0.8–1.5 mm long; main axis bracts narrowly oblong or spathulate, 6.5–8 × 0.8–1.3 mm in central portion of axis, glabrous; bracteoles linear, 2–3 mm long; flowers on short pedicels to 1–2.5 mm. Calyx lobes lanceolate, 2.7–3.2 mm long in flower, extending to 4.5 mm in those with fallen corollas, puberulous and glandular-pilose, later glabrescent. Only immature or poorly preserved corollas seen, c.12 mm long, white; pilose externally, glabrous within; tube campanulate. Only immature stamens seen; anther thecae superposed, parallel, each 1.5–1.75 mm long. Ovary with scattered minute papillose hairs. Capsule and seeds not seen.

Mozambique. Z: Gurué Dist., Chá Gurué, 5 km from Estação Pecuária, fl. 27.vii.1979, *de Koning* 7459 (BR, LMU).

Not known elsewhere. In dense forest close to river, understorey with *Cyathea;* c.1330 m.

Conservation notes: Known only from this specimen; Data Deficient.

The single collection seen lacks well-preserved mature flowers for dissection, making it difficult to draw conclusions on its identity. However, it does appear to be distinct within the Flora region, differing from *Isoglossa membranacea* in, for example, the longer bracts, the more densely hairy corolla and the larger anther thecae. It appears to be close to *I. laxa* Oliv. and allies from East Africa.

5. **Isoglossa multinervis** I. Darbysh. in Kew Bull. **64**: 422, fig.6 (2009). —Darbyshire in F.T.E.A., Acanthaceae **2**: 635, fig. 80,8 (2010). Type: Tanzania, Rungwe Dist., Ngozi crater, fl. 4.vii.1991, *Kayombo* 1105 (MO holotype, K).

Shrub 200–400 cm tall. Stems glabrous. Leaves ovate, (10.5)17–26 × (7)10.5–13.5 cm, base truncate or rounded with 1–2.5 cm long cuneate extension along upper petiole, or more evenly attenuate, apex acuminate or caudate, surfaces glabrous except for inconspicuous antrorse hairs on main veins above; lateral veins 17–25 pairs, these and scalariform tertiary venation conspicuous; petiole 45–110 mm long, glabrous. Inflorescence terminal, 1–3 developing from apex of each branch, spiciform or with 1–3 pairs of simple branches, main axis 9–28 cm long, cymules sessile, densely clustered towards apex, more widely spaced below; axis densely glandular-villose, hairs yellow, to 2–3.5 mm long, additionally puberulous; bracts and bracteoles minute and obscured by indumentum, triangular or lanceolate, 1–1.5 mm long; flowers subsessile, pedicels sometimes extending to 1.5 mm long in fruit. Calyx lobes lanceolate, 3.5–6.5 mm long in flower, barely extending in fruit, glandular-villose. Corolla 13.5–17.5 mm long, white or yellow, with deep red markings on palate; glabrous externally; tube strongly inflated from base, 6.5–8.5 mm long, 5–8 mm wide at mouth; upper lip broadly triangular, hooded, 6–8.5 mm long, lobes 0.8–1.5 mm; lower lip 7–9 mm long, lobes 1.5–2 mm, palate upraised and with conspicuous herring-bone patterning. Stamens attached in upper half of corolla tube; filaments free for 4–6.5 mm; anther thecae superposed, somewhat oblique, upper theca 2.4–3 mm long, lower theca 1.9–2.2 mm long. Capsule 29–38 mm long, glabrous. Seeds subdiscoid, 5–6 mm wide, rugose when immature, rather smooth at maturity except along rim.

Malawi. N: Chitipa Dist., Misuku Hills, Mugesse Forest, fr. 17.ix.1975, *Pawek* 10137 (K, MAL, MO, SRGH, UC).

Also in the Southern Highlands of Tanzania. Moist montane forest; c.1800 m.

Conservation notes: Localised distribution; assessed as Endangered under IUCN criterion B in F.T.E.A.

The very large fruits are particularly striking; fruits have not been seen in the closely allied *I. eliasbandae* but it is expected that they would be similarly large. This character alone would easily separate these two species from all other *Isoglossa* in our region.

This species has only been collected once in Malawi, when it was recorded as common. As in the closely related *I. eliasbandae*, it almost certainly has a plietesial life cycle; see discussion in Darbyshire (2009).

6. **Isoglossa eliasbandae** Brummitt in Kew Bull. **40**: 788, fig.1 (1985). Type: Malawi, Zomba Dist., Zomba Plateau, road 0.5 km below Ku Chawe, fl. 21.viii.1972, *Brummitt* 12926 (K holotype, BR, MAL, MO, SRGH).

Stout erect perennial herb to 200 cm tall. Stems crisped-pubescent in two opposite lines below nodes when young, later glabrescent. Leaves ovate or elliptic, to 14–16 × 8.5–9 cm, base asymmetric, abruptly attenuate, apex acuminate, antrorse-pubescent on main veins, with scattered hairs elsewhere above; lateral veins 11–14 pairs, conspicuous; petiole to 76 mm long, antrorse-pubescent in groove above. Inflorescence terminal, a paniculate thyrse with up to 5 pairs of ascending lateral branches, sometimes with secondary branching, main axis 12.5–19 cm long, cymules subsessile or shortly pedunculate, paired along axis or lateral branches, often 1-sided; axis densely glandular-villose, hairs yellow, 0.5–1.5 mm long, additionally spreading-puberulous; bracts minute, narrowly triangular to subulate, 0.7–1.7 × 0.2–0.5 mm; bracteoles to 0.8 mm long; pedicels to 2 mm. Calyx lobes lanceolate, 2.5–3.5 mm long in flower, glandular-villose. Corolla 14–17.5 mm long, white with purple markings on palate; pubescent externally particularly dorsally; tube strongly inflated from base, 7–9 mm long, 4.5–7.5 mm wide at mouth; upper lip broadly triangular, hooded, 6.5–8.5 mm long, lobes c.1 mm; lower lip 6.5–8.5 mm long, lobes 1.5–2 mm, palate raised, with conspicuous herringbone patterning. Stamens attached in upper half of corolla tube; filaments free for 4–4.5 mm; anther thecae superposed, subparallel, upper theca 3.4–3.8 mm long, lower theca 2.4–2.9 mm long. Capsule and seeds not seen.

Malawi. S: Zomba Dist., Zomba Plateau, road 0.5 km below Ku Chawe, fl. 21.viii.1972, *Brummitt* 12926 (K, BR, MAL, MO, SRGH).

Not known elsewhere. Margins of evergreen forest in gully; 1525 m.

Conservation notes: The only known locality is relatively undisturbed due to steepness. However, efforts to rediscover it by both Brummitt and Banda have failed. It is probably a plietesial species and so prone to large fluctuations in the number of mature individuals which, coupled with its extremely localised distribution, makes it vulnerable to habitat change; probably Critically Endangered under IUCN criterion B.

7. **Isoglossa ciliata** (Nees) Engl. in Bot. Jahrb. Syst. **10**: 265 (1889). —Clarke in Fl. Cap. **5**: 80 (1901) as *I. ciliata* Lindau. Type: South Africa, Uitenhage, Olifantshoek, Zwartkopsrivier and Bosjesmanrivier, n.d., *Ecklon & Zeyher* s.n. (holotype not traced, K, S).

Rhytiglossa ciliata Nees in Linnaea **15**: 364 (1841).

Ecteinanthus divaricatus T. Anderson in J. Linn. Soc., Bot. **7**: 45 (1863). Types: South Africa, Uitenhage and Albany, *Ecklon & Zeyher* s.n. (K syntype); Durban (Port Natal), *Gueinzius* s.n. (syntype, not traced).

Isoglossa sylvatica C.B. Clarke in Fl. Cap. **5**: 80 (1901). —Welman in Pl. Sthn. Afr., Strelitzia **14**: 99 (2003). Type: South Africa, Knysna Division, near Melville, fl. 10.v.1814, *Burchell* 5438 (K holotype).

Isoglossa densa N.E. Br. in Bull. Misc. Inform., Kew **1908**: 437 (1908). —Welman in Pl. Sthn. Afr., Strelitzia **14**: 99 (2003). Type: South Africa, near East London, fl.& fr. n.d., *Wood* in Herb. *Galpin* 3375 (K holotype, PRE).

Isoglossa grantii sensu Compton, Fl. Swaziland in J. S. Afr. Bot., suppl. **11**: 563 (1976) as regards *Compton* 30070, non C.B. Clarke.

Perennial herb or subshrub, 30–160 cm tall, often straggling. Stems retrorse-pubescent in two opposite furrows, with scattered crisped hairs elsewhere. Leaves sometimes immature at flowering, ovate or elliptic, 3.5–9.5 × 2–4.5 cm, base attenuate, apex acute to acuminate, sparsely antrorse-pubescent mainly on midrib, main veins and margin; lateral veins 4–6 pairs; petiole 10–44 mm long. Inflorescences terminal and axillary, fasciculate or shortly spiciform, 1–3 cm long, partially or wholly 1-sided, each fertile cymule single-flowered, sessile; bracts lanceolate, narrowly pandurate or oblanceolate, 7–11.5 × 1.5–4 mm, apex acuminate or spathulate, surface green with a narrow or (in our region) rather broad pale-hyaline margin in lower half, surface appressed-pubescent, margin densely pale-ciliate particularly on hyaline portion, sometimes also with short spreading glandular hairs; bracteoles lanceolate, 5.5–9.5 × 1.5–3 mm, hyaline margin ± broad. Calyx lobes linear-lanceolate, 4–7 mm long in flower and fruit, margin densely ciliate, glandular hairs usually absent. Corolla 12.5–14 mm long, white to mauve, often with red or purple markings on palate and/or throat; limb sparsely pubescent externally or glabrous except for a few hairs at lobe apex; tube 5–6 mm long, somewhat widened upwards, with an interrupted ring of hairs within around attachment point of stamens; upper lip hooded, 7–10 mm long, lobes to 1 mm; lower lip 7–8.5 mm long, lobes rounded, to 1 mm, palate upraised and with prominent herring-bone patterning. Stamens attached in upper half of corolla tube; filaments free for 4–5.5 mm; anther thecae superposed, held almost patent to filament or lower theca less strongly oblique, each 1–1.7 mm long. Capsule 5–8 mm long, glabrous, only 1–2 seeds maturing. Seeds 2.2–2.8 ×1.5–2.5 mm, rugose, rather smooth at maturity.

Mozambique. GI: Xai Xai Dist., Xai Xai (João Belo), Chipenhe, Chirindzene Forest, fl. 9.vi.1960, *Lemos & Balsinhas* 50 (BM, COI, K, LISC, LMA).

Also in South Africa (E & W Cape, KwaZulu-Natal) and Swaziland. In lowland forest understorey, often along forest margins or in clearings. Van Steenis (Bothalia **12**: 553, 1978) records this as a periodic mass-flowering, monocarpic species with a c.10 year cycle; c.20 m.

Conservation notes: Known only from a single collection in our region but locally abundant in the coastal forests of KwaZulu-Natal and the Cape, particularly favouring open forest and forest margins; Least Concern.

The single specimen seen from our region has lanceolate bracts with a broad hyaline margin in the lower half; some specimens from northern KwaZulu-Natal and Swaziland are similar. Plants from the Cape and southern KwaZulu-Natal tend to have narrower bracts and bracteoles with only a narrowly hyaline margin; the former are often oblanceolate or narrowly pandurate in shape. Our specimen lacks glandular hairs on the inflorescence; elsewhere these can be present or absent and this character appears not to hold any taxonomic value.

Oersted (1854) proposed the transfer of *Rhytiglossa ciliata* to *Isoglossa* in his generic protologue (Vidensk. Meddel. Dansk Naturhist. Foren. Kjobenhavn: 155, 1854), but did not formally make the combination; this was first validly published by Engler (1889).

8. **Isoglossa glandulosissima** K. Balkwill in Novon **23**: 79 (2014). Type: South Africa, Mpumulanga, farm Weltevreden 229JU (Pullen farm), c.34 km SE of Nelspruit, 23.ii.1989, *K. Balkwill & M.-J. Balkwill* 4287 (J holotype, E, K, MO, PRE).

> *Isoglossa stipitata* C.B. Clarke in Fl. Cap. **5**: 82 (1901) in part for *Flanagan* 2322, not of lectotype selected by Poriazis & Balkwill (2014), see note.
> *Isoglossa origanoides* sensu Clarke in Fl. Cap. **5**: 82 (1901) for *Drège* s.n., non (Nees) S. Moore.
> *Isoglossa sp. 1* sensu Poriazis & Balkwill in Bothalia **38**: 131–140 (2008).

Monocarpic perennial herb or subshrub 35–100 cm tall. Stems with numerous white retrorse, antrorse and/or subspreading hairs in two opposite lines, more widespread on young stems. Leaves sometimes absent at flowering/fruiting, mature leaves ovate-elliptic to narrowly ovate, 2–6.5 × 1–3 cm, base rounded to cuneate, apex acute or acuminate, shortly pubescent above, longer hairs on veins beneath; lateral veins 4–5 pairs; petiole 5–26 mm long. Inflorescences terminal on main and short lateral branches, often numerous at maturity, narrowly spiciform, 1.7–7.5 cm long, with decussately arranged bracts dense throughout, cymules single-flowered, sessile; axis antrorse- to spreading-pubescent with few to numerous spreading glandular hairs; bracts (ovate-)elliptic to lanceolate, 5.5–10.5 × 1.7–4.5 mm, base cuneate or attenuate, apex acute or attenuate, sometimes curved, densely (sparse when young) spreading glandular-pubescent with few to numerous short white appressed to ascending eglandular hairs at least along margin; bracteoles lanceolate, 4.5–8.5 × 0.8–1.2 mm. Calyx lobes linear-lanceolate, 4.5–8 mm long in flower, 7–12 mm in fruit, densely glandular-pubescent, with short eglandular hairs mainly along margin. Corolla 8.5–13 mm long, white with lilac or purple markings on palate; pubescent on limb and upper tube externally; tube subcylindrical, 4–5.5 mm long, with an interrupted ring of hairs 2–3 mm from base within; upper lip triangular, hooded, 5–8 mm long, lobes to 0.8 mm; lower lip 4.5–8.5 mm long, lobes rounded, 0.6–1.2 mm, palate upraised with prominent herringbone patterning. Stamens attached in upper half of corolla tube; filaments free for 3–4.5 mm; anther thecae superposed, held oblique or patent to filament, upper theca 1–1.3 mm long, lower theca 0.9–1.25 mm long. Capsule 13–15 mm long, glabrous. Seeds c.3 × 2.5 mm, rugose.

Mozambique. M: Moamba Dist., Ressano Garcia, fl. 18.ii.1955, *Exell, Mendonça & Wild* 475 (BM, LISC); Namaacha Dist., Goba to Fronteira da Goba, fl. 15.v.1975, *Marques* 2766 (DSM, LMU).

Also in Swaziland and South Africa (Eastern Cape, KwaZulu-Natal to North West and Mpumalanga). *Combretum–Acacia–Rhus* bushland; 50–200 m.

Conservation notes: Just extending into the Flora area but widespread and common in bushland habitats in South Africa; Least Concern.

This species forms part of a closely allied species complex from South Africa on which there has been much past confusion over taxonomy and nomenclature; the complex is currently under revision by Balkwill and Poriazis. Specimens have most often been named *Isoglossa stipitata* C.B. Clarke, but see discussions by Poriazis & Balkwill (2008, 2014).

9. **Isoglossa bracteosa** Mildbr. in Notizbl. Bot. Gart. Berlin-Dahlem **11**: 1086 (1934). —Darbyshire in F.T.E.A., Acanthaceae **2**: 638 (2010). Type: Tanzania, Kinole, Uluguru Mts, fl. 23.x.1932, *Schlieben* 2881 (B† holotype, BM, BR, G).

    Isoglossa imbricata Brummitt in Kew Bull. **40**: 790 (1985). Type: Tanzania, Iringa Dist., ridge above Sanje Falls, fl. 24.vii.1983, *Polhill, J.C. Lovett & J.M. Lovett* 5150 (K holotype, DSM, EA, MO).

    Isoglossa sp. aff. *substrobilina* sensu Dowsett-Lemaire in Bull. Jard. Bot. Belg. **55**: 317, 378 (1985).

    Isoglossa substrobilina sensu Phiri, Checklist Zambian Vasc. Pl.: 19 (2005); sensu Burrows & Willis, Pl. Nyika Plateau: 51 (2005), non C.B. Clarke.

Perennial herb or shrub 100–200 cm tall. Stems retrorse- or appressed-pubescent in two opposite lines or more widespread on young stems, hairs with conspicuous, brown(-orange) cell walls, later glabrescent. Leaves elliptic or somewhat obovate, 14.5–22 × (3.5)5–8.5 cm, base long-attenuate or cuneate, apex acuminate; surfaces pubescent or largely glabrous; lateral veins 7–13 pairs; petiole 18–53 mm long. Inflorescence terminal on principal or short lateral branches, spiciform or with single branch towards base, main axis 2–12.5 cm long, with decussately arranged bracts, imbricate when young, becoming more distant towards base at maturity, cymules single-flowered, sessile; axis villose in two opposite lines, sometimes also with spreading glandular hairs towards apex; bracts broadly ovate to elliptic, 5–10 × 4–7 mm, base attenuate, apex attenuate or acuminate, upper margins often reflexed at least when young, ± densely pilose along margins and on inner surface towards apex, outer surface largely glabrous or with ± dense glandular hairs; bracteoles lanceolate or narrowly elliptic, 4–7.5 mm long. Calyx lobes lanceolate, 4.5–7.5 mm long in flower, to 7–9 mm in fruit, shortly ciliate, with spreading glandular hairs externally. Corolla 10.5–15 mm long, white with pink or brown markings on palate; sparsely puberulous towards apex of lips externally; tube 4–5 mm long, somewhat widened upwards, with an interrupted ring of long hairs 2–3 mm from base within or these shorter and restricted to floor; upper lip narrowly triangular, hooded, 7.5–10 mm long, lobes 0.5–1.5 mm; lower lip 7–10 mm long, lobes rounded, 0.5–1.2 mm, palate upraised with prominent herring-bone patterning. Stamens attached in upper half of corolla tube; filaments free for 3.5–5.5 mm; anther thecae superposed, ± strongly oblique, one or both held almost patent to filament, upper theca 1–1.65 mm long, lower theca 1–1.45 mm long. Capsule c.11 mm long, glabrous. Immature seeds only seen, rugose.

Zambia. E: Isoka Dist., Kasoma Forest, fl. ix.1981, *Dowsett-Lemaire* 189 (BR, K). **Malawi.** N: Chitipa District, Misuku Hills, Mugesse Forest, fl. 17.ix.1975, *Pawek* 10129

(BR, K, MAL, SRGH, UC); Rumphi Dist., Uzumara Forest Reserve, fl.& fr. 28.ix.2009, *Chapama et al.* in MSB 996 (FRIM, K).

Also in S Tanzania. Moist montane and submontane forest; 1800–2000 m.

Conservation notes: Assessed as Vulnerable under IUCN criterion B in F.T.E.A.

Isoglossa bracteosa is closely related to *I. substrobilina* of Kenya, Uganda and Tanzania; it is most easily distinguished by the considerably larger leaves. In most specimens, including from the Misuku Hills, it is further separated by the spathe-like bracts with reflexed apex, but in *Dowsett-Lemaire* 189 even the immature bracts are held flat and erect. *I. milanjiensis* is also close but, in addition to the character given in the key, that species tends to have more widely branched paniculate inflorescences.

10. **Isoglossa milanjiensis** S. Moore in Trans. Linn. Soc., Bot. **4**: 33 (1894). —Clarke in F.T.A. **5**: 232 (1900). —Milne-Redhead in Mem. New York Bot. Gard. **9**: 25 (1953). Type: Malawi, Mulanje, fl.& fr. 1891, *Whyte* s.n. (BM holotype, K).

Isoglossa mossambicensis Lindau in Engler, Pflanzenw. Ost-Afrikas C: 372 (1895). —Clarke in F.T.A. **5**: 231 (1900). —Mapaura & Timberlake, Checklist Zimb. Vasc. Pl.: 14 (2004). —da Silva *et al.*, Prelim. Checklist Vasc. Pl. Mozamb.: 19 (2004). Type: Mozambique, Gorongosa, fl. 1884–1885, *Carvalho* 5617 [number not listed in protologue] (COI lectotype), lectotypified here.

Scrambling perennial herb or subshrub (15)100–350 cm tall. Stems retrorse-pubescent in two opposite lines and at nodes, later glabrescent. Leaves ovate or elliptic, 8–14.5(19.5) × 2.8–6(10) cm, base attenuate, apex acuminate, margin and principal veins pubescent or sparsely so, often also with scattered hairs on upper surface, midrib pilose towards leaf base when young, hairs extending onto young petioles; lateral veins 6–10 pairs; petiole 30–65 mm long. Inflorescence terminal, spiciform or often with 1–6 lateral branches at maturity, sometimes with secondary branching and forming a ± pyramidal panicle 4–15 cm long, cymules dense in upper portion of axis, more widely spaced below, sessile, single-flowered or one bract of a pair sometimes sterile on lateral branches; axis pale-pubescent, hairs retrorse towards base of spike, antrorse upwards, sometimes also glandular-pubescent; cymule bracts flabellate, elliptic or suborbicular in lower portion of axis, then 4–11 × 2–6.5 mm, becoming narrowly elliptic or lanceolate in upper portion, then 3–6 × 1–2.5 mm, base attenuate, apex acuminate, attenuate or acute, margin and often midrib ciliate, surfaces largely glabrous or sparsely to densely glandular-pubescent; bracts subtending inflorescence branches usually larger and ovate; bracteoles lanceolate, 3–6.5 × 1 mm. Calyx lobes lanceolate, 4.5–7.5 mm long in flower, 7.5–11 mm in fruit, indumentum as bracts. Corolla (12)13–16 mm long, white or tinged reddish, palate often with conspicuous purple to red-brown markings; limb sparsely puberulous externally; tube 3.7–5 mm long, somewhat widened upwards, with an interrupted ring of short hairs at attachment point of stamens within; upper lip oblong, hooded, (7.5)8.5–11.5 mm long, lobes 0.7–1.8 mm; lower lip (8)8.5–12 mm long, lobes rounded, 1–2.5 mm, palate upraised with prominent herring-bone patterning. Stamens attached in upper half of corolla tube; filaments free for 4–5 mm; anther thecae superposed, strongly oblique, at least lower theca held almost patent to filament, upper theca 1.35–1.7 mm long, lower theca 1.25–1.6 mm long. Capsule 10–14 mm long, glabrous. Seeds 3.5–4 × 2.8–3.5 mm, rugose when immature, inconspicuously so at maturity.

Zimbabwe. E: Mutare Dist., Vumba, Elephant Forest near Vumba Mts Hotel, fl.& fr. 25.iv.1959, *Chase* 7104 (BM, BR, K, LMA, SRGH). **Malawi**. S: Mulanje Mt, Lichenya Plateau, fl.& fr. 25.vi.1946, *Brass* 16422 (BR, K, NY, SRGH). **Mozambique**. MS: Gorongosa Dist., Mt Gorongosa, SE slopes, fl. 25.vii.1970, *Müller & Gordon* 1457 (K, LISC, SRGH); Sussundenga Dist., Serra Macuta, S slopes, fl. 2.vi.1971, *Pope* 439 (K, LMA, SRGH).

Not known elsewhere. Submontane and montane forest including margins and clearings, exposed riverbanks, mature *Cupressus* plantations, often dominant in forest understorey; (350)900–2000 m.

Conservation notes: Only known from Mts Mulanje, Gorongosa and mountains along the Zimbabwe–Mozambique border where it can be locally abundant, although some populations have been lost to forest clearance at lower altitudes. With only 9 locations known and inferred threats, considered Vulnerable under IUCN criterion B. Material seen from Mt Mulanje lacks glandular hairs on the inflorescence whilst those from the highlands of Mozambique and Zimbabwe develop a dense glandular indumentum as the inflorescence matures (the same situation having been documented by Poriazis & Balkwill for *Isoglossa* [sp. 1] *glandulosissima*; Bothalia **38**: 131–140, 2008). It may be possible to recognise two subspecies but field observations on Mulanje plants across a full flowering/fruiting season are needed before doing so.

11. **Isoglossa namuliensis** I. Darbysh. & T. Harris in Kew Bull. **66**: 244, fig.2 (2011). Type: Mozambique, Gurué Dist., Namuli Mt, slopes above Rio Licungo valley, fl. 16.xi.2007, *Harris* 324 (K holotype).

Weakly decumbent perennial herb to 15 cm high, stems to 70 cm long, but probably taller, rooting at lower nodes. Stems shortly crisped-pubescent when young, hairs mainly retrorse, later glabrescent. Leaves elliptic, 10–12.5 × 3.3–4.5 cm, base cuneate-attenuate, apex acuminate, main veins and margin shortly pubescent; lateral veins 6–8 pairs; petiole 18–40 mm long, antrorse-pubescent. Inflorescence terminal, spiciform, 7.5 cm long, cymules single-flowered, sessile; axis antrorse-pubescent, with scattered spreading glandular hairs increasing upwards; bracts flabellate, 9–11.5 × 4.5–7 mm, with pronounced outcurved acumen 4–4.5 mm long, margin with 1–5 prominent teeth below acumen, surface pubescent, hairs longest on margin and midrib, with interspersed short glandular hairs; bracteoles linear-lanceolate, 7–8.5 × 0.8–1.5 mm. Calyx lobes linear-lanceolate, 6.7–8.2 mm long in flower, shortly pubescent with interspersed short glandular hairs. Corolla 14.5–16.5 mm long, white with maroon markings on palate of lower lip and throat; limb sparsely pubescent externally; tube c.4 mm long, somewhat widened upwards, with a ring of hairs within immediately below attachment point of stamens; upper lip c.11.5 mm long, lobes 1.2–2.3 mm; lower lip 10–12.5 mm long, lobes oblong, 3.5–4 mm, palate upraised and with prominent herring-bone patterning. Stamens attached in upper half of corolla tube; filaments free for 4–4.5 mm; anther thecae superposed, upper theca 1.6–1.8 mm long, parallel or somewhat oblique to filament, lower theca 1.35–1.5 mm long, ± patent to filament. Capsule and seeds not seen.

Mozambique. Z: Gurué Dist., Namuli Mt, slopes above Rio Licungo valley, fl. 16.xi.2007, *Harris* 324 (K).

Not known elsewhere. Pathside in moist *Podocarpus–Syzygium* forest; c.1900 m.

Conservation notes: Assessed as Critically Endangered under IUCN criterion B in Harris, Darbyshire & Polhill (Kew Bull. **66**: 244, 2011).

12. **Isoglossa cataractarum** Brummitt & Feika in Kew Bull. **32**: 597, fig.2 (1978). Type: Zambia, Mbala Dist., Lunzua R. Falls, hydro-electric station, fl.& fr. 11.vii.1960, *Richards* 12861 (K holotype). FIGURE 8.6.**50**.

Shrub to 180–330 cm tall, often scandent, with widely divergent and ascending lateral branches, sometimes only one of a pair developing at each node. Stems shortly white antrorse-and/or appressed-pubescent particularly on two opposite sides, uppermost internodes sometimes with longer spreading or crisped hairs, later glabrescent. Leaves often absent at flowering / fruiting; mature leaves ovate or elliptic, 5.7–10 × 3–5.5 cm, base rounded to cuneate, apex attenuate or acuminate, pale-pubescent above and on veins beneath; lateral veins 6–10 pairs; petiole 3–30 mm long. Inflorescence terminal and in upper leafless axils of main and short lateral branches, densely fasciculate or shortly spiciform, 0.8–4 cm long, 1-sided with one bract of each pair along axis sterile, cymules subsessile, each 1–4-flowered; bracts imbricate, often caducous at fruiting, green with a pale base, elliptic or obovate, 4–11 × 1.5–6 mm, base attenuate, apex acuminate, puberulous and with few to numerous appressed pale hairs, sometimes only on margin and midrib, sometimes also with few short spreading glandular

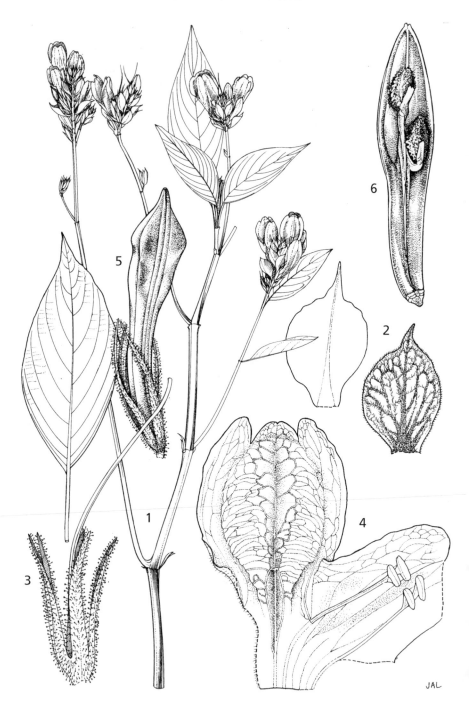

Fig. 8.6.**50**. ISOGLOSSA CATARACTARUM. 1, habit, with mature leaf (× ²/₃); 2, bract (× 5); 3, calyx with pistil, (× 5½); 4, dissected corolla with stamens, part of upper lip omitted (× 3); 5, capsule within calyx (× 4); 6, capsule valve with seeds (× 4). 1–4 from *Richards* 12861, 5 & 6 from *Richards* 12905. Drawn by Joanna Lowe. Reproduced from Kew Bulletin (1978).

hairs; bracteoles pale, (linear-)lanceolate or narrowly elliptic, 3.5–8 × 0.7–1.5 mm, indumentum as bracts but often with more glandular hairs. Calyx lobes linear-lanceolate, 5–8 mm long in flower, to 8.5–12 mm in fruit, spreading glandular-pubescent and puberulous, rarely mainly the latter, often with ascending hairs along lobe margins. Corolla 20–24 mm long, (pale) mauve to blue; pubescent externally, papillose within around attachment point of stamens; tube 11–13.5 mm long of which lower 5–7.5 mm cylindrical, c.2 mm wide, then throat abruptly widened laterally, 7–8 mm wide at mouth; upper lip triangular, 9–11 mm long, lobes 0.8–1.7 mm; lower lip elliptic, 9–12 mm long, lobes 1–2.5 mm, palate upraised with prominent herring-bone patterning. Stamens attached 6–7.5 mm from base of corolla; filaments free for 7.5–8.5 mm; thecae superposed, upper theca 2.1–2.4 mm long, patent to filament, lower theca 1.8–2.3 mm long, patent or oblique to filament. Capsule long-stipitate, 17–21.5 mm long, glabrous. Seeds 2.7–4 × 2.3–3 mm, rugose-tuberculate.

Zambia. N: Mbala Dist., Kalambo Falls, fl.& fr. 22.vii.1960, *Richards* 12905 (K). W: Ndola, fl.& fr. 4.xi.1970, *Fanshawe* 10958 (BR, EA, K, NDO).

Also in southern D.R. Congo (see note). Riverine forest and woodland, dense thicket on stream banks, amongst rocks in dry streambeds and by waterfalls; 900–1700 m.

Conservation notes: Known from only four localities (Lunzua Falls & Kalambo Falls in NE Zambia, Ndola in NC Zambia and Upemba Nat. Park in southern D.R. Congo) but unlikely to be threatened in the two former localities where it is recorded from thicket in ravines and by waterfalls; considered Data Deficient but may be threatened.

The branching pattern is unusual in this species and is more reminiscent of some of the scandent species of *Clerodendrum* (Lamiaceae). Often only a single branch is fully developed at each node, the opposite branch either being absent or much-reduced, the latter probably aiding climbing.

Specimens at BR from Katanga (Shaba), southern D.R. Congo have been labelled as *I. pyrophora* Champl. (e.g. *van Meel* 6473), but this name has not yet been published. This material appears to match the Zambian specimens of *I. cataractarum* and I consider them to be conspecific.

13. **Isoglossa pawekiae** Brummitt & Feika in Kew Bull. **32**: 595, fig.1 (1978). Type: Malawi, Chitipa Dist., Misuku Hills, Kalenga R., fl.& fr. 22.iv.1975, *Pawek* 9483 (K holotype, BR, MAL, MO, WAG).

Shrub to 150–300 cm tall. Young stems shortly crisped-pubescent on two opposite sides, hairs mainly antrorse; mature stems woody, glabrescent. Leaves (oblong-)elliptic or somewhat obovate, 10–18.5 × 4–6.3 cm, base of lower leaves long-attenuate, forming a narrow wing along midrib then abruptly truncate, apex acute to acuminate, upper surface glabrous or sparsely pubescent, principal veins beneath and margin shortly antrorse-pubescent; lateral veins 6–10 pairs, these and scalariform tertiary veins ± prominent beneath; petiole 0–2 mm long. Inflorescence terminal and often also in upper axils, narrowly spiciform, to 9–15.5 cm long, each cymule single-flowered, flowers sessile; axis pale crisped-puberulous and/or appressed-puberulous; bracts and bracteoles subequal, triangular to narrowly so, 2.5–6 × 1.5–3 mm, apex acute or attenuate, surface pale appressed-puberulous, margin pale-ciliate. Calyx tube 2–2.5 mm long, lobes lanceolate, 5–9 mm long in flower and fruit, imbricate towards base, margins narrowly hyaline, apex acute to acuminate, indumentum as bracts. Corolla 38–50 mm long, white, pale pink or pale purple; crisped-puberulous externally; tube 30–34 mm long, basal c.7 mm cylindrical, then broadly expanded into throat, with an interrupted ring of hairs within c.8 mm from base; upper lip broadly triangular, 14–15.5 mm long, lobes 3–3.5 mm; lower lip 15–16.5 mm long, lobes 3.5–4.5 mm, palate upraised with faint herring-bone patterning. Stamens attached c.14 mm from corolla base; filaments free for 12.5–14 mm; anther thecae parallel and held at subequal height, 3.2–4 mm long. Capsule 21–22 mm long, glabrous. Immature seeds only seen, rugose and with numerous minute processes.

Malawi. N: Karonga Dist., Stevenson Road, 21 km W of crossroads, 63 km E of Karonga, Vitakati R., fl. 26.iv.1977, *Pawek* 12698 (K, MAL, MO).

Not known elsewhere. Riverine thickets, termitaria in *Brachystegia* woodland, sometimes forming large colonies; 1000–1600 m.

Conservation notes: An extremely local species, apparently restricted to three river systems in the Misuku area of Malawi where it can be locally abundant; possibly threatened globally.

Balkwill *et al.* (S. Afr. J. Bot. **74**: 384, 2008) suggest that differences from *Isoglossa* sensu stricto in the macromorphology of this species may support its placement in a separate genus, but they do not explicitly state the characters they consider to be of significance. In my opinion this species falls well within the variation recorded in *Isoglossa* in eastern Africa; only the unusual leaf base and the imbricate calyx lobes are unique.

14. **Isoglossa floribunda** C.B. Clarke in F.T.A. **5**: 233 (1900). —Darbyshire in Kew Bull. **64**: 424, fig.7 (2009); in F.T.E.A., Acanthaceae **2**: 642, fig.82 (2011). Type: Mozambique, Chupanga, lower Zambezi, fl.& fr. viii.1858, *Kirk* s.n. (K holotype).

Schliebenia salviiflora sensu Vollesen in Opera Bot. **59**: 81 (1980), non Mildbr.

Isoglossa salviiflora (Mildbr.) Brummitt in Kew Bull. **40**: 786 (1985), in part as regards *Vollesen* in MRC 2323 & 3715.

Erect or decumbent perennial herb or subshrub 35–200 cm tall. Uppermost internodes eglandular- and glandular-pubescent, with or without occasional longer eglandular hairs, soon glabrescent. Leaves (ovate-)elliptic, 4.8–15 × 1.7–7 cm, base attenuate, apex acuminate, surfaces largely glabrous except for inconspicuous hairs on main veins; lateral veins 5–7 pairs; petiole 7–25 mm long. Inflorescence a series of wholly or partially 1-sided spiciform thyrses compounded into a ± large terminal panicle to 45 cm long, sometimes interrupted by ± reduced leaves; main axis of each thyrse 3–19 cm long, often curved, cymules densely clustered or more distant, sometimes strobilate; axis patent glandular- and eglandular-pubescent, usually with scattered long spreading eglandular hairs to 3 mm long; cymule bracts lanceolate to (oblong-) obovate, 1.5–9 × 1–6 mm, glandular-pubescent and eglandular-puberulent, with or without long spreading eglandular hairs particularly on margins; bracts subtending branches often foliaceous, ovate; bracteoles lanceolate to oblanceolate, 2–10.5 × 0.5–3.5 mm; flowers subsessile. Calyx lobes linear-lanceolate to narrowly oblanceolate, 3–10.5 mm long in flower, extending to 4.5–14 mm in fruit, indumentum as bracts. Corolla 19–33.5 mm long, blue to purple or rarely white, palate often paler, with or without darker mauve-blue guidelines; external surface sparsely pubescent, sometimes restricted to upper lip; tube 8.5–15 mm long, 2–2.5 mm wide at base, slightly upturned centrally then expanded to 3.5–5.5 mm wide at mouth, with a raised and interrupted ring of hairs 3–4 mm from base within; upper lip arcuate, hooded, 13.5–22.5 mm long, lobes minute; lower lip broad, 10.5–18.5 mm long, lobes 1–3 mm, palate upraised and with prominent herring-bone patterning. Stamens attached 3.5–5 mm below corolla mouth; filaments free for 11–17 mm, arcuate; anther thecae parallel, barely offset, upper theca 2.5–5 mm long, lower theca 2.3–4.5 mm long. Capsule 11.5–16.5 mm long, glabrous. Seeds 2.5–3 × 1.8–2.7 mm, rugose.

Subsp. **floribunda**. —Darbyshire in Kew Bull. **64**: 424, fig.7h,j (2009); in F.T.E.A., Acanthaceae **2**: 644 (2011).

Inflorescence axis either lax with clearly spaced cymules at least towards base, or more dense and clustered throughout but not strobilate; cymule bracts lanceolate to oblanceolate, 1.5–6 × 1–2 mm. Calyx lobes 3–10 mm long in flower. Corolla 19–29 mm long.

Zimbabwe. N: Hurungwe Dist., Chirundu, Zambezi valley, fl. 20.iii.1952, *Whellan* 644 (K, LISC, SRGH). S: Chiredzi Dist., Runde (Lundi) R., Fishan, fl.& fr. 28.iv.1962, *Drummond* 7795 (BR, K, LISC, SRGH). **Zambia**. C: Chongwe Dist., Chakwenga Copper prospect, fl. 1.iv.2008, *Bingham* 13463 (K). **Malawi**. S: Mulanje Dist., Sambani Hill, fl.& fr. 9.vi.1985, *la Croix* 2951 (K, MO). **Mozambique**. N: Chiúre Dist., 12 km from Chiúre (Novo) to Meloco, fl.& fr. 6.iv.1964, *Torre & Paiva* 11664 (BR, LISC, UPS, WAG). Z: Morrumbala Dist., slopes of Serra da Morrumbala, fl.& fr. 13.v.1943, *Torre* 5308 (BR, LISC, MO, PRE, WAG). T: Cahora Bassa Dist., Cabora Bassa, fl.& fr. 21.iv.1972, *Pereira*

& *Correia* 2214 (BR). MS: Sussundenga Dist., c.30 km N of Dombe, fl.& fr. 4.vi.1971, *Biegel* 3540 (K, LISC, SRGH). GI: Massangena Dist., R. Save, Mavue, near store, fl. 24.iv.1962, *Drummond* 7733 (BM, BR, K, LISC, SRGH).

Also in southern Tanzania. Riverine woodland, dry forest, grassland on margins of woodland; 100–650 m.

Conservation notes: Widespread but scattered and uncommon taxon; assessed as Data Deficient in F.T.E.A.

Subsp. *salviiflora* (Mildbr.) I. Darbysh., restricted to SE Tanzania, has broader, obovate cymule bracts and strobilate partial inflorescences.

15. **Isoglossa grandiflora** C.B. Clarke in F.T.A. **5**: 233 (1900). —Darbyshire in F.T.E.A., Acanthaceae **2**: 645, fig.80.11 (2011). Types: Malawi, Chikwawa (Shibisa) to Tshinsunze (Tshinmuze), fl.& fr. ix.1859, *Kirk* s.n. (K syntype); Manganja Hills, Mt Sochi, fl.& fr. ix.1861, *Meller* s.n. (K syntype).

Straggling subshrub 50–300 cm tall. Stems sparsely pubescent, hairs mainly antrorse or appressed, uppermost internodes also with short spreading glandular hairs, later glabrescent. Leaves (ovate-)elliptic, 6.4–22 × 3.2–10.5 cm, base attenuate or cuneate, apex acuminate, upper surface and veins beneath sparsely pubescent; lateral veins 6–9 pairs, prominent beneath; petiole 10–55(83) mm long. Inflorescence terminal (often with reduced inflorescences in uppermost leaf axils), spiciform or usually with 1–3(5) pairs of lax branches towards base, sometimes with secondary branching, main axis 7.5–38 cm long, cymules ± distant along axis, sessile, 1 to several-flowered, sometimes 1-sided along lateral branches; axis densely viscid glandular-pubescent, with appressed eglandular hairs, rarely the latter only; bracts ovate to lanceolate, 2.5–5.5(7.5) mm long, indumentum as axis; bracteoles somewhat narrower; flowers subsessile. Calyx lobes linear, (5)6–9 mm long in flower, to 6.5–12 mm in fruit, glandular-pubescent, with scattered eglandular hairs externally, rarely the latter only. Corolla 25–33.5 mm long, pink, purple or blue; pubescent externally; tube 10.5–14.5 mm long, 2–2.5(3) mm wide at base, abruptly upturned below centre then expanded to 4–6 mm wide at mouth, dorsally pubescent inside and with a ring of hairs 2.5–3.2 mm from base; upper lip arcuate, hooded, 15–20 mm long, lobes minute; lower lip 15–19 mm long, lobes 0.5–1.5 mm, palate upraised with prominent herring-bone patterning. Stamens attached 3.5–4.5 mm below corolla mouth; filaments free for 14–17 mm, arcuate; anther thecae barely offset, parallel, upper theca 2.6–3.4 mm long, lower theca 2.3–3 mm long. Capsule (11.5)14.5–21 mm long, glabrous. Seeds 3–4.2 × 2.5–3.1 mm, rugose.

Zambia. N: Mpika Dist., Mwamfushi R. gorge above Mpika boma, fl. 15.vi.1995, *Bingham* 10620 (K). **Malawi**. N: Rumphi Dist., South Rukuru gorge, 5 km E of Rumphi, fl.& fr. 21.v.1970, *Brummitt* 10962 (K, MAL). S: Blantyre Dist., Blantyre, Sunnyside suburb near Mudi R., fl. 3.viii.1970, *Brummitt & Harrison* 12403 (K, MAL, SRGH). **Mozambique**. T: Moatize Dist., Mt Zóbuè, fl.& fr. 3.x.1942, *Mendonça* 575 (BR, LISC).

Also in S Tanzania. Riverine woodland, thicket and forest, roadside banks, sometimes gregarious and locally abundant; 800–1600 m.

Conservation notes: Fairly common in Malawi; assessed as Least Concern in F.T.E.A.

Brummitt et al. 17166 from Njuli, S Malawi, is unusual in largely lacking the glandular element to the inflorescence, but otherwise resembles typical *Isoglossa grandiflora*; both glandular and eglandular forms are recorded in several other species of *Isoglossa* (see *I. milanjiensis* above).

16. **Isoglossa mbalensis** Brummitt in Kew Bull. **40**: 786 (1985). —Darbyshire in F.T.E.A., Acanthaceae **2**: 645 (2011). Type: Zambia, Mbala, fl. 3.iv.1960, *Fanshawe* 5640 (K holotype, LISC).

Erect perennial herb, 60–160 cm tall from a woody rootstock. Stems pubescent, hairs antrorse and retrorse, often with scattered longer hairs, uppermost internodes also with short spreading glandular hairs, later glabrescent. Leaves ovate, 7–12.5 × 4.3–7 cm, base shortly attenuate, apex acuminate, sparsely pubescent mainly on veins beneath, uppermost leaves also with glandular hairs towards base; lateral veins 5–6(7) pairs, prominent beneath; petiole 3–21 mm long. Inflorescence a terminal panicle 18–55 cm long, with 3–9 pairs of lax branches to 23 cm long, shortening upwards, sometimes with secondary branching, one of a pair sometimes barely developing, cymules widely spaced along axis, sessile; axis densely spreading glandular- and eglandular-pubescent; main axis bracts lanceolate, 5–12 mm long; cymule bracts ovate to lanceolate, 3–5.5(7) mm long, sometimes tinged purple; bracteoles as bracts but to 4 mm long; flowers sessile or pedicels to 2 mm long. Calyx lobes lanceolate, (2.5)3.5–5 mm long in flower, barely extending in fruit, glandular- and eglandular-pubescent. Corolla 23–29 mm long, deep blue-purple to mauve, palate sometimes brownish; external surface glabrous or pubescent; tube 11–13 mm long, 1.5–2.5(4) mm wide at base, abruptly upturned below centre then widened to 4–6.5 mm wide at mouth, an interrupted raised ring of hairs 2.5–3.5 mm from base within; upper lip arcuate, hooded, 14.5–18.5 mm long, lobes 0.8–1.5 mm; lower lip deflexed distally, 12.5–16 mm long, lobes ± 1.5 mm, palate upraised with prominent herring-bone patterning. Stamens attached at or just below corolla mouth; filaments free for 10–12.5 mm, arcuate; anther thecae barely offset, parallel, 3–4 mm long. Capsule (12.5)15.5–25 mm long, glabrous. Immature seeds only seen, flattened-ellipsoid, 4–5 × 2.8–3.1 mm, smooth.

Zambia. N: Mbala Dist., Lunzua Gorge, fl.& fr. 9.v.1936, *Burtt* 6271 (BM, BR, K).

Also in SW Tanzania. Dense herbage in miombo woodland, particularly at margins, riverine woodland, wooded grassland amongst tall grass, sometimes gregarious and locally abundant; 1300–1600 m.

Conservation notes: Very local but apparently abundant at some of its Zambian sites; assessed as Near Threatened in F.T.E.A.

Plants from southern Tanzania differ from most Zambian populations in the corollas being pubescent externally, not glabrous, although some Zambian specimens (e.g. *Richards* 11288) have sparse hairs on the upper tube.

This species is very close to *Duvernoia verdickii* De Wild. (combination in *Isoglossa* to be made by D. Champluvier) from the Katanga (Shaba) region of southern D.R. Congo, which differs primarily in having longer calyx lobes (5–8 mm long in flower) and larger fruits (28–30 mm long).

17. **Isoglossa eylesii** (S. Moore) Brummitt in Kew Bull. **40**: 787 (1985). Type: Zimbabwe, Mazowe, W side of Iron Mask Hill, fl. i.1909, *Eyles* 560 (BM holotype, SRGH). FIGURE 8.6.**51**.

    Adhatoda eylesii S. Moore in J. Bot. **48**: 253 (1910).

Suffruticose herb, 1 to several erect stems 50–150 cm tall from a woody rootstock. Stems crisped-pubescent in two opposite furrows and with a line of hairs at nodes. Leaves ovate(-elliptic), 8–12 × 3.3–6.5 cm, base rounded or lower leaves obtuse to acute, apex (sub)attenuate, surfaces sparsely pilose mainly on margin and main veins; lateral veins 6–9 pairs, prominent beneath; petiole of lower leaves 5–12 mm long. Inflorescence terminal, spiciform or with 1–2 pairs of lateral branches, main axis (5.5)10–39 cm long, subsessile pairs of cymules densely arranged along axis, lower cymules more distant, lowermost pair of bracts often sterile; axis glandular-villose, hairs 1–4 mm long, with numerous short subappressed pale eglandular hairs and scattered spreading long eglandular hairs to 5 mm; bracts ovate or lanceolate, 10–19.5 × 3–6.5 mm, apex attenuate to long-acuminate, indumentum as axis, short eglandular hairs mainly towards base; bracteoles (linear-)lanceolate, 9–17.5 × 1.5–4 mm; flowers sessile, pedicels to 2 mm long in fruit. Calyx lobes linear-oblanceolate, (7.5)10–13.5 mm long in flower to 13–18 mm in fruit, indumentum as bracts. Corolla (19)23–28 mm long, white to pale mauve, throat and palate often purple or brownish; pubescent externally, densely so on upper lip; tube inflated from immediately above base, floor strongly ventricose, (9)11–14 × 7–9 mm, irregularly bossed immediately above base within; upper lip strongly hooded and bulbous, (10)13–14.5 mm long, apex subentire; lower lip elliptic, 8–10 mm long, lobes 0.5–0.8 mm long, palate upraised with prominent herring-bone patterning, with scattered short-stalked glands. Stamens attached just below corolla mouth; filaments free for (6)9–10.5 mm; anther thecae subparallel and only slightly offset, each 3.5–4.5

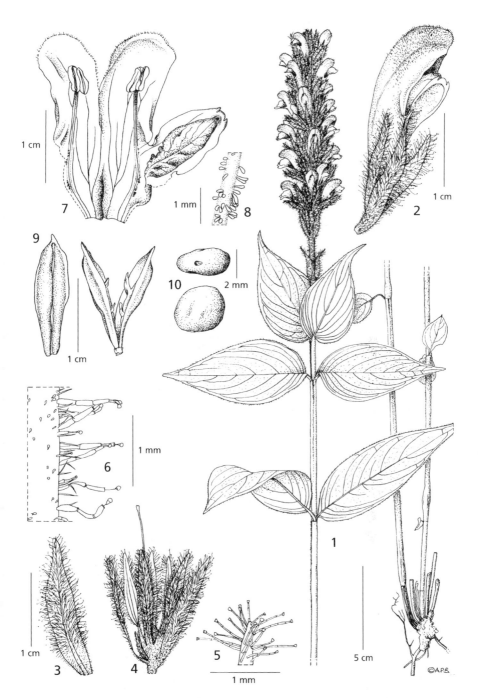

Fig. 8.6.**51**. ISOGLOSSA EYLESII. 1, habit, flowering branch and stem base; 2, flower, lateral view; 3, bract, outer face; 4, bracteoles and calyx with pistil; 5, detail of indumentum at apex of calyx lobe, outer surface, hydrated state; 6, detail of hairs on margin and inner surface of calyx lobe, dry state; 7, dissected corolla with stamens, upper lip split along the dorsal line; 8, detail of papillose hairs on adherent section of staminal filament; 9, capsule; 10, seed, ventral view and face. 1–6 from *Fanshawe* 2814, 7 & 8 from *Chase* 6322, 9 & 10 from *Fanshawe* 2296. Drawn by Andrew Brown.

mm long. Capsule (14.5)16–20.5 mm long, glabrous. Seeds glossy brown-black when mature, subellipsoid, 3.5–4.5 × 2.8–3.5 mm, smooth.

Zambia. N: Mpika, fl. 1.ii.1955, *Fanshawe* 1920 (BR, EA, K, NDO, SRGH). W: Kitwe, fr. 20.v.1955, *Fanshawe* 2296 (BR, EA, K, NDO). C: Chibombo Dist., Mutendere Game Ranch, Kamaila, fr. 1.iii.1996, *Bingham* 10951 (K). S: Kafue Dist., Munali Pass, Nega Nega Hills, fl.& fr. 26.ii.1964, *Angus* 3873 (BR, FHO, K). **Zimbabwe**. N: Bindura Dist., Bindura, fl. v.1940, *Hopkins* in GH 7660 (K, SRGH). C: Harare Dist., near Poti R., fl. 10.ii.1957, *Chase* 6322 (K, LMA, SRGH).

Also in D.R. Congo. Forming dense colonies in miombo woodland; 1100–1500 m.

Conservation notes: Widespread but scattered and apparently rather scarce in miombo woodland, can be the dominant herb in suitable habitat; Least Concern.

34. JUSTICIA L.[12]

Justicia L., Sp. Pl.: 15 (1753). —Graham in Kew Bull. **43**: 551–624 (1988).
    Dianthera L., Sp. Pl. **1**: 27 (1753).
    Adhatoda Miller, Gard. Dict. Abr., ed.4 (1754). —Clarke in F.T.A. **5**: 221 (1900).
    Gendarussa Nees in Wallich, Pl. Asiat. Rar. **3**: 76, 102 (1832).
    Rungia Nees in Wallich, Pl. Asiat. Rar. **3**: 77, 109 (1832). —Clarke in F.T.A. **5**: 252 (1900).
    Rhaphidospora Nees in Wallich, Pl. Asiat. Rar. **3**: 77, 115 (1832).
    Rostellularia Reichenb., Handb.: 190 (1837).
    Monechma Hochst. in Flora **24**: 374 (1841). —Clarke in F.T.A. **5**: 212 (1900).
    Tyloglossa Hochst. in Flora **26**: 72 (1843).
    Duvernoia Nees in De Candolle, Prodr. **11**: 322 (1847).
    Anisostachya Nees in De Candolle, Prodr. **11**: 368 (1847).
    Harnieria Solms-Laub. in Sitzber. Ges. Naturf. Fl. Tellur. **4**: 62 (1864).
    Nicoteba Lindau in Engler & Prantl, Nat. Pflanzenfam. **IV**, 3b: 329 (1895).
    Thamnojusticia Mildbr. in Notizbl. Bot. Gart. Berlin-Dahlem **11**: 825 (1933).

Herbs, shrubs or (rarely) small trees, erect to scrambling; cystoliths conspicuous or not. Leaves opposite, equal to distinctly anisophyllous, entire to crenate. Flowers in a wide variety of inflorescences from open dichasial cymes (sometimes aggregated into panicles) to subsessile or sessile dichasia aggregated or condensed into various types of racemoid or spiciform, sometimes secund, cymes with small or large sometimes strobilate bracts or in axillary clusters, or solitary; bracts persistent, prominent or not; bracteoles present, large or small. Calyx deeply divided into 5 subequal segments or with 1 segment reduced or almost absent. Corolla glabrous or (usually) hairy outside, often also with capitate glands, inside with band of hairs in upper part of tube; basal tube cylindric, widening upwards into a ± distinct throat; distinctly 2-lipped, upper lip shallowly 2-lobed, flat to distinctly hooded, lower lip shallowly to deeply 3-lobed, often with conspicuous pattern of transverse differently coloured lines ('herring-bone' pattern). Stamens 2, inserted in upper part of tube; no staminodes, anthers bithecous, thecae often unequal with upper smaller than lower, usually ± superposed, held parallel or at an angle to each other, lower theca with white sterile appendage (rarely apiculate). Style filiform, glabrous or hairy; stigma with two ellipsoid erect lobes. Capsule 2–4-seeded, clavate with a solid basal stalk, halves entire during dehiscence or with retinacula splitting from capsule wall or capsule walls splitting from base; retinacula strong, curved. Seed very variable, sphaeroid to discoid, reniform or cordiform, compressed or not, testa smooth or variously ornamented, rugulose, tuberculate, pubescent or echinate.

The largest genus in the Acanthaceae and often estimated at around 500 species, but almost certainly an underestimate. The present account alone lists 10 new species and detailed studies in other tropical regions (e.g. Wasshausen & Wood, Kew Bull.

---

[12] By Kaj Vollesen

58: 769, 2003; Vollesen in F.T.E.A., Acanthaceae **2**: 495, 2010) have also resulted in the description of many new species. With the recent inclusion of genera such as *Monechma* and *Rungia* a figure of 700 would seem more realistic.

Key to native species

1. Flowers solitary or in 2–4(6)-flowered dichasia which are sessile or subsessile in axils of vegetative leaves, occasionally internodes contracted towards apex of stem but all 'bracts' foliaceous (Fig. **58**.8, **59**) 2
  – Inflorescence not as above 27
2. Capsule 2-seeded; seeds kidney-shaped, testa smooth, glossy, glabrous (Sect. *Monechma*, in part) 3
  – Capsule 4-seeded; seeds not kidney-shaped, testa variously sculptured, rarely glabrous (Sect. *Harnieria*) 5
3. Calyx 4-lobed with no trace of 5th lobe; neither stems nor leaves densely whitish or greyish sericeous . . . . **54.** *divaricata*
  – Calyx 5-lobed with 5th lobe ¹/₂–²/₃ the length of other four; young stems and/or leaves densely whitish or greyish sericeous 4
4. Young stems densely whitish sericeous; leaves bright green, almost glabrous; bracteoles green, c.6 mm long, similar to calyx-lobes; corolla mauve or pale blue . . . . **55.** *australis*
  – Young stems greyish sericeous; leaves grey, densely sericeous on both surfaces; bracteoles greyish sericeous, 2–3 mm long, much shorter than calyx-lobes; corolla white. . . . . **56.** *incana*
5. Shrubby herbs or shrubs, at least basal part of stems woody 6
  – Annual or perennial herbs, basal part of stems not woody 11
6. Corolla lemon yellow to bright yellow . . . . **31.** *odora*
  – Corolla white to mauve or purple 7
7. Anther thecae yellow with distinct brown pigment patches at apex and base of upper theca and apex of lower theca . . . . **39.** *striata*
  – Anther thecae yellow to uniformly purple but never with distinct brown pigment patches at apex and base of upper theca and apex of lower theca 8
8. Upper internodes usually contracted forming a pseudo-racemose 'inflorescence'; bracts held erect, covering flowers . . . . **40.** *phyllostachys*
  – Upper internodes always well-spaced; bracts spreading, not covering flowers. . . 9
9. Corolla 10–17 mm long; capsule 12–19(24) mm long . . . . **32.** *capensis*
  – Corolla 4.5–9(11) mm long; capsule (4.5)5.5–7.5(9) mm long 10
10. Young stems, leaves, bracts and calyx minutely puberulous with hairs to 0.1 mm long (rarely also with scattered longer hairs) . . . .**36.** *rhodesiana*
  – Young stems, leaves, bracts and calyx puberulous, all hairs longer than 0.1 mm . . . . **35.** *protracta*
11. Corolla tube 4–7 mm longer than upper lip . . . . **33.** *striolata*
  – Corolla tube up to 2(3) mm longer than upper lip (rarely lip slightly longer than tube) 12
12. Annual herb 13
  – Perennial herb 20
13. Anther appendage broadly ellipsoid in outline, obtuse . . . . **47.** *mariae*
  – Anther appendage linear in outline, acute or bifid 14
14. Anther thecae yellow with distinct brown pigment patches at apex and base of upper theca and apex of lower theca . . . . **39.** *striata*
  – Anther thecae yellow to purple but never with distinct brown pigment patches at apex and base of upper theca and apex of lower theca. 15

15. 'Herring-bone' pattern on lower lip with raised ribs and with raised crenulate edge; anther appendage bent at 90° to theca. . . . .**46.** *richardsiae*
- 'Herring-bone' pattern on lower lip with raised ribs but without raised crenulate edge; anther appendage not bent relative to theca 16

16. Bract fused to pedicel making it appear as if flowers attached at apex of bract petiole; peduncles 1–2(3) mm long . . . .**43.** *mollugo*
- Bract not fused to pedicel, flowers attached at base of bract petiole; peduncles to 0.5 mm long 17

17. Upper internodes strongly contracted forming a pseudo-raceme; lateral branches spreading at almost right angles to main stem . . . . **44.** *syncollotheca*
- Upper internodes not contracted; lateral branches leaving main stem at angles of 45–60° 18

18. Leaves 4–10 times as long as wide; corolla 4–5 mm long, lower lip 2–3 × 2.5–4 mm; capsule 3–4 mm long, obtuse. . . . .**45.** *obtusicapsula*
- Leaves up to 3 times as long as wide; corolla 4–9(11) mm long, lower lip 3–6(9) × 3–6.5(9) mm; capsule 4–7.5 mm long, acute 19

19. Calyx lobes 3–6.5(7.5) mm long, not with long pilose hairs, basal part white and spongiform in fruit; corolla 4–7 mm long; capsule 4–6 mm long. . **34.** *heterocarpa*
- Calyx lobes 1.5–3.5 mm long, usually with long (1–2 mm) pilose hairs, basal part not white or spongiform in fruit; corolla (4.5)6–9(11) mm long; capsule (4.5)5.5–7.5 mm long. . . . . **35.** *protracta*

20. Appendage on lower theca ellipsoid or broadly so in outline, bent 90° relative to theca, broadly rounded; bracteoles 2–7 mm long 21
- Appendage on lower theca linear in outline, not bent relative to theca, acute or often bifid; bracteoles less than 2 mm long (very rarely to 2.5 mm) 22

21. Young shoots covered with small succulent orbicular or reniform leaves; flowers 2–4 per axil; peduncles 1–5(6) mm long; corolla 11–16 mm long . . **42.** *elegantula*
- Young shoots not covered with small succulent orbicular or reniform leaves; flowers all single; peduncles less than 1 mm long; corolla 8–10 mm long . . . .**41.** *lithospermoides*

22. Young stems, leaves, bracts and calyx minutely puberulous with hairs to 0.1 mm long, rarely also with scattered longer hairs . . . .**36.** *rhodesiana*
- Young stems, leaves, bracts and calyx puberulous to pubescent or pilose, all hairs longer than 0.1 mm 23

23. Anther thecae yellow with distinct brown pigment patches at apex and base of upper theca and apex of lower theca . . . . **39.** *striata*
- Anther thecae yellow to uniformly purple but never with distinct brown pigment patches at apex and base of upper theca and apex of lower theca 24

24. Leaves less than twice as long as wide, base usually truncate to cordate, more rarely cuneate; young stems puberulous to tomentose . . . .**37.** *hedrenii*
- Leaves more than twice as long as wide, base attenuate to rounded; young stems puberulous to pubescent or sparsely so 25

25. Leaves distinctly discolorous, much darker above, upper 4–12 (rarely 2–3) times longer than wide. . . . . **38.** *whytei*
- Leaves not or only very slightly discolorous, all 2–3 times as long as wide 26

26. Upper internodes usually contracted forming a pseudo-racemose 'inflorescence'; bracts held erect, covering flowers; calyx lobes 4–6 mm long; corolla 8–14 mm long . . . . **40.** *phyllostachys*
- Upper internodes always well-spaced; bracts spreading, not covering flowers; calyx lobes 1.5–3.5 mm long; corolla (4.5)6–9(11) mm long . . . . **35.** *protracta*

27. Flowers in axillary secund (only 1 bract at each node supporting a flower) spiciform cymes; bracts small (up to 5 mm long), not strobilate; bracteoles small (Sect. *Ansellia*, Fig. **57**) 28
- Inflorescence not as above 34
28. Perennial herb with rootstock or rhizome 29
- Annual herb without rootstock or rhizome 32
29. Corolla 8–14 mm long 30
- Corolla 5–7.5 mm long 31
30. Spikes 2(3)-flowered; capsule 9–13 mm long; stem hairs with purple transverse walls . . . .**27.** *nuttii*
- Spikes (2)3–5-flowered; capsule 14–16 mm long; stem hairs with colourless transverse walls . . . .**26.** *crassiradix*
31. Stem with longitudinal band of hairs; capsule 6–8.5 mm long, hairy in upper half, more rarely glabrous; seeds c.1.5 mm in diameter; in a variety of woodland, bushland and grassland habitats, but not on grey to black clay . . **25.** *anagalloides*
- Stem glabrous apart from thin transverse band at nodes; capsule (8)9–12 mm long, glabrous (very rarely hairy at apex); seeds c.2 mm in diameter; in grassland or bushland on grey to black clay . . . . **29.** *anselliana*
32. Stem glabrous apart from thin transverse band at nodes; corolla 5.5–7.5 mm long; in grassland or bushland on grey to black clay . . . . **29.** *anselliana*
- Stem with longitudinal band of hairs; corolla 3.5–5 mm long; in a variety of woodland, bushland and grassland, but not on grey to black clay 33
33. Capsule 4–6 mm long; seeds c.1 mm in diameter; most or all leaves widest at or near middle . . . . **28.** *exigua*
- Capsule 8–12 mm long; seeds 1.5–2 mm in diameter; most or all leaves widest below middle . . . .**30.** *matammensis*
34. Flowers in 1–3(15)-flowered subsessile dichasia aggregated into clearly defined terminal (or also axillary from upper axils) racemoid cymes; bracts never strobilate (Sect. *Tyloglossa*, Fig. **54**) 35
- Inflorescence not as above 41
35. Plant with all leaves in a basal rosette and ± appressed to ground . . **14.** *fittonioides*
- Plant with well-defined leafy stems 36
36. Corolla bright yellow 37
- Corolla white, pink, blue or mauve, often with dark veins or with darker markings on lower lip 38
37. Main axis bracts and cyme bracts lanceolate to narrowly ovate, 1–2(3) mm wide; dorsal calyx lobe only slightly longer than others; upper corolla lip not edged with brown . . . . **15.** *flava*
- Main axis bracts and cyme bracts elliptic, 3–4 mm wide; dorsal calyx lobe amost twice as long as others; upper corolla lip edged with brown . . . . . **16.** *gorongozana*
38. Annual herb; corolla with conspicuous blue to purple venation on both lips . . . . **20.** *kirkiana*
- Perennial herb; corolla without conspicuous darker venation but usually with darker markings on lower lip 39
39. Corolla blue to mauve, 11–14 mm long; calyx 7–9 mm long; capsule glabrous . . . . **17.** *petiolaris*
- Corolla white to pink or mauve, 6–12 mm long; calyx 3–7 mm long; capsule puberulous 40
40. Erect, ascending or scrambling herb from creeping rhizome, often rooting at lower nodes; bracts with long ciliate hairs and long-stalked capitate glands; largest leaf (1.5)2–5 cm wide; calyx 3–5 mm long; corolla 6–8.5 mm long; capsule 5–7.5 mm long . . . . **19.** *nyassana*

- Pyrophytic herb usually with several erect stems from woody rootstock; bracts without pilose hairs, with short-stalked capitate glands; largest leaf 0.3–1 cm wide; calyx 4–7 mm long; corolla 8–12 mm long; capsule 9–11 mm long . . . . **18.** *linearispica*

41. Calyx of 4 equal sepals 2.5–3 mm long plus a 5th reduced to a small tooth; corolla 2.5–3 mm long (Sect. *Anisostachya*, Fig. **56**) . . . . **24.** *tenella*
- Calyx of 5 equal linear to lanceolate sepals 3–21 mm long; corolla 6–31 mm long 42

42. Flowers in dense strobilate, spiciform, secund (only 1 bract at each node supporting a flower) or quadrangular cymes; bracteoles large and bract-like; capsule 4-seeded (Fig. **55**) 43
- Inflorescences various, dense strobilate, spiciform cymes or not; bracteoles not bract-like; capsule 2–4-seeded 45

43. Cymes quadrangular (all bracts supporting a flower); bracts without (rarely with) white margin (Sect. *Vascia*) . . . . **21.** *ruwenzoriensis*
- Cymes secund (only 1 bract at each node supporting a flower); bracts with white margin (Sect. *Betonica*) 44

44. Bracts (5.5)7–16 mm long; sepals 3.5–7 mm long; corolla white, 8–14 mm long . . . . **22.** *betonica*
- Bracts 20–25 mm long; sepals c.10 mm long; corolla pale yellow, c.20 mm long. . . . . **23.** *sp. A*

45. Capsule 2-seeded (not known in sp. 51) 46
- Capsule 4-seeded; seed not kidney-shaped, testa variously sculptured (rarely smooth); inflorescences various 59

46. Seeds orbicular, testa densely tuberculate; flowers in loose elongated terminal racemiform cymes or loose panicles; bracts small . . . . **12.** *bequaertii*
- Seeds kidney-shaped, testa smooth, glossy, glabrous (rarely hairy); flowers in dense terminal and axillary racemiform cymes; bracts usually large, strobilate or not; bracteoles small or large (Sect. *Monechma*, in part, Fig. **60**). 47

47. Annual herb 48
- Perennial herb, shrubby herb or shrub 51

48. Calyx 8–13 mm long, longer than the 7–8 mm long corolla; all bracts foliaceous; bracteoles 1–2 cm long; capsule glabrous, 9–11 mm long. . . . . **48.** *ciliata*
- Calyx 3.5–7.5 mm long, shorter than the 5.5–11 mm long corolla; bracts (apart from lowermost ones) clearly different from leaves; bracteoles minute (to 1.5 mm long); capsule hairy, 4.5–7 mm long 49

49. Corolla c.11 mm long, lower lip c.8 mm long; anther thecae c.1 mm long; bracts finely ciliate with ciliae less than 0.5 mm long . . . . **51.** *sp. B*
- Corolla 5.5–10.5 mm long, lower lip 3–7 mm long; anther thecae 0.5–0.75 mm long; bracts with ciliae 1–2 mm long (rarely without) 50

50. Bract apex subacute to truncate, recurved, with distinct 'shoulders'. **49.** *bracteata*
- Bract apex acute to acuminate (rarely subacute), straight if acute and recurved if acuminate, usually without 'shoulders' . . . . **50.** *monechmoides*

51. Flowers mostly in axillary subsessile racemoid cymes; calyx 4–6.5 mm long; corolla 10–14 mm long; capsule 7–9 mm long . . . . **53.** *eminii*
- Flowers in terminal racemoid cymes (or also a few lateral from upper axils); calyx 6–19(21) mm long; corolla 11–25 mm long. 52

52. Bracts, bracteoles and calyx-lobes with 3 strong rib-like longitudinal veins from base to apex; capsule glabrous. . . . . **57.** *tricostata*
- Bracts, bracteoles and calyx-lobes with a single strong central vein and weak lateral veins 53

53. Bracteoles wider than bracts, emarginate with a small apiculus, distinctly ciliate with broad glossy hairs; capsule glabrous . . . . **52.** *sp. C*

- Bracteoles often same length as bracts but always distinctly narrower, acute to rounded, not distinctly ciliate; capsule (where known) hairy 54
54. Flowers in 0.5–1.5 cm long terminal and axillary cymes with non-strobilate bracts; bracts 4–6 mm long; bracteoles 3–6 mm long . . . . **58.** *fanshawei*
- Flowers in longer terminal cymes with strobilate bracts; bracts and bracteoles 7–25 mm long 55
55. Bracts and bracteoles 7–10 mm long; only one bract per node supporting a flower; corolla c.11 mm long . . . . **59.** *varians*
- Bracts 12–25 mm long; bracteoles 8–25 mm long; all bracts supporting a flower; corolla 11–25 mm long. 56
56. Bracts, bracteoles and calyx-lobes with broad pale hyaline margins; corolla 11–15 mm long . . . . **60.** *scabrida*
- Bracts, bracteoles and calyx-lobes without broad pale hyaline margins; corolla 15–25 mm long 57
57. Calyx 7–10 mm long; corolla 22–25 mm long; bracteoles 8–13 mm long; anthers hairy on connective . . . . **61.** *kasamae*
- Calyx 11–19(21) mm long; corolla 15–20 mm long; bracteoles 12–25 mm long; anthers glabrous 58
58. Leaf-base truncate to cuneate (rarely shortly attenuate); largest leaf (3)4–7 × 1.5–3.5 cm, up to twice as long as wide; corolla 15–18 mm long, upper lip only slightly longer than tube. . . . . **62.** *subsessilis*
- Leaf-base attenuate; largest leaf 8–9 × 1.7–2.2 cm, more than 3 times as long as wide; corolla 17–20 mm long, upper lip more than twice as long as tube . . **63.** *attenuifolia*
59. Seed testa smooth and shiny, glabrous 60
- Seed testa variously sculptured; flowers in various types of open inflorescence (dichasia, racemoid cymes, panicles) with distinct peduncles and non-strobilate bracts (Sect. *Rhaphidospora* and Sect. *Justicia*, Figs **52**, **53**) 61
60. Flowers in a large open panicle 10–30 cm long; corolla 2.7–3.1 cm long . . . .
. . . . **6.** *salvioides*
- Flowers in a condensed racemoid cyme 4–7 cm long; corolla c.1.7 cm long . . . . .
. . . . **64.** *tetrasperma*
61. Corolla 2.7–3.5 cm long (measured along upper lip); upper corolla lip bent backwards forming an angle with tube; flowers in racemoid cymes aggregated into large panicles 62
- Corolla 6–18 mm long (measured along upper lip); dorsal edge of corolla tube and upper lip forming a straight line; flowers in axillary dichasial cymes or in axillary and/or terminal racemoid cymes 63
62. Inflorescence axis and calyx with dense stalked capitate glands; young stems glabrous or with scattered stalked capitate glands; leaf base attenuate; calyx 8–10 mm long; corolla puberulous, 2.7–3.1 cm long; capsule glandular, 2.8–3.4 cm long, seed c.5 mm long. . . . . **6.** *salvioides*
- Inflorescence axis sericeous, calyx ciliate, no stalked capitate glands; young stems sparsely antrorsely sericeous; leaf base truncate to subcuneate; calyx 10–14 mm long; corolla glabrous, 3.3–3.5 cm long; capsule glabrous, c.3.7 cm long; seed c.7 mm long . . . . **7.** *niassensis*
63. Flowers in 1–2-flowered axillary cymes with peduncle 1.5–4 cm long; petiole 0.5–1 mm long; leaf base truncate to subcordate; bracts, bracteoles and calyx with white hyaline margins. . . . . **5.** *dalaënsis*
- Flowers in axillary dichasia with more than 2 flowers or in terminal and/or axillary racemoid cymes (if solitary then peduncle shorter); petiole longer; leaf base cuneate to attenuate; bracts, bracteoles and calyx without white hyaline margins 64

64. Flowers in axillary dichasial cymes 65
   – Flowers in terminal and/or axillary racemoid cymes 68
65. Corolla 12–14 mm long, white; dichasia 3–7-flowered . . . . **3.** *asystasioides*
   – Corolla 6–11.5 mm long; some or all dichasia with more than 7 flowers . . . .
   66
66. Corolla white with reddish streaks on lower lip. **4.** *rodgersii*
   – Corolla pale yellow to yellow or pale green, with or without markings 67
67. Leaves less than three times as long as wide; flowers in dichasial cymes throughout;
   calyx finely puberulous with short stubby non-capitate glandular hairs; corolla 6–9
   mm long; capsule 10–15 mm long, puberulous; seeds glochidiate . . . . **1.** *scandens*
   – Leaves more than three times as long as wide; cymes becoming ± racemoid
   upwards; calyx glabrous; corolla 9–11.5 mm long; capsule c.16 mm long,
   glandular; seeds tuberculate . . . . **2.** *mbalaensis*
68. Corolla tube c.4 mm longer than upper lip; capsule c.1 cm long, with placentae
   separating from outer fruit wall and rising elastically . . . . **13.** *gendarussa*
   – Corolla tube and upper lip ± same length (less than 2 mm difference); capsule
   1.4–2.7 cm long, placentae not separating from outer fruit wall and not rising
   elastically 69
69. Corolla 16–18 mm long, white or greenish white; peduncle 4–8.5 cm long . . . . .
   . . . . **8.** *aconitiflora*
   – Corolla 8–12 mm long; inflorescence subsessile or with peduncle to 4 cm long. .
   70
70. Corolla greenish white; capsule 15–19 mm long . . . . **11.** *lenticellata*
   – Corolla salmon pink to dull brownish or greyish to greenish purple 71
71. Cymes 4–32 cm long, pedunculate; capsule 1.4–1.8 cm long, with sculpturing of
   raised partly reticulate ridges; seeds c.2 mm in diameter . . . **10.** *stachytarphetoides*
   – Cymes 1–3(4.5) cm long, subsessile; capsule 1.9–2.7 cm long, without sculpturing
   of raised partly reticulate ridges; seeds c.4 mm in diameter . . . . . . **9.** *francoiseana*

Sect. **Justicia**; Graham in Kew Bull. **43**: 595 (1988) and Sect. **Rhaphidospora** (Nees) T.
Anderson in J. Linn. Soc., Bot. **7**: 43 (1863). —Graham in Kew Bull. **43**: 587 (1988).

Flowers in various types of open inflorescence (dichasia, racemoid cymes, sometimes
aggregated into panicles), peduncles distinct; bracts not imbricate; bracteoles small. Calyx with 5
equal lobes. Appendage on lower anther theca short, linear, flat or triangular. Capsule 4-seeded.
Seeds tuberculate to echinulate (rarely glochidiate) or smooth. Species 1–13.

I find it impossible to keep these two groups separate.

1. **Justicia scandens** Vahl, Symb. Bot. **2**: 7 (1791). —Vollesen in F.T.E.A., Acanthaceae
2: 513 (2010). Type: India, "Malabar", n.d., *Koenig* s.n. (C holotype). FIGURE
8.6.52.

   Justicia glabra Roxb., Fl. Ind. **1**: 132 (1820). —Clarke in F.T.A. **5**: 208 (1900). —Binns,
   Checklist Herb. Fl. Malawi: 14 (1968). —Graham in Kew Bull. **43**: 589 (1988). —Mapaura
   & Timberlake, Checklist Zimb. Pl.: 14 (2004). —da Silva *et al.*, Prelim. Checklist Vasc. Pl.
   Mozamb.: 19 (2004). —Phiri, Checklist Zamb. Vasc. Pl.: 19 (2005). —Setshogo, Checklist Pl.
   Botswana: 18 (2005). Type: as for *J. scandens*.

   Rhaphidospora glabra (Roxb.) Nees in Wallich, Pl. Asiat. Rar. **3**: 115 (1832); in De Candolle,
   Prodr. **11**: 499 (1847).

Erect to scandent perennial (rarely annual) herb with 1 to several stems from a creeping
rhizome; stems to 1.3 m long, subglabrous or bifariously or uniformly puberulous (rarely
pubescent or pilose); plant not drying black. Leaves subglabrous to puberulous along midrib
and veins; petiole to 6(8) cm long; lamina ovate to elliptic or broadly so, largest (5)7–20 ×

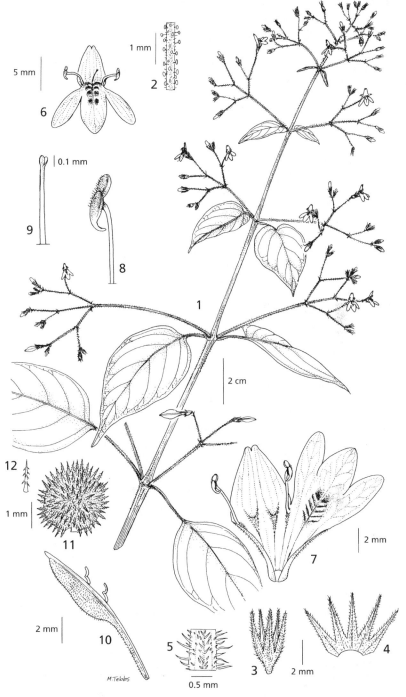

Fig. 8.6.**52**. JUSTICIA SCANDENS. 1, habit; 2, detail of pedicel indumentum; 3, calyx; 4, calyx opened up; 5, detail of calyx lobe; 6, corolla, front view; 7, corolla, opened up; 8, apical part of filament and anther; 9, apical part of style and stigma; 10, capsule; 11, seed; 12, glochidia from seed. 1, 2 & 8 from *Gillett* 13507, 3–7 & 9 from *Newbould* 3339, 1–12 from *Luke* 2690. Drawn by Margaret Tebbs. Reproduced from Flora of Tropical East Africa (2010).

3.5–10.5 cm; apex acuminate to cuspidate with an acute to obtuse tip; base attenuate to truncate. Flowers in lax axillary dichasia, often forming loose panicles towards apex; peduncles, axes and pedicels subglabrous to puberulous, with or without sparse to dense stalked capitate glands; peduncles 1–4.5 cm long; branches to 2.5 cm long; main axis bracts foliaceous, smaller, sessile and subcordate to cordate upwards; secondary bracts and bracteoles linear, 1–2 mm long. Calyx 3–5 mm long, finely puberulous or sparsely so with short stubby non-capitate glandular hairs, usually ciliate; lobes linear-lanceolate, acute to acuminate. Corolla white to pale yellow or yellowish green with red to mauve markings at base of lower lip, 6–9 mm long, puberulous; tube 4–5 mm long and c.2 mm diameter, slightly curved ventrally; upper lip (2)3–5 mm long, flat or slightly hooded; lower lip deflexed, 4–5 mm long, deeply 3-lobed. Filaments 2–3 mm long; thecae 0.5–1 mm long, oblong or broadly so, glabrous or hairy, lower with acute curved appendage c.0.75 mm long. Capsule 10–15 mm long, uniformly puberulous or sparsely so. Seeds circular in outline, with dense glochidiate hairs, 2–3 mm diameter, glochidiae c.0.25 mm long.

Botswana. N: Chobe Dist., Kachikau, Shaile Camp, 950 m, fl.& fr. 21.iv.1975, *Edwards* 4364 (K, PRE). **Zambia**. B: Sesheke Dist., Masese, fl.& fr. 14.iii.1961, *Fanshawe* 6433 (K, LISC). N: Kaputa Dist., Lake Mweru Wantipa, 1050 m, fl.& fr. 10.iv.1957, *Richards* 9123 (K). W: Mpongwe Dist., Lake Kashiba, Mpongwe, fl.& fr. 5.vi.1957, *Fanshawe* 3322 (EA, K). C: Mpika Dist., Mfuwe, Luangwa Valley, 650 m, fl.& fr. 10.iv.1969, *Astle* 5694 (K). S: Livingstone Dist., Victoria Falls, Boiling Pots, 850 m, fl., 21.ii.1997, *Luwiika et al.* 459 (K, MO). **Zimbabwe**. N: Hurungwe Dist., Kariba Gorge, fl. 25.ii.1953, *Wild* 4091 (K, LISC, SRGH). E: Chipinge Dist., between Chibunje and Mangazi, 450 m, fl. 30.i.1975, *Biegel et al.* 4873 (K, SRGH). S: Chiredzi Dist., Lundi R., Fishans, fl., 28.iv.1962, *Drummond* 7793 (K, LISC, SRGH). **Malawi**. N: Karonga Dist., Kondowe to Karonga, 1225 m, fl.& fr. vii.1896, *Whyte* s.n. (K). C: Lilongwe Dist., Lilongwe Nature Sanctuary, Lingazi R., fl.& fr. 3.v.1984, *Patel et al.* 1480 (K, MAL). S: Zomba Dist., Changalume escarpment, 11 km W of Zomba, 850 m, fl. 12.iii.1977, *Brummitt & Patel* 14815 (K, MAL). **Mozambique**. T: Mágoè Dist., Mphende (Mágoè), along Zambezi R., fl.& fr. iv.1972, *Costa Martins* s.n. (LMU). MS: Sussundenga Dist., 20 km N of Dombo, fl.& fr. 4.vi.1971, *Biegel & Pope* 3542 (K, LISC, SRGH). GI: Chokwe Dist., Chokwe (Guijá), along Limpopo R., fl. 9.vi.1947, *Pedrógão* 273 (K).

Widespread in tropical Africa, Madagascar and Tropical Asia. Drier types of lowland to montane forest, usually on margins, riverine forest, scrub and woodland, termite mounds; 20–1250(1850) m.

Conservation notes: Widespread; not threatened.

2. **Justicia mbalaensis** Vollesen, sp. nov. Differs from *J. scandens* in the narrow leaves (more than 3 times as long as wide), the cymes being racemoid upwards, the glabrous (not puberulous) calyx, the larger (9–11.5 mm not 6–9 mm long) corolla, the larger (c.16 mm not 10–15 mm long) capsule and in the tuberculate (not glochidiate) seed. Type: Zambia, Mbala Dist., Mpulungu road, Inono stream, 27.iv.1952, *Richards* 1527 (K holotype).

Erect to scandent or scrambling perennial herb with a single stem (sometimes branched near base) from a creeping branched rhizome; stems to 0.5(1) m long, glabrous apart from thin line of hairs at nodes when young, distinctly longitudinally ridged; plant not drying black. Leaves slightly anisophyllous, pale green beneath, lamina elliptic or narrowly so, largest 8–14.5 × 2.5–4 cm, more than 3 times as long as wide; apex acute; base long attenuate, often to stem. Flowers in lax terminal panicles, usually becoming racemoid towards apex; peduncles, axes and pedicels glabrous; peduncles 0.5–2(6) cm long; branches to 1(1.5) cm long; main axis bracts absent apart from single foliaceous pair towards base of inflorescence; secondary bracts and bracteoles linear, 1–2.5(3) mm long. Calyx 3–4(5) mm long, glabrous; lobes lanceolate to narrowly ovate, acuminate. Corolla pale green to greenish mauve with brown lines, 9–11.5 mm long, puberulous; tube 4–6 mm long and c.2 mm diameter, slightly curved ventrally; upper lip 5–6 mm long, slightly hooded; lower lip deflexed, 4–5 mm long, deeply 3-lobed. Filaments c.3 mm long; thecae held at right angles to each other, c.1 mm long, oblong, glabrous, lower with acute straight appendage c.1 mm long. Capsule c.16 mm long, sparsely uniformly puberulous and with short-stalked capitate glands. Seeds oblong-ellipsoid in outline, densely tuberculate, c.4 mm long.

Zambia. N: Mbala Dist., Mpulungu road, Inono stream, 1225 m, fl. 5.x.1954, *Richards* 1952 (K); Mbala Dist., Lake Chila, fl. 13.vii.1964, *Mutimushi* 905 (K).

Also in Tanzania. Dry evergreen thicket and riverine thicket; 1225–1700 m.

Conservation notes: Restricted distribution and in a moderately threatened habitat; possibly Near Threatened.

3. **Justicia asystasioides** (Lindau) M.E. Steiner in Kew Bull. **44**: 709 (1989). —White *et al.*, Everg. For. Fl. Malawi: 112 (2001). —Vollesen in F.T.E.A., Acanthaceae **2**: 515 (2010). Types: Tanzania, Lushoto Dist., W Usambara Mts, Sakare, 25.ix.1902, *Engler* 932 & 1008 (B† syntypes). Neotype: Lushoto Dist., W Usambara Mts, Baga Forest Reserve, 2.iii.1984, *Borhidi et al.* 84434 (UPS neotype, ETH, NHT).

Duvernoia asystasioides Lindau in Bot. Jahrb. Syst. **38**: 72 (1905).

Thamnojusticia amabilis Mildbr. in Notizbl. Bot. Gart. Berlin-Dahlem **11**: 826 (1933). Type: Tanzania, Mahenge Dist., Sali, 16.v.1932, *Schlieben* 2202 (B† holotype, BM, BR, HBG).

Thamnojusticia grandiflora Mildbr. in Notizbl. Bot. Gart. Berlin-Dahlem **12**: 100 (1934). Type: Tanzania, Morogoro Dist., Nguru Mts, 9.vii.1933, *Schlieben* 4086 (B† holotype, BM, BR, HBG, S).

Justicia amabilis (Mildbr.) V.A.W. Graham in Kew Bull. **43**: 595 (1988).

Justicia mildbraedii V.A.W. Graham in Kew Bull. **43**: 596 (1988). Type: As for *Thamnojusticia grandiflora*.

Erect shrub to 3 m tall; young stems sparsely to densely antrorsely sericeous or sericeous-puberulous with broad glossy hairs with purple walls; older branches with straw-coloured bark; cystoliths inconspicuous. Leaves subequal to slightly anisophyllous, sericeous to puberulous or sparsely so on midrib and veins, distinctly ciliate towards base; petiole 0.5–2 cm long; lamina ovate to elliptic, largest 6–18 × 2.8–6 cm; apex acute to subacuminate; base attenuate, not decurrent. Flowers in 3–7-flowered axillary dichasia; peduncles, axes and pedicels glabrous to sparsely puberulous, no capitate glands; peduncle 1.5–5 cm long; branches 0.5–2 cm long; secondary bracts and bracteoles linear-lanceolate, c.2 mm long. Calyx 5–7 mm long, subglabrous to sparsely puberulous, no capitate glands; lobes lanceolate, acute to acuminate. Corolla white with red streaks on lower lip, 12–14 mm long, glabrous; tube 8–10 mm long and c.3.5 mm diameter; upper lip 4–5 mm long, flat to slightly hooded; lower lip deflexed, 5–7 mm long, deeply 3-lobed, middle lobe much longer and wider. Filaments c.3 mm long, connective enlarged and thecae held at right angle to each other; thecae c.1 mm long, oblong to ellipsoid, parallel, glabrous or sparsely hairy, lower apiculate or with short appendage c.0.3 mm long. Capsule glabrous, 20–25 mm long. Seeds strongly verrucose, 4–5 mm diameter.

Malawi. N: Rumphi Dist., North Viphya, Uzumara Forest, 1700–1900 m, fl. 18.vi.1983, *Dowsett-Lemaire* 806 (K). **Mozambique.** Z: Lugela Dist., Mt Mabu, 1200 m, fr. 16.x.2008, *Harris et al.* 592 (K, LMA, LMU).

Also in Tanzania. Wet evergreen montane forest; 1000–2000 m.

Conservation notes: Local distribution but habitat not under particular threat; not threatened.

4. **Justicia rodgersii** Vollesen in F.T.E.A., Acanthaceae **2**: 519 (2010). Type: Tanzania, Iringa Dist., Udzungwa Mts Nat. Park, Mt Luhomero, 2.x.2000, *Luke et al.* 6835 (EA holotype, K).

Justicia sp. 1 sensu White *et al.*, Evergr. For. Fl. Malawi: 116 (2001).

Shrubby herb or shrub to 2 m tall; young stems bifariously sericeous-puberulous, cystoliths not visible; older branches with longitudinally ribbed glossy brown bark; plant drying black. Leaves with conspicuous cystoliths, slightly to strongly anisophyllous, sericeous-puberulous on midrib and veins and distinctly ciliate on margins; petiole 1–5 cm long; lamina ovate, largest 9–11 × 4.5–6 cm; apex subacuminate with acute tip; base cuneate to attenuate, equal to slightly unequal sided; margin crenate. Flowers in few-flowered axillary dichasia, 1 per node; peduncles, axes and pedicels puberulous, with conspicuous cystoliths; peduncles 1–4 mm long; branches 3–7 mm long (longer than peduncle); secondary bracts and bracteoles linear, 2–4 mm long. Calyx 6.5–9 mm long, puberulous; lobes linear-lanceolate, cuspidate. Corolla white with reddish streaks on lower lip, 9–10 mm long, glabrous; tube 4.5–5 mm long and c.2 mm diameter, slightly pouched ventrally; upper lip 4.5–5 mm long, hooded; lower lip deflexed, c.5 mm long, deeply 3-lobed. Filaments c.3 mm long; thecae c.1 mm long, oblong, parallel, about 50% overlapping, glabrous, lower with minute acute appendage c.0.25 mm long. Capsule densely puberulous, sometimes breaking irregularly, 13–14 mm long. Seeds circular in outline, tuberculate, c.3 mm diameter.

Malawi. N: Nkhata Bay Dist., South Viphya, Chamambo Forest, 1750 m, fl. 23.v.1983, *Dowsett-Lemaire* 741 (K).

Also in Tanzania. Evergreen montane forest; 1750 m.

Conservation notes: Known only from this and four collections from SW Tanzania; possibly Vulnerable.

5. **Justicia dalaënsis** Benoist in Bol. Soc. Brot., sér.2 **24**: 28 (1950). Types: Angola, Lunda, Dala, Luma-Cassai, Cuxi R., 28.iv.1937, *Exell & Mendonça* 1375 (COI lectotype, BM); Lunda, Dala, 23.iv.1937, *Exell & Mendonça* 1150 (BM, COI syntype), lectotypified here.

Suffrutescent pyrophytic herb with 2 to several erect unbranched or sparsely branched stems from a large woody rootstock; stems to 1 m long, distinctly longitudinally ridged, glabrous to sparsely puberulous. Leaves distinctly glandular punctuate, ciliate or sparsely so, otherwise glabrous to sparsely puberulous on veins; petiole 0.5–1 mm long; lamina ovate, largest 2.5–6 × 1.2–2.7 cm; apex acute; base truncate to subcordate. Flowers in 1–2-flowered axillary cymes, sometimes 2 per axil; peduncle 1.5–4 cm long, glabrous or sparsely puberulous; bracts and bracteoles narrowly triangular, 5–8 mm long, with strong rib-like veins and white scarious margins; pedicels strongly ribbed, 0.5–1 mm long. Calyx 6–8(10) mm long, with ciliate lobes, otherwise glabrous or sparsely puberulous; lobes lanceolate with strongly raised midrib and conspicuous scarious white margins, acuminate. Corolla whitish green with mauve markings on lower lip, 10–11 mm long, puberulous; tube 5–7 mm long and 2–3 mm diameter; upper lip 4–5 mm long, hooded; lower lip deflexed, 4–5 mm long, deeply 3-lobed, lobes elliptic-oblong. Filaments held in upper lip; thecae c.1 mm long, oblong, held at right angles to each other, hairy, lower with linear appendage c.1 mm long. Capsule uniformly puberulous, 15–18 mm long. Seeds not compressed, ovoid-cordiform, densely tuberculate, c.3 mm long.

Zambia. W: Kitwe, fl., 25.x.1969, *Mutimushi* 3819 (K); Chingola Dist., fl.& fr. 9.xii.1972, *Fanshawe* 11698 (K).

Also in Angola and D.R. Congo. *Brachystegia* woodland; appearing after burning or just before the rains; 1000–1500 m in Angola and Congo.

Conservation notes: Widespread; not threatened.

6. **Justicia salvioides** Milne-Redh. in Bull. Misc. Inform., Kew **1936**: 488 (1936). — Vollesen in F.T.E.A., Acanthaceae 2: 522 (2010). Types: Tanzania, Massaini, n.d., *Fischer* 506 (K lectotype); Tanzania, Usagara, Mlali, 2.vi.1890, *Stuhlmann* 207 (B† syntype), lectotypified here. FIGURE 8.6.53.

Duvernoia salviiflora Lindau in Bot. Jahrb. Syst. **20**: 42 (1894).

Justicia salviiflora (Lindau) C.B. Clarke in F.T.A. **5**: 205 (1900), non Kunth.

Much-branched erect shrubby herb or shrub to 2(?4) m tall; young stems rounded with numerous longitudinal shallow pale ribs and green furrows, glabrous or with scattered (rarely dense) stalked capitate glands on uppermost node below inflorescence. Leaves often absent at time of flowering, glabrous; petiole to 2 mm long; lamina ovate to elliptic, largest 4–5 × 1.5–2.5 cm; apex acute to subacuminate with a blunt to acute tip; base attenuate, decurrent. Flowers

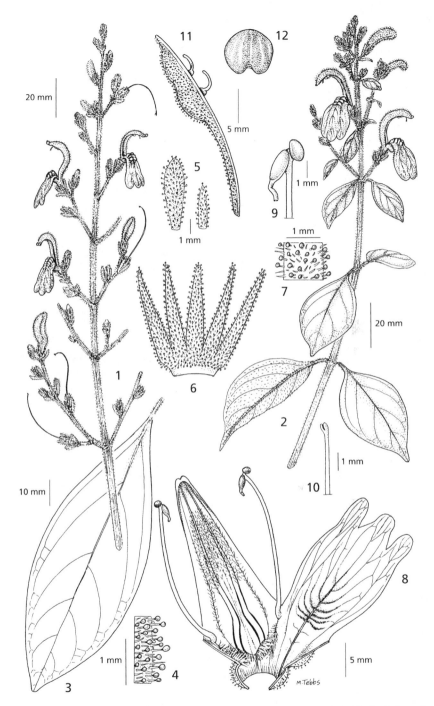

Fig. 8.6.**53**. JUSTICIA SALVIOIDES. 1, habit, leafless flowering plant; 2, habit, leafy flowering plant; 3, large leaf; 4, detail of inflorescence indumentum; 5, bract (left) and bracteole (right); 6, calyx opened up; 7, detail of calyx indumentum; 8, corolla opened up; 9, apical part of filament and anther; 10, apical part of style and stigma; 11, capsule; 12, seed. 1 & 4–10 from *Bidgood et al.* 7481, 2 from *Congdon* 194, 3 from *Burtt* 2505, 11 & 12 from *Congdon* 499. Drawn by Margaret Tebbs. Reproduced from Flora of Tropical East Africa (2010).

solitary (rarely in 3s) in racemiform cymes congested into large terminal panicles, 10–30 cm long; axes, branches and pedicels densely glandular puberulous (sticky) and with a layer of fine non-glandular hairs; branches 3–8 cm long; main bracts foliaceous towards base, soon falling; secondary bracts ovate-elliptic or narrowly so, to 8 mm long; bracteoles linear-lanceolate, 3–6 mm long. Calyx 8–10 mm long, with dense stalked capitate glands and sparse to dense fine non-glandular hairs; lobes lanceolate, acute. Corolla white with purple markings on lower lip, tube pink on outside, 2.7–3.1 cm long, puberulous with mixture of glands and hairs; tube 0.9–1.1 cm long and 4–6 mm diameter, gibbose ventrally; upper lip 1.6–2 cm long, bent backwards and strongly curved and hooded; lower lip deflexed, 1.2–2 cm long, deeply 3-lobed, middle lobe wider. Filaments held in upper lip; thecae 1.5–2.5 mm long, oblong, parallel or at an angle to each other on a broadened receptacle, with few to many long pilose hairs, lower with bifurcate appendage c.1.5 mm long. Capsule (from East African material) 2.8–3.4 cm long, glandular puberulous. Seeds ('*Monechma*'type) kidney-shaped, not compressed, smooth, c.5 × 6 mm.

Zambia. N: Kaputa Dist., Sumbu to Bulaya, 750 m, fr. 22.ix.1956, *Richards* 6260 (K); Kaputa Dist., Bulaya, fl. 14.ix.1958, *Fanshawe* 4828 (K).

Also in Tanzania. Deciduous thicket (muteshe, itigi) on grey clay soil; 750–950 m.

Conservation notes: Local distribution and habitat type under some threat; possibly Vulnerable.

7. **Justicia niassensis** Vollesen, sp. nov. Differs from *Justicia salvioides* in the non-glandular indumentum on stems, inflorescences and calyces, truncate to subcordate (not attenuate) leafbase, longer (10–14 vs. 8–10 mm) calyx, larger (3.3–3.5 vs. 2.7–3.1 cm) corolla which is glabrous (not puberulous), larger (c.3.7 vs. 2.8–3.4 cm) glabrous (not glandular-puberulous) capsule and larger (c.7 vs. c.5 mm long) seed. Type. Mozambique, Pemba Dist., road to Maringanha Lighthouse, 21.iii.1960, *Gomes e Sousa* 4553 (K holotype).

Shrub to 2 m tall; young stems finely sericeous, older with pale yellow warty bark. Leaves present at time of flowering, glabrous; petiole to 3 cm long; lamina broadly ovate, largest 8–11.5 × 6–7 cm; apex subacuminate; base truncate to subcuneate, not decurrent. Flowers solitary or in 2–3-flowered cymules in racemiform cymes congested into a large terminal panicle, 10–20 cm long; axes, branches and pedicels finely antrorsely sericeous, no capitate glands; branches to 7 cm long; main bracts foliaceous to near apex of panicle, soon falling; secondary bracts lanceolate to narrowly ovate, 8–12 mm long; bracteoles linear-lanceolate, 5–8 mm long. Calyx 10–14 mm long puberulous-ciliate, otherwise glabrous; lobes lanceolate, cuspidate. Corolla "white or violet", 3.3–3.5 cm long, glabrous; tube gibbose ventrally and widened upwards, 1–1.3 cm long and 3.5–5 mm diameter in basal cylindric part; upper lip 2.3–2.7 cm long (from apex of cylindric tube), bent backwards and strongly curved and hooded; lower lip spreading to deflexed, 2–2.5 cm long, deeply 3-lobed, middle lobe wider. Filaments held in upper lip; thecae 2–2.5 mm long, oblong, held at right angles to each other on a broadened receptacle, glabrous, lower with linear appendage c.0.5 mm long. Capsule c.3.7 cm long, glabrous. Seeds oblong, flattened, smooth, c.7 × 5 mm.

Mozambique. N: Pemba Dist., road to Maringanha Lighthouse, fl. 21.iii.1960, *Gomes e Sousa* 4553 (K); Nacala Dist., Praia de Nacala Porta, fl.& fr. 28.vii.1979, *Macitela* 63 (K, LMA).

Not known elsewhere. Habitat not indicated, but probably coastal bushland; below 25 m.

Conservation notes: Endemic to coastal parts of N Mozambique and known only from these 2 collections; assessed as EN B1ab+B2ab by the Southern African Plant Specialist Group (2014).

Closest to *J. salvioides* which has a glabrous stem, attenuate leafbase, densely glandular inflorescence and calyx, shorter calyx, puberulous corolla, and a smaller glandular capsule and seed.

8. **Justicia aconitiflora** (Meeuse) Vollesen, comb. nov. Type: South Africa, Mpumalanga, Barberton, 12 km S of Komatipoort, Komati R., 21.iv.1953, *Codd* 7778 (PRE holotype, K).

Duvernoia aconitiflora Meeuse in Fl. Pl. Afr. **31**: pl. 1216 (1956). —Dyer *et al.*, Wild Fl. Transvaal: 320, pl.160 (1962). —Manning & Getliffe Norris in F.S.A. **30**(3,1): 15 (1995). —Schmidt *et al.*, Trees Shrubs Mpumalanga: 610 (2002).

Shrub to 2.5 m tall; young branches distinctly longitudinally ridged, sparsely antrorsely sericeous. Leaves glandular punctuate, glabrous or sparsely ciliate along midrib; petiole to 2 cm long; lamina narrowly ovate, largest 8.5–10.5 × 2.5–3 cm; apex obtuse; base attenuate, decurrent. Flowers in axillary racemoid cymes from upper leaf axils; peduncle 4–8.5 cm long, sparsely and finely sericeous and with subsessile glands; bracts and bracteoles falling early, triangular, subacute, to 3 mm long. Calyx 3–4 mm long, fused about halfway up, puberulous with short stubby glossy hairs; lobes triangular, acute or subacute. Corolla white or greenish white with maroon markings on lower lip, 16–18 mm long, puberulous; tube c.7 mm long and c.2.5 mm diameter; upper lip 9–11 mm long, strongly hooded; lower lip deflexed with recoiled lobes, 9–11 mm long, deeply 3-lobed, lobes ovate-elliptic. Filaments held in upper lip; thecae c.2.5 mm long, oblong-ellipsoid, parallel, glabrous, lower with small mucro-like appendage c.0.25 mm long. Capsule 18–22 mm long, densely and finely puberulous, with sculpturing of partly reticulate ridges. Seeds not compressed, reniform, reticulately sculptured, c.5 mm long.

Mozambique. M: Namaacha Dist., Namaacha, Pedreiras de Movene, fl. 14.iii.1958, *Barbosa & Lemos* 8265 (EA, K); Namaacha Dist., Mahlavatimuke, Rio Movene, fr. 21.viii.1980, *Zunguze & Mafumo* 250 (K, LMU).

Also in Swaziland and South Africa (Mpumulanga). Riverbanks and floodplains with clay soil, *Acacia–Sclerocarya–Combretum* bushland on clay; 50–1200 m.

Conservation notes: Known from less than 10 collections in a rather restricted area; possibly threatened.

9. **Justicia francoiseana** Brummitt in Kew Bull. **40**: 792 (1985). —White *et al.*, Evergr. For. Fl. Malawi: 116 (2001). —Vollesen in F.T.E.A., Acanthaceae **2**: 528 (2010). Type: Uganda, Toro/Mubende Dist., Lake Albert, Mizizi R., 4.xii.1906, *Bagshawe* 1332 (BM holotype, BM).

Adhatoda bagshawei S. Moore in J. Bot. **45**: 333 (1907), non *Justicia bagshawei* S. Moore. *Justicia bagshawei* (S. Moore) Eyles in Trans. Roy. Soc. S. Afr. **5**: 486 (1916), non S. Moore.

Shrub or small tree to 5 m tall, rarely scandent; young branches longitudinally striate, glabrous or sericeous-puberulous. Leaves glabrous or sericeous-puberulous on veins and midrib; petiole to 3 cm long; lamina ovate or narrowly so, largest 9–14(18) × 3–5.8(9.3) cm; apex acuminate or subacuminate; base attenuate, decurrent. Flowers in short axillary and/or terminal subsessile racemiform cymes 1–3(4.5) cm long; rachis finely sericeous-puberulous; bracts and bracteoles triangular or narrowly so, 2–3 mm long, acute to acuminate; pedicels to 1 mm long. Calyx 3–4.5 mm long, puberulous, without capitate glands; lobes triangular, acute. Corolla dull brownish or greenish purple, 10–12 mm long, puberulous; tube 6–7 mm long and 2–3 mm diameter, not gibbose; upper lip 4–6 mm long, strongly hooded; lower lip deflexed, 5–6 mm long, deeply 3-lobed, lobes ovate-elliptic. Filaments held in upper lip; thecae c.1 mm long, oblong, parallel, glabrous, lower with small appendage c.0.25 mm long. Capsule 19–27 mm long, finely puberulous. Seeds verrucose or reticulately sculptured, reniform, not compressed, c.4 mm long.

Zimbabwe. E: Chipinge Dist., Chirinda Forest, 1100 m, fl.& fr. x.1964, *Goldsmith* 43/64 (EA, K, SRGH). **Malawi.** S: Nsanje Dist., Malawi Hills Forest Reserve, 650–750 m, fl. 13.viii.1983, *Dowsett-Lemaire* 917 (K). **Mozambique.** MS: Gorongosa Dist., Gorongosa Nat. Park, Sangarassa Forest, Chitengo, fl.& fr. v.1972, *Tinley* 2599 (K, SRGH).

Also in Uganda. Lowland and intermediate evergreen or semi-evergreen forest; 30–1250 m.

Conservation notes: Widespread but in a threatened habitat; possibly Near Threatened.

10. **Justicia stachytarphetoides** (Lindau) C.B. Clarke in F.T.A. **5**: 194 (1900). —
Vollesen in F.T.E.A., Acanthaceae **2**: 527 (2010). Type: Tanzania, Uzaramo Dist.,
Madessa, 17.v.1894, *Stuhlmann* 8121 (B† holotype, K).

Duvernoia stachytarphetoides Lindau in Engler, Pflanzenw. Ost-Afr. **C**: 372 (1895).

Erect to scrambling shrubby herb or shrub to 1.25 m tall; young branches puberulous or
sparsely so. Leaves sparsely puberulous on midrib and larger veins; petiole to 2.5 cm long; lamina
ovate to elliptic, largest 6–12.5 × 2.5–6.5 cm; apex subacute to shortly acuminate; base cuneate
to attenuate. Flowers all solitary or in 3-flowered sessile dichasia aggregated into loose terminal
racemiform cymes 4–32 cm long, usually with lateral cymes from lower nodes thus forming a loose
panicle to 50 × 15 cm; peduncle to 4 cm long; peduncle and rachis sparsely to densely puberulous;
bracts and bracteoles lanceolate to narrowly elliptic, 4–6 mm long, uniformly puberulous to
densely so. Calyx 4–5 mm long, densely puberulous; lobes lanceolate, acuminate. Corolla salmon
pink to greyish purple, with pink to purple markings on lower lip, 9–12 mm long, densely
puberulous; tube 5–7 mm long and 1–2 mm diameter; upper lip 4–5 mm long, strongly hooded;
lower lip deflexed, 4–5 mm long, deeply 3-lobed into one broad triangular middle lobe and two
strap-shaped recoiled and twisted lateral lobes. Filaments held in upper lip; thecae c.1 mm long,
oblong, at an angle to each other, upper hairy, lower glabrous, lower with curved appendage c.0.5
mm long. Capsule with longitudinal and transverse ridges, 14–18 mm long, puberulous or densely
so. Seeds reticulately sculptured with central ridge, reniform, not compressed, c.2 × 2.5 mm long.

Mozambique. N: Ancuabe Dist., 20 km N of Ancuabe, Mt Miquita, 350 m, fl.
2.viii.1983, *Groenendijk & Dungo* 463 (LMA, LMU). Z: Lugela Dist., Namagoa, fl.
31.viii.1948, *Faulkner* 282 (K). MS: Gorongosa Dist., Gorongosa Nat. Park, fl. v.1972,
Tinley 2624 (K, LISC, SRGH).

Also in Somalia, Kenya and Tanzania. Evergreen and semi-deciduous lowland
forest, riverine forest, moist mopane woodland; 30–350 m.

Conservation notes: Widespread; not threatened.

11. **Justicia lenticellata** Champl. in Pl. Ecol. Evol. **146**: 110 (2013). Type: D.R. Congo,
Katanga, Upemba Nat. Park, Kanonga, ii.1949, *de Witte* 5615 (BR holotype, CAS,
K, MO).

Shrub to 3 m tall; young branches longitudinally striate, finely puberulous or sericeous-
puberulous. Leaves finely puberulous or sericeous-puberulous on midrib and veins; petiole to
2.5 cm long; lamina ovate, largest 5.5–9.3 × 2.1–4.2 cm; apex subacuminate; base attenuate,
decurrent. Flowers single (rarely in 2-flowered cymules) in terminal subsessile racemiform cymes
to 7 cm long; rachis finely puberulous, with or without scattered stalked capitate glands; lower
bracts foliaceous, upper bracts and bracteoles lanceolate, 2–3 mm long, falling early; pedicels to
1 mm long. Calyx 4–6(9 in fruit) mm long, finely and sparsely puberulous, with scattered capitate
glands; lobes lanceolate, distinctly 3-veined, acuminate. Corolla greenish white, no markings
on lower lip, 8–9 mm long, finely puberulous; tube 4–5 mm long and c.2 mm diameter, not
gibbose; upper lip c.4 mm long, hooded; lower lip deflexed, 3–4 mm long, deeply 3-lobed, lobes
oblong-elliptic. Filaments held in upper lip; thecae c.1 mm long, oblong, parallel, glabrous,
lower with linear appendage c.0.25 mm long. Capsule 15–19 mm long, minutely puberulous.
Seeds tuberculate, ovate-reniform, not compressed, c.4 mm long.

Zambia. N: Kasama Dist., Luombe R., Chishimba Falls, 1225 m, fl. 31.iii.1955,
Richards 5254 (K). W: Mwinilunga Dist., Kalene Hill, fr. 18.v.1969, *Mutimushi* 3201 (K).

Also in D.R. Congo. Riverine forest (mushitu), damp woodland bordering riverine
forest; 1200–1400 m.

Conservation notes: Fairly restricted distribution and habitat threatened; possibly
Near Threatened.

12. **Justicia bequaertii** De Wild. in Bull. Jard. Bot. Brux. **4**: 429 (1914). Type: D.R.
Congo, Shaba, Bukama, 16.vi.1911, *Bequaert* 108 (BR holotype).

Monechma praecox Milne-Redh. in Bull. Misc. Inform., Kew **1937**: 430 (1937). Type: Zambia, Solwezi Dist., Solwezi Dambo, 20.ix.1930, *Milne-Redhead* 1150 (K holotype, BR). *Justicia praecox* (Milne-Redh.) Milne-Redh. in Kew Bull. **8**: 444 (1953).

Suffrutescent pyrophytic herb with several erect unbranched stems (mostly inflorescences) to 50 cm tall from a large woody rootstock, flowering when leafless, rarely with very young leaves; stems distinctly longitudinally striate and furrowed, subglabrous to puberulous, rarely pubescent. Leaves sparsely pubescent with broad glossy hairs when young, soon glabrescent, glandular punctate; petiole to 1 cm long; lamina narrowly ovate, narrowly elliptic or oblong, largest c.6.5 × 2.2 cm (only one mature leaf seen); apex acute to obtuse; base cuneate to attenuate and then decurrent. Flowers solitary or in 2(3)-flowered cymules in elongated terminal racemiform cymes, unbranched or with lateral cymes from lower nodes thus forming a loose panicle to 35 cm long, lateral cymes to 20 cm long; peduncle and nodes to 5(10) cm long, subglabrous to puberulous; bracts and bracteoles lanceolate to narrowly triangular, 3–4 mm long (sometimes to 9 mm and slightly foliaceous near base), acute to acuminate; cymule peduncles to 2(7) mm long; pedicels 1–3 mm long. Calyx 3.5–6 mm long, ciliate or sparsely so to uniformly puberulous; lobes lanceolate, acuminate. Corolla white, rarely with purple patches on lower lip, 8–11 mm long, puberulous; tube 4–6 mm long and 2–3 mm diameter; upper lip 4–5 mm long, hooded; lower lip deflexed, 4–5 mm long, deeply 3-lobed, lobes elliptic-oblong. Filaments held in upper lip; thecae 1–1.5 mm long, oblong, held at right angles to each other, glabrous, lower with linear appendage c.1 mm long. Capsule 2-seeded, 3.2–3.5 cm long, glabrous. Seeds not compressed, reniform, densely tuberculate, c.8 mm long.

Zambia. N: Kaputa Dist., Mporokoso to Mkupa, fl.& fr. 25.x.1949, *Bullock* 1385 (K). W: Kitwe, fl.& fr. 6.xi.1966, *Fanshawe* 9826 (K, LISC). C: Kabwe Dist., Kabwe (Broken Hill), fl. vii.1920, *Rogers* 26153 (K).

Also in D.R. Congo. *Brachystegia* and chipya woodland and dambos, appearing after burning; 1000–1500 m.

Conservation notes: Fairly widespread; not threatened.

As pointed out by Milne-Redhead (1953) this species is intermediate with '*Monechma*'. It has a 2-seeded capsule but at the same time sculptured seeds.

13. **Justicia gendarussa** Burm. f., Fl. Indica: 10 (1768). —Clarke in F.T.A. **5**: 203 (1900). —Vollesen in F.T.E.A., Acanthaceae **2**: 529 (2010). Type: India and Indonesia, "Malabar, Amboina and Java", *Herb. Burman* s.n. (G holotype).

Shrubby herb or shrub to 1.5(?3) m tall; young branches purple, glabrous or with a thin band of hairs at nodes. Leaves glabrous; petiole to 7 mm long; lamina narrowly ovate or narrowly elliptic, largest 7.3–9.5 × 1.3–1.5 cm; apex obtuse; base attenuate, not decurrent. Flowers in 2–3(7)-flowered cymules at base and single upwards in terminal (or also from upper leaf axils) subsessile racemiform cymes to 12 cm long; rachis glabrous or sparsely puberulous upwards; lower bracts foliaceous, upper bracts and bracteoles linear to lanceolate, to 1 cm (bracts) or 4 mm (bracteoles) long, falling early; pedicels to 0.5 mm long. Calyx 4–7 mm long, sparsely sericeous-puberulous; lobes linear-lanceolate, cuspidate. Corolla white to purple, with darker markings on lower lip, 14–16 mm long, glabrous; tube 9–10 mm long and c.2 mm diameter; upper lip 5–6 mm long, flat or slightly hooded; lower lip spreading, 7–8 mm long, deeply 3-lobed, lobes oblong-elliptic. Filaments held under upper lip; thecae c.1 mm long, oblong, parallel, almost superposed, glabrous, lower with small appendage c.0.25 mm long. Capsule c.10 mm long, with placentae separating from outer fruit wall and rising elastically, minutely puberulous. Immature seeds densely reticulate-tuberculate.

Mozambique. N: Mogovolas Dist., Nametil, fl. 12.vii.1948, *Pedro & Pedrógão* 4436 (EA). MS: Gorongosa Dist., Gorongosa, n.d., *Carvalho* (cited by Clarke (1900), not seen).

Widespread in India and SE Asia, usually cultivated and naturalised. In the Flora area grown as a hedge plant, sometimes persisting in abandoned gardens and fields; 20–200 m.

Conservation notes: Introduced; not threatened.

Many authors and collectors comment on the fact that ripe fruits are hardly ever seen. The plant is invariably propagated by cuttings. It is widely cultivated and has been so for centuries. The exact area of origin is unknown but is probably somewhere in India or SE Asia. It was almost certainly brought to East Africa with Arab traders many centuries ago.

Sect. **Tyloglossa** (Hochst.) Lindau in Engler & Prantl, Nat. Pflanzenfam. IV **3b**: 349 (1895). —Graham in Kew Bull. **43**: 590 (1988).

Inflorescence of subsessile 1–3(15)-flowered dichasia aggregated into clearly defined terminal (or also axillary from upper axils) pseudo-racemoid cymes; bracts not imbricate; bracteoles small. Calyx with 5 equal lobes. Appendage on lower anther theca linear, entire or bifurcate. Capsule 4-seeded. Seeds verrucose to tuberculate. Species 14–20.

This is almost certainly not a natural group.

14. **Justicia fittonioides** S. Moore in J. Bot. **16**: 134 (1878). —Clarke in F.T.A. **5**: 189 (1899). —Vollesen in F.T.E.A., Acanthaceae **2**: 530 (2010). Type: Kenya, Mombasa Dist., near Mombasa, i.1878, *Wakefield* s.n. (K holotype, K).

Justicia dolichopoda Mildbr. in Notizbl. Bot. Gart. Berlin-Dahlem **12**: 721 (1935). Type: Tanzania, Lindi Dist., Lake Lutamba, 9.i.1935, *Schlieben* 5847 (B† holotype, BM, K, LISC).

Acaulescent perennial herb with thick creeping rhizome or with a short puberulous stem to 2 cm long. Leaves in a rosette, often appressed to ground, puberulous or sericeous-puberulous on petiole, midrib and larger veins; petiole 1.5–8 cm long; lamina oblong-elliptic or broadly so to ovate-cordiform or reniform, largest 6–19 × 4–15 cm; apex broadly rounded, rarely retuse; base cuneate to cordate, rarely shortly attenuate, usually with one lobe larger than the other. Flowers in 3–15-flowered dichasia; cymes 1.5–8 cm long, sometimes with 2 short lateral cymes from lowermost node; peduncle 4–17 cm long; peduncle and rachis puberulous or retrorsely sericeous-puberulous; main axis bracts broadly elliptic to orbicular or reniform, 6–9 × 3.5–8.5 mm, subacute to broadly rounded and apiculate, puberulous or sparsely so, distinctly 5-veined from base; dichasia bracts and bracteoles lanceolate to elliptic or obovate, 4–6 mm long. Calyx 3–5 mm long, subglabrous to minutely puberulous; lobes linear-lanceolate, acute to cuspidate. Corolla white, with pink markings on lower lip, upper lip greenish outside, 7–11.5 mm long, glabrous to sparsely puberulous; tube 4–6 mm long; upper lip 4–7 mm long; lower lip spreading or deflexed, 4–6 mm long, deeply 3-lobed, middle lobe much wider than laterals. Filaments 2–4 mm long; thecae 1.5–2 mm long, oblong, ± 50% overlapping, upper hairy, both thecae with acute appendage 0.5 (upper) to 1 (lower) mm long. Capsule 15–20 mm long, glabrous to sparsely and minutely puberulous. Seeds (immature) densely verrucose-reticulately sculptured, ovoid, c.4 mm long.

Mozambique. N: Macomia Dist., Quirimbas Nat. Park, 12°07'S 40°12'E, 100 m, fl. 6.xii.2003, *Luke et al.* 9895 (K); Macomia Dist., W of Quiterajo, Namacubi Forest, 75 m, fl.& fr. 29.xi.2008, *Goyder et al.* 5070 (K, LMA, LMU, P).

Also in Kenya and Tanzania. Dry deciduous or semi-evergreen coastal forest and thicket, termite mounds; 50–100 m.

Conservation notes: Restricted distribution within the Flora area, and in a threatened habitat; possibly Vulnerable.

15. **Justicia flava** (Vahl) Vahl, Symb. Bot. **2**: 15 (1791). —Clarke in F.T.A. **5**: 190 (1899); in Fl. Cap. **5**: 58 (1901). —Bandeira, Bolnick & Barbosa, Fl. Nat. Sul Mozam.: 195 (2007). —Vollesen in F.T.E.A., Acanthaceae **2**: 532 (2010). Type: Yemen, Al Hadiyah, iii.1763, *Forsskål* 394 (C lectotype). FIGURE 8.6.54.

Dianthera flava Vahl, Symb. Bot. **1**: 5 (1790).

Adhatoda fasciata Nees in De Candolle, Prodr. **11**: 402 (1847). Type: South Africa, KwaZulu-Natal, Umgeni R., 1837, *Drège* s.n. (K lectotype), lectotypified here.
Adhatoda petiolaris Nees in De Candolle, Prodr. **11**: 402 (1847) for *Forbes* s.n., Delagoa Bay, non sensu stricto.
Justicia fasciata (Nees) T. Anderson in J. Linn. Soc., Bot. **7**: 39 (1863).
Justicia palustris (Hochst.) T. Anderson var. *dispersa* Lindau in Bot. Jahrb. Syst. **20**: 72 (1894). Type: Tanzania, Kilimanjaro, xi.1884, *Johnston* 59 (K lectotype, BM, K).

Perennial herb with 1 to several erect, ascending or scrambling stems from a creeping rhizome or woody rootstock; stems to 0.6(2 in scrambling plants) m long, subglabrous to retrorsely sericeous, rarely pubescent or antrorsely sericeous. Leaves sericeous-pubescent or sparsely so on midrib and larger veins, rarely also on lamina; petiole to 2(3.5) cm long; lamina narrowly to broadly ovate to elliptic, largest 4–14.5 × 1.5–6.5 cm; apex acuminate to rounded; base cuneate to attenuate, rarely truncate. Flowers in 1–3-flowered dichasia; cymes 2–15(22) cm long; peduncle to 12(16) cm long; peduncle and rachis with indumentum as stems; main axis bracts sometimes tinged purple, lanceolate to narrowly ovate, 7–18 × 1–2(3) mm, sparsely puberulous to pubescent and with scattered to dense stalked capitate glands, usually conspicuously ciliate with hairs to 2 mm long; dichasia bracts and bracteoles linear-lanceolate to narrowly ovate, 7–14 mm long, puberulous and with capitate glands; bracteoles longer than calyx. Calyx 3–6(8) mm long, sericeous-puberulous at least on veins and with long pilose ciliae, without (rarely with) capitate glands; lobes lanceolate to narrowly triangular, acute to acuminate. Corolla yellow, lower lip with 3 and upper lip with 2 conspicuous reddish brown streaks (absent in C & N Mozambique, Malawi and N & E Zambia), 7–14 mm long, puberulous; tube 4–8 mm long and 3–4 mm diameter; upper lip 3–6 mm long; lower lip spreading or deflexed, 6–11 mm long. Filaments 2–5 mm long, bending out of flower after anthesis; thecae 1–1.5 mm long, oblong, 25–33% overlapping, upper hairy, lower with linear appendage c.1 mm long. Capsule 7–11 mm long, uniformly puberulous or densely so. Seeds black, densely tuberculate, ovoid, 1.5–2 mm long.

Botswana. SE: Central Dist., Tuli Block, Seleka Ranch, 800 m, fl.& fr. 22.ii.1977, *Hansen* 3050 (C, GAB, K, PRE, SRGH). **Zambia.** B: Zambezi Dist., Zambezi (Balovale), fl.& fr. 25.ii.1964, *Fanshawe* 8341 (K). N: Isoka Dist., Lupita to Makulu, fl.& fr. 20.viii.1965, *Lawton* 1269 (K). W: Kabompo Dist., Lusangwa, 1100 m, fl.& fr. 26.i.1975, *Brummitt et al.* 14133 (K). S: Livingstone, fl.& fr. n.d., *Gairdner* 560 (K). **Zimbabwe.** W: Bulawayo, fl.& fr. xii.1897, *Rand* 101 (BM). C: Charter Dist., 32 km W of Chivu (Enkeldoorn), fl.& fr. xi.1957, *Miller* 4755 (K, SRGH). E: Chipinge Dist., Save R., Rupisi Hot Springs, 500 m, fl.& fr. 29.i.1948, *Wild* 2392 (K, SRGH). S: Mwenezi Dist., Nuanetsi Expt. Station, fl.& fr. 7.ii.1976, *Takaindisa* 10 (K, SRGH). **Malawi.** N: Chitipa Dist., Chisenga, 1550 m, fl.& fr. 11.vii.1970, *Brummitt* 12019 (K, MAL). C: Lilongwe Dist., Dzalanyama Forest Reserve, 1550 m, fl.& fr. 26.iii.1977, *Brummitt et al.* 14929 (K, MAL). S: Mwanza Dist., Thambani Forest Reserve, fl.& fr. 29.vii.1984, *Patel & Nachamba* 1512 (C, K, MAL). **Mozambique.** T: Moatize Dist., Zóbuè, 900 m, fl.& fr. 17.vi.1947, *Hornby* 2761 (K, SRGH). MS: Maringua Dist., S bank of Save R., 175 m, fl.& fr. 30.vi.1950, *Chase* 2463 (BM, LISC, SRGH). GI: Bilene Dist., Chissano to Licilo, fl.& fr. 5.viii.1958, *Barbosa & Lemos* 8295 (K). M: Maputo Dist., fl.& fr. 7.iv.1947, *Hornby* 2638 (K, PRE, SRGH).

Widespread in tropical Africa, South Africa and Arabia. Riverine forest and scrub, usually on margins and in clearings, and in a wide range of woodland, bushland and grassland from high rainfall to semi-desert conditions, often in disturbed or secondary vegetation; 20–2000 m.

Conservation notes: Widespread; not threatened.

The material from N & C Mozambique, Malawi and N & E Zambia (as well as material from C and S Tanzania) differ in the absence of the conspicuous reddish brown streaks on the corolla, but no other differences can be found. At the moment I am keeping all material as one taxon although separating this out at subspecific rank is tempting. Occasionally faint lines are visible at the base of the lower lip, and a few collections from S Zimbabwe, well outside the area of this 'form', also lack the streaks.

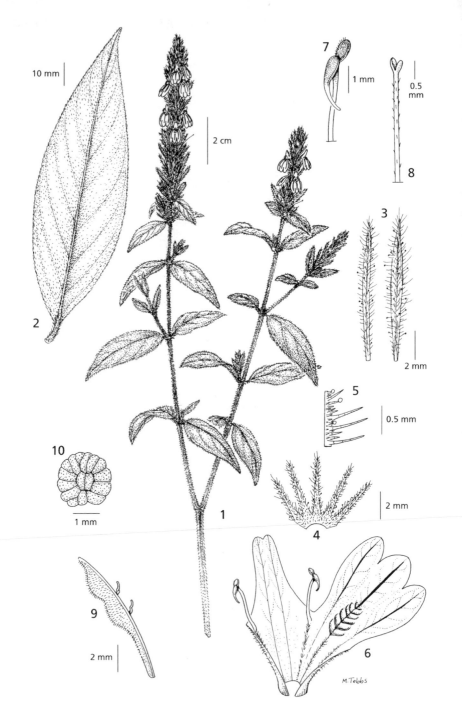

Fig. 8.6.54. JUSTICIA FLAVA. 1, habit; 2, leaf; 3, bract (right) and bracteole (left); 4, calyx opened up; 5, detail of calyx lobe; 6, corolla opened up; 7, apical part of filament and anther; 8, apical part of style and stigma; 9, capsule; 10, seed. 1 & 3–6 from *Drummond & Hemsley* 4122, 2 & 7–8 from *Bidgood et al.* 109, 9–10 from *Gereau* 4256. Drawn by Margaret Tebbs. Reproduced from Flora of Tropical East Africa (2010).

16. **Justicia gorongozana** Vollesen, sp. nov. Differs from *J. flava* in having elliptic (not lanceolate to narrowly ovate) 3–4 mm (not 1–2(3) mm) wide bracts, the dorsal calyx lobe being almost twice as long as the rest, and in having a brown-edged upper corolla lip. Type: Mozambique, Gorongosa Dist., Gorongosa Nat. Park, 4 km NNW of Conduè, 10.vi.2012, *Bester* 11132 (K holotype, PRE).

Perennial herb, basal part of stem creeping, rooting at nodes, apical part erect, to 30 cm long, retrorsely sericeous when young, soon glabrescent. Leaves antrorsely sericeous on midrib, larger veins and margins, with scattered hairs on lamina; petiole to 2 cm long; lamina ovate to elliptic, largest 5.5–8.5 × 2.5–4 cm; apex acute to subacuminate with a blunt tip; base cuneate to attenuate. Flowers in 1–3-flowered dichasia; cymes 2–3.5 cm long; peduncle to 1 cm long; peduncle and rachis sericeous; main axis bracts and dichasia bracts elliptic, 10–15 × 3–4 mm, sparsely puberulous and with scattered stalked capitate glands, conspicuously ciliate with hairs to 1.5 mm long; bracteoles linear to narrowly elliptic, c.8 mm long, longer than calyx, indumentum similar. Calyx 3–4 (ventral and lateral lobes)–7 (dorsal lobe) mm long, sparsely puberulous, ciliate at apex of lobes, without capitate glands; lobes linear-lanceolate, cuspidate. Corolla yellow, lower lip without reddish brown streaks, upper lip edged with brown, 11–12 mm long, puberulous; tube c.8 mm long and 3 mm diameter; upper lip 3–4 mm long, slightly hooded; lower lip spreading, 5–8 mm long. Filaments c.4 mm long, bending out of flower after anthesis; thecae 1–1.5 mm long, oblong, 25–33% overlapping, hairy on connective, lower with linear appendage c.1 mm long. Capsule 7–8 mm long, uniformly puberulous. Seeds black, densely tuberculate, ovoid, 1.5–2 mm long.

Mozambique. N: Macomia Dist., Messalo (Msalu) R., fl.& fr. ii.1912, *Allen* 131 (K). MS: Cheringoma Dist., Gorongosa Nat. Park, W edge of Cheringoma Plateau, Nhamfisse R. Gorge, 125 m, fl. 12.vi.2012, *Burrows* 12785 (BNRH, K).

Not known elsewhere. Lowland forest and riverine forest, on limestone; 10–300 m.

Conservation notes: Endemic to lowland forests in N & C Mozambique, which are under threat; possibly Vulnerable.

Known only from these three collections. The Allen collection is a long way north of the two recent collections but matches them in all characters.

17. **Justicia petiolaris** (Nees) T. Anderson in J. Linn. Soc., Bot. **7**: 39 (1863). —Clarke in Fl. Cap. **5**: 59 (1901). —Immelman in Fl. Pl. Afr. **48**: pl.1897 (1984); in F.S.A. **30**(3,1): 29 (1995). —Bandeira *et al.*, Fl. Nat. Sul Mozamb.: 194 (2007). Types: South Africa, KwaZulu-Natal, Umzimvubu R., 1837, *Drège* s.n. (K lectotype, P); Umgeni, 1837, *Drège* s.n. (P syntype), lectotypified by Immelman (1995).

Adhatoda petiolaris Nees in De Candolle, Prodr. **11**: 402 (1847).

Justicia incerta C.B. Clarke in Fl. Cap. **5**: 66 (1901). —Meeuse in Fl. Pl. Afr. **31**: pl.1237 (1956). Type: South Africa, Mpumalanga, Boshweld, between Elands R. and Klippan, 1875–1880, *Rehmann* 5058 (K holotype, Z).

Justicia petiolaris subsp. *incerta* (C.B. Clarke) Immelman in Bothalia **16**: 40 (1986); in F.S.A. **30**(3,1): 30 (1995).

Justicia petiolaris subsp. *bowiei* sensu da Silva *et al.*, Prelim. Checklist Vasc. Pl. Mozamb.: 19 (2004), non (C.B. Clarke) Immelman.

Perennial herb with erect, decumbent or scrambling stems from creeping rhizome, often rooting at lower nodes; stem to 1 m long, sparsely retrorsely sericeous to puberulous. Leaves sparsely sericeous on midrib and larger veins; petiole to 2 cm long; lamina ovate to elliptic, largest 2.8–8.5 × 1.5–4 cm; apex acute to obtuse; base attenuate, decurrent. Flowers in 1–3-flowered dichasia; cymes 3–13 cm long, dense, terminal, upper leaf axils often with clusters of flowers; peduncle to 6.5 cm long; peduncle and rachis sparsely to densely sericeous-puberulous, no capitate capitate glands; main axis bracts, dichasia bracts and bracteoles ovate to elliptic or narrowly so to obovate, 8–12 mm long, sparsely sericeous on edges and midrib, sometimes with a few capitate glands with ellipsoid heads. Calyx 7–9 mm long, with scattered long curly hairs and scattered long capitate glands, sparsely pubescent-ciliate; lobes lanceolate to narrowly ovate, with white edges, acute to acuminate. Corolla blue to mauve with pink 'mouth' and darker blue veins,

11–14 mm long, sparsely puberulous; tube 5–7 mm long and 2–3 mm diameter; upper lip 6–7 mm long; lower lip spreading or deflexed, 6–8 mm long. Filaments 3–4 mm long, bending out of flower after anthesis; thecae c.1 mm long, oblong, about a third overlapping, upper sparsely hairy, lower with linear acute appendage c.1 mm long. Capsule 9–10 mm long, glabrous. Seeds not seen.

Botswana. SE: Central Dist., Shoshong, 1883, *Holub* s.n. (K). **Mozambique**. GI: Xai-Xai Dist., Xai Xai (João Belo), Chipenhe, Chiconela, fl. 1.iv.1959, *Barbosa & Lemos* 8445 (K, LISC). M: Maputo Dist, fl.& fr. 7.iv.1947, *Hornby* 2637 (K, PRE, SRGH).

Also in Swaziland and South Africa (Mpumalanga, KwaZulu-Natal, E Cape). Lowland and riverine forest; 10–1000 m.

Conservation notes: Widespread, although main habitat under threat; not threatened.

The third collection cited in the original description (*Forbes* s.n., Delago Bay) is *Justicia flava*.

The Botswana record is significantly distant from the main distribution, and at a higher altitude.

18. **Justicia linearispica** C.B. Clarke in F.T.A. **5**: 192 (1899). —Vollesen in F.T.E.A., Acanthaceae **2**: 539 (2010). Type: Zambia, Isoka/Mbala Dist., Stevenson Road, xi.1893, *Scott Elliot* 8267 (K holotype, BM).

Perennial pyrophytic herb with 1 to several erect (rarely scrambling) stems from a large often branched woody rootstock; stems to 60 cm long, glabrous to sparsely and finely bifariously puberulous, usually only below nodes, rarely uniformly puberulous or densely so. Leaves glabrous, rarely puberulous on midrib and larger veins, sessile; lamina lanceolate to narrowly elliptic, largest 1.5–6 × 0.3–1 cm; apex acute or subacute; base attenuate, decurrent. Flowers single; cymes 1.5–11 cm long, terminal, loose at base, denser upwards; peduncle 0.5–7 cm long, glabrous (towards base) to sparsely and finely puberulous, towards apex sometimes with subsessile capitate glands; rachis finely puberulous and with (rarely without) subsessile stalked capitate glands; bracts and bracteoles ovate-triangular or narrowly so, 3.5–6 mm long, puberulous and with dense stalked capitate glands. Calyx 4–7 mm long, with similar indumentum; lobes lanceolate to narrowly triangular, acute. Corolla white with pink 'herring-bones' on lower lip, 8–12 mm long, minutely puberulous and with subsessile glands; tube 4–6 mm long and 2–3 mm diameter; upper lip 4–6 mm long; lower lip spreading to deflexed, 6–8 mm long. Filaments 3–4 mm long, bending out of flower after anthesis; thecae dark purple, c.1.5 mm long, oblong, only opening for about half their length, about a third overlapping, upper hairy, lower with obtuse or furcate appendage 0.5–0.75 mm long. Capsule 9–11 mm long, puberulous. Seeds verrucose-spinulose, subcircular, c.3 mm long.

Zambia. N: Mbala Dist., Kawimbe, 1750 m, fl.& fr. 6.x.1959, *Richards* 11496 (K, LISC). E: Isoka Dist., Nyika Plateau, near Rest House, 2150 m, fl.& fr. 21.x.1958, *Robson & Angus* 193 (BM, K, LISC). **Malawi**. N: Rumphi Dist., Nyika Nat. Park, Chelinda Bridge, 2400 m, fl. 22.i.1992, *Goyder et al.* 3546 (K, MAL). C: Ntchisi Dist., Mt Chipata, 1350 m, fl. 2.v.1980, *Brummitt et al.* 15592 (C, K, MAL).

Also in Tanzania and Angola. Montane grassland, often in seasonally waterlogged soil but also in drier types, woodland, persisting in plantations and cultivations; 1300–2400 m.

Conservation notes: Moderately widespread; not threatened.

19. **Justicia nyassana** Lindau in Bot. Jahrb. Syst. **20**: 66 (1894). —Clarke in F.T.A. **5**: 192 (1899). —Vollesen in F.T.E.A., Acanthaceae **2**: 539 (2010). Type: Malawi, no locality, 1891, *Buchanan* 290 (B† holotype, BM, K).

  Justicia ulugurica Lindau in Bot. Jahrb. Syst. **22**: 126 (1895). —Clarke in F.T.A. **5**: 194 (1900). Type: Tanzania, Morogoro Dist., Uluguru Mts, Ukami, 18.x.1894, *Stuhlmann* 8866 (B† holotype).

Perennial herb with erect, ascending or scrambling stems from a creeping rhizome, often rooting at lower nodes; stems to 1.5 m long, sparsely sericeous to puberulous or pubescent, rarely densely pubescent. Leaves sparsely sericeous to pubescent on midrib and larger veins or also on lamina; petiole to 1.5(2.5) cm long; lamina narrowly to broadly ovate to elliptic, largest (4)5–14.5 × (1.5)2–5 cm; apex acute to acuminate; base attenuate, decurrent, rarely cuneate. Flowers in 1-flowered dichasia (sometimes 2–3-flowered in basal part); cymes 1–13 cm long, terminal, dense to rather open; peduncle to 11(15) cm long, indumentum as stem, towards apex often with stalked capitate glands; rachis densely whitish to yellowish puberulous, with scattered long pilose hairs and with (rarely without) stalked capitate glands; main axis bracts linear to narrowly ovate, (4)5–10 × 1–2(3) mm, puberulous or sparsely so and with stalked capitate glands, distinctly ciliate with hairs to 2 mm long; dichasia bracts and bracteoles linear to lanceolate, 5–7(8) mm long, both with similar indumentum. Calyx 3–5 mm long, with similar indumentum; lobes filiform to linear, cuspidate. Corolla white to mauve or pink, with pink to purple 'herring-bones' on lower lip, 6–8.5 mm long, sparsely puberulous on lobes, glabrous on whole or basal part of tube, with stalked capitate glands; tube 4–6.5 mm long and 1–1.5 mm diameter; upper lip 2–3 mm long; lower lip spreading, 4–6 mm long. Filaments 1–2 mm long, not bending out of flower after anthesis, thecae 0.5–1 mm long, narrowly oblong, about a half overlapping, upper hairy, lower with linear acute appendage c.0.5 mm long. Capsule 5–7.5 mm long, puberulous. Seeds black, densely tuberculate, tubercles warty, subcircular, 1–1.5 mm diameter.

Zambia. N: Mbala Dist., Lunzua Power Station, 900 m, fl.& fr. 22.v.1962, *Richards* 16491 (K). W: Mwinilunga Dist., Matonchi R., fl.& fr. 9.ii.1938, *Milne-Redhead* 4513 (K). **Zimbabwe**. E: Mutare Dist., Vumba, Tumbling Waters, 1375 m, fl.& fr. 15.v.1955, *Chase* 5578 (BM, K, LISC, SRGH). **Malawi**. N: Rumphi Dist., Njakwa Gorge, 1075 m, fl.& fr. 20.v.1973, *Pawek* 6748 (K, MO). C: Dedza Dist., Mua Livulezi Forest, fl.& fr. 31.v.1962, *Adlard* 471 (K, LISC, SRGH). S: Blantyre Dist., Mt Ndirande, 1225 m, fl. 3.v.1970, *Brummitt* 10345 (K, LISC, MAL). **Mozambique**. N: Malema Dist., Mutuali, fl. 12.v.1948, *Pedro & Pedrógão* 3342 (EA). Z: Gurué Dist., Malema R., Marope, 1150 m, fl.& fr. 1.viii.1979, *de Koning* 7528 (LISC, LMU, WAG). MS: Sussundenga Dist., Chimanimani Mts, Musapa Gap, 975 m, fl.& fr. 6.x.1950, *Sturgeon & Panton* GH30653 (EA, K, SRGH). GI: Massingir Dist., Massingir (Machisugu), 25 m, fl. 10.ii.1898, *Schlechter* 12115 (K).

Also in Central African Republic, Sudan, Ethiopia, D.R. Congo, Burundi, Uganda, Kenya and Tanzania. Lowland to montane forest, usually on edges, paths and clearings, riverine forest, damp places in woodland, termite mounds and roadsides; 20–1500 m.

Conservation notes: Widespread; not threatened.

20. **Justicia kirkiana** T. Anderson in J. Linn. Soc., Bot. **7**: 39 (1863). —Clarke in F.T.A. **5**: 192 (1899). —Burrows & Willis, Pl. Nyika Plateau: 54 (2005). —Vollesen in F.T.E.A., Acanthaceae **2**: 543 (2010). Type: Mozambique, Tete, ii.1859, *Kirk* s.n. (K holotype).

Justicia woodsiae S. Moore in J. Bot. **67**: 271 (1929). Type: Zambia, Mazabuka, 17.ii.1921, *H.S. Woods* 23 (BM holotype).

Annual herb with single (but often branched) erect stem from small taproot, rarely decumbent or straggling; stems to 1 m long, puberulous to retrorsely sericeous or sparsely so. Leaves sparsely sericeous-puberulous on midrib and larger veins; petiole to 1 cm long; lamina lanceolate to narrowly ovate or narrowly elliptic (rarely ovate or elliptic), largest 5.5–16.5 × 0.5–3.2 cm; apex acute to acuminate; base attenuate, decurrent. Flowers in 1–3-flowered dichasia, sometimes all 1-flowered; cymes 1.5–11.5(14) cm long, terminal; peduncle to 6.5 cm long; peduncle and rachis puberulous or sericeous-puberulous to pubescent and sometimes with stalked capitate glands; main axis bracts oblanceolate to narrowly obovate-spathulate, 6–15 × 1–1.5 mm, puberulous with moniliform glossy hairs and with long stalked capitate glands, distinctly ciliate; dichasia bracts narrowly obovate to narrowly spathulate, 6–11(13) mm long; bracteoles linear to oblanceolate, slightly shorter, both with similar indumentum. Calyx 3–4 mm long, puberulous and with scattered capitate glands; lobes lanceolate to narrowly triangular, with white edges, cuspidate. Corolla greenish white to pale bluish green or pale yellow, with distinct blue to purple venation on both lips and a yellow patch at base of lower lip, 7–9 mm long, sparsely puberulous on lobes, glabrous on tube; tube 4–6 mm long and 2–3 mm diameter; upper lip 3–4 mm long; lower lip spreading or deflexed, 5–8 mm long. Filaments 2–3 mm long, not bending out of flower after anthesis, thecae c.1 mm long, oblong, about 25% overlapping, upper hairy, lower with linear acute to obtuse appendage c.0.5 mm long. Capsule 5–8 mm long, subglabrous to sparsely sericeous-puberulous. Seeds black, tuberculate, round, c.1.5 mm diameter.

Botswana. N: Okavango Delta, Qoroqwe Is., fl.& fr. 22.iv.1984, *P.A. Smith* 4420 (K, SRGH). **Zambia**. B: Sesheke Dist., Masese, fl.& fr. 12.iii.1960, *Fanshawe* 5472 (K). C: Chibombo Dist., Kamaila, Mutendere Game Ranch, 1175 m, fl.& fr. 1.iii.1996, *Bingham* 10952 (K). E: Katete Dist., Mpangwe Hills, 1100 m, fl.& fr. 18.iii.1956, *Wright* 89 (K). S: Sinazongwe Dist., Chete Is. Nat. Park, 500 m, fl.& fr. 1.iii.1997, *Zimba et al.* 1007 (K, MO). **Zimbabwe**. N: Darwin Dist., Nyatandi R., 825 m, fl.& fr. 27.i.1960, *Phipps* 2425 (K, SRGH). W: Hwange Dist., Matetsi Safari Area, 925 m, fl.& fr. 16.iii.1981, *Gonde* 366 (K, SRGH). C: Hurungwe Dist., Hurungwe Nat. Park, S of Chirundu, 325 m, fl.& fr. 14.ii.1981, *Philcox et al.* 8527 (K). E: Mutare Dist., Maranke, fl. 14.i.1970, *Mavi* 1090 (K, SRGH). S: Bikita Dist., Devuli R. Bridge, 600 m, fl.& fr. 15.iv.1963, *Chase* 7989 (K, LISC, SRGH). **Malawi**. N: Kasungu Dist., Chipala Hill, 1000 m, fl.& fr. 14.i.1959, *Robson & Jackson* 1175 (BM, K, LISC). C: Lilongwe Dist., Chitedze, 1150 m, fl.& fr. 22.iii.1955, *Exell et al.* 1119 (BM, LISC). S: Mangochi Dist., Cape Maclear, 550 m, fl.& fr. 7.iv.1991, *Bidgood & Vollesen* 2177 (C, EA, K, MAL). **Mozambique**. T: Mutarara Dist., Lower Zambesi R., Lupata, fl.& fr. 20.iv.1860, *Kirk* s.n. (K). MS: Chemba Dist., Chiramba (Shiramba), fr. 13.iv.1860, *Kirk* s.n. (K).

Also in Tanzania and Uganda. Woodland, bushland and grassland, rocky hills, also in disturbed vegetation; 100–1400 m.

Conservation notes: Widespread; not threatened.

There is a collection (*Pimenta* 34102 at LMA) which has the locality "Sul do Save – arredores do L. Marques". This could be either Mozambique GI or M, but since the locality is so vague and from far outside the rest of the distribution area it has not been included in the citations.

Sect. **Vascia** Lindau in Engler & Prantl, Nat. Pflanzenfam. IV **3b**: 395 (1895). — Graham in Kew Bull. **43**: 584 (1988).

Flowers single, subsessile, in dense 4-sided spiciform cymes with both bracts at each node supporting a flower; bracts large, strobilate; bracteoles large, bract-like. Calyx with 5 equal lobes. Appendage on lower anther theca short and stubby or reduced to a small tooth (rarely elongated). Capsule 4-seeded. Seeds reticulately sculptured. Species 21.

21. **Justicia ruwenzoriensis** C.B. Clarke in F.T.A. **5**: 185 (1899). —Vollesen in F.T.E.A., Acanthaceae **2**: 543 (2010). Type: Tanzania, Bukoba Dist., Karagwe, ix.1893, *Scott Elliot* 8120 (K holotype, BM). FIGURE 8.6.55, 12–13.

Justicia andongensis C.B. Clarke in F.T.A. **5**: 185 (1899). Type: Angola, Pungo Andongo, n.d., *Welwitsch* 5111 (BM lectotype, K, P), lectotypified here.

Justicia gossweileri S. Moore in J. Bot. **44**: 28 (1906). Type: Angola, Luanda, 8.vii.1903, *Gossweiler* 1071 (BM holotype, COI, K).

Perennial herb with several erect to decumbent stems from woody rootstock, often pyrophytic; stems to 50 (usually less than 30) cm long, puberulous or densely so. Leaves puberulous; petiole to 7 mm long; lamina ovate to elliptic, largest 1.5–5 × 0.8–3 cm; apex rounded to acute; base cuneate to attenuate, decurrent. Spikes 2–7 cm long, terminal, sometimes interrupted towards base; peduncle to 3.5 cm long; rachis densely puberulous; lower pair of bracts often foliaceous; middle and upper pale to dark green, with or without dark reticulate venation, without (rarely with) white scarious margin, ovate to elliptic, 6–9 × 3–6 mm, puberulous, subacute to acuminate at apex, cuneate to truncate at base; bracteoles like bracts. Calyx 3–4.5 mm long, finely puberulous with short stubby capitate glandular hairs, usually with longer hairs on edges and midrib; lobes acute to acuminate. Corolla white, with or without faint purple markings in throat, 8–11 mm long, puberulous or retrorsely sericeous puberulous; tube 4–6 mm long and 2–3.5 mm in diameter apically; upper lip slightly hooded, 3.5–5 mm long; lower lip deflexed, 3–6 mm long, deeply 3-lobed with middle lobe much wider. Anthers purple; thecae c.1 (upper) and c.1.5 (lower) mm long, lower with an acute curved appendage c.1 mm long, connective glabrous. Capsule 13–15 mm long, densely uniformly puberulous. Seeds not seen.

Zambia. N: Mbala Dist., near Mbala, 1500 m, fl. 18.viii.1960, *Richards* 13118 (K, LISC). W: Solwezi Dist., fl.& fr. 29.ix.1973, *Fanshawe* 12106 (K, NDO).

Also in Sudan, Ethiopia, Rwanda, Uganda, Kenya, Tanzania, Angola. Fireswept dambo grassland; 1200–1500 m.

Conservation notes: Widespread; not threatened.

Sect. **Betonica** (Nees) T. Anderson in J. Linn. Soc., Bot. **7**: 38 (1863). —Graham in Kew Bull. **43**: 585 (1988).

Flowers single, rarely with 2 additional non-developing lateral buds, subsessile, in dense 1-sided spiciform cymes with only one bract at each node supporting a flower, all flowers turned to the same side; bracts large, strobilate; bracteoles large, bract-like. Calyx with 5 equal lobes. Appendage on lower anther theca with elongated acute appendage. Capsule 4-seeded. Seeds tuberculate. Species 22–23.

22. **Justicia betonica** L., Sp. Pl.: 15 (1753). —Clarke in F.T.A. **5**: 184 (1899); in Fl. Cap. **5**: 57 (1901). —Meyer in Merxmüller, Prodr. Fl. SW Afr. **130**: 35 (1968). —Immelman in F.S.A. **30**(3,1): 23 (1995). —Bandeira, Bolnick & Barbosa, Fl. Nat. Sul Mozamb.: 195 (2007). —Vollesen in F.T.E.A., Acanthaceae **2**: 549 (2010). Type: Rheede, Hort. Mal. **2**: t.21 (1679), iconotype. FIGURE 8.6.55, 1–11.

Adhatoda betonica (L.) Nees in Wallich, Pl. Asiat. Rar. **3**: 103 (1832); in De Candolle, Prodr. **11**: 385 (1847).

Adhatoda lupulina Nees in De Candolle, Prodr. **11**: 385 (1847). Type: South Africa, KwaZulu-Natal, Umlozi R., 1837, *Drège* s.n. (K lectotype), lectotypified here.

Adhatoda variegata Nees var. *pallidior* Nees in De Candolle, Prodr. **11**: 385 (1847). Type: South Africa, Gauteng, Aapies R., n.d., *Burke* 514 (K holotype, PRE).

Adhatoda cheiranthifolia Nees in De Candolle, Prodr. **11**: 387 (1847). Type: South Africa, Gauteng, Magaliesberg, n.d., *Burke* 327 (K holotype).

Justicia betonicoides C.B. Clarke in F.T.A. **5**: 184 (1899); in Fl. Cap. **5**: 58 (1901. —Compton, Fl. Swaziland: 564 (1967). Type: Sudan, Jur, Seriba Ghattas, 8.iv.1869, *Schweinfurth* 1423 (K lectotype), lectotypified here.

Justicia cheiranthifolia (Nees) C.B. Clarke in Fl. Cap. **5**: 58 (1901). —Compton, Fl. Swaziland: 564 (1976).

Justicia pallidior (Nees) C.B. Clarke in Fl. Cap. **5**: 58 (1901).

Justicia pallidior var. *cooperi* C.B. Clarke in Fl. Cap. **5**: 58 (1901). Type: South Africa, "British Kaffraria", 1860, *Cooper* 3114 (K holotype).

Justicia uninervis S. Moore in J. Bot. **46**: 74 (1908). Type: South Africa, Mpumulanga, Nelspruit, 6.xi.1905, *Rogers* 269 (BM holotype).

Annual, perennial or shrubby herb with 1 to several erect, ascending or decumbent stems from a creeping rhizome, woody rootstock or taproot; stems to 1(1.5) m long, glabrous (with band of hairs at nodes) to sparsely uniformly puberulous. Leaves glabrous to puberulous; petiole 0–2.5 cm long; lamina linear-lanceolate to ovate or elliptic, largest 1.5–18(22) × (0.3)0.5–6.5 cm; apex rounded to acuminate; base cuneate to attenuate, decurrent. Spikes 2–15(18) cm long, terminal and axillary; peduncle to 4(5.5) cm long; rachis puberulous to pubescent or sparsely so; bracts white to green with dark green venation, more rarely uniformly green or tinged purplish,

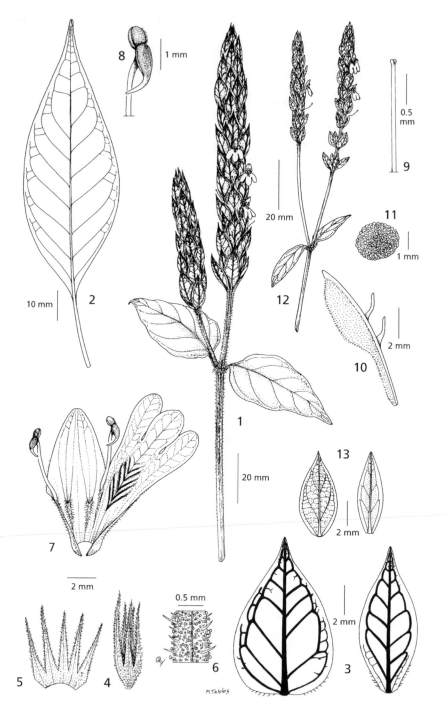

Fig. 8.6.55. JUSTICIA BETONICA. 1, habit; 2, leaf; 3, bract (left) and bracteole (right); 4, calyx; 5, calyx opened up; 6, detail of calyx lobe; 7, corolla opened up; 8, apical part of filament and anther; 9, apical part of style and stigma; 10, capsule; 11, seed. JUSTICIA RUWENZORIENSIS. 12, habit; 13, bract (left) and bracteole (right). 1 from *Richards* 19975, 2, 3 & 7–8 from *Verdcourt* 1675, 4–6 & 9 from *Harris* 1613, 10 & 11 from *Drummond & Hemsley* 2289, 12–13 from *Lye* 2067. Drawn by Margaret Tebbs. Reproduced from Flora of Tropical East Africa (2010).

ovate or broadly so, rarely lanceolate, (5.5)7–16 × (2)3.5–10 mm, glabrous to puberulous, acute to acuminate and apiculate at apex, cuneate to truncate at base; bracteoles 7–13 × 2–5 mm. Calyx 3.5–7 mm long, finely puberulous with stubby non-capitate glandular hairs and with longer hairs on edges (rarely without); lobes acute to acuminate. Corolla white with purple spots on lower lip, 8–14 mm long, densely puberulous and with capitate glands; tube 4.5–8 mm long and 2–3 mm in diameter apically; upper lip slightly hooded, 3.5–6 mm long; lower lip deflexed, 3–5.5 × 4–6 mm, central lobe wider and longer than laterals. Anthers purplish, upper theca 0.5–1 mm long, lower 1–1.5 mm long, with an acute appendage c.1 mm long, connective hairy. Capsule 10–20 mm long, uniformly puberulous or densely so. Seeds densely tuberculate, c.3 mm long.

Botswana. N: Ngamiland Dist., Maun, fl.& fr. v.1967, *Lambrecht* 189 (K, LISC, SRGH). SE: Central Dist., Mahalapye Village, fl.& fr. 16.xii.1963, *Yalala* 397 (K, LISC, SRGH). **Zambia.** B: Senanga West Dist., between Kaanja and Mutomena, Mbume, fl.& fr. 10.iv.1994, *Bingham* 10061 (K). N: Mbala Dist., 20 km from Kawimbe, Kara Village, 1675 m, fl. 4.v.1957, *Richards* 9560 (K). W: Ndola, fl.& fr. 27.iii.1954, *Fanshawe* 1027 (EA, K). C: Mumbwa Dist., 70 km W of Lusaka, Mukulaikwa Agric. Station, fl. 22.iii.1963, *Angus* 3603 (FHO, K). E: Chipata Dist., Luangwa Valley, Mkania, 600 m, fr. 30.v.1968, *Astle* 5176 (K). S: Choma Dist., Choma, fl.& fr. 16.iv.1963, *van Rensburg* 1956 (K). **Zimbabwe.** N: Mutoko Dist., Mudzi Dam, 1225 m, fl.& fr. 16.ii.1962, *Wild* 5674 (K, SRGH). W: Hwange Nat. Park, Shapi road, 1025 m, fl.& fr. 28.ii.1967, *Rushworth* 290 (K, LISC, SRGH). C: Gweru Dist., 28 km S of Kwekwe, Mlezu School Farm, 1275 m, fl. 4.iv.1966, *Biegel* 1076 (K, SRGH). E: Chimanimani Dist., Heathfield, fl. 7.vi.1966, *Loveridge* 1709 (K, LISC, SRGH). S: Chiredzi Dist., Save–Runde (Sabi–Lundi) junction, Chuhonja (Chuhanje) ridge, 375 m, fl.& fr. 9.vi.1950, *Wild* 3466 (K, LISC, SRGH). **Malawi.** N: Chitipa Dist., Kaseye Mission, 1250 m, fl.& fr. 5.iv.1969, *Pawek* 1930 (K). C: Lilongwe Dist., Lilongwe Nature Sanctury, fl.& fr. 5.v.1984, *Patel et al.* 1497 (K, MAL). S: Blantyre Dist., Matenje road, 2 km N of Limbe, 1175 m, fl.& fr. 12.iv.1970, *Brummitt* 9811 (K, LISC, MAL). **Mozambique.** MS: Sussundenga Dist., E of Makurupini R., 400 m, fl. 11.vi.1971, *Biegel & Pope* 3584 (K, LISC, SRGH). M: Boane Dist., Umbeluzi, between Quinta do Umbeluzi and Boane, 15 m, fl. 20.xi.1946, *Gomes e Sousa* 3468 (K).

Widespread in tropical and South Africa and in tropical Asia, introduced in America. In a wide range of grassland, bushland and woodland, riverine forest and scrub, margins of wet forest, also in secondary vegetation and as a weed; 10–1750(1950) m.

Conservation notes: Widespread; not threatened.

A widespread and very variable species as shown by the lengthy synonymy. But the variation – though sometimes fairly distinct within certain regions – when viewed over the total area of the species does not allow for any infraspecific taxa to be separated.

23. **Justicia sp. A** (= *J.E. & S.M. Burrows* 9902).

Shrubby herb or shrub to 2 m tall; young branches quadrangular, sparsely to densely retrorsely sericeous-puberulous on edges or uniformly so, glabrescent. Leaves pubescent when young, soon glabrescent; petiole to 7 cm long; lamina ovate or elliptic, largest 25–30 × 9–10 cm; apex acuminate; base attenuate, decurrent. Spikes c.10 cm long, axillary; peduncle c.10 cm long; peduncle and rachis puberulous; bracts pale green with darker venation, elliptic, 2–2.5 × 1.3–1.5 cm, sparsely puberulous, densest on veins, acute at apex, subcuneate at base; bracteoles similar to bracts but narrower. Calyx c.10 mm long, puberulous with stalked capitate glands and hairs, divided to c.2 mm from base; lobes narrowly elliptic, acute. Corolla pale yellow with greenish upper lip, with brownish markings on lower lip, c.2 cm long, sparsely puberulous and with scattered stalked capitate glands; tube c.1 cm long, basal part cylindrical, c.4 mm in diameter, apical part widened, c.7 mm in diameter; upper lip strongly hooded, almost semiglobose in side view, c.10 × 8 mm; lower lip deflexed, deeply 3-lobed, c.10 mm long, central lobe c.8 × 5 mm, oblong, lateral lobes c.8 × 3 mm. Thecae oblong, c.2.5 mm long, lower with minute appendage, connective hairy. Capsule and seeds not seen.

Mozambique. MS: Gorongosa Dist., Mt Gorongosa, 1200 m, fl. 7.iii.2007, *J.E. & S.M. Burrows* 9902 (BNRH).

Not known elsewhere. Wet evergreen montane forest; 1200 m.

Conservation notes: Endemic to Mt Gorongosa in C Mozambique, and probably also on other similar mountains; possibly Near Threatened.

Closely related to *Justicia engleriana* Lindau from East Africa, and the record in F.T.E.A. of *J. engleriana* from Mozambique refers to this species.

A second non-flowering collection (Mozambique Z: Gurué Dist., Mt Namuli, Mukocha Forest, 1725 m, 28.v.2007, *Timberlake et al.* 5051 (K, LMA)) is possibly this species.

Sect. **Anisostachya** (Nees) Benth. in Bentham & Hooker, Gen. Pl.: 1110 (1876).

Sect. *Betonica* (Nees) T. Anderson subsect. *Anisostachya* (Nees) V.A.W. Graham in Kew Bull. **43**: 586 (1988).

Flowers single, subsessile, in dense spiciform 1-sided racemoid cymes with only one bract at each node supporting a flower and all flowers turned to the same side; bracts strobilate; bracteoles small, lanceolate. Calyx with 4 equal lobes, the 5th lobe reduced to a small tooth; divided to near base. Lower anther theca with flattened, obtuse appendage. Capsule 4-seeded. Seed sparsely hairy or echinate. Species 24.

24. **Justicia tenella** (Nees) T. Anderson in J. Linn. Soc., Bot. **7**: 40 (1863). —Clarke in F.T.A. **5**: 187 (1899. —Vollesen in F.T.E.A., Acanthaceae **2**: 556 (2010). Type: Madagascar, Emirna, n.d., *Bojer* s.n. (K lectotype), lectotypified here. FIGURE 8.6.**56**.

Rostellularia tenella Nees in De Candolle, Prodr. **11**: 369 (1847).

Anisostachya tenella (Nees) Lindau in Engler & Prantl, Nat. Pflanzenfam. IV **3b**: 329 (1895).

Annual herb, basal part of stem erect or (more often) creeping and rooting at nodes, apical part always erect; stems to 35 cm long, puberulous or sparsely so in two bands with broad curly hairs. Leaves with scattered hairs along veins beneath, above with uniformly scattered hairs; petiole to 3 cm long; lamina ovate to elliptic or broadly so, largest 2–5.5 × 1–2.8 cm; apex subacute to broadly rounded; base attenuate, decurrent. Spikes 0.5–2 cm long, axillary; peduncle to 1.8 cm long; rachis with indumentum as stem; bracts pale green, broadly obcordiform, 3–4.5 × 2.5–4 mm, glabrous to sparsely puberulous-ciliate, truncate to retuse at apex, attenuate at base; bracteoles 1.5–2 mm long. Calyx glabrous or finely and sparsely puberulous-ciliate; 4 long lobes 2.5–3 mm long, cuspidate, the 5th lobe 0.5–1 mm long. Corolla pure white or with purple markings on lower lip, 2.5–3 mm long, glabrous; tube 1.5–2 mm long and c.1 mm in diameter apically; upper lip c.1 mm long; lower lip deflexed, c.1 mm long. Stamens exserted; thecae similar, c.0.3 mm long, lower with flattened obtuse appendage c.0.5 mm long. Capsule 2.5–3 mm long, with a few hairs near apex. Seeds sparsely hairy, c.0.75 mm long.

Zimbabwe. E: Nyanga Dist., Honde Valley, Pungwe R., 750 m, fl.& fr. 16.ix.1964, *Loveridge* 1122 (K, SRGH). **Malawi**. N: Nkhata Bay Dist., Chinteche, 475 m, fl.& fr. 11.v.1970, *Brummitt* 10616 (K, LISC, MAL). S: Mulanje Dist., Great Ruo Gorge, Lufiri Power Station, 775 m, fl.& fr. 18.vi.1962, *Richards* 16750 (K). **Mozambique**. MS: Manica Dist., Mt Bandula, 700 m, fl.& fr. 9.vii.1957, *Chase* 6590 (K, LISC, SRGH).

Also in Senegal, Guinée, Mali, Sierra Leone, Liberia, Ivory Coast, Togo, Ghana, Nigeria, Bioko, Cameroon, Gabon, Equatorial Guinea, Central African Republic, Sudan, D.R. Congo, Burundi, Tanzania and Madagascar. Lowland evergreen and riverine forest, also secondary forest and a weed in cleared forest areas; 200–800(1050) m.

Conservation notes: Widespread; not threatened.

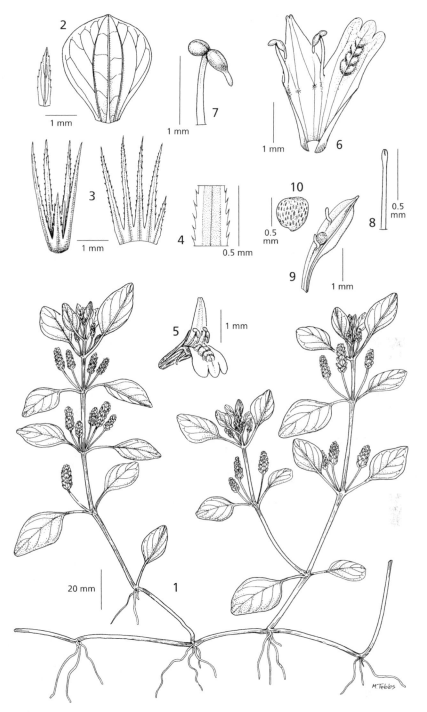

Fig. 8.6.**56**. JUSTICIA TENELLA. 1, habit; 2, bract (right) and bracteole (left); 3, calyx, whole and opened up; 4, detail of calyx lobe; 5, corolla; 6, corolla opened up; 7, apical part of filament and anther; 8, apical part of style and stigma; 9, capsule; 10, seed. 1 & 9–10 from *Mhoro* 170, 2–4 & 8 from *Richards* 9851, 5–7 from *Vaughan* 607. Drawn by Margaret Tebbs. Reproduced from Flora of Tropical East Africa (2010).

Sect. **Ansellia** C.B. Clarke in F.T.A. **5**: 183 (1899); in Fl. Cap. **5**: 57 (1901). —Ensermu in Symb. Bot. Ups. **29**(2): 51 (1990).

    Sect. *Rostellaria* T. Anderson subsect. *Ansellia* (C.B. Clarke) V.A.W. Graham in Kew Bull. **43**: 598 (1988).

Flowers small, subsessile, in axillary secund racemoid cymes; with only one bract at each node supporting a flower and all flowers turned to the same side, but alternate sterile bracts often present; bracteoles minute. Three dorsal calyx lobes longer than two ventral and middle dorsal lobe longer than lateral and sometimes also slightly wider. Appendage on lower anther theca lower square-cut at apex. Capsule 4-seeded. Seeds double reticulate, flattened or not. Species 25–30.

25. **Justicia anagalloides** (Nees) T. Anderson in J. Linn. Soc., Bot. **7**: 42 (1863). —Clarke in Fl. Cap. **5**: 65 (1901). —Graham in Kew Bull. **43**: 598 (1988). — Ensermu in Symb. Bot. Ups. **29**(2): 52, figs.20, 21 (1990). —Vollesen in F.T.E.A., Acanthaceae **2**: 556 (2010). Type: South Africa, Gauteng, Aapies R., n.d., *Burke* s.n. (K holotype). FIGURE 8.6.57.

    Adhatoda anagalloides Nees in De Candolle, Prodr. **11**: 403 (1847).

    Justicia uncinulata Oliv. in Trans. Linn. Soc. **29**: 130, t.129a (1875). —Clarke in F.T.A. **5**: 210 (1900). —Graham in Kew Bull. **43**: 598 (1988). Type: Tanzania, Mpwapwa/Kilosa Dist., Rubeho Mts, xii.1860, *Speke & Grant* s.n. (K holotype).

    Justicia crassiradix C.B. Clarke var. *hispida* C.B. Clarke in F.T.A. **5**: 210 (1900). Type: Malawi, Shire Highlands, vii.1885, *Buchanan* 461 (K lectotype), lectotypified by Ensermu (1990).

    Justicia psammophila Mildbr. in Notizbl. Bot. Gart. Berlin-Dahlem **11**: 1089 (1934). — Graham in Kew Bull. **43**: 598 (1988). Type: Tanzania, Morogoro, 28.i.1933, *Schlieben* 3309 (B lectotype, BM, BR, G, HBG, LISC, P, S), lectotypified by Ensermu (1990).

    Justicia anselliana sensu Clarke in F.T.A. **5**: 208 (1900) for *Buchanan* 483, 876 & 1385, non (Nees) T. Anderson.

Perennial herb with several unbranched or branched erect, decumbent or creeping (then sometimes rooting) stems from a vertical or horizontal woody rootstock; stems to 45 cm long, bifariously puberulous with retrorse hairs when young or uniformly pubescent to tomentose, sometimes also with longer (to 2 mm) broad curly glossy many-celled hairs with white walls. Leaves glabrous (finely ciliate towards apex) or with scattered to dense long (to 2.5 mm) broad glossy many-celled hairs on margins and sometimes also on veins, rarely uniformly pubescent; petiole 0–5 mm long; lamina linear-lanceolate to ovate or elliptic, largest 1.2–6.2 × 0.2–2 cm; apex subacute to rounded; base cuneate to attenuate. Spikes 3–5(6)-flowered; peduncle 0.3–4(4.5) cm long, subglabrous to bifariously or uniformly sericeous-puberulous, rarely also with scattered long many-celled hairs; rachis 7–30 mm long, with similar indumentum; bracts and bracteoles 1.5–3.5 mm long, subulate, glabrous. Calyx glabrous or minutely hispid-ciliate towards apex; lobes subulate to linear, apiculate, two lower 3–5 mm long, three upper 5–9 mm long. Corolla white, with or without faint pink to mauve 'herring-bones' on lower lip and two pink lines on upper lip, 5–7 mm long; tube 2–3 mm long; upper lip 3–4 mm long; lower lip 3–4 mm long, middle lobe 2–3 × 1–2 mm. Filaments 2–3 mm long; anthers c.1.5 mm long, thecae c.0.75 mm long. Capsule 6–8.5 mm long, hairy in upper half, more rarely glabrous. Seeds c.1.5 mm in diameter.

    Zambia. N: Isoka Dist., 25 km S of Chitipa, 5 km into Zambia, facing Mafinga Hills, 1075 m, fl. 27.xii.1972, *Pawek* 6159 (K, MAL, MO). S: Namwala Dist., 3 km from Namwala, near Kafue Nat. Park road, fl., 10.xii.1962, *van Rensburg* ZM42 (EA). **Zimbabwe**. N: Lomagundi Dist., Darwendale, Audley End Farm, 1300 m, fl. 24.i.1968, *Biegel* 2863 (K, SRGH). C: Harare Dist., Goromonzi, 1375 m, fl.& fr. 7.xii.1951, *Wild* 3698 (K, LISC, SRGH). E: Mutare Dist., Vumba Mts, Norseland, 1525 m, fl. iii.1949, *Wild* 2838 (K, SRGH). S: Mberengwa Dist., Mt Bukwa, 1350 m, fl.& fr. 29.iv.1973, *Pope* 1013 (K, SRGH). **Malawi**. N: Chitipa Dist., Kapoka Crossroads, 1275 m, fl.& fr. 29.xii.1977, *Pawek* 13454 (K, MAL, MO, SRGH). C: Mchinji Dist., 7 km W of Namitete, fl.& fr. 29.iv.1989, *Radcliffe-Smith et al.* 5796 (C, K). S: Zomba Dist., Zomba, Mbidi Estate, fl.& fr. 11.i.1960, *Banda* 363 (BM, K, SRGH). **Mozambique**. N: Macomia Dist.,

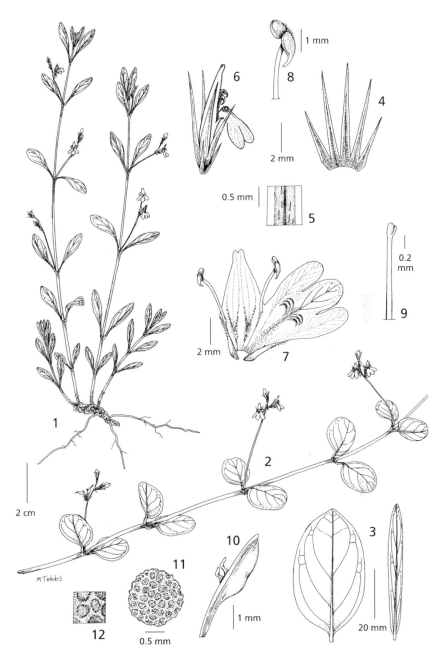

Fig. 8.6.57. JUSTICIA ANAGALLOIDES. 1, habit, erect form; 2, habit, trailing form; 3, variation in leaf shape; 4, calyx opened up; 5, detail of calyx lobe; 6, calyx and corolla; 7, corolla opened up; 8, apical part of filament and anther; 9, apical part of style and stigma; 10, capsule; 11, seed; 12, detail of seed. 1 & 4–5 from *Richards* 24171, 2 from *Faden & Ng'weno* 74/546, 3 (narrow leaf) from *Drummond & Hemsley* 1776 and (broad leaf) *Vollesen* MRC3614, 6–9 from *Faulkner* 507, 10–12 from *Bidgood et al.* 3969. Drawn by Margaret Tebbs. Reproduced from Flora of Tropical East Africa (2010).

Quiterajo, Namacubi Forest, Messalo R., 25 m, fl. 23.xi.2009, *Crawford* 696 (K, LMA, LMU, P). Z: Gilé Dist., 72 km from Gilé on road to Mulela R., fl. 10.x.1949, *Barbosa & Carvalho* 4365 (K). T: Angónia Dist., 8 km N of Mlangeni, Msese Hill, 1475 m, fl. 18.ii.1970, *Brummitt* 8612 (K). MS: Sussundenga Dist., Manica, Zonue R., 750 m, fl.& fr. 4.iii.1907, *Johnson* 256 (K).

Also in Ethiopia, Somalia, Uganda, Kenya, Tanzania, D.R. Congo, Rwanda, Burundi, Swaziland, South Africa. In a wide variety of grassland, bushland and woodland on a broad range of soils; (20)250–1600 m.

Conservation notes: Widespread; not threatened.

26. **Justicia crassiradix** C.B. Clarke in F.T.A. **5**: 210 (1900). —Graham in Kew Bull. **43**: 598 (1988). —Ensermu in Symb. Bot. Ups. **29**(2): 61, figs.24, 25 (1990). —Vollesen in F.T.E.A., Acanthaceae **2**: 560 (2010). Type: Zambia, Mbala Dist., Urungu, Fwambo, ix.1863, *Carson* 107 (K lectotype), lectotypified by Ensermu (1990).

Perennial herb with 1–2 erect unbranched (rarely branched) stems from a creeping, sometimes branched rhizome with fleshy roots; stems to 35 cm long, glabrous (with fringe of hairs at nodes) to pubescent with broad glossy many-celled hairs with white walls. Leaves sessile, glabrous to pubescent with similar hairs; lamina linear to lanceolate, largest 1.5–8 × 0.2–0.8 cm; apex acute to acuminate; base cuneate to attenuate; lateral veins absent or very inconspicuous. Spikes (2)3–5-flowered; peduncle 2–8.5 cm long, glabrous to pubescent with white hairs; rachis (0.5)1–3 cm long, with similar indumentum; bracts and bracteoles linear to narrowly triangular, 2–6 mm long, acuminate, glabrous or finely ciliate. Calyx glabrous, rarely sparsely puberulous, and finely hispid-ciliate; lobes subulate to linear, two lower 4–7 mm long, three upper 7–10 mm long. Corolla white with pink to pale mauve 'herring-bones' on lower lip (more rarely pure white), 10–14 mm long; tube 5–7 mm long; upper lip 5–7 mm long; lower lip 6–9 mm long , middle lobe 4–6 × 3–5 mm, lateral lobes narrower. Filaments 4–5.5 mm long; anthers 2–3 mm long, thecae 1–1.5 mm long. Capsule 14–16 mm long, glabrous. Seeds 2.5–3 mm in diameter.

Zambia. N: Mbala Dist., Sumbawanga road, Zombe Plain, 1500 m, fl.& fr. 29.xii.1964, *Richards* 19397 (EA, K); Isoka Dist., 45 km on Tunduma–Mbala road, 1350 m, fl. 10.i.1975, *Brummitt & Polhill* 13710 (K).

Also in Tanzania. Dambos, valley grassland; 1200–1750 m.

Conservation notes: Local distribution and in a somewhat threatened habitat; possibly Near Threatened.

27. **Justicia nuttii** C.B. Clarke in F.T.A. **5**: 210 (1900). —Graham in Kew Bull. **43**: 598 (1988). —Ensermu in Symb. Bot. Ups. **29**(2): 66, figs.28,29 (1990). —Burrows & Willis, Pl. Nyika Plateau: 51 (2005). —Vollesen in F.T.E.A., Acanthaceae **2**: 561 (2010). Type: Tanzania, between Lake Tanganyika and Lake Rukwa, 1896, *Nutt* s.n. (K holotype).

Justicia nuttii var. *blantyrensis* C.B. Clarke in F.T.A. **5**: 210 (1900). —Binns, Checklist Herb. Fl. Malawi: 14 (1968). Type: Malawi, Shire Highlands, Blantyre, 6.vii.1879, *Buchanan* 20 (K lectotype), lectotypified by Ensermu (1990).

Justicia goetzei Lindau in Bot. Jahrb. Syst. **28**: 484 (1900). —Clarke in F.T.A. **5**: 514 (1900). —Graham in Kew Bull. **43**: 598 (1988). Type: Tanzania, Iringa Dist., Uhehe, Ukano Mts, 1898, *Goetze* 685 (K lectotype, BM, BR, E, P), lectotypified by Ensermu (1990).

Justicia schliebenii Mildbr. in Notizbl. Bot. Gart. Berlin-Dahlem **11**: 411 (1932). Type: Tanzania, Njombe Dist., Upembe, Likanga, 28.iii.1931, *Schlieben* 458 (G lectotype, BR, P), lectotypified by Ensermu (1990).

Perennial herb with 1 to several erect to decumbent unbranched or branched stems from a creeping, sometimes branched rhizome; stems to 40 cm long, sparsely to densely pubescent with broad glossy many-celled hairs with coloured walls. Leaves glabrous (rarely) to pubescent with similar hairs, above uniformly so, beneath mainly along veins; petiole 1–5 mm long; lamina ovate to elliptic or oblong or narrowly so, rarely lanceolate, largest 1.5–5 × 0.7–2.5 cm; apex acute to broadly rounded; base cuneate to cordate. Spikes 2(3)-flowered; peduncle 1.5–6 cm long, subglabrous to pubescent with broad hairs with coloured walls; rachis 3–11(15) mm long, puberulous (sometimes only in a narrow band) to pubescent; bracts and bracteoles subulate to linear, 1–7 mm long, glabrous to puberulous. Calyx glabrous to puberulous; lobes subulate to linear, apiculate, two lower (3.5)4.5–9 mm long, three upper (4)6–14 mm long. Corolla white, rarely with faint pale mauve 'herring-bones' on lower lip, 8–13 mm long; tube 4–6 mm long; upper lip 4–7 mm long; lower lip 5–8 mm long, middle lobe 3.5–6.5 × 3–5 mm, lateral lobes narrower. Filaments 4–6 mm long; anthers 2–3 mm long, thecae 1–1.5 mm long. Capsule 9–13 mm long, pubescent in upper half with broad hairs with coloured walls, rarely glabrous. Seeds 2.5–3 mm in diameter.

Zambia. N: Mbala Dist., Saisi R., 1500 m, fl.& fr. 27.ii.1957, *Richards* 8367 (K). **Malawi**. N: Mzimba Dist., Mzuzu, 1375 m, fl.& fr. 11.ii.1974, *Pawek* 8080 (EA, K, MAL, MO, SRGH). C: Dedza Dist., Chiwao Hill, fl. 3.ii.1967, *Salubeni* 542 (K, LISC, SRGH) S: Shire Highlands, Blantyre, fl.& fr. n.d., *Buchanan* 165 (K). **Mozambique**. T: Angónia Dist., Ulónguè, Serra Dómuè, 1500 m, fl. 19.xii.1980, *Macuácua* 1477 (K, LISC).

Also in D.R. Congo and Tanzania. Montane grassland and bushland, dambos and valley grassland in woodland, more rarely in woodland; 1200–2200 m.

Conservation notes: Widespread; not threatened.

28. **Justicia exigua** S. Moore in J. Bot. **38**: 204 (1900). —Clarke in F.T.A. **5**: 514 (1900); in Fl. Cap. **5**: 66 (1901). —Graham in Kew Bull. **43**: 598 (1988). —Ensermu in Symb. Bot. Ups. **29**(2): 71, figs.32, 33 (1990). —Vollesen in F.T.E.A., Acanthaceae **2**: 562 (2010). Type: Zimbabwe, Bulawayo, v.1898, *Rand* 389 (BM holotype).

Justicia matammensis sensu Clarke in F.T.A. **5**: 209 (1900) for *Holub* 1208–1210 & in Fl. Cap. **5**: 66 (1901). —Meyer in Merxmüller, Prodr. Fl. SW Afr. **130**: 36 (1968), non (Schweinf.) Oliv.

Justicia anselliana sensu Immelman in Fl. Pl. Afr. **49**: pl.1932 (1986) in part. —Setshogo, Checklist Pl. Botswana: 18 (2005), non (Nees) T. Anderson.

Annual herb with much-branched erect, ascending or decumbent (then often rooting) stems to 45 cm long; stems bifariously puberulous and with few to many long broad curly many-celled hairs to 2 mm long. Leaves glabrous or with scattered long broad curly many-celled hairs along midrib, more rarely also on lamina and edges towards base; petiole 1–7 mm long; lamina elliptic or narrowly so, widest near middle, largest 0.8–3.8(5) × 0.3–1.5(2.5) cm; apex subacute to broadly rounded; base cuneate to attenuate. Spikes 3–6(7)-flowered; peduncle 0.2–2.5 cm long, bifariously puberulous or sparsely so and (towards base) often with long broad curly many-celled hairs; rachis 0.5–2.2 cm long, bifariously puberulous or sparsely so; bracts and bracteoles subulate to linear, 0.5–1.5 mm long, glabrous. Calyx glabrous; lobes linear, two lower 2–4 mm long, three upper 3–6 mm long. Corolla white with faint pink to mauve 'herring-bones' on lower lip and two pink lines on upper lip, 3.5–5 mm long; tube 1.5–2.5 mm long; upper lip 2–3 mm long; lower lip 2.5–3.5 mm long, middle lobe 1.5–2.5 × 1.5–2 mm. Filaments 1.5–2 mm long; anthers c.1 mm long, thecae c.0.5 mm long. Capsule 4–6 mm long, with scattered hairs in apical half (rarely almost to base), sometimes with purple lines and patches. Seeds c.1 mm in diameter.

Botswana. N: Ngamiland, Okavango, Xudum R., 925 m, fl.& fr. 15.iii.1961, *Richards* 14733 (K). SE: Central Dist., 15 km on Mosolotschame–Serowe road, 1150 m, fl.& fr. 22.iv.2005, *Darbyshire et al.* 454 (K). **Zambia**. B: Sesheke Dist., Masese, fl.& fr. 12.iii.1960, *Fanshawe* 5479 (K). S: Choma Dist., 11 km S of Mapanza, 1075 m, fl.& fr. 20.iii.1955, *Robinson* 1134 (K, LISC). **Zimbabwe**. W: Matobo Dist., Besna Kobila Farm, 1500 m, fl. xii.1956, *Miller* 4007 (K, LISC, SRGH). C: Gweru Dist., 30 km SSE of Kwekwe (Que Que), Mlezu School Farm, 1275 m, fl. 20.i.1966, *Biegel* 812 (K, SRGH). S: Mberengwa Dist., Belingwe to West Nicholson road, Zingedzi R. Bridge, fl.& fr. 4.v.1972, *Pope* 633 (K, LISC, SRGH). **Mozambique**. M: Magude Dist., near Magude, Ungabana, fl.& fr. 5.xii.1980, *Jansen & Nuvunga* 7657 (K, LMA).

Also in Sudan, Ethiopia, D.R. Congo, Rwanda, Burundi, Uganda, Kenya, Tanzania, Angola, Namibia, South Africa. In a wide variety of woodland, bushland and grassland, often in disturbed or secondary vegetation or as a weed in cultivated areas, on a wide range of soils; 40–1500 m.

Conservation notes: Widespread; not threatened.

29. **Justicia anselliana** (Nees) T. Anderson in J. Linn. Soc., Bot. **7**: 44 (1863). —Clarke in F.T.A. **5**: 208 (1900), in part. —Graham in Kew Bull. **43**: 598 (1988). —Ensermu in Symb. Bot. Ups. **29**(2): 74, figs.34, 35 (1990). —Immelman in F.S.A. **30**(3,1): 41 (1995). —Vollesen in F.T.E.A., Acanthaceae **2**: 564 (2010). Type: Liberia, Cape Palmas, 1842, *Ansell* s.n. (K lectotype), lectotypified by Graham (1988).

Adhatoda anselliana Nees in De Candolle, Prodr. **11**: 403 (1847).

Dianthera anselliana (Nees) Bentham in Gen. Pl. **2**: 1114 (1876).

Justicia kapiriensis De Wild. in Bull. Jard. Bot. État **5**: 12 (1915). Type: D.R. Congo, Kapiri Valley, ii.1913, *Homblé* 1089 (BR lectotype), lectotypified by Ensermu (1990).

Short-lived perennial herb with one to several erect, ascending or decumbent (then often rooting) unbranched or sparsely branched stems to 50 cm from a short creeping rhizome; stems glabrous except for a thin band at nodes. Leaves minutely hispid-ciliate, otherwise glabrous or with a few hairs at base (rarely also along midrib); petiole 0–5 mm long; lamina lanceolate to narrowly elliptic or narrowly obovate, largest 1.5–5.5 × 0.3–1.2 cm; apex acute to acuminate (rarely rounded); base cuneate to attenuate. Spikes (1)2–5-flowered; peduncle 0.5–3.5 cm long, glabrous; rachis (0.5)1–2.5 cm long, glabrous; bracts and bracteoles subulate to linear, 1–2.5 mm long, glabrous; sterile bracts to 2 mm long. Calyx minutely scabrid-ciliate; lobes linear-lanceolate, cuspidate, two lower 3–5 mm long, three upper 4–6(7) mm long. Corolla white with faint pink to mauve 'herring-bones' on lower lip and two pink lines on upper lip, 5.5–7.5 mm long; tube 2.5–3.5 mm long; upper lip 3–4 mm long; lower lip 3–5 mm long, middle lobe 3–4 × 2–3 mm. Filaments 2.5–3.5 mm long; anthers 1–1.5 mm long, thecae 0.5–0.75 mm long. Capsule (8)9–12 mm long, glabrous (very rarely hairy at apex), sometimes with purple lines. Seeds 1.5–2 mm in diameter.

Caprivi. Kakumba Is., 925 m, fl. 14.i.1959, *Killick & Leistner* 3417 (K, PRE). **Zambia**. B: Mongu Dist., Mongu to Lealui, fl.& fr. 9.xi.1959, *Drummond & Cookson* 6265 (K, LISC, SRGH). W: Solwezi Dist., Solwezi, 1350 m, fl.& fr. 9.iv.1960, *Robinson* 3476 (K). C: Lusaka Dist., 30 km NNE of Lusaka, N of Kasisi, Chongwe R., 1150 m, fl. 29.ix.1972, *Strid* 2235 (C, K). S: Namwala Dist., Namwala, near Kafue R., 1000 m, fl.& fr. 9.i.1957, *Robinson* 2084 (K).

Also in Liberia, Ghana, Togo, Benin, Nigeria, Central African Republic, Sudan, D.R. Congo, Rwanda, Uganda, Kenya, Tanzania, Angola, Namibia. Grassland and *Acacia* bushland on grey to black clay in dambos and river valleys; 900–1400 m.

Conservation notes: Widespread; not threatened.

30. **Justicia matammensis** (Schweinf.) Oliv. in Trans. Linn. Soc. Lond. **29**: 130 (1875). —Clarke in F.T.A. **5**: 209 (1900). —Meyer in Merxmüller, Prodr. Fl. SW Afr. **130**: 36 (1968). —Graham in Kew Bull. **43**: 598 (1988). —Ensermu in Symb. Bot. Ups. **29**(2): 78, figs.36, 37 (1990). —Immelman in F.S.A. **30**(3,1): 43 (1995). —Vollesen in F.T.E.A., Acanthaceae **2**: 565 (2010). Type: Sudan/Ethiopia, Matamma, 29.vii.1865, *Schweinfurth* 130C (K lectotype, BM), lectotypified by Ensermu (1990).

Adhatoda matammensis Schweinf. in Verh. Königl. Zool.-Bot. Ges. Wien **18**: 464 (1868).

Annual herb with unbranched or branched erect, ascending or decumbent (then often rooting) stems to 50 cm long; stems bifariously puberulous and usually with few to many long glossy curly many-celled hairs to 3 mm long. Leaves with sparse to dense long curly hairs on midrib and lower part of margin (rarely absent), sometimes also on lamina; petiole 2–18 mm long; lamina ovate to elliptic (most or all widest below middle), largest 2.5–9 × 1–3.5 cm; apex subacute to broadly rounded; base cuneate to attenuate. Spikes 3–8-flowered; peduncle 0.8–4(5) cm long, bifariously puberulous and with (rarely without) few to many long glossy curly many-celled hairs; rachis 0.5–2.5 cm long, bifariously puberulous; bracts and bracteoles subulate, 1–2.5 mm long, glabrous. Calyx glabrous; lobes linear-lanceolate, two lower 2–3.5 mm long, three upper 3–4.5 mm long. Corolla white with faint pink to mauve 'herring-bones' on lower lip and two pink lines on upper lip, 4–5 mm long; tube 2–2.5 mm long; upper lip 2–3 mm long; lower lip 3–4 mm long, middle lobe 1.5–3 × 1–2 mm. Filaments 2–2.5 mm long; anthers 1–1.5 mm long, thecae 0.5–0.75 mm long. Capsule 8–10.5(12) mm long, hairy in apical half, sometimes with purple lines. Seeds c.2 mm in diameter.

Botswana. N: Central Dist., 120 km from Francistown, Mosetse R., 1025 m, fl.& fr. 8.iii.1961, *Richards* 14579 (K). **Zambia**. E: Chipata Dist., between Mushoro (Msoro) and Great East Road, Mwangazi R., 900 m, fl. 6.i.1959, *Robson* 1045 (BM, K, LISC). S: Choma Dist., Mapanza Mission, 1075 m, fl. 14.i.1953, *Robinson* 46 (K). **Zimbabwe**. N: Hurungwe Dist., Kariba Gorge, 500 m, fl. 25.ii.1953, *Wild* 4032 (K, SRGH). W: Nkayi Dist., Gwampa Forest Reserve, 925 m, fl.& fr. ii.1956, *Goldsmith* 89/56 (K, LISC, SRGH). E: Chimanimani Dist., Haroni R., fl. 26.iv.1973, *Mavi* 1443 (BM, K, LISC, SRGH). S: Chiredzi Dist., Runde (Lundi) R., Chipinda Pools, 300 m, fl. 12.i.1961, *Goodier* 52 (K, PRE, SRGH). **Malawi**. N: Karonga Dist., 25 km SSE of Karonga, 500 m, fl.& fr. 31.i.1992, *Goyder et al.* 3591 (K). C: Lilongwe Dist., Mchenzi R., fl. 16.i.1953, *Jackson* 1032 (K). S: Blantyre Dist., Mpatamanga Gorge, E Bank of Shire R., 225 m, fl.& fr. 9.ii.1970, *Brummitt* 8488 (K, MAL). **Mozambique**. T: Changara Dist., Boroma, Marenga, fl. 1890, *Menyharth* 544 (C). MS: Chemba Dist., Chiou, CICA Expt. Farm, fl.& fr. 5.iv.1962, *Balsinhas & Macuácua* 558 (BM, K, LISC).

Also in Central African Republic, Sudan, Ethiopia, D.R. Congo, Burundi, Uganda, Kenya, Tanzania, South Africa. In a wide variety of woodland, bushland and grassland (including secondary types), riverine forest, often a weed in cultivation, on a wide variety of soils; 100–1100 m.

Conservation notes: Widespread; not threatened.

Sect. **Harnieria** (Solms-Laub.) Benth. in Bentham & Hooker, Gen. Pl. **2**: 1109 (1876). —Hedrén in Kew Bull. **43**: 349 (1988). —Graham in Kew Bull. **43**: 591 (1988). —Hedrén in Bull. Jard. Bot. Belg. **58**: 129 (1988); in Bot. J. Linn. Soc. **103**: 263 (1990); in Nordic J. Bot. **10**: 357 (1990).

Flowers single or in 2–4(6)-flowered cymules in axils of upper leaves; flowers almost always developing on both sides; bracts foliaceous; bracteoles small, subulate to linear or narrowly triangular. Calyx with 5 equal lobes. Lower anther theca appendage acute and often bifid at apex, more rarely broad and bent (transversally ellipsoid). Capsule 4-seeded. Seeds black, tuberculate, with a central ridge. Species 31–47.

The species of the *Justicia heterocarpa* group (species 34–40) often develop spiny indehiscent 1-seeded fruits (see Fig. 8.6.59).

31. **Justicia odora** (Forssk.) Lam., Encycl. Meth. Bot. **1**(2): 629 (1785). —Clarke in F.T.A. **5**: 201 (1900). —Meyer in Merxmüller, Prodr. Fl. SW Afr. **130**: 36 (1968). —Hedrén in Symb. Bot. Ups. **29**(1): 66, fig.25 (1989). —Immelman in F.S.A. **30**(3,1): 34 (1995). —Vollesen in F.T.E.A., Acanthaceae **2**: 565 (2010). Type: Yemen, Surdud, ii.1763, *Forsskål* 384 (C holotype). FIGURE **8.6.58**.

Dianthera odora Forssk., Fl. Aegypt.-Arab.: 8 (1775).

Adhatoda odora (Forssk.) Nees in De Candolle, Prodr. **11**: 399 (1847).

Justicia polymorpha Schinz in Verh. Bot. Vereins Prov. Brandenburg **31**: 203 (1890). Type: Namibia, Amboland, Ohando, 14.iii.1886, *Schinz* 32 (Z lectotype, K), lectotypified here.

Justicia fischeri Lindau in Bot. Jahrb. Syst. **20**: 65 (1894). —Clarke in F.T.A. **5**: 202 (1900). Type: Tanzania, Lushoto Dist., Usambara Mts, Kwa Mshuza, 5.viii.1893, *Holst* 8897 (K lectotype), lectotypified here.
Justicia lycioides Schinz in Vierteljahr. Naturf. Ges. Zürich **61**: 440 (1916). Type: Namibia, Ondonga to Uukuambi, 26.ii.1901, *Rautanen* 785 (Z holotype).

Much-branched shrubby herb or shrub to 1 m tall; young stems pale green with straw-coloured ribs, glabrous, older branches with first dark brownish then pale grey bark. Leaves occasionally with scattered hairs on margins near base, otherwise glabrous, with few to many yellow to orange subsessile capitate glands, often only or mostly along midrib; petiole 0–3 mm long; lamina narrowly ovate to ovate or narrowly elliptic to elliptic, largest 1.6–7.5 × 0.8–3 cm; apex acute to rounded, often apiculate; base cuneate to attenuate, rarely cuneate. Flowers single or 2–4 per axil of upper leaves; bracts 0.5–2 cm long; bracteoles 0.5–2 mm long; peduncles and pedicels 0–1(2) mm long. Calyx puberulous-ciliate, otherwise glabrous to puberulous, divided to 1–1.5 mm from base; lobes 3–9 mm long. Corolla lemon yellow to bright yellow, upper lip sometimes with red to purple veins and a purple patch at base, lower lip with well-developed 'herring-bones', 6–14(16) mm long; tube 2–6(7) mm long, widening immediately above base, 2–5 mm diameter apically; upper lip 4–9(10) mm long (upper lip 2–3 mm longer than tube); lower lip 4–13(15) × 5–12 mm. Filaments 3.5–6 mm long; thecae yellow, with glandular exudate, 1–1.5 mm long, lower with bifid appendage 0.5–1 mm long. Capsule 10–15 mm long, glabrous. Seeds 2.5–3 mm diameter.

Botswana. N: Ngamiland Dist., Motopi Irrigation Scheme, fl.& fr. 19.xii.1978, *P.A. Smith* 2602 (K, PRE, SRGH). SE: Central Dist., Ilalamabele–Mosu area, Sua Pan, fl.& fr. 14.i.1974, *Ngoni* 328 (K, SRGH). **Zimbabwe**. W: Hwange Nat. Park, 25 km SW of Sinamatella Camp, Mandavu Dam to Shumba Camp, 1025 m, fl. 25.ii.1967, *Rushworth* 222 (K, LISC, SRGH).

Also in Egypt, Saudi Arabia, Yemen, Oman, Sudan, Eritrea, Ethiopia, Djibouti, Somalia, Uganda, Kenya, Tanzania, Angola, Namibia and South Africa. Bushland on sandy to loamy or stony soil, rocky hillsides, riverine scrub, mopane woodland; 900–1200 m.

Conservation notes: Widespread; not threatened.

32. **Justicia capensis** Thunb., Prodr. Fl. Cap. **2**: 104 (1800). —Clarke in Fl. Cap. **5**: 60 (1901). —Hedrén in Kew Bull. **43**: 349, fig.1 (1988); in Symb. Bot. Ups. **29**(1): 72 (1989). —Immelman in F.S.A. **30**(3,1): 35 (1995). —Vollesen in F.T.E.A., Acanthaceae **2**: 568 (2010). Type: South Africa, no locality, n.d., *Herb. Thunberg* 362 (UPS lectotype), lectotypified by Hedrén (1988).
Gendarussa capensis (Thunb.) Nees in Linnaea **15**: 366 (1841).
Adhatoda capensis (Thunb.) Nees in De Candolle, Prodr. **11**: 391 (1847).
Justicia sansibarensis Lindau in Bot. Jahrb. Syst. **20**: 71 (1894) —Clarke in F.T.A. **5**: 202 (1900). Type: Zanzibar, Kidoti, xi.1873, *Hildebrandt* 983 (COR lectotype, BM), lectotypified by Hedrén (1988).

Shrubby herb or shrub to 1(2) m tall, sparsely to densely branched; stems pale yellow to dark green, puberulous or sparsely with with short erect or bent hairs, sometimes with capitate glands near nodes. Leaves glabrous or sparsely hairy on midrib and larger veins and conspicuously ciliate, with few to many yellow to orange subsessile capitate glands; petiole to 3 cm long; lamina ovate to elliptic or obovate, largest 2.5–11(13) × 1.2–4(6) cm; apex subacute to rounded; base attenuate. Flowers usually 2 per axil of upper leaves, rarely some single; bracts 0.3–1(1.5) cm long. Calyx puberulous or sericeous-puberulous or sparsely so, rarely also with a few stalked capitate glands; lobes 2.5–8(10.5) mm long. Corolla white to purple with white tube, upper lip without darker veins, lower lip with well-developed 'herring-bones', 10–17 mm long; tube (4)5–9 mm long, 1.5–4 mm diameter apically; upper lip (4)5–9 mm long, tube and lip same length; lower lip (4)6–12 × 5.5–12 mm. Filaments (3)5–8 mm long; thecae yellow or brown, c.1 mm long, lower with bifid appendage 0.5–1 mm long. Capsule 12–19(24) mm long, glabrous. Seeds 2–3 mm diameter, densely tuberculate.

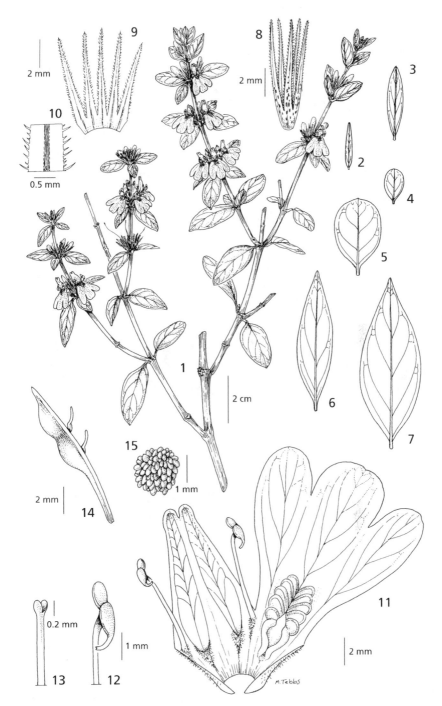

Fig. 8.6.**58**. JUSTICIA ODORA. 1, habit; 2–7, series of leaves; 8, calyx; 9, calyx opened up; 10, detail of calyx lobe; 11, corolla opened up; 12, apical part of filament and anther; 13, apical part of style and stigma; 14, capsule; 15, seed. 1 & 7 from *Bidgood et al.* 1029, 2 from *Greenway* 13828, 3 & 14–15 from *Tweedie* 4019, 4 from *Bally* 14651, 5 & 8–10 from *Richards* 24219, 6 from *Pearce & Vollesen* 942, 11–13 from *Bidgood et al.* 1277. Drawn by Margaret Tebbs. Reproduced from Flora of Tropical East Africa (2010).

Mozambique. GI: Massinga Dist., Pomene, fl.& fr. 20.vi.1980, *de Koning* 8201 (LMA, LMU). M: Boane Dist., Umbeluzi (Umbolosi), 12.xii.1897, *Schlechter* 11719 (K, Z). Also in Kenya, Tanzania, Zanzibar and South Africa. Coastal bushland, sand dunes; c.10 m.

Conservation notes: Widespread, although more local in the Flora area; probably not threatened.

33. **Justicia striolata** Mildbr. in Notizbl. Bot. Gart. Berlin-Dahlem **11**: 411 (1932). —Hedrén in Symb. Bot. Ups. **29**,1: 99, fig.53 (1989). —Vollesen in F.T.E.A., Acanthaceae **2**: 576 (2010). Type: Tanzania, Njombe Dist., Lupembe, Ruhudje, iv.1931, *Schlieben* 678 (B† holotype, BM, EA, K).

   Justicia phyllostachys sensu Clarke in F.T.A. **5**: 188 (1899) for *Whyte* s.n. (Mt Chiradzulu), non sensu stricto.

Annual or short-lived perennial herb, basal part of stems creeping and rooting, apical part erect or ascending, sparsely branched, to 1m long, sparsely to densely sericeous-puberulous with curly or bent hairs. Leaves subglabrous or with scattered hairs on midrib and larger veins, occasionally also on lamina; petiole to 4 cm long, sericeous, often also with long pilose hairs to 2 mm long and scattered capitate glands; lamina ovate to elliptic, largest 2.5–9.5 × 1.7–4.8 cm; apex acuminate to acute; base attenuate to rounded. Flowers single or 2 in axils of upper leaves or 2–4 per leaf in short spiciform cymes on short axillary branches; bracts (sometimes absent in solitary flowers) ovate to elliptic or broadly so, to 9 mm long, conspicuously ciliate with long white hairs to 1.5 mm long. Calyx subglabrous or with scattered to dense long straight hairs, often only towards apex but sometimes all over, sometimes with stalked capitate glands; lobes 3–5 mm long. Corolla white to pale mauve, upper lip with darker veins, lower lip with well-developed 'herring-bones', 12–20 mm long; tube 9–12(14) mm long, c.1 mm diameter in middle, 4–7 mm longer than upper lip; upper lip 3–8 mm long; lower lip 5–10 × 5–11 mm. Filaments 3–5 mm long; thecae white to pale yellow, 0.5–1 mm long, lower with entire acute appendage c.0.25 mm long. Capsule sometimes 2-seeded by abortion, 6.5–8(9) mm long, with scattered hairs near apex. Seeds 1.5–2 mm long, densely tuberculate.

Zambia. N: Mbala Dist., old Boma gardens, 1525 m, fl. 10.v.1955, *Richards* 5634 (K). **Malawi**. N: Chitipa Dist., Mughese, 1575 m, fl.& fr. 1.viii.1977, *Phillips* 2707B (K, MO). C: Dedza Dist., Mt Dedza, 2050 m, fl.& fr. 24.iv.1970, *Brummitt* 10121 (C, EA, K, LISC, MAL). S: Mulanje Dist., Mchese Forest Reserve, above Ulolo, 1000 m, fl. 14.ii.1992, *Goyder & Paton* 3650 (K). **Mozambique**. N: Lichinga Dist., Lichinga, 6.viii.1934, *Torre* 186 (COI, LISC).

Also in D.R. Congo and Tanzania. Medium altitude to montane moist evergreen forest and forest margins, persisting in plantations and old fields; (650)1000–2100 m.

Conservation notes: Widespread; not threatened.

34. **Justicia heterocarpa** T. Anderson in J. Linn. Soc., Bot. **7**: 41 (1863). —Clarke in F.T.A. **5**: 200 (1900). —Hedrén in Symb. Bot. Ups. **29**(1): 105 (1989); in Nordic J. Bot. **10**: 374, figs.27, 28 (1990). —Vollesen in F.T.E.A., Acanthaceae **2**: 580 (2010). Type: Ethiopia, Gageros, 7.ix.1854, *Schimper* in *Hohenacker* 2300 (K lectotype, BM), lectotypified by Hedrén (1990).

   Justicia leptocarpa Lindau in Bot. Jahrb. Syst. **20**: 70 (1894). —Clarke in F.T.A. **5**: 200 (1900). Type: Tanzania, Lushoto Dist., Mashewa, vii.1893, *Holst* 8799 (K lectotype), lectotypified by Hedrén (1990).

Subsp. **dinteri** (S. Moore) Hedrén in Symb. Bot. Ups. **29**(1): 106 (1989); in Nordic J. Bot. **10**: 378, figs.27, 28 (1990). —Immelman in F.S.A. **30**(3,1): 33 (1995). —Vollesen in F.T.E.A., Acanthaceae **2**: 582 (2010). Type: Namibia, Otjitua, 8.v.1907, *Dinter* 87 (BM holotype, K).

Justicia dinteri S. Moore in J. Bot. **57**: 246 (1919). —Meyer in Merxmüller, Prodr. Fl. SW Afr. **130**: 35 (1968).

Annual herb with a single unbranched to much-branched stem from a taproot; stems erect (rarely straggling) to 0.8(1) m long, when young puberulous or sparsely so and with long (1–2 mm) pilose hairs, with or without capitate glands. Leaves sparsely puberulous or sparsely pubescent, densest on midrib and margins towards base, usually also with stalked capitate glands; petiole to 4.5 cm long; lamina ovate or elliptic, largest 3.7–9.3 × 1.4–4.5 cm; apex acuminate to subacute; base attenuate to truncate. Flowers 2–4(6) in axils of upper leaves; bracts absent or lanceolate to suborbicular or reniform, to 0.8 cm long, sometimes with long (2(3) mm) glossy curly hairs, with or without stalked capitate glands. Calyx puberulous or densely so, usually with long pilose hairs to 2 mm long and stalked capitate glands; lobes 3–6.5(7.5) mm long, basal part white and spongiform in fruit. Corolla pale or dark mauve, rarely white, lower lip with distinct 'herring-bone' pattern, upper lip with dark veins, 4–7 mm long; tube 3–4.5 mm long, widening immediately above base, 1–2 mm longer than upper lip; upper lip 1.5–3 mm long, teeth to 1 mm; lower lip 3–5 × 3–6 mm. Filaments 1–2 mm long; thecae yellow, c.0.5 mm long, lower with acute appendage 0.2–0.3 mm long. Capsule 4–6 mm long, glabrous or sparsely puberulous in upper half; indehiscent fruits often present, 3–4 mm long, with 4 subentire to (usually) dissected wings. Seeds c.1.5 mm long.

Caprivi. E Caprivi, Katima Mulilo, Mpacha, fl.& fr. 27.ii.1975, *Edwards* 4338 (K, PRE). **Botswana**. N: Ngamiland Dist., Maun, fl.& fr. 31.iii.1975, *P.A. Smith* 1339 (K, LISC, SRGH). SE: Kgatleng Dist., Mochudi, Phutodikobo Hill, 925–1050 m, fl.& fr. 10.iii.1967, *Mitchison* 37 (K). **Zambia**. B: Kalabo Dist., Lukona, 1150 m, fl.& fr. 20.iii.1996, *Luwiika et al.* 394 (K, MO). C: Mumbwa Dist., Blue Lagoon Nat. Park, Nakeenda, 975 m, fl.& fr. 25.iv.2000, *Bingham* 12211 (K). S: Sesheke Dist., Nanga Nat. Forest, 30 km NW of Mulolezi Boma, 900 m, fl.& fr. 6.iii.1996, *Harder et al.* 3649 (K, MO). **Zimbabwe**. N: Gokwe Dist., Sengwa Research Station, fl. 16.iii.1976, *Guy* 2402 (K, SRGH). W: Matopos Nat. Park, 1350 m, fl.& fr. 24.ii.1981, *Philcox & Leppard* 8811 (C, K). C: Chegutu Dist., Poole Farm, Mupfure (Umfuli) R., 1225 m, fl.& fr. 18.iv.1948, *Hornby* 2889 (K, SRGH). **Malawi**. S: Mangochi Dist., Thumbi Is., c.500 m, fl.& fr. 12.iv.1989, *Steiner* 639 (C, K, UPS).

Also in Tanzania, Angola, Namibia and South Africa. Riverine forest, woodland and grassland, mopane and *Baikiaea* woodland and forest, *Acacia* and *Combretum* woodland, weed; 450–1500 m.

Conservation notes: Widespread; not threatened.

There are two other subspecies in NE and E Africa one of which extends to India. For differences between the three subspecies, see Vollesen (2010).

35. **Justicia protracta** (Nees) T. Anderson in J. Linn. Soc., Bot. **7**: 41 (1863). — Immelman in Bothalia **16**: 40 (1986). —Hedrén in Symb. Bot. Ups. **29**(1): 108 (1989); in Nordic J. Bot. **10**: 391, fig.36 (1990). —Immelman in F.S.A. **30**(3,1): 30 (1995). Type: South Africa, Cape Prov., Zwartkops R., n.d., *Ecklon* 456 (BOL lectotype, MEL), lectotypified here.

Gendarussa protracta Nees in Linnaea **15**: 371 (1841).
Gendarussa prunellaefolia Hochst. in Flora **28**: 71 (1845). Type: South Africa, Durban Bay, 1840, *Krauss* 304 (K lectotype), lectotypified here.
Gendarussa mollis Hochst. in Flora **28**: 71 (1845). Type: South Africa, KwaZulu-Natal, between Umlazi R. and Durban Bay, 1840, *Krauss* 61 (K lectotype), lectotypified here.
Adhatoda protracta (Nees) Nees in De Candolle, Prodr. **11**: 390 (1847).
Adhatoda microphylla Klotzsch in Peters, Naturw. Reise Mossamb. **6**: 216 (1861). Type: Mozambique, Mozambique Is., n.d., *Peters* s.n. (B† holotype).
Justicia microphylla (Klotzsch) Lindau in Engler & Prantl, Nat. Pflanzenfam. IV **3b**: 349 (1895).
Justicia filifolia sensu Clarke in F.T.A. **5**: 198 (1900) for *Kirk* s.n., Shupanga & *Kirk* s.n., Shire R., non Lindau.

Justicia pulegioides C.B. Clarke in Fl. Cap. **5**: 62 (1901), excl. syn. *Chaetacanthus personii.*
Type: South Africa, KwaZulu-Natal, Durban Flats, 1887, *Wood* 1019 (K lectotype, BOL, PRE,
SAM), lectotypified here.
Justicia kraussii C.B. Clarke in Fl. Cap. **5**: 62 (1901). Type: South Africa, KwaZulu-Natal,
Inanda, vi.1879, *Wood* 423 (K lectotype, PRE), lectotypified here.
Justicia kraussii var. *florida* C.B. Clarke in Fl. Cap. **5**: 63 (1901). Type: South Africa,
KwaZulu-Natal, Inanda, x.1879, *Wood* 566 (K holotype, BM).
Justicia woodii C.B. Clarke in Fl. Cap. **5**: 64 (1901). Type: South Africa, KwaZulu-Natal,
Noodsberg, v.1879, *Wood* 112 (K holotype).
Justicia striata (Klotzsch) Bullock subsp. *melampyrum* (S. Moore) J.K. Morton in Kew Bull.
32: 443 (1978), for *Chase* 5224, non sensu stricto.
Justicia striata var. *filifolia* (Lindau) J.K. Morton in Kew Bull. **32**: 443 (1978) for *Hutchinson
& Gillett* 3289, non sensu stricto.
Justicia syncollotheca sensu Hedrén in Bull. Jard. Bot. Belg. **58**: 148 (1988) for *Pawek* 11669,
non Milne-Redh.

Annual, perennial or shrubby herb with one to several unbranched to much-branched
stems, sometimes from a woody rootstock; stems erect to scrambling, to 1 m long, when young
puberulous or sparsely so (sometimes only near nodes) with hairs 0.2–0.5 mm long, rarely with
longer pilose hairs to 1 mm long, usually with scattered to dense stalked capitate glands. Leaves
sparsely puberulous or sparsely pubescent, densest on midrib and margins, usually with capitate
glands towards base and on petiole, same colour on both sides; petiole to 2(4) cm long; lamina
narrowly ovate to ovate, rarely lanceolate or elliptic, largest 1–9(13) × 0.3–3.5(4.5) cm, more than
twice as long as wide; apex subacuminate to obtuse; base attenuate to rounded. Flowers single
or 2–4 in axils of upper leaves; bracts elliptic or broadly so, to 1 cm long, minutely puberulous,
usually with long (to 2 mm) pilose hairs and with stalked capitate glands; bracteoles to 1(2.5) mm
long. Calyx sparsely puberulous to puberulous, rarely subglabrous with hairs 0.1–0.5 mm long,
sometimes with stalked capitate glands; lobes 1.5–3.5 mm long. Corolla white to pale mauve,
lower lip with (rarely without) distinct 'herring-bone' pattern, upper lip without darker veins,
(4.5)6–9(11) mm long; tube 2.5–5.5 mm long, widening immediately above base, 0.5–2 mm
longer than upper lip; upper lip 2–3.5(5.5) mm long, teeth 0.5–1 mm long; lower lip 3–6(9) ×
3–6.5(9) mm. Filaments 1–3 mm long; thecae yellow (rarely with purple sides), 0.5–1 mm long,
lower with acute appendage c.0.3 mm long. Capsule (4.5)5.5–7.5 mm long, uniformly puberulous
(hairs more than 0.1 mm long) all over, more rarely subglabrous or glabrous; indehiscent fruits
sometimes present, 3–4 mm long, with 4 entire to dissected wings. Seeds 1–1.5 mm long.

Zimbabwe. N: Hurungwe Dist., 18 km ESE of Chirundu Bridge, Mensa Pan, 450
m, fl.& fr. 29.i.1958, *Drummond* 5333 (EA, K, LISC, SRGH). E: Mutare Dist., 10 km
S of Mutare (Umtali), Dora R., fl.& fr. 20.v.1935, *Eyles* 8427 (K). S: Chiredzi Dist.,
Runde (Lundi) R., Save–Runde junction, 250 m, fl.& fr. 7.vi.1950, *Wild* 3443 (K,
LISC, SRGH). **Malawi.** S: Chikwawa Dist., Livingstone Falls, W bank of Shire R., 100
m, fl.& fr. 21.iv.1970, *Brummitt* 9998 (K, LISC, MAL). **Mozambique.** MS: Guro Dist.,
31 km on Changara–Catandica (Gouveia) road, 600 m, fl.& fr. 24.v.1971, *Torre &
Correia* 18589 (LISC, LMA, LMU, UPS, WAG). GI: Panda Dist., between Mangorro
and Panda, Inhassune R., fl. 7.iv.1959, *Barbosa & Lemos* 8526 (K, LISC). M: Namaacha
Dist., Namaacha, fl.& fr. 20.vii.1957, *Marques* 2059 (LMU).
Also in South Africa and Swaziland. Dry forest (including riverine), woodland,
bushland and grassland, roadsides; 50–1300 m.
Conservation notes: Widespread; not threatened.

36. **Justicia rhodesiana** S. Moore in J. Bot. **51**: 188 (1913). Type: Zimbabwe, Bulawayo,
18.ii.1912, *Rogers* 5740 (BM lectotype, K), lectotypified by Hedrén (1990).

Justicia protracta (Nees) T. Anderson subsp. *rhodesiana* (S. Moore) Immelman in Bothalia
16: 40 (1986). —Hedrén in Symb. Bot. Ups. **29**(1): 109 (1989); in Nordic J. Bot. **10**: 393
(1990). —Immelman in F.S.A. **30**(3,1): 31 (1995). —Mapura & Timberlake, Checklist Zimb.
Vasc. Pl.: 14 (2004). —Setshogo, Checklist Pl. Botswana: 18 (2005).

Justicia protracta sensu Meyer in Merxmüller, Prodr. Fl. SW Afr. **130**: 37 (1968), non (Nees) T. Anderson.

Shrubby herb or shrub or pyrophytic suffrutescent herb with several unbranched or sparsely branched stems from a creeping woody rootstock; stems erect to decumbent, then often rooting at lower nodes, to 1 m long (to 30 cm in pyrophytic plants), when young uniformly minutely puberulous or densely so with hairs to 0.1 mm long, rarely with a few hairs to 0.25 mm long, older branches greyish with slightly corky bark. Leaves sparsely minutely puberulous, usually densest on midrib and margins, very rarely with scattered longer hairs; petiole to 0.5 cm long; lamina narrowly ovate to ovate, rarely elliptic, largest 0.7–2.5(3.2) × 0.3–1.2(1.6) cm; apex acute to broadly rounded base attenuate to rounded. Flowers single or 2–4 in axils of upper leaves; bracts ovate to elliptic, to 0.8 cm long, minutely puberulous, with stalked capitate glands; bracteoles to 1(2.5) mm long. Calyx uniformly minutely puberulous with hairs to 0.1 mm long, with or without stalked capitate glands; lobes (1.5)2–4.5(5.5) mm long. Corolla white, rarely tinged mauve or purple, lower lip without distinct 'herring-bone' pattern, upper lip sometimes with faintly mauve veins, 5–9(10) mm long; tube 3–5(6) mm long, widening immediately above base, 0–1 mm longer than upper lip; upper lip 2–5 mm long, teeth 0.5–1 mm long; lower lip 3–6 × 3–6 mm. Filaments 1–3 mm long; thecae yellow, rarely purple tinged, 0.5–1 mm long, lower with acute appendage c.0.3 mm long. Capsule 6–7.5(9) mm long, uniformly minutely puberulous (hairs to 0.1 mm long); indehiscent fruits not seen. Seeds c.1 mm long.

Botswana. SE: Central Dist., Mahalapye, Lephephe, 975 m, fl. 13.i.1958, *de Beer* 540 (K, SRGH). **Zimbabwe**. W: Matobo Dist., 32 km S of Antelope, Noel Nickel Mine, fl.& fr. 10.vi.1968, *Wild* 7727 (K, LISC, SRGH). C: Makonde Dist., Rusape, fl.& fr. iv.1953, *Dehn* R87/53 (K, SRGH). E: Chimanimani Dist., Rocklands, 1225 m, fl. 7.x.1950, *Panton* GH30512 (K, SRGH). S: Beitbridge Dist., Fort Tuli, fl.& fr. 23.iii.1959, *Drummond* 5957 (K, SRGH). **Mozambique**. GI: Guijá Dist., Caniçado to Massingir, fl.& fr. 8.viii.1973, *Correia & Marques* 3221 (LMU).

Also in Namibia and South Africa. *Acacia–Commiphora* bushland on rocky hills and mopane bushland; 20–1600 m.

Conservation notes: Widespread; not threatened.

I have decided to resurrect this species. It is difficult to accept two subspecies (*Justicia protracta* subsp. *protracta* and subsp. *rhodesiana*) having more or less the same distribution areas and ecology. The choice is then to either sink them into one species or accept two closely related species; as they usually have slightly different facies I have followed the latter course.

Justicia rhodesiana has occasionally been reported from soils with high content of heavy metals, but the label data are generally not good enough to tell whether this is regularly the case.

37. **Justicia hedrenii** Vollesen, sp. nov. Differs from *Justicia protracta* in the wider (less than twice as long as wide) leaves with a truncate to cordate base, the denser stem indumentum and the larger capsules. Type: Zimbabwe, Mutare Dist., Mutaranpanda, 4.ii.1956, *Chase* 5993 (K holotype, BM, SRGH).

Justicia sp. B sensu Hedrén in Symb. Bot. Ups. **29**(1): 111 (1989); in Nordic J. Bot. **10**: 394, fig.38 (1990).

Perennial herb with several unbranched or sparsely branched stems from a woody rootstock; stems erect to decumbent, to 0.6 m long, when young puberulous to pubescent or tomentose with hairs to 2 mm long, sometimes with scattered stalked capitate glands. Leaves puberulous to pubescent or sparsely (rarely densely) so, occasionally with capitate glands towards base, same colour on both sides; petiole to 3(8) mm long; lamina ovate to cordiform or broadly so, largest 1–2.6 × 0.8–1.6 cm, less than twice as long as wide; apex subacute to rounded; base truncate to cordate. Flowers all single or some in 2s in axils of upper leaves; bracts absent (usually) or elliptic, to 0.5 cm long, indumentum as leaves. Calyx puberulous or sparsely so with hairs to 0.5(1) mm long (over 0.1 mm); lobes 2.5–4.5 mm long. Corolla white to mauve, lower lip with or without distinct 'herring-bone' pattern, upper lip with or without darker veins, 7–9.5(10.5) mm long; tube 4–5.5 mm long, widening immediately above base, (0.5)1–2 mm longer than upper lip; upper lip 3–4(5) mm long, teeth c.0.5 mm long; lower lip 4.5–7 × 4–6 mm. Filaments 1–3 mm long; thecae yellow, 0.5–1 mm long, lower with acute appendage c.0.3 mm long. Capsule (6)7–9 mm long, uniformly puberulous (hairs more than 0.1 mm long); indehiscent fruits not seen. Seed not seen.

Zimbabwe. E: Mutare Dist., Mt Inyamatshira (Inyamakabira), 1425 m, fl.& fr. 4.ii.1956, *Chase* 5992 (K, LISC, SRGH); Nyanga Dist., Mutarazi Falls, fl.& fr. 12.ii.1961, *Goodier* 1039 (K, LISC, SRGH).

Not known elsewhere. Montane grassland, rocky outcrops, roadsides; 1400–2300 m.

Conservation notes: Endemic to the Mutare–Nyanga area of E Zimbabwe but in a widespread habitat; possibly Near Threatened.

A high altitude species closely related to *Justicia protracta*. It differs in the wider truncate to cordate leaves, the usually denser stem indumentum and usually larger capsules.

38. **Justicia whytei** S. Moore in Trans. Linn. Soc. London, Bot. **4**: 32 (1894). —Clarke in F.T.A. **5**: 198 (1900), in part. —Hedrén in Symb. Bot. Ups. **29**(1): 112 (1989); in Nordic J. Bot. **10**: 394, fig.37 (1990). Type: Malawi, Mt Mulanje, x.1891, *Whyte* 136 (BM holotype).

   Justicia filifolia sensu Clarke in F.T.A. **5**: 198 (1900) for *Buchanan* 51, non Lindau.
   Justicia melampyrum sensu Clarke in F.T.A. **5**: 199 (1900) in part, non S. Moore.
   Justicia striata subsp. *striata* var. *striata* sensu Morton in Kew Bull. **32**: 441 (1978), non (Klotzsch) Bullock.

Pyrophytic perennial herb with several unbranched to much-branched stems from a woody rootstock; stems erect, ascending or decumbent, to 1 m long, when young puberulous to pubescent or sparsely so with hairs to 1 mm long, rarely with scattered stalked capitate glands. Leaves sparsely pubescent, densest on midrib and margins, usually with capitate glands towards base and on petiole, dark green above, much paler beneath; petiole to 7 mm long; lamina lanceolate to ovate, always lanceolate towards tip of stems, largest 2.3–6.5 × 0.4–1.3(2) cm, lower 2–3 times as long as wide, upper 4–12 times; apex acute to rounded; base attenuate to rounded. Flowers 2–4 in axils of upper leaves; bracts lanceolate to ovate or elliptic, to 1.1 cm long, indumentum as leaves. Calyx puberulous or sparsely so with hairs to 0.5 mm long (over 0.1 mm); lobes 2–4 mm long. Corolla mauve to bright purple, lower lip with distinct 'herring-bone' pattern, upper lip usually with darker veins, 8–11 mm long; tube 5–6.5 mm long, widening immediately above base, 1.5–2 mm longer than upper lip; upper lip 3.5–5 mm long, teeth to 1 mm long; lower lip 6–8 × 8–12 mm. Filaments 2–4 mm long; thecae yellow, rarely pale purple, c.1 mm long, lower with acute appendage c.0.5 mm long. Capsule 7–9.5 mm long, uniformly puberulous (hairs more than 0.1 mm long); indehiscent fruits not seen. Seeds c.1 mm long.

Zimbabwe. E: Nyanga Dist., Pungwe Hills, 1825 m, fl.& fr. 31.i.1939, *Hopkins* GH 7204 (K, SRGH). **Malawi**. C: Dedza Dist., foot of Golomoti escarpment, 950 m, fl.& fr. 18.ii.1979, *Brummitt* 15408 (K, MAL). S: Zomba Dist., lower slopes of Mt Malosa, 900 m, fl.& fr. 6.xi.1977, *Brummitt & Dudley* 15047 (K, MAL). **Mozambique**. Z: Milange Dist., Serra Chiperone, Marrega, 775 m, fl.& fr. 24.i.1972, *Correia & Marques* 2274 (LMU). MS: Nhamatanda? Dist., between Beira and Manica (Massikessi), 250 m, fl. xi-xii.1899, *Cecil* 26 (K).

Not known elsewhere. *Brachystegia* woodland, persisting in secondary woodland and grassland after clearing, also in plantations; 250–1900 m.

Conservation notes: Endemic to the Flora area. Widespread; not threatened.

39. **Justicia striata** (Klotzsch) Bullock in Bull. Misc. Inform., Kew **1932**: 502 (1932). —Morton in Kew Bull. **32**: 441 (1978). —Hedrén in Symb. Bot. Ups. **29**(1): 110 (1989); in Nordic J. Bot. **10**: 385, fig.35 (1990). —Burrows & Willis, Pl. Nyika Plateau: 51 (2005). —Vollesen in F.T.E.A., Acanthaceae **2**: 584 (2010). Type: Mozambique, Zambézia, Namacurra to Mocuba, 26.iii.1943, *Torre* 4994 (LISC neotype, LMA, LMU, WAG), selected by Hedrén (1990). FIGURE 8.6.**59**.

Adhatoda striata Klotzsch in Peters, Naturw. Reise Mossamb. **6**: 216 (1861).

Justicia filifolia Lindau in Bot. Jahrb. Syst. **20**: 70 (1894). —Clarke in F.T.A. **5**: 198 (1900) in part. Type: Malawi, Shire Highlands, 1887, *Last* s.n. (K lectotype), lectotypified by Hedrén (1990).

Justicia melampyrum S. Moore in Trans. Linn. Soc. London, Bot. **4**: 32 (1894). —Clarke in F.T.A. **5**: 199 (1900). Type: Malawi, Mt. Mulanje, 1891, *Whyte* 135 (BM holotype).

Justicia dyschoristeoides C.B. Clarke in F.T.A. **5**: 197 (1900). Type: Tanzania, Kilimanjaro, 1884, *Johnston* s.n. (K lectotype, BM), lectotypified by Hedrén (1990).

Justicia whytei sensu C.B. Clarke in F.T.A. **5**: 198 (1900) for *Buchanan* 304, non S. Moore.

Justicia lindaui C.B. Clarke in F.T.A. **5**: 199 (1900). Type: Tanzania, Usambara Mts, Kwa Mshuza, ix.1893, *Holst* 4324 (K lectotype, COI, HBG, M, P, W, Z), lectotypified by Hedrén (1990).

Justicia forbesii S. Moore in J. Bot. **44**: 219 (1906). Type: "Madagascar", n.d., *Forbes* 22 (BM holotype), see notes.

Justicia infirma Mildbr. in Notizbl. Bot. Gart. Berlin-Dahlem **11**: 410 (1932). Type: Tanzania, Mahenge, Massagati, 5-12.vi.1931, *Schlieben* 1145 (B† holotype, BM, BR, G, M, P, S, Z).

Justicia striata subsp. *melampyrum* (S. Moore) J.K. Morton in Kew Bull. **32**: 443 (1978), in part.

Justicia striata subsp. *striata* var. *filifolia* (Lindau) J.K. Morton in Kew Bull. **32**: 443 (1978).

Justicia striata subsp. *striata* var. *dyschoristeoides* (C.B. Clarke) Hedrén in Symb. Bot. Ups. **29**(1): 111 (1989); in Nordic J. Bot. **10**: 390 (1990).

Justicia striata subsp. *austromontana* Hedrén in Symb. Bot. Ups. **29**(1): 111 (1989), invalidly published; in Nordic J. Bot. **10**: 390 (1990). Type: Tanzania, Mbeya Dist., Mbeya, 22.iii.1985, *Hedrén et al.* 248 (UPS holotype, B, BR, DSM, EA, K, LISC, MAL, MO, NHT, P, SRGH, WAG).

Annual or perennial herb with erect, ascending (sometimes rooting at lower nodes) or scrambling stems from a taproot (when annual) or from a creeping often branched rhizome, rarely a scrambling subshrub with woody basal stems; stems to 0.75 m long in erect plants, to 2 m when scrambling, when young subglabrous to sericeous, puberulous or pubescent. Leaves subglabrous to sparsely sericeous, puberulous or pubescent, densest on midrib and margins, with or without stalked capitate glands on margins towards base and on petioles; petiole to 2.5(3.5) cm long; lamina linear-lanceolate to ovate (rarely broadly so), narrower towards tip of stems, largest (1.5)2–10(12) × 0.4–4(5.5) cm; apex acuminate to subacute; base attenuate to subcuneate. Flowers 2–6 in axils of upper leaves, rarely some single; bracts lanceolate to broadly ovate or broadly elliptic, to 1.3 cm long, with or without long (to 1.5 mm) pilose hairs, with or without stalked capitate glands. Calyx subglabrous to puberulous with straight hairs to 0.5 mm long, sometimes with stalked capitate glands; lobes 3–6.5 mm long. Corolla white to mauve or purple, lower lip with distinct 'herring-bone' pattern, upper lip without dark veins, 7–14 mm long; tube 4–7 mm long, widening immediately above base, same length as upper lip or to 2 mm longer, very rarely with lip to 2(4) mm longer than tube; upper lip 3–7(9) mm long, teeth to 1 mm long; lower lip 4–11 × 5–12 mm. Filaments 2–6 mm long; thecae yellow with brown pigment patches at apex and base of upper theca and at apex of lower theca, 0.5–1 mm long, lower with acute appendage 0.3–0.5 mm long. Capsule 5–8 mm long, acute, glabrous to sparsely puberulous in upper half; indehiscent fruits 3–4 mm long, 4-winged. Seeds 1–1.5 mm long.

Zambia. N: Mbala Dist., 45 km on Mbala–Tunduma road, 1500 m, fl.& fr. 1.vi.1962, *Richards* 16502 (EA, K). W: Kitwe Dist., Kitwe, 1250 m, fl.& fr. 13.ii.1961, *Linley* 77 (K, SRGH). C: Serenje Dist., Kundalila Falls, fl.& fr. 29.iii.1984, *Brummitt et al.* 16970 (K). S: Choma Dist., Mapanza East, Ngongo R., 1075 m, fl.& fr. 10.v.1953, *Robinson* 210 (EA, K). **Zimbabwe**. N: Hurungwe Dist., Magunge, Gwambara R., fl.& fr. 27.iv.1955, *Shiff* 10 (K, SRGH). W: Masvingo Dist., Masvingo (Victoria), fl. 1909, *Monro* 896 (BM).

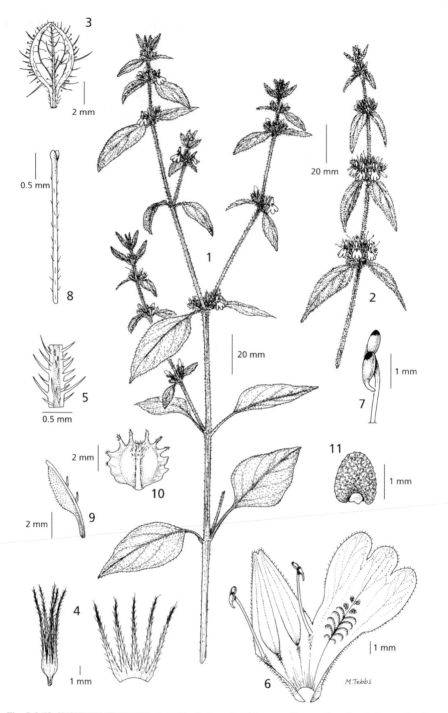

Fig. 8.6.**59**. JUSTICIA STRIATA. 1, habit; 2, branch with many-flowered nodes; 3, bract; 4, calyx whole and opened up; 5, detail of calyx lobe; 6, corolla opened up; 7, apical part of filament and anther; 8, apical part of style and stigma; 9, capsule; 10, indehiscent fruit; 11, seed. 1 & 9–10 from *Drummond & Hemsley* 3829, 2, 4–6 & 8 from *Bidgood et al.* 703, 3 from *Faulkner* 3995, 7 from *Tweedie* 1614, 11 from *Milne-Redhead & Taylor* 7281. Drawn by Margaret Tebbs. Reproduced from Flora of Tropical East Africa (2010).

C: Chegutu Dist., Poole Farm, Mupfure (Umfuli) R., 1225 m, fl. 12.iii.1950, *Hornby* 3123 (K, SRGH). E: Chipinge Dist., Chirinda, 1175 m, fl.& fr. 22.iv.1950, *Whellan* 419 (K, SRGH). S: Mberengwa Dist., Mt Buchwa, 1000 m, fl. 27.iv.1973, *Pope* 941 (K). **Malawi**. N: Nkata Bay Dist., above Chikale Beach, 600 m, fl. 27.iv.1970, *Pawek* 3457 (K). C: Dedza Dist., S of Dedza, Chincherare Hill, 1675 m, fl.& fr. 24.iv.1971, *Pawek* 4705 (K, MAL). S: Zomba Dist., Nselema village, fl.& fr. 23.iv.1955, *Banda* 71 (BM, K). **Mozambique**. N: Ribáuè Dist., 60 km W of Nampula, fl.& fr. 23.v.1961, *Leach & Rutherford-Smith* 10977 (K, LISC, SRGH). Z: Gurué Dist., Namuli Peaks, near Gurué (Vila Junqueiro), 975 m, fl. 25.vii.1962, *Leach & Schelpe* 11455 (K, LISC, SRGH). T: Macanga Dist., Mt. Furancungo, 1250–1450 m, fl. 17.iii.1966, *Pereira et al.* 1863 (LMU). MS: Nhamatanda Dist., between Gondola and Nhamatanda (Vila Machado), fl. 22.iii.1960, *Wild & Leach* 5217 (K, SRGH). GI: Vilankulo Dist., Benguerua Is., fl. 8.xi.1958, *Mogg* 28889 (K).

Also in Ethiopia, Somalia, Kenya, Tanzania, D.R. Congo and Angola. Lowland to montane forest, usually on edges or in grassy glades, riverine forest, grassland, bushland and woodland (including *Acacia* and *Brachystegia*), rocky hills, roadsides, often becoming weedy and surviving in disturbed vegetation; 20–1900 m.

Conservation notes: Widespread; not threatened.

Hedrén (1990) divides the Eastern and Southern African material of *Justicia striata* into two subspecies, and one of these into two varieties. I have decided not to maintain these taxa here. There are numerous intermediates between the more or less glabrous small-flowered lowland form (subsp. *striata* var. *striata*) and the more hairy large-flowered form (subsp. *striata* var. *dyschoristeoides*). Some specimens of the glabrous to hairy large-flowered highland form (subsp. *austromontana*) do look very distinct with their large purple corollas, but closer study shows a total gradation towards the coastal forms.

The type of *Justicia forbesii* is said to be from Madagascar but, as explained by Hedrén (1990), Forbes also collected in Mozambique. As no species from this section have ever been collected from Madagascar since, it seems reasonable to assume that the type was collected in Mozambique and wrongly labelled. The specimen is typical of coastal *J. striata*.

Subsp. *occidentalis* J.K. Morton from Ghana to Central African Republic and W Congo is not considered here; it may or may not be worthy of subspecific rank.

40. **Justicia phyllostachys** C.B. Clarke in F.T.A. 5: 188 (1900). —Hedrén in Symb. Bot. Ups. **29**(1): 115 (1989); in Nordic J. Bot. **10**: 396, fig.39 (1990). —Burrows & Willis, Pl. Nyika Plateau: 51 (2005). —Vollesen in F.T.E.A., Acanthaceae 2: 586 (2010). Type: Malawi, Nyika Plateau, vi-vii.1896, *Whyte* 118 (K lectotype, Z), lectotypified by Hedrén (1990).

Justicia umbratilis S. Moore in J. Bot. **51**: 216 (1913). Type: D.R. Congo/Zambia, Lake Mweru, 18.v.1908, *Kässner* 2804 (BM holotype).

Justicia whytei sensu Clarke in F.T.A. 5: 198 (1900) for *Whyte* s.n., Masuku Plateau, non S. Moore.

Justicia melampyrum sensu Clarke in F.T.A. 5: 199 (1900) for *Johnson* s.n. & *Whyte* s.n., Masuku Plateau, non S. Moore.

Justicia striata sensu Morton in Kew Bull. **32**: 443 (1978) in part, non (Klotzsch) Bullock.

Perennial herb, often pyrophytic, with one to several erect, ascending, decumbent or scrambling usually unbranched or little-branched stems from a woody rootstock, basal part of stems sometimes subshrubby; stems to 75(100) cm long, when young subglabrous to sericeous, puberulous or pubescent with hairs to 1.5 mm long. Leaves subglabrous to sparsely puberulous or sparsely pubescent, densest on midrib and margins, without stalked capitate glands; petiole to 0.5(1.3) cm long; lamina lanceolate to ovate (rarely broadly so), usually lanceolate towards stem tips and in 'inflorescence', largest 1.2–6.5 × 0.6–2.5 cm; apex subacuminate to rounded, base shortly attenuate to truncate. Flowers 2–6 in axils of upper leaves, sometimes single, upper internodes usually contracted thus creating a pseudo-racemose inflorescence; bracts lanceolate to ovate or elliptic, to 1.3 cm long, held erect (usually covering flowers), often ciliate with long (to 2 mm) pilose hairs, without stalked capitate glands. Calyx ciliate or sparsely so on edges and midrib, more rarely subglabrous, hairs to 1 mm; lobes 4–6 mm long. Corolla from pale pink or pale mauve to dark purple, very rarely white, lower lip with distinct 'herring-bone' pattern, upper lip with dark veins, 8–14 mm long; tube 4–7 mm long, widening immediately above base; same length as upper lip or 2 mm longer; upper lip 4–7 mm long, teeth to 1 mm long; lower lip 6–10 × 6–12 mm. Filaments 2–5 mm long; thecae yellow without brown pigment patches, 0.5–1 mm long, lower with acute appendage 0.3–0.5 mm long. Capsule 6.5–9(10) mm long, acute, glabrous to sparsely puberulous in upper half; indehiscent fruits absent. Seeds 1–1.5 mm long.

Zambia. N: Kasama Dist., Misamfu (Misamfwa), fl.& fr. 10.ix.1958, *Fanshawe* 4773 (K). W: Mufulira, 1225 m, fl.& fr. 7.xi.1948, *Cruse* 423 (K). C: Lusaka, Stuart Park, 1225 m, fl. 1.x.1961, *Lusaka Nat. Hist. Club* 58 (K). E: Chama Dist., Nyika Nat. Park, Zambian Govt. Rest House, fl.& fr. 17.iv.1986, *Philcox et al.* 9979 (K). **Zimbabwe**. N: Hurungwe Dist., Mwami (Miami), 1375 m, fl. 29.xi.1948, *Hopkins* GH 24516 (K, SRGH). C: Harare Dist., Warren Hill, fl. 29.x.1967, *Biegel* 2303 (K, SRGH). **Malawi**. N: Rumphi Dist., Nyika Nat. Park, Makunguru Hills, 15 km W of Chelinda, 2100 m, fl. 21.i.1992, *Goyder et al.* 3536 (K). **Mozambique**. N: Lago Dist.?, "east of Lake Nyasa", fl. 1884, *Johnson* s.n. (K).

Also in Burundi, D.R. Congo, Tanzania and Angola. Montane grassland and bushland (including secondary types), forest margins and glades, higher altitude wetter *Brachystegia* and *Uapaca* woodland, weed in old fields, persisting in plantations, roadsides, rarely in seasonally wet valley grassland; 1200–2300 m.

Conservation notes: Widespread; not threatened.

41. **Justicia lithospermoides** Lindau in Fries, Wiss. Ergebn. Schwed. Rhod.-Kongo-Exped. **1**: 308 (1916). —Hedrén in Symb. Bot. Ups. **29**(1): 114 (1989); in Bot. J. Linn. Soc. **103**: 272, fig.7 (1990). —Vollesen in F.T.E.A., Acanthaceae 2: 588 (2010). Type: Zambia, Mansa Dist., Mansa (Fort Roseberry), 14.ix.1911, *R.E. Fries* 532b (UPS lectotype), lectotypified by Hedrén (1990).

Justicia anselliana sensu Clarke in F.T.A. **5**: 209 (1900) for *Whyte* s.n., Chitipa (Fort Hill), non (Nees) T. Anderson.

Peristrophe usta C.B. Clarke in F.T.A. **5**: 244 (1900) for *Carson* s.n., non sensu stricto.

Perennial herb with several erect to ascending or decumbent (rarely rooting), unbranched or sparsely branched stems from a woody rootstock; stems to 50 cm long, when young subglabrous (hairy below nodes) to puberulous, rarely pubescent or densely so. Leaves (sometimes only partly developed when flowering) glabrous to pubescent; petiole 0–1.5 mm long; lamina linear to ovate, elliptic or slightly obovate, largest 0.7–4 × 0.1–0.6 cm; apex subacute to broadly rounded; base attenuate. Flowers single in axils of upper leaves, sometimes only one per node; bracts absent; bracteoles 2–7 mm long; pedicels 0.5–1.5 mm long. Calyx puberulous or sparsely so, rarely also with stalked capitate glands; lobes 4–7(9.5 in fruit) mm long. Corolla bluish or reddish purple, mouth almost closed with stamens hidden under upper lip, lower lip with distinct 'herring-bone' pattern, upper lip with dark veins, 8–10 mm long; tube 5–6 mm long, widening immediately above base, 1–3 mm longer than lip; upper lip 3–4 mm long, teeth c.0.5 mm long; lower lip 5–8 × 5–8 mm. Filaments 2–4 mm long; thecae yellow or tinged purple, c.1 mm long, lower with appendage c.0.5 mm long, broadly ellipsoid in outline, bent 90° relative to theca, obtuse. Capsule 5.5–7.5 mm long, glabrous or sparsely puberulous near apex. Seeds c.1.5 mm long.

Zambia. B: Zambezi Dist., Zambezi (Balovale), fl. 29.xi.1964, *Fanshawe* 8909 (K). N: Samfya Dist., 5 km W of Samfya, Kasamba dambo, 1300 m, fl.& fr. 21.iv.1989, *Goyder et al.* 3107 (K). W: Kitwe, fl. 26.x.1955, *Fanshawe* 2555 (K). C: Serenje Dist., fl.& fr. 12.viii.1965, *Fanshawe* 9245 (K). **Malawi**. N: Chitipa Dist., Chitipa (Fort Hill), 1200 m, fl.& fr. vii.1896, *Whyte* s.n. (K). C: Dedza Dist., Mt Dedza, fl., 23.x.1956, *Banda* 308 (BM, K, SRGH). **Mozambique**. N: Mandimba Dist., Mandimba, fl.& fr. 25.xi.1941, *Hornby* 3500 (K).

Also in Burundi, D.R. Congo and Tanzania. Appearing shortly after burning in short-grass dambos and valley grassland, on sandy-peaty to clay soils; 700–1700 m.

Conservation notes: Widespread; not threatened.

42. **Justicia elegantula** S. Moore in J. Bot. **38**: 204 (1900). —Clarke in F.T.A. **5**: 513 (1900). —Plowes, Wild Fl. Rhodesia: pl.169 (1976). —Immelman in Bothalia **16**: 41 (1986). —Hedrén in Symb. Bot. Ups. **29**(1): 114 (1989); in Bot. J. Linn. Soc. **103**: 271, fig.6 (1990). Type: Zimbabwe, Harare, ix.1898, *Rand* 508 (BM holotype).

Justicia elegantula var. *elatior* S. Moore in J. Bot. **38**: 204 (1900). —Clarke in F.T.A. **5**: 513 (1900). —Phiri, Checklist Zamb. Vasc. Pl.: 18 (2005). Type: Zimbabwe, Harare, ix.1898, *Rand* 642 (BM holotype).

Justicia elegantula var. *repens* S. Moore in J. Bot. **38**: 204 (1900). —Clarke in F.T.A. **5**: 513 (1900). Type: Zimbabwe, Bulawayo, xii.1897, *Rand* 179 (BM holotype).

Justicia brycei C.B. Clarke in Fl. Cap. **5**: 67 (1901). —Guillarmod, Fl. Lesotho: 217 (1971). —Immelman in Bothalia **16**: 41 (1986). Type: Lesotho, "Basoutoland, Mt. Machache", v.1896, *Bryce* s.n. (K holotype), see note.

Perennial herb with several erect to decumbent or creeping (and then sometimes rooting), unbranched or sparsely branched stems from a woody rootstock; stems to 50 cm long, when young subglabrous (hairy below nodes) to densely pubescent; young shoots covered with small succulent orbicular or reniform leaves, pale yellow and tinged purple. Radical leaves (sometimes only partly developed when flowering) glabrous to sparsely pubescent; petiole 0–3 mm long; lamina linear-lanceolate to broadly ovate, broadly elliptic or obovate, largest 1–4.5(6) × 0.15–2.5 cm; apex obtuse to broadly rounded; base attenuate. Flowers 2–4 in axils of upper leaves, sometimes only one per node; bracts oblanceolate to obovate, to 1.2(1.7) cm long; bracteoles 2–5 mm long; peduncle 1–5(6) cm long; pedicels 1–2(4) mm long. Calyx puberulous or sparsely so, rarely subglabrous; lobes 6–9(12 in fruit) mm long. Corolla bluish or reddish purple, mouth almost closed with stamens hidden under upper lip, lower lip with distinct 'herring-bone' pattern, upper lip with dark veins, 11–16 mm long; tube 5–8 mm long, widening immediately above base, 0–2(3) mm longer than lip, rarely lip to 1 mm longer than tube; upper lip 5–8 mm long, teeth c.0.5 mm long; lower lip 5–11 × 5–13 mm. Filaments 3–5 mm long; thecae yellow, c.1 mm long, lower with appendage c.0.4 mm long, broadly ellipsoid in outline, bent 90° relative to theca, obtuse. Capsule 6–7.5 mm long, glabrous or sparsely puberulous. Seeds c.1.5 mm long.

Zambia. B: Sesheke Dist., 14 km NW of Masese, fl. 11.viii.1947, *Brenan & Keay* 7676 (EA, K). N: Luwingu Dist., Luwingu, fl.& fr. 15.v.1958, *Angus* 1963 (K). W: Luanshya Dist., fl. 27.vii.1954, *Fanshawe* 1400 (K). C: Mkushi Dist., 20 km on Kapiri Mposhi–Serenje road, fl.& fr. 28.iii.1984, *Brummitt et al.* 16931 (K). E: Chipata Dist., 14 km from Chipata (Fort Jameson) on Malawi road, 1075 m, fl.& fr. 25.ix.1960, *Wright* 276 (K). S: Kalomo Dist., Mwezi and Machile rivers, 975 m, fl. 19.viii.1963, *Bainbridge* 892 (K). **Zimbabwe**. N: Makonde Dist., Silverside Mine, fl. 4.xi.1965, *Wild* 7465 (K, SRGH). W: Matobo Dist., Matopo Hills, 1400 m, fl.& fr. xi.1902, *Eyles* 1101 (BM). C: Harare Dist., near Harare, fl., 2.ix.1956, *Angus* 1396 (K). E: Nyanga Dist., fl. 8.xi.1930, *T.C.E. Fries et al.* 2784 (K). **Malawi**. C: Kasungu Dist., Kasungu Nat. Park, 925 m, 8.ix.1972, *Pawek* 5701 (K).

Also in D.R. Congo. There is a dry season after-burn form in short-grass dambos and woodland with a rainy season form in wetter grassland; 900–1700 m.

Conservation notes: Widespread; not threatened.

The small fleshy leaves covering the young shoots give this species a very characteristic appearance. The dry season form with its short erect shoots is superficially very different from the wet season form with long creeping stems, but the former often has the long, partly burnt stems of the previous season attached.

Guillarmod (1971), Immelman (1986) and Hedrén (1990) discuss the provenance of the type of *Justicia brycei* and conclude that it almost certainly came from Macheke in Zimbabwe and not from Lesotho.

Hedrén (1989 & 1990) cites this species as occurring in Mozambique, but he doesn't cite any specimens nor does his distribution map show it as occurring there.

43. **Justicia mollugo** C.B. Clarke in F.T.A. **5**: 200 (1900). —Hedrén in Bull. Jard. Bot. Belg. **58**: 140, fig.5 (1988); in Distrib. Pl. Afr. **35**: map 1166 (1988); in Symb. Bot. Ups. **29**(1): 117 (1989). —Burrows & Willis, Pl. Nyika Plateau: 51 (2005). — Vollesen in F.T.E.A., Acanthaceae **2**: 589 (2010). Type: Malawi, no locality, 1891, *Buchanan* s.n. (K lectotype, BM, G, US), lectotypified by Hedrén (1988).

Justicia leptocarpa sensu Lindau in Bot. Jahrb. Syst. **20**: 70 (1894) for *Buchanan* s.n., non sensu stricto.

Justicia heterocarpa sensu Clarke in F.T.A. **5**: 200 (1900) for *Nutt* s.n., non T. Anderson.

Single-stemmed annual herb, stem erect (rarely decumbent), unbranched or sparsely branched from near base, side branches at angle of less than 45°; to 20(30) cm long but usually less than 10 cm, when young bifariously sericeous-puberulous or sparsely so. Leaves sparsely puberulous on midrib and margins, often with stalked glands; petiole to 5 mm long; lamina ovate to elliptic or narrowly so, sometimes lanceolate towards apex, largest 0.4–2.3 × 0.15–1.4 cm, usually (1.5)2–3 times as long as wide; apex acute to obtuse; base attenuate to cuneate. Flowers 2–4 in axils of upper leaves or some (rarely all) single; bracts lanceolate to narrowly elliptic, to 8 mm long, sparsely puberulous, not pilose-ciliate or with scattered ciliae to 0.5 mm long; bract fused to pedicel looking as if flowers attached at apex of petiole; peduncles 1–2(3) mm long. Calyx sparsely puberulous-ciliate; lobes yellowish green to green, 2–4 mm long. Corolla white to mauve, lower lip with indistinct to distinct 'herring-bone' pattern with raised ribs but without crenulate edges, upper lip with faintly darker veins, 3–5 mm long; tube white, 2–3 mm long, widening immediately above base, c.1 mm longer than lip; upper lip 1–2 mm long, teeth minute; lower lip 1.5–3 × 2–3.5 mm. Filaments 1–2 mm long; anthers exserted, thecae with a purplish band on sides, 0.25–0.5 mm long, lower with acute appendage c.0.2 mm long. Capsule 3–4.5(5) mm long, acute, sparsely puberulous near apex, more rarely glabrous. Seeds c.1 mm long.

Zambia. N: Mbala Dist., Kawimbe, 1675 m, fl.& fr. 23.iii.1957, *Richards* 8846 (K). W: Mwinilunga Dist., Matonchi Farm, fl.& fr. 22.i.1938, *Milne-Redhead* 4284 (BM, K). C: Serenje Dist., Kundalila Falls, 1400 m, fl.& fr. 13.iii.1975, *Hooper & Townsend* 713 (K). **Malawi**. N: Chitipa Dist., 90 km SE of Chisenga, 1550 m, fl.& fr. 21.iv.1976, *Pawek* 11184 (K, MAL, MO). C: Ntchisi Dist., Ntchisi Forest Reserve, 1400–1600 m, fl.& fr. 25.iii.1970, *Brummitt* 9373 (K, MAL). S: Mulanje Dist., Mt Mchese Forest Reserve, above Ulolo, 1000 m, fl.& fr. 14.ii.1992, *Goyder & Paton* 3647 (K). **Mozambique**. N: Lichinga Dist., Lichinga (Vila Cabral), Massangulo ridge, *Torre & Paiva* 10751 (LISC, cited in Hedrén).

Also in Burundi, D.R. Congo, Tanzania and Angola. *Brachystegia* woodland and grassland, rocky hills, also in disturbed woodland, roadsides, pine plantations, sometimes a weed; (650) 1000–2400 m.

Conservation notes: Widespread; not threatened.

This species must be a strong contender for the "smallest Acanthaceae in the World"; I have seen fully-developed flowering and fruiting specimens smaller than a matchstick.

44. **Justicia syncollotheca** Milne-Redh. in Bull. Misc. Inform., Kew **1937**: 429 (1937). —Hedrén in Bull. Jard. Bot. Belg. **58**: 148, fig.9 (1988); in Distrib. Pl. Afr. **35**: map 1168 (1988); in Symb. Bot. Ups. **29**(1): 117 (1989). Type: Zambia, Solwezi, 11.vi.1930, *Milne-Redhead* 489 (K holotype).

Annual herb, single-stemmed but much-branched from base with lateral branches spreading ± at right angles to main stem; main stem erect, lateral branches decumbent (sometimes rooting at lower nodes) to ascending, stems to 70 cm long, when young uniformly puberulous or sericeous-puberulous or sparsely so, sometimes with stalked glands. Leaves often almost absent when flowering, sparsely pubescent on midrib and margins, densest towards base, hairs to 1 mm long; petiole to 3(8) mm long; lamina ovate to elliptic or narrowly so, narrower towards apex, largest 1.3–3.2(4) × 0.4–1.3 cm, usually 2–3 times as long as wide; apex acute to obtuse; base attenuate. Flowers 2(4) or some single in axils of upper leaves, upper internodes strongly contracted forming a 'pseudo-raceme'; bracts oblanceolate to narrowly obovate, to 6 mm long, not pilose-ciliate, not fused to pedicel. Calyx uniformly puberulous and with stalked glands (rarely without); lobes 3.5–5(6 in fruit) mm long. Corolla mauve or pink to purple, lower lip with very distinct dark purple and white 'herring-bone' pattern with raised ribs but without crenulate edges, upper lip with darker veins, 4–6.5(9) mm long; tube white, 2.5–3.5(6) mm long, widening immediately above base, c.1(3) mm longer than lip; upper lip 1.5–3 mm long, teeth c.0.25 mm long; lower lip 2–3.5(5) × 2.5–4 mm. Filaments 1–2 mm long; thecae yellow, 0.5 mm long, lower with linear straight acute appendage 0.5 mm long. Capsule 3–4.5 mm long, acute, puberulous or sparsely so all over or in apical half only. Seeds 1–1.5 mm long.

Zambia. N: Mbala Dist., Mbala, 1600 m, fl.& fr. 2.vi.1967, *Richards* 22296 (K). W: Solwezi Dist., Solwezi, fl.& fr. 11.vi.1930, *Milne-Redhead* 489 (K). C: Lusaka, Stuart Park, 1225 m, fl.& fr. 2.vii.1961, *Lusaka Nat. Hist. Club* 47 (K). **Malawi**. N: Chitipa Dist., Chisenga, 1550 m, fl.& fr. 12.vii.1970, *Brummitt* 12025 (EA, K, MAL).

Also in D.R. Congo and Tanzania. *Brachystegia* woodland, grassland (sometimes secondary) and lawns; 1200–1600 m.

Conservation notes: Widespread; not threatened.

A rarely collected species with a fairly wide distribution. It flowers in the dry season while the related *Justicia mollugo* always flowers in the rainy season. It seems to have adapted to growing in fairly closely cut lawns; about half of known collections are from such places.

45. **Justicia obtusicapsula** Hedrén in Bull. Jard. Bot. Belg. **58**: 151, fig.10 (1988); in Distrib. Pl. Afr. **35**: map 1169 (1988); in Symb. Bot. Ups. **29**(1): 118 (1989). — Vollesen in F.T.E.A., Acanthaceae **2**: 590 (2010). Type: Zambia, Kasama Dist., Mungwi to Kasama, 30.iv.1962, *Richards* 16419 (K holotype, LISC).

Erect single-stemmed annual herb, much-branched from about 10 cm above ground level, lateral branches spreading at 45–60° to main stem; stems to 60 cm long, when young uniformly sericeous-puberulous. Leaves finely and sparsely pubescent, densest on midrib and margins towards base; petiole to 3 mm long; lamina lanceolate to narrowly elliptic or narrowly obovate, often linear towards apex, largest 3–5.5 × 0.3–1.1 cm, 4–10 times as long as wide; apex subacute to obtuse; base attenuate. Flowers 2–4 in axils of upper leaves, upper internodes not contracted; bracts lanceolate to narrowly elliptic or oblanceolate, to 4(8) mm long, not pilose-ciliate, not fused to pedicel. Calyx finely uniformly puberulous, sometimes with subsessile capitate glands; lobes 2–3 mm long. Corolla pale mauve or bright purple, lower lip with indistinct 'herring-bone' pattern with raised ribs but without crenulate edges, upper lip with darker veins, 4–5 mm long; tube white, 2.5–3 mm long, widening immediately above base, c.1 mm longer than lip; upper lip 1.5–2 mm long, teeth 0.25 mm long; lower lip 2–3 × 2.5–4 mm. Filaments 1–2 mm long; thecae yellow, 0.5 mm long, lower with linear straight acute appendage 0.3 mm long. Capsule 3–4 mm long, obtuse, puberulous all over. Seeds c.1 mm long.

Zambia. N: Mbala Dist., Chilongwelo, 1225 m, fl.& fr. 12.iv.1952, *Richards* 1385 (K); Mbala Dist., Kalala Village, 1500 m, fl.& fr. 25.iv.1968, *Richards* 23224 (K).

Also in Tanzania. *Brachystegia* woodland on sandy to loamy soil, roadsides; 1200–1600 m.

Conservation notes: Restricted distribution but probably not threatened.

46. **Justicia richardsiae** Hedrén in Bull. Jard. Bot. Belg. **58**: 133, fig.2 (1988); in Distrib. Pl. Afr. **35**: map 1163 (1988); in Symb. Bot. Ups. **29**(1): 116 (1989). Type: Zambia, Mbala Dist., Ndundu, 10.iii.1955, *Richards* 4870 (K holotype).

Single-stemmed annual herb, basal part of stem often creeping and rooting, apical part erect to scrambling; stem unbranched or sparsely branched, to 40 cm long, sparsely to densely puberulous when young, hairs to 1 mm long, sometimes with stalked glands. Leaves sparsely puberulous or pubescent on midrib and edges, densest towards base, hairs to 1 mm long; petiole to 1 cm long; lamina ovate or broadly so (narrowly so towards apex of stems), largest 1.2–3 × 0.7–1.8 cm, less than twice as long as wide; apex obtuse; base attenuate. Flowers single or 2 (rarely 4) in axils of upper leaves, upper internodes not contracted; bracts lanceolate to narrowly ovate, to 7 mm long, pilose-ciliate with broad glossy hairs to 1.5 mm long, sometimes with stalked capitate glands, not fused to pedicel. Calyx puberulous-ciliate or sparsely so with broad glossy hairs to 1 mm long; lobes uniformly green, 1.5–2.5(3.5 in fruit) mm long. Corolla a rich mauve or purple, lower lip with very distinct dark purple and white 'herring-bone' pattern with strongly raised crenulate edges, upper lip with dark veins and darker towards tip, 7–9 mm long; tube white, 4–5 mm long, widening immediately above base, 1–2 mm longer than lip; upper lip 3–4 mm long, teeth 0.5 mm long; lower lip 5–7.5 × 6–8.5 mm. Filaments c.3 mm long; anthers hidden in throat, thecae yellow, 0.75–1 mm long, lower with linear appendage 0.4 mm long, bent at 90° to theca. Capsule 5–6.5 mm long, acute, glabrous or with a few hairs near apex. Seeds 1–1.5 mm long.

Zambia. N: Mbala Dist., Isanya road, 1525 m, fl.& fr, 12.i.1952, *Richards* 446 (K); Ndundu, 1750 m, fl. 6.iii.1965, *Richards* 19737 (K).

Also in D.R. Congo. In tall grass in swamps and on riverbanks; 1400–1800 m.

Conservation notes: Restricted distribution and habitat somewhat threatened; possibly Near Threatened.

47. **Justicia mariae** Hedrén in Bull. Jard. Bot. Belg. **58**: 135, fig.3 (1988); in Distrib. Pl. Afr. **35**: map 1164 (1988); in Symb. Bot. Ups. **29**(1): 117 (1989). —Vollesen in F.T.E.A., Acanthaceae 2: 591 (2010). Type: Zambia, Kaputa Dist., Mweru Wantipa, road to Bulaya, 12.iv.1957, *Richards* 9173 (K holotype).

Single-stemmed annual herb, basal part of stem creeping and rooting, apical part erect; stem unbranched or sparsely branched, to 35 cm long, puberulous or sparsely so when young. Leaves subglabrous to puberulous on midrib and edges, densest towards base, hairs to 1 mm long; petiole 0–2 mm long; lamina lanceolate to narrowly elliptic, largest 1.5–3.5 × 0.3–0.8 cm, more than 3 times as long as wide; apex subacute to obtuse; base attenuate. Flowers single or 2 in axils of upper leaves, upper internodes contracted; bracts lanceolate, to 1 cm long, pilose-ciliate with broad glossy hairs to 1.5 mm long, sometimes with stalked capitate glands. Calyx apically pubescent with broad glossy hairs to 1 mm long, glabrous towards base; lobes dark green towards apex, yellowish towards base, 3.5–5(6 in fruit) mm long. Corolla a rich bluish purple, lower lip with distinct 'herring-bone' pattern with raised crenulate edges, upper lip with dark veins, 8–9 mm long; tube white, 4–4.5 mm long, widening immediately above base, tube and lip same length; upper lip 4–4.5 mm long, teeth 0.5 mm long; lower lip 4–5 × 5–6 mm. Filaments c.3 mm long; thecae yellow, 0.5 mm long, lower with appendage 0.3 mm long, broadly ellipsoid in outline, straight, obtuse. Capsule c.5.5 mm long, obtuse, sparsely puberulous near apex. Seeds c.1 mm long.

Zambia. N: Mbala Dist., Chinakila, 1375 m, fl. 11.iv.1956, *Vesey-Fitzgerald* 1039 (K); Kaputa Dist., Mweru Wantipa, road to Bulaya, 1050 m, fl.& fr. 12.iv.1957, *Richards* 9173 (K).

Also in Tanzania. Swampy grassland; 1000–1400 m.

Conservation notes: Restricted distribution and habitat somewhat threatened; possibly Near Threatened.

This species has a broadly ellipsoid obtuse anther appendage like *Justicia elegantula* and *J. lithospermoides*, but here the appendage is straight and not bent 90° relative to the theca.

Sect. **Monechma** (Hochst.) T. Anderson in J. Linn. Soc. Bot. **7**: 43 (1863). —Hedrén in Nordic J. Bot. **10**: 151 (1990).

Flowers single or in 3(5)-flowered cymules, widely spaced or aggregated into terminal racemoid cymes, or also with cymes from upper axils; bracts foliaceous or differentiated upwards; bracteoles small or similar to bracts. Calyx with 5 equal lobes (rarely one lobe reduced to a small tooth). Lower anther theca with linear entire or bifurcate appendage. Capsules 2(4)-seeded. Seeds smooth, kidney-shaped, glossy, glabrous (rarely hairy). Species 48–64.

48. **Justicia ciliata** Jacq., Hort. Bot. Vindob. **2**: 47, t.104 (1772). —Vollesen in F.T.E.A., Acanthaceae **2**: 592 (2010). Type: Illustration t.104 in Hort. Bot. Vindob. **2** (1772), from plant cultivated in Vienna. See note.

Justicia ciliaris L.f., Suppl. Pl.: 84 (1781). Type as for *J. ciliata*.
Monechma hispidum Hochst. in Flora **24**: 375 (1841). —Clarke in F.T.A. **5**: 213 (1900). — Binns, Checklist Herb. Fl. Malawi: 15 (1968). Type: Sudan, Kordofan, Mt Kohn, 6.xi.1839, *Kotschy* 239 (TUB holotype, HBG, K, M, S, US, WAG).
Monechma ciliatum (Jacq.) Milne-Redh. in Bull. Misc. Inform., Kew **1934**: 304 (1934); in Kew Bull. **5**: 381 (1951).

Erect to decumbent often much-branched annual herb; stems to 40 cm long, sparsely to densely pubescent or pilose. Leaves sparsely to densely pubescent or pilose on veins and margins; petiole to 8 mm long; lamina linear-lanceolate to narrowly elliptic, largest 5–9 × 0.5–1 cm, at least 5 times longer than wide; apex subacute to rounded; base cuneate to attenuate, not decurrent. Flowers solitary in axils, upwards usually congested into terminal pseudo-racemes, sometimes with short lateral branches; bracts foliaceous, as leaves or slightly smaller, distinctly white-ciliate in basal part with ciliae (1.5)2–4 mm long; bracteoles linear or linear-lanceolate, (1)1.2–2 cm long, with similar ciliae. Calyx 8–13 mm long, with similar ciliae; lobes linear, acuminate, with a single broad central vein and with broad hyaline white margins. Corolla lower lip white with purple to mauve lines, upper lip white to pale green with brownish veins, 7–8 mm long, shorter than calyx, sparsely sericeous; tube 3–4 mm long with a conspicuous central pouch and 2 smaller lateral pouches; upper lip c.4 mm long, hooded; lower lip c.5 mm long, broadly obovate, lobes c.2 mm long. Filaments c.2.5 mm long; anthers hairy, thecae superposed, upper smaller and held at angle to lower, c.0.5 (upper) and 1 (lower) mm long, lower with bifurcate appendage c.1 mm long. Capsule 9–11 mm long, glabrous. Seeds black, 3–3.5 × 4.5–5 mm, with two large tufts of thick moniliform hairs.

Zambia. S: Kalomo Dist., fl. 16.ii.1965, *Fanshawe* 9114 (K). **Malawi**. N: Karonga Dist., 30 km on Karonga–Stevenson road, 750 m, fl, 16.iv.1976, *Pawek* 11075 (K, MAL, MO).

Also in Gambia, Senegal, Guinea, Sierra Leone, Mali, Burkina Faso, Ivory Coast, Ghana, Togo, Niger, Nigeria, Cameroon, Chad, Central African Republic, D.R. Congo, Sudan, Ethiopia, Rwanda, Burundi, Uganda and Tanzania. Grassland on black clay soils; 750–1200 m.

Conservation notes: Widespread, especially outside the Flora area; not threatened.

In Linnaeus' Suppl. Pl.: 84 (1781) the type is said to come from Sri Lanka (Ceylon), but there is nothing to indicate that this species occurs outside Africa. Jacquin says its origin is unknown and as his plate is a perfect match for the species it can be used as the type. The seed from which the plant illustrated was grown almost certainly came from West Africa.

49. **Justicia bracteata** (Hochst.) Zarb, Cat. Spec. Bot. Pfund: 32 (1879). —Vollesen in F.T.E.A., Acanthaceae **2**: 593 (2010). Type: Sudan, Kordofan, Tejara, 19.xi.1839, *Kotschy* 261 (TUB holotype, GZU, HBG, K, M, MPU, S, STU, US). FIGURE 8.6.**60**.

Monechma bracteatum Hochst. in Flora **24**: 375 (1841). —Nees in De Candolle, Prodr. **11**: 411 (1847). —Clarke in F.T.A. **5**: 214 (1900); in Fl. Cap. **5**: 68 (1901).
Monechma angustifolium Nees in De Candolle, Prodr. **11**: 412 (1847). Type: South Africa, KwaZulu-Natal, Durban, n.d., *Drège* s.n. (K holotype).

Justicia heterostegia T. Thoms. in Speke, Nile Append.: 643 (1863). Type: Tanzania, Bukoba Dist., Karagwe, ii.1862, *Speke & Grant* 433 (K holotype).
Justicia debilis (Forssk.) Vahl var. *angustifolia* (Nees) Oliv. in Trans. Linn. Soc. Lond. **29**: 129 (1873). —Clarke in F.T.A. **5**: 215 (1900). —Binns, Checklist Herb. Fl. Malawi: 15 (1968).
Monechma bracteatum var. *non-strobilifera* C.B. Clarke in F.T.A. **5**: 215 (1900). Type: Malawi, Kondowa to Karonga, vii.1896, *Whyte* s.n. (K lectotype); Eritrea, Habab, n.d., *Hildebrandt* 442 (B† syntype), lectotypified here.
Monechma bracteatum var. *angustifolium* (Nees) C.B. Clarke in F.T.A. **5**: 215 (1900). — Binns, Checklist Herb. Fl. Malawi: 15 (1968).
Monechma debile sensu Schinz in Mém. Herb. Boissier **10**: 64 (1900). —Binns, Checklist Herb. Fl. Malawi: 15 (1968). —Meyer in Merxmüller, Prodr. Fl. SW Afr. **130**: 45 (1968). — Setshogo, Checklist Pl. Botswana: 19 (2005) in part, non (Forssk.) Nees.

Erect (rarely ascending to decumbent) unbranched or little branched annual herb; stems to 1(1.5) m long, finely sericeous or densely so when young with downwardly directed appressed hairs or sericeous-puberulous with hairs bending down apically, no capitate glands. Leaves sparsely sericeous-puberulous along veins and edges, with or without sparse yellow subsessile glands; petiole to 2(4) cm long; lamina ovate to elliptic or narrowly so (rarely lanceolate or obovate), largest (3)4–12 × (0.8)1–5.5(6.5) cm, usually 2–4 times as long as wide; apex acute or subacute (rarely rounded); base attenuate, decurrent. Flowers solitary or paired, usually in numerous axillary racemoid cymes, long and many-flowered downwards, gradually shorter upwards, sometimes ± fusing to form an elongated terminal pseudo-raceme; main axis bracts foliaceous, raceme bracts imbricate, oblong-elliptic or broadly so (rarely obovate), lowermost pairs usually orbicular to reniform, 4–9(11) × 3–7 mm in middle of racemes, subacute to truncate with recurved tip, always with distinct 'shoulders', lower with rounded to subcordate base, finely puberulous and with few to many stalked yellow capitate glands, distinctly ciliate (rarely not) with broad glossy ciliae 1–2 mm long; bracteoles linear or acicular, 1–1.5 mm long. Calyx 4–7 mm long, finely sericeous-puberulous with a mixture of broad non-glandular hairs, short non-capitate glands and scattered stalked yellow capitate glands; lobes linear to filiform, cuspidate, with a single central vein. Corolla white to purple, with or without purple lines on lower lip, puberulous and with scattered capitate glands, 5.5–10 mm long; tube 3.5–6 mm long; upper lip 2–5 mm long, flat or slightly hooded; lower lip 3–6 mm long, broadly obovate, lobes 1–2 mm long. Filaments c.2 mm long; anthers hairy, thecae 0.5–0.75 mm long, lower with simple or bifurcate appendage 0.5 mm long. Capsule 4.5–7 mm long, densely sericeous-puberulous, hairs spreading near apex, deflexed near base. Seeds mottled grey and black, 2–2.5 × 2.5–3 mm, glabrous.

Botswana. N: Ngamiland Dist., Santsarr–Seronga road, 7 km E of Masoko Pan, fl. 18.iii.1973, *P.A. Smith* 456D (K, SRGH). SE: Kweneng Dist., 25 km NW of entry to Khutse Game Reserve via Salajwe, fl.& fr. 7.iii.1977, *Skarpe* 147 (K). **Zambia**. B: Kalabo Dist., Lukona, 1150 m, fl.& fr. 20.iii.1996, *Luwiika et al.* 375 (K, MO). N: Mbala Dist., Chilongwelo, 1425 m, fl. 4.v.1955, *Richards* 5514 (EA, K). W: Kitwe, fl. 3.iii.1965, *Mutimushi* 1195 (K). C: Lusaka Dist., 8 km E of Lusaka, fl. 21.iii.1957, *Noak* 172 (K, SRGH). E: Chipata Dist., Chizombo, 600 m, fl. 20.ii.1969, *Astle* 5511 (K). S: Choma Dist., Mapanza Mission, 1050 m, fl.& fr. 5.iv.1953, *Robinson* 157 (K). **Zimbabwe**. N: Hurungwe Dist., Mwami (Miami), 1375 m, fl. 5.iii.1947, *Wild* 1830 (K, SRGH). W: Hwange Dist., Matetsi Safari Area, 1000 m, fl.& fr. 6.iv.1978, *Gonde* 138 (K, SRGH). C: Gweru Dist., 30 km S of Kwekwe (Que Que), Mlezu School Farm, 1275 m, fl. iii.1966, *Biegel* 967 (K, SRGH). **Malawi**. N: Mzimba Dist., 19 km on Rumphi–Ekwendeni road, Emanyeleni, 1050 m, fl.& fr. 21.v.1970, *Brummitt* 10984 (K, MAL). C: Dedza Dist., Ntakataka, fl.& fr. 11.iv.1969, *Salubeni* 1320 (K, MAL, SRGH). S: Nsanje Dist., 5 km NW of Nsanje, Nyamadzere Rest House, 175 m, fl.& fr. 28.v.1970, *Brummitt* 11143 (K, MAL). **Mozambique**. N: Nampula Dist., 23 km on Nampula–Meconta road, fl. 6.v.1948, *Pedro & Pedrógão* 3191 (EA, K). MS: Chimoio Dist., 15 km N of Vanduzi, 600 m, fl.& fr. 27.iv.1962, *Chase* 7696 (EA, K, SRGH).

Also in India(?), Yemen, Oman, Sudan, Eritrea, Ethiopia, Somalia, D.R. Congo, Uganda, Kenya, Tanzania, Angola, Botswana, South Africa and Namibia. Woodland,

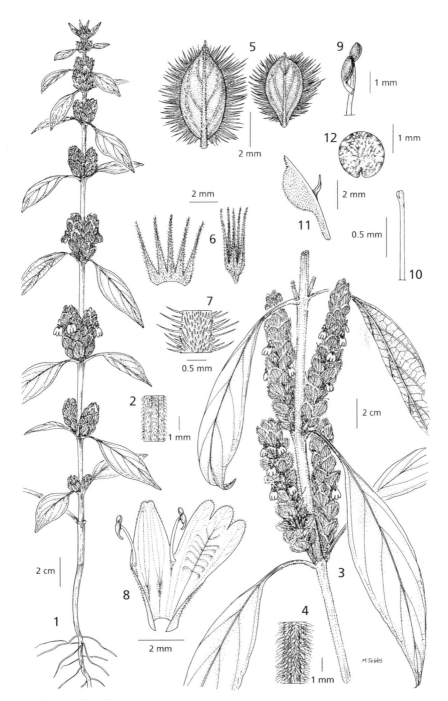

Fig. 8.6.**60**. JUSTICIA BRACTEATA. 1, habit; 2, detail of stem indumentum; 3, branch with many-flowered inflorescences; 4, detail of stem indumentum; 5, bract, lower and middle; 6, calyx whole and opened up; 7, detail of calyx lobe; 8, corolla opened up; 9, apical part of filament and anther; 10, apical part of style and stigma; 11, capsule; 12, seed. 1 & 2 from *Richards* 20425, 3 & 4 from *Bullock* 3923, 5–12 from *Greenway & Kanuri* 12083. Drawn by Margaret Tebbs. Reproduced from Flora of Tropical East Africa (2010).

bushland and grassland on a wide range of soils, commonly in old cultivation, on roadsides and as a weed, seems to thrive in areas with slight to moderate disturbance; 100–1450 m.

Conservation notes: Widespread; not threatened.

A few collections from Botswana (*Lambrecht* 76, *Richards* 14846) have upwardly directed hairs on the stems or a mixture of upwardly and downwardly directed hairs, but are otherwise completely typical.

50. **Justicia monechmoides** S. Moore in J. Bot. **18**: 311 (1880). Type: Angola, Luanda, Imbondeiro dos Lobos, n.d., *Welwitsch* 5140 (BM lectotype, K, LISU, P), lectotypified here.

   Monechma welwitschii C.B. Clarke in F.T.A. **5**: 216 (1900). —Binns, Checklist Herb. Fl. Malawi: 15 (1968), illegitimate name. Type: As for *Justicia monechmoides*.

   Monechma tettense C.B. Clarke in F.T.A. **5**: 216 (1900). —Binns, Checklist Herb. Fl. Malawi: 15 (1968). Type: Mozambique, Lower Zambesi, opposite Tete, Kaimba Is., 1860, *Kirk* s.n. (K lectotype), lectotypified here.

   Monechma monechmoides (S. Moore) Hutch., Botanist Sthn. Afr.: 524 (1946). —Binns, Checklist Herb. Fl. Malawi: 15 (1968).

   Monechma debile sensu Munday in F.S.A. **30**(3,1): 59 (1995) in part, non (Forssk.) Nees.

Erect to decumbent, unbranched to much-branched annual herb; stems to 1.25 m long, when young sparsely puberulous to tomentose or sericeous with spreading or downwardly bent hairs, with (rarely without) sparse to dense stalked capitate glands. Leaves sparsely sericeous-puberulous along veins and edges, sometimes with stalked glands towards base; petiole to 2(3) cm long; lamina narrowly ovate to ovate or narrowly elliptic (rarely lanceolate), largest 5–12(15) × 0.7–4(7) cm, (2)3–5 times as long as wide; apex acute to acuminate, base attenuate, decurrent. Flowers solitary or paired, in numerous axillary loose to dense (imbricate bracts) racemoid cymes, long and many-flowered downwards, gradually shorter upwards, sometimes fusing to form an elongated terminal pseudo-raceme; either all bracts supporting flowers or ± 1-sided; main axis bracts foliaceous, raceme bracts ovate or elliptic to reniform (lower), (4)5–10 × 2–6 mm in middle of racemes, acute to acuminate (rarely subacute), straight if acute, recurved and sometimes twisted if acuminate, with or usually without distinct 'shoulders', lower with truncate to subcordate base, sparsely to densely puberulous to pubescent, usually with dense stalked yellow capitate glands, often sticky, often with long (over 1 mm) broad glossy hairs on underside, distinctly ciliate with broad glossy ciliae to 1.5 mm long; bracteoles linear or acicular, 1–1.5 mm long. Calyx 3.5–7.5 mm long, puberulous with non-glandular hairs and usually numerous stalked capitate glands, also with (rarely without) long broad non-glandular hairs to 1 mm long; lobes linear to filiform, acuminate to cuspidate, 1-veined. Corolla white to mauve, with pink lines on lower lip, 8–10.5 mm long, sparsely sericeous-puberulous and with scattered capitate glands; tube (3.5)4.5–6.5 mm long; upper lip (2.5)3–5 mm long, flat or slightly hooded; lower lip 3–7 mm long, broadly obovate, lobes 1–2 mm long. Filaments 2–3 mm long; anthers with a few hairs, thecae 0.5–0.75 mm long, lower with acute appendage c.0.5 mm long. Capsule 5–7 mm long, densely sericeous-puberulous, hairs spreading near apex, deflexed near base. Seeds mottled grey and black, glabrous, c.2 × 3 mm.

Botswana. N: Ngamiland Dist., Maun, Thamalakane R., fl.& fr. 9.vi.1994, *Cole* 835 (K, PRE). SE: Central Dist., 60 km NW of Serowe, fl.& fr. 24.iii.1965, *Wild & Drummond* 7285 (K, SRGH). **Zambia**. N: Mpika Dist., North Luangwa Nat. Park, 600 m, fl. 22.iv.1994, *P.P. Smith* 522 (K). C: Kafue Dist., 15 km S of Lusaka, Mt Makulu Research Station, fl.& fr. 23.viii.1958, *Simwanda* 121 (K, SRGH). E: Petuake Dist., Minga, fl.& fr. 4.vi.1958, *Fanshawe* 4520 (K). S: Mazabuka Dist., Mazabuka, fl., 26.iii.1963, *van Rensburg* 1816 (K). **Zimbabwe**. N: Gokwe Dist., Tare R., fl.& fr. 10.iv.1962, *Bingham* 223 (K, SRGH). W: Hwange Dist., 80 km NE of Kamatiri, Sebungwe R., fl.& fr. 9.v.1955, *Plowes* 1794 (K, SRGH). C: Chegutu Dist., Hartley A Safari Area, 925 m, fl.& fr. 4.vii.1974, *Müller* 2164 (K, SRGH). E: Chimanimani Dist., Changadzi R., 600 m, fl. 15.iv.1963, *Chase* 7993 (K, SRGH). S: Beitbridge Dist., 40 km on Beitbridge–Chiturupadzi road, Nuli Hills, fl. 20.iii.1967, *Mavi* 259 (K, SRGH). **Malawi.** C: Salima Dist., Lifidzi Khola, fl. 18.v.1985, *Salubeni et al.* 4208 (K, MAL). S: Blantyre Dist., Shire R., opposite Matope Mission, 475 m, fl. 10.iv.1984, *Brummitt et al.* 17146 (K, MAL). **Mozambique.** T: Zumbu Dist., Msusa, Mucanha (Nkanga) R., 200 m, fl.& fr. 25.vii.1950, *Chase* 2838 (BM, EA, K, SRGH). MS: Guro Dist., 6 km on Mungári–Nhacolo (Tambara) road, 200 m, fl. 12.v.1971, *Torre & Correia* 18375 (BR, LISC, LMA, LMU, WAG). M: Maputo Dist., 4 km from Catembe, Tembe R., fl.& fr. 6.v.1981, *de Koning & Boane* 8682 (BM, K, LISC, LMU).

Also in Angola, Namibia and South Africa. Woodland, bushland and grassland, often in disturbed or secondary vegetation or roadsides, often weedy and favoured by moderate disturbance; (5)150–1200 m.

Conservation notes: Widespread; not threatened.

51. Justicia sp. B (= *Bester* 11112).

Erect branched annual herb; stems to 10 cm long, finely retrorsely puberulous when young, glabrescent, no capitate glands. Leaves with a few hairs on midrib and finely antrorsely ciliate; petiole to 2 mm long; lamina lanceolate, largest 5–6 × 0.7–1 cm, more than 5 times longer than wide; apex subacute to rounded, base attenuate, decurrent. Flowers paired, in terminal dense (imbricate bracts) racemoid cymes to 3 cm long; only one bract per node supporting flowers; raceme bracts broadly ovate to orbicular, lower tinged purple, c.8 × 6 mm in middle of racemes, subacute to rounded with a small apiculus, without distinct 'shoulders', lower with truncate base, finely and sparsely puberulous and with scattered capitate glands, finely ciliate with ciliae less than 0.5 mm long; bracteoles narrowly triangular, c.0.5 mm long. Calyx 5–6 mm long, puberulous and with capitate glands, without long broad non-glandular hairs; lobes filiform, cuspidate, 1-veined. Corolla white tinged with purple, with purple 'herring-bones' on lower lip, c.11 mm long, sparsely sericeous-puberulous and with scattered glands; tube c.6 mm long; upper lip c.5 mm long, flat; lower lip c.8 mm long, broadly obovate, lobes 2–3 mm long. Filaments 2–3 mm long; anthers densely hairy, thecae c.1 mm long, lower with acute appendage c.1 mm long. Capsule and seeds not seen.

Mozambique. MS: Cheringoma Dist., Gorongosa Nat. Park, 23 km SW of Inhaminga and 86 km ENE of Gorongosa, Muzamba R. Gorge, 225 m, fl. 9.vi.2012, *Bester* 11112 (K, PRE).

Not known elsewhere. Open bushland on limestone cliffs; 200–250 m.

Conservation notes: Known only from one collection on the Cheringoma plateau in a restricted habitat; probably Vulnerable.

This species is close to *Justicia monechmoides* from which it differs in the broad, obtuse, finely ciliate bracts and the larger flower with larger lower lip. The differences in flower size may seem slight, but considering the large amount of material seen of *J. monechmoides* I consider it – together with the differences in bracts – enough to warrant recognition. Limestone habitats are well known for harbouring local endemics.

52. Justicia sp. C (= *Sanane* 630).

Perennial herb with 1 to few erect unbranched stems from a woody rootstock; stems to 50 cm long, with 2 bands of puberulous hairs running down from nodes. Leaves finely scabrid-ciliate, otherwise glabrous; petiole absent; lamina narrowly ovate-elliptic, largest 5–6.5 × 0.6–1.1 cm; apex acute, base attenuate, decurrent to stem. Flowers single, in 6–8 cm long terminal racemoid cymes; peduncle not clearly defined; lowermost 1–2 pairs of bracts foliaceous, middle and upper narrowly oblong-elliptic, 8–12 × 3–5 mm, obtuse, finely puberulous, with dense short-stalked capitate glands, distinctly ciliate with broad glossy ciliae 1–2 mm long; bracteoles broadly oblong-elliptic, wider than bracts, 9–11 × 5–8 mm, emarginate with a small apiculus, 1–1.5 mm long, with distinct broad glossy ciliae 1–2 mm long. Calyx subequally 5-lobed, c.7 mm long, finely puberulous, with subsessile capitate glands; lobes linear-lanceolate, cuspidate, with single central vein. Corolla mauve, without darker lines on lower lip, c.14 mm long, finely puberulous; tube c.8 mm long; upper lip c.6 mm long, slightly hooded; lower lip c.8 mm long, broadly obovate, lobes c.2 mm long. Filaments c.5 mm long; anthers hairy, thecae 0.75 (upper) to 1.25 (lower) mm long, lower with acute appendage 0.75 mm long. Capsule (immature) c.8.5 mm long, finely sericeous-puberulous, hairs spreading near apex, deflexed near base. Seeds (immature) mottled grey and black, glabrous.

Zambia. N: Mbala Dist., Kambole escarpment, 1925 m, fl. 22.iv.1969, *Sanane* 630 (K). Not known elsewhere. *Brachystegia* woodland on rocky hill; c.1900 m.

Conservation notes: Known only from one collection, but habitat widespread; possibly Vulnerable.

This differs most conspicuously from *Justicia bracteata* in its perennial habit, large wide bracteoles and large corolla.

53. **Justicia eminii** Lindau in Bot. Jahrb. Syst. **20**: 68 (1894). —Clarke in F.T.A. **5**: 187 (1900). —Hedrén in Symb. Bot. Ups. **29**(1): 73, fig.29 (1989). —Vollesen in F.T.E.A., Acanthaceae **2**: 596 (2010). Type: Uganda, Ankole Dist., Mpororo, Ruhanga, 20.iv.1891, *Stuhlmann* 2086 (B† holotype, K).

> *Monechma bracteatum* Hochst. var. *eciliata* C.B. Clarke in F.T.A. **5**: 215 (1900). Type: Malawi, Nyika Plateau, vi.1896, *Whyte* 135 (K lectotype), lectotypified here.
>
> *Monechma debile* (Forssk.) Nees var. *eciliata* (C.B. Clarke) Chiov. in Nuovo Giorn. Bot. Ital., n.s. **26**: 161 (1919).

Erect shrubby herb or shrub to 1.25 m tall, usually sparsely branched; young stems sericeous to sericeous-puberulous or puberulous with upwardly directed (or spreading and bent up apically) hairs. Leaves sericeous or sericeous-puberulous or sparsely so, densest on veins; petiole to 1.5(2) cm; lamina ovate to elliptic, largest 3–10(12) × 1.5–3.5 cm, usually 2–3 times as long as wide; apex acute to rounded, base attenuate, decurrent. Flowers solitary or in 3-flowered cymules usually aggregated into a terminal racemoid cymes, also with usually numerous axillary racemoid cymes; main axis bracts foliaceous, raceme bracts ± imbricate, ovate-elliptic or broadly so, 7–11 × 4–6 mm in middle of racemes, acute to rounded and apiculate, finely puberulous and with scattered to dense subsessile or short-stalked capitate glands, not ciliate; bracteoles like reduced bracts or minute, acicular to triangular. Calyx 4–6.5 mm long, puberulous and with broad long-stalked capitate glands; lobes linear, acuminate to cuspidate, with single strong central vein. Corolla purple or dark purple with white lines on lower lip, 10–14 mm long, puberulous and with scattered capitate glands; tube 6–8 mm long; upper lip 4–6 mm long, slightly hooded; lower lip 6–9 mm long, broadly obovate, lobes 2–3 mm long. Filaments 2–4 mm long; anthers subglabrous to densely hairy, thecae with purple band near lower edge, c.1 mm long, lower with simple or bifurcate appendage 0.5–0.75 mm long. Capsule 7–9 mm long, densely puberulous. Seeds mottled grey and black, glabrous, c.3 × 3 mm.

Zambia. N: Mbala Dist., Mt Sunzu, 2400 m, fl. 23.iv.1962, *Richards* 16364 (EA, K). W: Mwinilunga Dist., Chingola, fl.& fr. 24.vi.1967, *Fanshawe* 10132 (K). C: Mpika Dist., Mutinondo Wilderness Area, fl. 12.v.2012, *Merrett* 790 (K). **Malawi**. N: Mzimba Dist., 12 km W of Mzuzu, 1225 m, fl.& fr. 20.vi.1976, *Pawek* 11378 (K, MAL, MO, SRGH).

Also in Uganda, D.R. Congo, Rwanda, Burundi and Tanzania. Woodland and bushland, often on rocky hills, montane grassland and forest margins, riverine woodland; 1200–2400 m.

Conservation notes: Widespread; not threatened.

54. **Justicia divaricata** (Nees) T. Anderson in J. Linn. Soc., Bot. **7**: 42 (1863). Type: South Africa, Nothern Cape, Groote Poort, n.d., *Hoffmansegg* s.n. (B lectotype), lectotypified here. FIGURE 8.6.**61**.

> *Adhatoda divaricata* Nees in De Candolle, Prodr. **11**: 391 (1847).
>
> *Justicia mossamedea* S. Moore in J. Bot. **18**: 312 (1880). Type: Angola, Namibe (Moçamedes),

Mossamedes to Carvalheiros, vii.1859, *Welwitsch* 5003 (BM lectotype, K, P); Namibe, Mossamedes to Carvalheiros, vii.1859, *Welwitsch* 5004 (BM syntype, K), lectotypified here.
Justicia nepeta S. Moore in J. Bot. **18**: 312 (1880). Type: Angola, Luanda, Boa Vista, n.d., *Welwitsch* 5185 (BM holotype, K, P).
Justicia namaensis Schinz in Verh. Bot. Verein. Prov. Brandenburg **31**: 202 (1890). Type: Namibia, Namaqualand, Gamosab, iv.1885, *Schinz* 25 (Z holotype, K).
Monechma floridum C.B. Clarke in F.T.A. **5**: 219 (1900). Type: Angola, Banza do Libongo, ix.1858, *Welwitsch* 5120 (BM lectotype, K, LISU); Benguela, Bumbo, n.d., *Welwitsch* 5032 (BM syntype, LISU), lectotypified here.
Monechma nepeta (S. Moore) C.B. Clarke in F.T.A. **5**: 219 (1900).
Monechma spissum C.B. Clarke in F.T.A. **5**: 219 (1900). Type: Angola, Luanda, Teba to Quicuxe, iii.1854, *Welwitsch* 5066 (BM holotype).
Monechma fimbriatum C.B. Clarke in Fl. Cap. **5**: 72 (1901). —Compton, Fl. Swaziland: 566 (1976). Type: South Africa, "Natal and Zululand", n.d., *Gerrard* 1269 (K holotype).
Monechma divaricatum (Nees) C.B. Clarke in Fl. Cap. **5**: 72 (1901). —Meyer in Merxmüller, Prodr. Fl. SW Afr. **130**: 45 (1968). —Munday in F.S.A. **30**(3,1): 55 (1995). —da Silva *et al.*, Prelim. Checklist Vasc. Pl. Mozambique: 19 (2004). —Setshogo, Checklist Pl. Botswana: 19 (2005).
Monechma namaense (Schinz) C.B. Clarke in Fl. Cap. **5**: 73 (1901).
Monechma nepetoides C.B. Clarke in Fl. Cap. **5**: 73 (1901). Type: South Africa, Northern Cape, Griqualand West, Hay, Klipfontein, n.d., *Burchell* 2149 (K lectotype); Kalahari Region, Prieska, Keikams Poort (Modder Gat Poort), n.d., *Burchell* 1616 (K syntype), lectotypified here.
Monechma angustissimum S. Moore in J. Bot. **41**: 137 (1903). Type: South Africa, Northern Cape, between Schmidt's Drift and Griquatown, 1902, *Barrett-Hamilton* s.n. (BM holotype, MO).
Monechma eremum S. Moore in J. Bot. **45**: 231 (1907). Type: Namibia, Damaraland, 1879, *Een* s.n. (BM holotype).
Monechma terminale S. Moore in J. Bot. **46**: 75 (1908). Type: South Africa, Mpumalanga, Komatipoort, 14.vi.1906, *Rogers* 893 (BM holotype, GRA, PRE).

Erect, straggling or decumbent, usually much-branched annual, perennial or shrubby herb; stems to 1 m long, subglabrous to densely puberulous when young or sericeous, often also with longer pilose hairs and/or stalked capitate glands. Leaves subglabrous to sparsely puberulous or minutely puberulous; petiole not clearly defined; lamina linear to lanceolate or oblanceolate to narrowly obovate, largest 1–5.2 × 0.1–0.8(1) cm; apex subacute to rounded, base attenuate, decurrent to stem. Flowers solitary in axils of foliaceous bracts, sometimes congested towards stem tips; bracteoles large, green with conspicuous hyaline edges, similar to calyx lobes or slightly larger, oblong to slightly obovate, 5–10(13) mm long, sparsely to densely puberulous, with or without long pilose hairs and stalked capitate glands. Calyx 4-lobed without trace of 5th lobe, green with conspicuous hyaline edges, 4–6(8) mm long, with similar indumentum; lobes linear-lanceolate to slightly oblanceolate, acute or subacute, with single central vein. Corolla pale mauve to purple (rarely white), with darker lines on lower lip, 7–13 mm long, puberulous; tube 4–7 mm long; upper lip 3–6 mm long, flat; lower lip 5–8 mm long, broadly obovate, lobes 2–3 mm long, rounded. Filaments 2–3 mm long; anthers hairy, thecae 0.5 (upper) to 1 (lower) mm long, lower with acute appendage 0.5 mm long. Capsule 5.5–9(11) mm long, sparsely puberulous or sericeous-puberulous in upper half or near apex only. Seeds mottled grey and black, glabrous, 2–3 × 2–2.5 mm.

Caprivi. E Caprivi, Singalamwe, Kwando Camp, fl.& fr. vii.1974, *Pienaar & Vahrmeijer* 457 (K, PRE). **Botswana**. N: Ngamiland Dist., SE of Tsau, fl. 14.iv.1967, *Lambrecht* 131 (K, SRGH). SW: Ghanzi Dist., 4 km N of Dondong, fl. 29.iv.1976, *Skarpe* 65 (K). SE: Mosetse Dist., 120 km from Francistown, Mosetse R., 1025 m, fl. 9.iii.1961, *Richards* 14609 (K). **Zambia**. B: Senanga Dist., 40 km W of Nangweshi, 1000 m, fl.& fr. 6.viii.1952, *Codd* 7414 (BM, K, PRE). C: Kafue Dist., 30 km W of Lusaka, Mzimbili Ranch, 1200 m, fl. 10.iv.1998, *Bingham* 11650 (K). **Zimbabwe**. N: Hurungwe Dist., Zambezi Valley, Rifa R., 500 m, fl. 24.ii.1953, *Wild* 4073 (K, SRGH). E: Chimanimani Dist., Umvumvumvu,

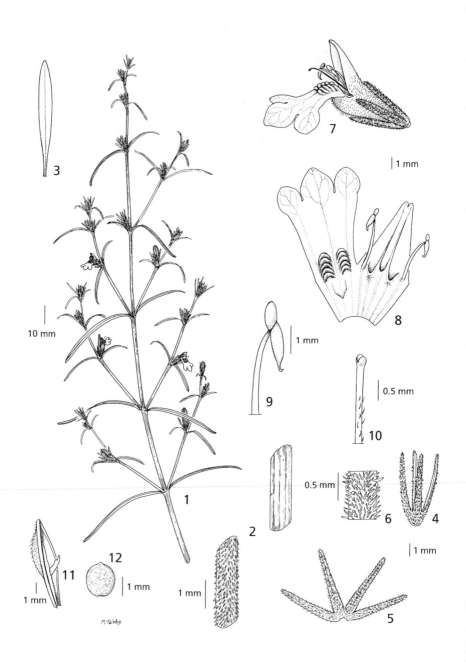

Fig. 8.6.**61**. JUSTICIA DIVARICATA. 1, habit; 2, details of stem indumentum, glabrous and hairy plants; 3, large leaf; 4, calyx; 5, calyx, opened up; 6, detail of calyx indumentum; 7, corolla; 8, corolla, opened up; 9, apical part of filament and anther; 10, apical part of style and stigma; 11, capsule; 12, seed. 1, 2 (glabrous) from *Richards* 14609, 2 (hairy) & 7–10 from *Skarpe* 427, 3 from *De Hoogh* 105, 4–6 & 11–12 from *Mitchison* 9. Drawn by Margaret Tebbs.

fl. ii.1931, *Myres* 727 (K). S: Beitbridge Dist., Limpopo Ranches road, Umzingwane R., off Bulawayo road, fl. 25.ii.1959, *Drummond* 6029 (K, SRGH). **Mozambique.** GI: Massingir Dist., Mahosi (Mabosi), fl. 25.vi.1967, *Pedro & Pedrógão* 2069 (K, PRE). M: Moamba Dist., Ressano Garcia, 300 m, fl. 24.xii.1897, *Schlechter* 11912 (K).

Also in Swaziland, South Africa, Namibia and Angola. *Acacia* and mopane bushland, often more or less disturbed or secondary, roadsides, a weed on a wide range of soils; 100–1200 m.

Conservation notes: Widespread; not threatened.

As indicated by the lengthy list of synonyms, this is an extremely variable species. Taking the large amount of South African material into consideration, I see no way of dividing it into meaningful taxa. It is easily recognised by the complete absence of the 5th calyx lobe.

55. **Justicia australis** (P.G. Mey.) Vollesen, comb. nov. Type: Namibia, Gibeon Dist., Habies Farm, 7.iv.1956, *Volk* 12244 (M holotype).

> *Monechma australe* P.G. Mey. in Mitt. Bot. Staatssaml. München **3**: 602 (1960); in Merxmüller, Prodr. Fl. SW Afr. **130**: 41 (1968).
> *Monechma genistifolium* (Engl.) C.B. Clarke subsp. *australe* (P.G. Mey.) Munday in F.S.A. **30**(3,1): 54 (1995).
> *Monechma hereroense* sensu Clarke in F.T.A. **5**: 220 (1900) in part. —Meyer in Mitt. Bot. Staatssaml. München **2**: 303 (1957), non sensu stricto.

Usually much-branched shrublet to 50 cm tall, often forming cushions; young branches densely white sericeous with downwardly directed appressed hairs. Leaves bright green, glabrous or sparsely sericeous on midrib and edges, sessile; lamina lanceolate, largest c.2.5 × 0.3 cm; apex acute, base attenuate, decurrent to stem. Flowers solitary in axils of foliaceous bracts, not congested towards stem tips; bracteoles large, green, with narrow hyaline edges towards base, similar to calyx lobes but shorter, lanceolate, c.6 mm long, sparsely minutely puberulous, with short capitate glands and scattered longer non-glandular hairs. Calyx 5-lobed, 5th lobe shorter and often only half length of other four, green with narrow hyaline edges towards base, with similar indumentum, c.6.5 mm long; lobes lanceolate, acute, with single central vein. Corolla mauve or pale blue (?rarely white) with darker lines on lower lip, c.13 mm long, puberulous; tube c.7 mm long; upper lip c.6 mm long, hooded; lower lip c.7 mm long, broadly obovate, lobes c.2 mm long, rounded. Filaments c.4 mm long; anthers hairy, thecae 1 (upper) to 1.5 (lower) mm long, lower with bifurcate appendage c.1 mm long. Capsule c.8 mm long, glabrous or sparsely puberulous at the very tip. Seeds not seen.

Botswana. SW: Kgalagadi Dist., Gemsbok Park, Ooi Kolle (Ooikalk), 5 km E of Nossop, fl.& fr. 13.iii.1969, *Blair Rains & Yalala* 29 (K, SRGH).

Also in South Africa and Namibia. Open grassland and bushland on alkaline hardpans; c.900 m.

Conservation notes: Habitat not threatened; not threatened.

56. **Justicia incana** (Nees) T. Anderson in J. Linn. Soc., Bot. **7**: 42 (1863). Type: South Africa, Northern Cape, between Beaufort West and Rhinoster Kop, n.d., *Drège* s.n. (K lectotype), lectotypified here.

> *Gendarussa incana* Nees in Linnaea **15**: 367 (1841).
> *Adhatoda incana* (Nees) Nees in De Candolle, Prodr. **11**: 393 (1847).
> *Monechma incanum* (Nees) C.B. Clarke in Fl. Cap. **5**: 69 (1901). —Meyer in Merxmüller, Prodr. Fl. SW Afr. **130**: 46 (1968). —Munday in F.S.A. **30**(3,1): 50 (1995). —Setshogo, Checklist Pl. Botswana: 19 (2005).

Usually much-branched shrublet to 50 cm tall, often forming cushions; young branches densely white sericeous with downwardly directed appressed hairs. Leaves grey, densely sericeous over both surfaces; petiole 0–2 mm; lamina lanceolate to oblanceolate or narrowly oblong, largest

1.3–2.2 × 0.3–0.5(0.7) cm; apex subacute to rounded, base attenuate, usually decurrent to stem. Flowers solitary in axils of foliaceous bracts, not congested towards stem tips; bracteoles small, grey, lanceolate, 2–3 mm long, densely greyish sericeous. Calyx 5-lobed with 5th lobe 1/2–2/3 the other 4, 5–8 mm long, densely greyish sericeous; lobes lanceolate, obtuse, with single central vein. Corolla white, sometimes with faint mauve lines on lower lip, 11–14 mm long, puberulous; tube 6–8 mm long; upper lip 5–7 mm long, slightly hooded; lower lip 6–9 mm long, oblong-obovate, lobes 2–3(4) mm long, rounded. Filaments 3–4 mm long; anthers hairy, thecae only opening about half their length, 1 (upper) to 1.5 (lower) mm long, lower with acute appendage c.0.5 mm long. Capsule c.12 mm long, glabrous or sparsely puberulous at the very tip. Seeds not seen.

Botswana. SW: Kgalagadi Dist., 70 km S of Tshane, fl. 10.iii.1969, *Blair Rains & Yalala* 17 (K). SE: Kweneng Dist., 28 km NW of entry to Khutse Game Reserve via Salajwe, fl. 7.iii.1977, *Skarpe* 168 (K).

Also in South Africa and Namibia. Grassland and bushland on sandy to clay soil, often on hardpans, sometimes dominant; c.1100 m.

Conservation notes: Widespread and apparently common; not threatened.

57. **Justicia tricostata** Vollesen in F.T.E.A., Acanthaceae 2: 598 (2010). Type: Tanzania, Sumbawanga Dist., Tatanda Mission, 24.ii.1994, *Bidgood et al.* 2431 (K holotype, BR, C, CAS, DSM, EA, K, MO, NHT, P, WAG).

Brittle-stemmed shrubby herb or shrub, sometimes (always?) from a woody rootstock; stems to 1.5 m long, sericeous-puberulous or sericeous-pubescent with upwardly directed hairs. Leaves sparsely sericeous-puberulous on midrib and veins; petiole ill-defined, to 3 mm long; lamina narrowly ovate or narrowly elliptic (rarely elliptic-obovate), largest 3.5–10 × 0.5–2.3(2.8) cm; apex acute to rounded, base attenuate, decurrent; venation of 1–2 pairs of strong veins from below middle curving up and running to near apex, sometimes also 1–2 pairs of similar veins from above middle. Flowers single (rarely in 3-flowered cymules), in (1.5)2.5–11(15) cm long racemoid cymes, terminal and often also from upper leaf axils, usually dense but occasionally interrupted in basal part (rarely almost to apex); peduncle not clearly defined; lowermost 1–2 pairs of bracts foliaceous, middle and upper narrowly ovate to ovate, acute with straight or recurved tip, without broad distinct hyaline margins, 6–12(15) × 2–4 mm, subglabrous to sericeous-puberulous, not ciliate, with 3(5) strongly raised rib-like longitudinal veins from base to apex; bracteoles similar in shape, venation and indumentum, shorter (basal) to longer (apical) than bracts, 7–12(14) mm long. Calyx subequally 5-lobed, 8–13 mm long, with similar indumentum; lobes lanceolate to narrowly ovate, with 3 strong rib-like veins from base to apex, often pale green between veins and on edges. Corolla white to pale mauve, with or without darker lines on lower lip, 17–20 mm long, sparsely puberulous; tube 8–10 mm long; upper lip 8–11 mm long, slightly hooded; lower lip 8–13 mm long, elliptic-obovate, lobes 2–3 mm long. Filaments 8–11 mm long; anthers glabrous, thecae c.2 mm long, lower with bifurcate appendage c.1.5 mm long. Capsule 10–13 mm long, glabrous. Seeds mottled grey and black, c.3.5 × 4 mm, glabrous.

Zambia. N: Mbala Dist., Kalambo Falls, 1050 m, fl. 3.iii.1962, *Richards* 16217 (K, LISC). W: Chingola Dist., fl. 19.iii.1968, *Mutimushi* 2560 (K). C: Chongwe Dist., 125 km E of Lusaka, Chakwenga headwaters, fl. 14.ii.1965, *Robinson* 6379 (K). S: Choma Dist., Sinazongwe–Choma road, 1200 m, fl. 10.iii.1997, *Luwiika et al.* 598 (K, MO).

Also in Tanzania. *Brachystegia* woodland, usually on stony soils, stony ridges and outcrops; 750–1600 m.

Conservation notes: Widespread distribution; not threatened.

58. **Justicia fanshawei** Vollesen, sp. nov. Related to *Justicia tricostata* from which it differs in the numerous short (0.5–1.5 vs. (1.5)2.5–11(15) cm long) mostly axillary racemoid cymes, in the 1-veined (not 3-veined) bracts and calyx lobes, and in the larger (13–17 vs. 10–13 mm long) capsule. Type: Zambia, Mansa (Fort Rosebery) Dist., 5.v.1964, *Fanshawe* 8576 (K holotype, NDO).

Shrubby herb or shrub with several stems from a woody rootstock; stems to 1.5(2) m long, whitish puberulous to sericeous or densely so with upwardly directed hairs. Leaves glabrous to sparsely sericeous-puberulous on midrib and veins; petiole ill-defined, to 3 mm long; lamina narrowly to broadly ovate to elliptic, largest 2.5–4.5 × 0.7–2.3 cm; apex subacute to rounded, base attenuate, decurrent; venation pinnate. Flowers single, in 0.5–1.5 cm long terminal (sometimes absent) and axillary racemoid cymes with less than 10 flowers; peduncle clearly defined, 0.1–1.3 cm long; lowermost pair of bracts foliaceous, middle and upper not imbricate, elliptic-obovate (lower) to lanceolate (upper), 4–6 × 1–3 mm, subacute to rounded, rarely truncate with a 3-toothed apex, finely sericeous or densely so, with a single raised central vein; bracteoles lanceolate or oblanceolate, acute or with a 3-toothed apex, indumentum similar, 3–6 mm long. Calyx subequally 5-lobed, (7)8–11 mm long, finely and densely puberulous (rarely sericeous), with non-capitate shiny glandular hairs and scattered longer capitate glands; lobes linear-lanceolate, acute to acuminate, rarely truncate with 3-toothed apex, with single central rib-like vein. Corolla white to pale mauve, without darker lines on lower lip, 15–18 mm long, puberulous and with scattered capitate glands; tube 7–10 mm long; upper lip 7–9 mm long, slightly hooded; lower lip 11–14 mm long, elliptic-obovate, lobes 3–6 mm long. Filaments 6–9 mm long; anthers hairy, thecae c.2 (upper) to 3 (lower) mm long, lower with sharply bent acute appendage c.1.5 mm long. Capsule 13–17 mm long, finely puberulous. Seeds mottled grey and black, c.4 × 3.5 mm, glabrous.

Zambia. N: Mporokoso Dist., Kalungwishi R., Kundabwika Falls, 1125 m, fl. 14.iv.1961, *Phipps & Vesey-Fitzgerald* 3173 (K, LISC, SRGH); Kawambwa Dist., fl.& fr. 23.viii.1957, *Fanshawe* 3527 (K).

Not known elsewhere. Rocky outcrops in *Brachystegia* woodland, riverine fringes; 900–1300 m.

Conservation notes: Endemic to NE Zambia but habitat not threatened; probably not threatened.

59. **Justicia varians** (C.B. Clarke) Vollesen, comb. nov. Types: Malawi, Nyika Plateau, vi.1896, *Whyte* s.n. (K syntype); Karonga Dist., Kondowa to Karonga, vii.1896, *Whyte* s.n. (K lectotype), lectotypified here.

Monechma varians C.B. Clarke in F.T.A. 5: 216 (1900). —Binns, Checklist Herb. Fl. Malawi: 15 (1968). —Burrows & Willis, Pl. Nyika Plateau: 51 (2005).

Perennial herb with 1–5 erect unbranched or sparsely branched stems from a woody rootstock; stems to 35 cm long, puberulous or sparsely so, uniformly so or in two bands. Leaves subglabrous to sparsely puberulous, densest along veins; petiole 0(2) mm long; lamina narrowly ovate or narrowly elliptic, largest 3–5 × 0.7–1.3 cm, more than 3 times longer than wide; apex subacute to rounded, base cuneate to shortly attenuate; venation pinnate. Flowers in 2–5 cm long dense racemoid cymes, terminal or also from upper leaf axils, 1-sided with only one bract at each node supporting a flower; peduncle not clearly defined; lowermost bracts foliaceous, upwards gradually differentiated, middle and upper ones green, without hyaline margins, narrowly oblong-elliptic, 7–10 × 2–3.5 mm, finely puberulous, with scattered subsessile capitate glands; bracteoles similar in shape, size and indumentum. Calyx 5-lobed, with 4 lanceolate to oblong subacute lobes and one shorter linear-lanceolate acuminate lobe, 6–7.5 (long lobes) and 3–4.5 (short lobe) mm long, finely puberulous and with subsessile capitate glands; lobes without hyaline margins, with one strong central vein. Corolla white, without markings on lower lip, c.11 mm long, puberulous; tube c.5 mm long; upper lip c.6 mm long, slightly hooded; lower lip c.8 mm long, broadly obovate, lobes c.2 mm long. Filaments c.4 mm long; anthers black, hairy, thecae 0.75 mm long, lower with obtuse appendage 0.5 mm long. Capsule and seeds not seen.

Malawi. N: Chitipa Dist., Nyika Plateau, Sawi Valley, 1700 m, fl. 11.ix.1972, *Brummitt & Synge* WC437 (K).

Not known elsewhere. High altitude *Brachystegia* woodland and grassland; 1700–2200 m.

Conservation notes: Endemic to the Nyika Plateau, but probably not threatened.

A very rare species known only from four collections from N Malawi. The apparent rarity may be due to the fact that it flowers from July to September, i.e. at the height of the dry season. Two collections are from 1896 and one from 1972.

60. **Justicia scabrida** S. Moore in J. Bot. **18**: 310 (1880). Types: Angola, Cuanza Norte, Pungo Andongo, Condo, 11.iii.1857, *Welwitsch* 5085 (BM syntype, K, LISU); Pungo Andongo, Condo, 11.iii.1857, *Welwitsch* 5092 (BM lectotype, K, LISU), lectotypified here.

    Justicia marginata Lindau in Bot. Jahrb. Syst. **20**: 73 (1894). Type: Angola, Malanje, iii.1879, *Buchner* 33 (B† holotype, K).

    Monechma scabridum (S. Moore) C.B. Clarke in F.T.A. **5**: 217 (1900).

Perennial or shrubby herb with several erect or ascending stems from a woody rootstock; stems to 1.5 m long, puberulous to pubescent or sericeous-puberulous (rarely subglabrous), usually with a mixture of short appressed and longer spreading hairs. Leaves dark green, sparsely sericeous to puberulous, densest along veins, often also with longer spreading hairs; petiole 4–9(12) mm long; lamina narrowly ovate or narrowly elliptic, largest 7–14.5 × 1.5–4 cm, more than 3 times longer than wide; apex acute to subacuminate, base truncate to shortly attenuate. Flowers in 3–15(20) cm long racemoid cymes, terminal and from upper leaf axils, usually dense but occasionally interrupted in basal part; peduncle clearly defined, 1–5(7) cm long; lowermost bracts foliaceous or not, upwards differentiated, middle and upper yellowish green, with broad distinct hyaline margins, elliptic or (usually) narrowly so, 12–18 × 2.5–4 mm, finely puberulous or sericeous-puberulous or sparsely so, with scattered subsessile capitate glands, distinctly ciliate; bracteoles similar or narrower, 12–15 mm long, with similar indumentum. Calyx subequally 5-lobed, 6–7(9) mm long, finely puberulous or sparsely so and sometimes with a few longer hairs on midribs, not ciliate; lobes lanceolate, acuminate to cuspidate, usually with conspicuous hyaline margins, with strong central vein. Corolla white or creamy white, with mauve lines on lower lip, 11–15 mm long, finely or minutely puberulous; tube 5–6 mm long; upper lip 6–9 mm long, slightly hooded; lower lip 6–9 mm long, broadly oblong, lobes c.1 mm long. Filaments 5–7 mm long; anthers glabrous or with a few hairs, thecae c.1.5 mm long, lower with acute or bifurcate appendage c.1.5 mm long. Capsule 8–12 mm long, densely and finely puberulous. Seeds mottled grey and black, 3–3.5 × 4–4.5 mm, whitish sericeous-puberulous on edges, sparser on flanks, glabrescent.

    Zambia. W: Kitwe, fl.& fr. 27.iv.1966, *Mutimushi* 1392 (K, NDO); Kitwe, 3.iii.1968, *Mutimushi* 2510 (K, NDO).

    Also in D.R. Congo and Angola. *Brachystegia* woodland or tall grassland at dambo edges; 1200–1500 m.

    Conservation notes: Although of restricted distribution in the Flora area, probably not threatened globally.

    Justicia scabrida has occasionally been united with the W African *J. depauperata* T. Anderson which has smaller corolla and glabrous capsule.

61. **Justicia kasamae** Vollesen, sp. nov. Differs from *Justicia fanshawei* in the longer terminal inflorescences, the longer bracts and bracteoles and the larger corolla. Type: Zambia, Kasama, 1.vii.1964, *Fanshawe* 8774 (K holotype).

Sparsely branched shrubby herb or shrub to 1(2) m tall; stems erect, brittle, puberulous or densely so. Leaves puberulous or sparsely so, densest along veins; petiole ill-defined, to 2 mm long; lamina elliptic, largest 2–3.8 × 1.3–2.5 cm; apex broadly rounded, base attenuate, decurrent. Flowers in terminal 3–8 cm long racemoid cymes, dense or interrupted in basal part; peduncle not clearly defined; lowermost bracts foliaceous, upwards pale green, middle and upper elliptic or broadly so or obovate, without hyaline margins, 13–18 × 9–13 mm, truncate to retuse and apiculate, rounded at base, puberulous and with scattered subsessile pale yellow capitate glands, indistinctly ciliate; bracteoles narrowly elliptic or elliptic, 8–13 mm long, with similar indumentum. Calyx subequally 5-lobed, 7–10 mm long, puberulous, no capitate glands; lobes lanceolate, acute to sub-acuminate, with strong central and 2 weaker lateral veins. Corolla pale mauve (rarely white), with mauve lines on lower lip, 22–25 mm long, densely puberulous; tube 11–13 mm long; upper lip 11–13 mm long, slightly hooded; lower lip c.15 mm long, obovate, lobes c.8 mm long. Filaments c.14 mm long; anthers hairy, thecae c.2 mm long, lower with acute sharply bent appendage c.1.5 mm long. Capsule 12–15 mm long, densely and finely puberulous. Seeds mottled grey and black, c.3.5 × 4 mm, whitish sericeous or densely so on edges, sparser on flanks, glabrescent.

Zambia. N: Kasama Dist., Mungwi–Kasama road, 1250 m, fl.& fr. 30.iv.1962, *Richards* 16434 (K); Kasama Dist., Lua Lua, fr. 14.vii.1949, *Hornby* 3061 (EA, K).

Not known elsewhere. Rocky hills in *Brachystegia* woodland; 1200–1300 m.

Conservation notes: Apparently endemic to Kasama District in NE Zambia but general habitat not under threat; possibly Near Threatened.

62. **Justicia subsessilis** Oliv. in Trans. Linn. Soc. London **29**: 129, t.129b (1875). — Vollesen in F.T.E.A., Acanthaceae **2**: 599 (2010). Type: Tanzania, Bukoba Dist., Karagwe, 2.xii.1861, *Speke & Grant* 213 (K holotype).

Justicia simplicispica C.B. Clarke in F.T.A. **5**: 188 (1900). Type: "East Africa" (Kenya or Tanzania), no locality, n.d., *Scott Elliot* s.n. (K holotype).

Monechma subsessile (Oliv.) C.B. Clarke in F.T.A. **5**: 216 (1900).

Monechma nemoralis S. Moore in J. Bot. **47**: 296 (1909). Type: D.R. Congo, Niembe R., 27.v.1908, *Kässner* 3010 (BM holotype, K).

Perennial herb with several erect or ascending stems from a large woody rootstock; stems to 75 cm long, puberulous to densely pubescent or sericeous-puberulous, often with mixture of short appressed and longer spreading hairs. Leaves subglabrous to pubescent or sericeous-pubescent, densest along veins; petiole 1–7 mm long; lamina ovate to broadly so or elliptic, largest (3)4–7 × 1.5–3.5 cm; apex subacute to rounded, base truncate to rounded, rarely attenuate. Flowers in terminal 3–13 cm long racemoid cymes, usually dense but occasionally interrupted in basal part; peduncle not clearly defined; lowermost bracts foliaceous, differentiated upwards, middle and upper pale or yellowish green to green, with or without indistinct hyaline margins, elliptic or narrowly so, 12–22 × 4–10 mm, puberulous to pubescent or sparsely so, with scattered to dense subsessile pale yellow capitate glands, distinctly ciliate; bracteoles similar or longer and narrower, 12–22(25) mm long, with similar indumentum. Calyx 5-lobed, lobes subequal or 5th lobe about ⅔ the other 4, 11–19(21) mm long, with similar indumentum; lobes linear-lanceolate, cuspidate, with strong central and 2 weaker lateral veins. Corolla white to yellow or yellowish green, with or without pink lines on lower lip, 15–18 mm long, puberulous; tube 6–7 mm long; upper lip 9–11 mm long, slightly hooded; lower lip 8–11 mm long, obovate, lobes c.2 mm long. Filaments 6–9 mm long; anthers glabrous, thecae c.2 mm long, lower with bifurcate appendage 1.5–2 mm long. Capsule 10–14 mm long, densely and finely sericeous-puberulous, hairs spreading near apex, deflexed near base. Seeds mottled grey and black, 3–3.5 × 4–4.5 mm, whitish sericeous or densely so on edges, sparser on flanks, glabrescent.

Zambia. N: Mbala Dist., Katwe road, 1525 m, fl. 28.ii.1955, *Richards* 4714 (K). W: Solwezi Dist., 7 km ENE of Kamlezi School, Jirundu Bot. Reserve, 1475 m, fl. 18.ii.1995, *Nawa et al.* 67 (K, MO). C: Mpika Dist., South Luangwa Nat. Park, 525 m, fr. 21.v.2004, *Bingham* 12765 (K). S: Mazabuka Dist., Mazabuka, fl. 23.i.1960, *White* 6349 (FHO, K). **Zimbabwe**. N: Makonde Dist., Hunyani, fl. 3.i.1932, *Stent* GH 5624 (BM, K, SRGH). C: Harare Dist., Twentydales road, fl. 3.ii.1961, *Rutherford-Smith* 488 (K, SRGH). E: Nyanga Dist., Juliasdale, Mt Dombo road, 1775 m, fl. 15.i.1987, *Bayliss* 10650 (K, MO).

Also in D.R. Congo, Rwanda, Burundi, Uganda, Kenya and Tanzania. Grassland, woodland and bushland with short grass cover; 500–1800 m.

Conservation notes: Widespread; not threatened.

63. **Justicia attenuifolia** Vollesen in F.T.E.A., Acanthaceae **2**: 599 (2010). Type: Tanzania, Tunduru Dist., 95 km on Masasi–Tunduru road, 19.iii.1963, *Richards* 17968 (K holotype).

Perennial herb with several erect or ascending stems from a woody rootstock; stems to 75 cm long, sericeous-puberulous when young. Leaves sparsely sericeous-puberulous on midrib, otherwise glabrous; petiole to 3 mm long; lamina narrowly ovate or narrowly elliptic, largest 8–9 × 1.7–2.2 cm, more than 3 times longer than wide; apex subacute to rounded, base attenuate, decurrent. Flowers in 4–9 cm long racemoid cymes, terminal and from upper leaf axils, dense or interrupted basally; peduncle clearly defined, 3–6.5 cm long; lowermost bracts usually foliaceous, differentiated upwards, middle and upper ones green, without hyaline margins, narrowly elliptic, 18–25 × 4–6 mm, sericeous-puberulous and with scattered subsessile pale yellow capitate glands, distinctly ciliate; bracteoles similar or narrower, 20–25 mm long, with similar indumentum. Calyx 5-lobed with subequal lobes or 5th lobe ¾ of other 4, 17–19 mm long, puberulous; lobes linear-lanceolate, cuspidate, with strong central and 2 weaker lateral veins. Corolla cream or yellowish green, without(?) pink lines on lower lip, 17–20 mm long, sparsely puberulous; tube 5–6 mm long; upper lip 12–14 mm long, slightly hooded; lower lip 10–12 mm long, oblong-obovate, lobes c.2 mm long. Filaments 8–9 mm long; anthers glabrous, thecae c.2.5 mm long, lower with bifurcate appendage c.2 mm long. Capsule c.17 mm long, densely and finely sericeous-puberulous, hairs spreading near apex, deflexed near base. Seeds not seen.

Mozambique. N: Cuamba Dist., 75 km NW of Cuamba (Nova Freixo) on Mandimba road, 700 m, fl.& fr. 25.v.1961, *Leach & Rutherford-Smith* 11009 (K, SRGH); Mecula Dist., Mbatamila, fr. 7.vi.2003, *Golding, Timberlake & Clarke* 8 (K, LMA, SRGH).

Also in Tanzania. Open *Brachystegia* woodland on sandy soil; 700 m.

Conservation notes: Restricted distribution but widespread habitat; not threatened.

Known only from these and two other collections from S Tanzania. Differs from *Justicia subsessilis* in the large attenuate leaves, large corolla with long upper lip and large capsules. *J. scabrida* from W Zambia and Angola has similar leaves but the corolla and capsule are smaller than *J. subsessilis*.

64. **Justicia tetrasperma** Hedrén in Nordic J. Bot. **10**: 151, fig.1 (1990). —Vollesen in F.T.E.A., Acanthaceae **2**: 600 (2010). Type: D.R. Congo, Karanga (Shaba), Marungu, n.d., *Vilain* 60 (BRLU holotype).

Shrubby perennial herb with several erect unbranched stems from a woody rootstock; stems to 1 m long, densely puberulous. Leaves puberulous on midrib and veins, sparser on lamina; petiole to 6 mm long; lamina narrowly ovate, largest c.8.5 × 2.8 cm; apex subacute, base attenuate. Flowers in 4–7 cm long dense terminal racemoid cymes; peduncle not clearly defined; lowermost bracts foliaceous, differentiated upwards, middle and upper ones green, without hyaline margins, ovate-elliptic, 14–18 × 7–9 mm, sparsely puberulous, with longer hairs on veins and distinctly ciliate; bracteoles small, narrowly triangular, c.2 × 0.5 mm long. Calyx subequally 5-lobed or 5th lobe about ²/₃ of other 4, c.11 mm long, puberulous and pilose-ciliate; lobes linear-lanceolate, cuspidate, with 1 strong central vein. Corolla pure white, c.17 mm long, puberulous; tube c.9 mm long; upper lip c.8 mm long, slightly hooded; lower lip c.10 mm long, broadly obovate, lobes c.6 mm long, middle much wider than laterals. Filaments c.7 mm long; anthers glabrous, thecae 1.5–2 mm long, lower with bifurcate appendage c.1.5 mm long. Capsule 4-seeded, 11–13 mm long, densely puberulous; seeds mottled grey and black, c.3 × 3 mm, glabrous.

Zambia. N: Mbala Dist., fl. 10.ix.1969, *Fanshawe* 10625 (K).

Also in D.R. Congo and Tanzania. Moist grassland in mushitu (swamp forest); c.1800 m.

Conservation notes: Local distribution and threatened habitat; Near Threatened.

Hedrén (1990) discusses at length the affinities of this species and how it breaks down the characters differentiating *Justicia* from *Monechma*.

This collection differs from the Tanzania and Congo material in its smaller calyx (c.16 mm in Tanzania) and smaller white corolla (c.19 mm long and purple in Tanzania). But the total variation in a small amount of material is no bigger than seen in other widespread species.

35. RHINACANTHUS Nees [13]

Rhinacanthus Nees in Wallich, Pl. Asiat. Rar. **3**: 76 (1832). —Nees in De Candolle, Prodr. **11**: 442 (1847). —Darbyshire & Harris in Kew Bull. **61**: 401–418 (2006).

Perennial herbs or subshrubs; cystoliths numerous, ± conspicuous. Stems 6-angular or terete, often with short pale retrorse and/or antrorse hairs. Leaves evergreen or rarely deciduous, opposite-decussate, petiolate, pairs equal to somewhat anisophyllous; blade margin entire or obscurely repand. Inflorescences axillary and/or terminal, fasciculate, spiciform or paniculate; bracts and bracteoles inconspicuous, then linear to lanceolate or the former leaf-like; flowers subsessile, in clusters or solitary. Calyx divided almost to base; lobes 5, subequal, linear-lanceolate. Corolla bilabiate, pubescent externally except towards base of tube where it is glabrous; tube narrowly long-cylindrical, usually longer than limb, roof with 2 lines of hairs within extending onto upper lip; upper lip ovate to lanceolate, apically notched or entire; lower lip divided in upper half or third into 3 subequal rounded to oblong lobes. Stamens 2; filaments attached near mouth of corolla tube, shortly exserted; anthers bithecous, thecae slightly to strongly offset, ± oblique, lower theca sometimes with a short appendage (tail). Disk shallowly cupular, with V-shaped slit. Ovary bilocular, 2 ovules per locule; style filiform; stigma bifid, lobes unequal. Capsule clavate, long-stipitate, apex shortly beaked. Seeds held on retinacula, lenticular with a hilar excavation, brown to black, ± tuberculate.

A genus of c. 25 species, confined to the Old World tropics and subtropics. In addition to the four species here, *R. latilabiatus* (K. Balkwill) I. Darbysh. from South Africa and Swaziland is likely to occur in southern Mozambique – see also note under *Dicliptera cernua* (Nees) I. Darbysh.

1. Corolla 21–43 mm long including tube 12.5–30 mm long; upper lip narrowly lanceolate 2
– Corolla (9)12–19 mm long including tube (6)7.5–14.5 mm; upper lip ovate or triangular-ovate. 4
2. Corolla tube 12.5–17 mm long, ratio of tube:lower lip 1.3–1.65:1 . . . . . . **3.** *sp. A*
– Corolla tube 17–30 mm long, ratio of tube:lower lip 1.8–3:1 3
3. Cauline leaves* ovate or lanceolate, with apex acute or attenuate, length/ width ratio (1.8)2–4(4.8):1; young leaves rather sparingly and inconspicuously pubescent beneath . . . . **1.** *zambesiacus*
– Cauline leaves broadly ovate with apex obtuse, rounded or shortly and abruptly acuminate, length/width ratio 1–1.5:1; young leaves ± densely and conspicuously pubescent beneath . . . . **2.** *xerophilus*
4. Thecae of each anther offset but overlapping, slightly oblique; lower theca shortly tailed . . . . **4.** *virens*
– Thecae of each anther superposed and separated by portion of filament, highly oblique, upper theca almost patent to filament; lower theca not tailed. . . . .
. . . .**5.** *submontanus*

* Excluding lowermost leaf pairs which are often elliptic in both species.

1. **Rhinacanthus zambesiacus** I. Darbysh. in Kew Bull. **67**: 768 (2012). Type: Malawi, Dedza Dist., c.5 km E of Linthipe on Chongoni road, fl. 30.iv.1989, *Goyder, Pope & Radcliffe-Smith* 3144 (K holotype, MAL, PRE). FIGURE 8.6.**62**.

    Rhinacanthus communis sensu Clarke in F.T.A. **5**: 225 (1900) in part for *Buchanan* 72 & 863, *Kirk* s.n., *Last* s.n., *Scott-Elliot* 8655 & 8598 & *Whyte* s.n., non Nees.

───────────────────────

[13] By Iain Darbyshire

Rhinacanthus gracilis sensu auct., non Klotzsch. —da Silva *et al.*, Prelim. Checklist Vasc. Pl. Mozamb.: 19 (2004). —Mapaura & Timberlake, Checklist Zimb. Vasc. Pl.: 14 (2004). —Phiri, Checklist Zamb. Vasc. Pl.: 19 (2005). —Darbyshire & Harris in Kew Bull. **61**: 408 (2006). Neotype rejected by Darbyshire in Kew Bull. **67**: 768 (2012) – see Excluded species. *Rhinacanthus rotundifolius* sensu da Silva *et al.*, Prelim. Checklist Vasc. Pl. Mozamb.: 19 (2004), non C.B. Clarke (see note).

Erect or straggling suffruticose herb, 10–90(150) cm tall; stems subangular with dark longitudinal stripes, rather densely spreading antrorse- and retrorse-pubescent. Leaves ovate, lanceolate or lowermost pairs elliptic, (1.9)5–13.5(17) × (0.7)1.5-6 cm, base of lower cauline pairs acute to attenuate, becoming rounded to shallowly cordate in uppermost pairs, apex acute to attenuate, upper surface and nerves beneath finely pubescent, hairs ± antrorse; lateral veins (4)5–6(7) pairs, pale and prominent beneath; petiole to 10(16) mm long but uppermost leaves sessile. Inflorescences mostly terminal, laxly spiciform to paniculate, 5-40 × 0.8-25 cm; flowers subsessile, usually spaced along axis singly or in opposite pairs, sometimes clustered at ends of inflorescence branches; axis puberulous with increasing numbers of short capitate glandular hairs distally, sometimes also with scattered longer eglandular hairs; bracts (excluding leafy pairs sometimes present towards base of main axis) and bracteoles linear-lanceolate, 1.5–3(3.5) mm long. Calyx lobes (2.7)3.5–5(5.5) mm long, puberulous with blunt-tipped spreading hairs and capitate glandular hairs externally, the latter rarely sparse. Corolla (24)30–43 mm long, pubescent, with scattered short glandular hairs externally; tube white to pink, (18)22–30 mm long, straight or shallowly curved; limb white, cream or pale yellow within, with or without purple markings at mouth, sometimes pink-tinged externally; upper lip lanceolate, (5.5)6–8 mm long, apex shallowly notched; lower lip (6)8–13.5(14.5) mm long, lobes oblong-ovate, median lobe (4)5–8(9.5) × 3.3–5(5.7) mm, apex rounded. Staminal filaments (2)2.5–5 mm long, glabrous; anther thecae offset and oblique, slightly overlapping, lower theca 0.9–1.35 mm long, base apiculate or with minute pale tail, upper theca 0.65–1 mm long. Ovary pubescent at apex; style pubescent towards base. Capsule 13–20 mm long, pubescent and with scattered short glandular hairs. Seeds black, 2.2–3.2 mm wide, tuberculate.

Botswana. N: Central Dist., dirt track W immediately left of road after cattle checkpoint NW of Tutume, fl. 28.iv.2005, *Darbyshire* 459 (GAB, K). **Zambia**. C: Mpika Dist., 4.8 km S of Lubi R., South Luangwa Nat. Park (Game Reserve), fl. 19.iv.1967, *Prince* 508 (K). E: Lundazi Dist., Lukusuzi Nat. Park, fl.& fr. 14.iv.1971, *Sayer* 1197 (SRGH). S: Sinazongwe Dist., Siamambo, fl.& fr. 17.v.1961, *Fanshawe* 6572 (K, LISC, NDO, SRGH). **Zimbabwe**. N: Gowke Dist., near Gwave R., fl. 16.iv.1962, *Bingham* 230 (K, SRGH). W: Hwange Dist., Victoria Falls, near Mtuzi Bridge, fl. 14.iii.1979, *Mshasha* 186 (K, SRGH). C: Chegutu Dist., Poole Farm, fl.& fr. 7.iv.1954, *Wild* 4557 (K, LISC, SRGH). E: Mutare Dist., commonage, fl.& fr. 23.iii.1959, *Chase* 7075 (BM, K, LMA, SRGH). S: Chivi Dist., near Madzivire Dip, c.6.4 km N of Lundi R., fl. 4.v.1962, *Drummond* 7931 (K, PRE, SRGH). **Malawi**. C: Dedza Dist., c.5 km from Linthipe on Chongoni road, fl. 30.iv.1989, *Pope, Radcliffe-Smith & Goyder* 2238 (BR, K, LISC, MAL, SRGH). S: Nsanje Dist., 5 km NW of Nsanje, near Nyamadzere Rest House, fl.& fr. 28.v.1970, *Brummitt* 11133 (K, MAL, SRGH). **Mozambique**. N: Mandimba Dist., Mandimba, fl. 1.xii.1941, *Hornby* 3513 (K). Z: Lugela Dist., Namagoa, 200 km inland from Quelimane, fl.& fr. 1944, *Faulkner* PRE 319 (EA, K, LMA, SRGH). T: Cahora Bassa Dist., Songo, perimeter of fence by S margin of Zambezi, by dam, fl.& fr. 8.iii.1972, *Macêdo* 5023 (LISC, LMA, LMU). MS: ?Gondola Dist., Chimoio (Vila Pery)–Mutarara (Matarara), fl. 8.vi.1949, *Pedro & Pedrógão* 6310 (LMA).

Not known elsewhere. Miombo and mopane woodland, riverine woodland, in shade on rocky hillslopes, wooded termitaria, sometimes in disturbed or recently burnt woodland; 50–1400 m.

Conservation notes: Endemic to the Flora area. Widespread and often common; Least Concern.

Plants from the northern part of the range (including some apparently pyrophytic populations) tend to be smaller in stature and often have an unbranched spiciform

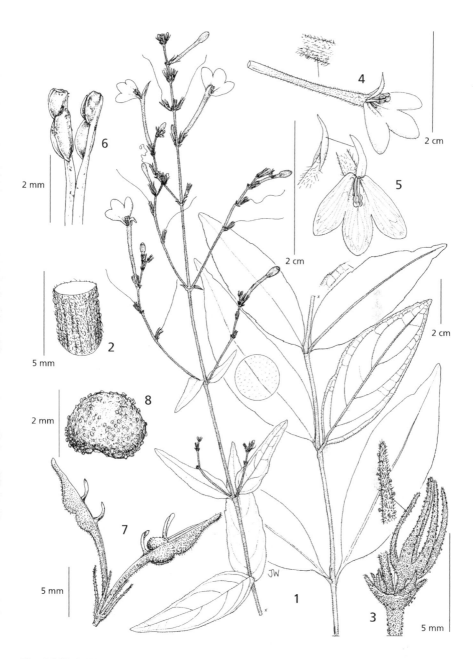

Fig. 8.6.**62**. RHINACANTHUS ZAMBESIACUS. 1, habit: inflorescence and cauline leaves (×
²/₃), with detail of leaf indumentum; 2, detail of stem indumentum (× 5); 3, cymule bracts,
bracteoles and calyx (× 6), with detail of calyx indumentum; 4, corolla, lateral view (× 1½), with
detail of external indumentum; 5, corolla limb, face view, and with dorsal view of upper lip (×
2); 6, detail of stamens (× 10); 7, capsule with persistent calyx (× 3); 8, seed (× 10). 1 & 3 from
Patel et al. 1506, 2 & 6–8 from *Brummitt* 11133, 4 & 5 from *Bidgood & Vollesen* 2174. Drawn by
Juliet Williamson.

inflorescences. Those from the south and west are generally more robust with a paniculate inflorescence, especially in Zambian populations where the inflorescence can be very widely branched. However, there is much overlap and the variation appears clinal.

The name *Rhinacanthus rotundifolius* has previously been misapplied to specimens of *R. xerophilus* in South Africa, but in the Mozambique checklist, da Silva *et al.* (2004) apply this name to specimens of *R. zambesiacus* from Mozambique MS. True *R. rotundifolius* C.B. Clarke is a very distinctive species from arid NE Africa.

2. **Rhinacanthus xerophilus** A. Meeuse in Bothalia **7**: 410 (1960). —Balkwill in F.S.A. **30**(3,1): 13 (1995). —Darbyshire in Kew Bull. **67**: 768, map 1 (2012). Type: South Africa, Pietersburg Dist., Soekmekaar, fl.& fr. 28.i.1954, *Meeuse* 9221 (PRE holotype, K, SRGH).

Erect or straggling suffruticose herb or shrublet, 15–40(70) cm tall, woody towards base; stems pale green with darker stripes, weakly 6-angular, densely pubescent with white retrorse or mixed retrorse and antrorse hairs, rarely also with interspersed spreading glandular hairs. Leaves somewhat leathery, most cauline leaves broadly ovate, 1.5–4.2 × 1.5–4 cm, base shallowly cordate or truncate, apex obtuse or rounded to shortly acuminate, surfaces pubescent with white antrorse or spreading hairs, rather dense beneath when young, particularly along veins; lateral veins 4–6(7) pairs, pale and prominent beneath; lowermost cauline leaves often more elliptic(-obovate) with cuneate base; petiole to 7(13) mm but upper cauline leaves subsessile. Inflorescences mostly terminal, spiciform to laxly (often narrowly) paniculate, 1.5–12(20) × 0.7–4(9) cm, flowers sessile in opposite single-flowered cymules spaced along axis, sometimes more clustered; axis with short antrorse and/or spreading white hairs, usually with numerous capitate stalked glands at least in distal portion; main axis bracts towards inflorescence base often foliaceous, typically 3–11 × 1.5–9 mm, rapidly reducing in size up axis, bracts in distal portion and bracteoles linear-lanceolate, 1.5–3 mm long. Calyx lobes 2.5–6 mm long, puberulous with short, blunt-tipped spreading hairs, usually with numerous capitate stalked glands externally, margin with longer white eglandular hairs. Corolla white, with or without purple markings on lower lip, 24–33.5 mm long, pilose, with short glandular hairs externally; tube 17–24 mm long, staight; upper lip lanceolate, 4.5–8 × 1.5–2.3 mm, apex recuved, shallowly notched; lower lip 6–10 mm long, lobes elliptic-rounded, median lobe 3–6.3 × 3–3.5 mm, palate with few minute glandular hairs. Staminal filaments 2.5–3.5 mm long, glabrous; anther thecae offset and barely overlapping, parallel to slightly oblique, lower theca 0.8–1.2 mm long, base apiculate or muticous, upper theca 0.65–0.85 mm long. Ovary with pale hairs and short-stalked glands towards apex; style pubescent towards base. Capsule 14–16.5 mm long, pilose, with scattered capitate stalked glands. Seeds black, c.2 mm wide, tuberculate.

Botswana. SE: Central Dist., NE Tuli Block Game Reserve, Farm Uitspan 3-MS, fl.& fr. 8.i.1993, *Baytopp* 228 (PRE). **Zimbabwe.** S: Beitbridge Dist., Marangutzi Mt, SW slopes, c.19 km W of R.S.E.S. base camp at Bubye crossing, c.32 km W of Mateke Hills, bud 10.v.1958, *Boughey* 2863 (SRGH). **Mozambique.** GI: Massingir Dist., Massingir, rio dos Elefantes, 81 km to NW of Lagoa Nova, fl.& fr. 19.iii.1973, *Lousã & Rosa* 355 (LMA). M: Moamba Dist., Ressano Garcia, fl. 22.xii.1897, *Schlechter* 11881 (BM, K).

Also in South Africa (Limpopo, Mpumalanga). Dry bushland on sandy or stony soils, sometimes amongst boulders on or at base of rocky slopes; 50–600 m.

Conservation notes: Locally common in NE South Africa but scarce in the Flora area; Least Concern.

Whilst most populations of *Rhinacanthus xerophilus* are easily separable from the largely allopatric *R. zambesiacus*, a few collections appear somewhat intermediate (Map 1 in Darbyshire 2012). For example, *Wild* 4361 from Mtilikwe R. (Zimbabwe S) and *Rand* 76 from Francistown (Botswana N) are both close to *R. xerophilus* but are more robust with larger leaves that are more acute at the apex than normal.

3. **Rhinacanthus sp. A** (= *Barbosa* 1880).

Suffruticose herb producing precocious flowering shoots 10–15 cm tall from a woody base, with basal remains of previous year's vegetative shoots often present; young stems slender, 1–1.5 mm diameter, subangular, densely pubescent with white spreading hairs; mature stem much more robust, 2.5–4.5 mm diameter, softly woody with pale surface. Leaves immature or absent at flowering, mature leaves not seen; immature leaves subsessile, ± broadly ovate(-elliptic) or suborbicular, 0.9–1.5 × 0.5–0.9 cm, base rounded, apex rounded with minute acumen or more obtuse to acute, pubescent along veins and margin, lower surface puberulous between veins with mixed glandular and eglandular hairs; with 3–4 pairs of lateral veins. Inflorescence terminal, spiciform, sometimes with 3 spikes terminating a branch, each 4.5–12 cm long, flowers sessile in opposite single-flowered cymules spaced along axis; axis densely glandular- and eglandular-puberulous, the glandular hairs capitate; main axis bracts towards infloresence base often foliaceous, broadly ovate, 5–7.5 × 2.5–6 mm, rapidly reducing in size up axis, bracts in distal portion lanceolate or linear-lanceolate, 2.5–4 mm long; bracteoles similar but 1.8–3 mm long. Calyx lobes 3.8–6 mm long, glandular- and eglandular-puberulous. Corolla white or white-pink with brown-red markings on lower lip, 21–27.5 mm long, retrorse-pubescent, with short-stalked glandular hairs externally; tube 12.5–17 mm long, staight or shallowly curved; upper lip lanceolate, 6.5–9.5 × c.2.3 mm, apex shallowly notched; lower lip 8–12 mm long, lobes elliptic-rounded, median lobe 2.5–4.3 mm long, palate with few minute glandular hairs. Staminal filaments 4–5.7 mm long, glabrous; anther thecae offset and barely overlapping, slightly oblique, lower theca 0.9–1.2 mm long, base with minute pale tail, upper theca 0.7–0.8 mm long. Ovary with few pale hairs and short-stalked glands towards apex; style pubescent towards base. Capsule 14.5–17 mm long, pubescent, with interspersed short glandular hairs; seeds black, c.2.5 mm wide, tuberculate.

Mozambique. N: Eráti Dist., between Lúrio and Namapa, old Posto de Nivete, fl.& fr. 17.viii.1948, *Pedro & Pedrógão* 4842 (EA, LMU); between Pemba (Porto Amélia) and Ancuabe, fl.& fr. 24.viii.1948, *Barbosa* 1880 (LISC).

Not known elsewhere. Habitat not recorded; 100–300 m.

Conservation notes: Known only from two collections, it is clearly highly restricted and may be threatened due to habitat loss. Barbosa recorded it as gregarious; possibly Vulnerable.

This species is closely allied to *Rhinacanthus xerophilus* and *R. zambesiacus* but separated from both by the short corolla tube relative to the limb. The young leaves are rather broad and hairy as in *R. xerophilus* but the species differs in the growth habit (*R. xerophilus* flowers from mature leafy stems) and the shortly tailed lower anther thecae (tail not or barely developed in *R. xerophilus*). Some plants of *R. zambesiacus* from the northern part of its range can have a similar growth habit to *sp. A*, but differ in the markedly longer corolla tube, the less conspicuously hairy stem and the narrower, less hairy immature leaves. However, *Torre & Paiva* 11542 from Mogovolas in N Mozambique has mature foliage matching *R. zambesiacus* but a short corolla tube as in *sp. A*. More material is needed before drawing firm conclusions on this taxon's status.

4. **Rhinacanthus virens** (Nees) Milne-Redh. in Exell, Suppl. Cat. Vasc. Pl. S. Tomé: 37 (1956). —Heine in F.W.T.A. ed.2 **2**: 425 (1963), excl. var. *obtusifolius*. —Darbyshire & Harris in Kew Bull. **61**: 414 (2006). —Darbyshire in F.T.E.A., Acanthaceae 2: 603 (2010). Type: Gabon, "Gaboon coast", fr. 1787, *Middleton* s.n. (K holotype).

Leptostachya virens Nees in De Candolle, Prodr. **11**: 378 (1847).

Rhinacanthus dewevrei De Wild. & T. Durand in Compt. Rend. Soc. Bot. Belg. **38**: 105 (1899). Type: D.R. Congo, Bokakata, fl. 10.iii.1896 (6.ii.1896 on K sheet), *Dewèvre* 804 [no. not listed in protologue] (BR holotype, K), but see note.

Siphonoglossa rubra S. Moore in J. Bot. **44**: 88 (1906). Type: Uganda, Entebbe, fl. 15.ix.1905, *Bagshawe* 750 (BM holotype).

Rhinacanthus minimus S. Moore in J. Bot. **58**: 47 (1920). Type: D.R. Congo, Bukala, fl.& fr. 1914, *Vanderyst* 4972 (BM holotype, BR).

Rhinacanthus communis sensu auct., non Nees. —Clarke in F.T.A. **5**: 224, 514 (1900), in part. —Heine in F.W.T.A. ed.1 **2**: 266 (1931), in part for W African specimens.

Weak-stemmed perennial herb, 15–50 cm tall, often decumbent and rooting at lower nodes; stems (sub) 6-angular, pale antrorse- and/or retrorse-pubescent mostly on angles. Leaves ovate to lanceolate, 2.3–11.5 × 0.6–4 cm, base attenuate, cuneate or obtuse, apex acute to acuminate, surfaces glabrous or antrorse-pubescent particularly on veins beneath; lateral veins 4–6 pairs; petiole 2–8 mm. Inflorescence axillary in upper leaf axils, spiciform or in a lax, few-branched panicle, 1.5–8.5(12.5) × 0.5–5.5 cm, sometimes compounded to form loose leafy terminal panicles; flowers in subsessile clusters of 2–4 along and at apex of axis; primary peduncles antrorse-pubescent, usually with scattered glandular hairs, axis of inflorescence branches puberulous and with more numerous short glandular hairs; cymule bracts and bracteoles linear-lanceolate, 1.3–4 mm long, bracts subtending inflorescence branches sometimes more foliaceous but much reduced. Calyx lobes 3–4(5.5) mm long, eglandular- and glandular-puberulous. Corolla 14–19 mm long, white in our region (elsewhere pink to mauve), pubescent, with few short glandular hairs on limb externally; tube 10–14.5 mm long; limb only slightly spreading; upper lip triangular-ovate, 2.5–3 mm long and wide, apex attenuate or notched; lower lip 4–5.5 mm long, glabrous within, lobes rounded, median lobe 2–3 × 2–2.5 mm. Staminal filaments 1.7–2 mm long, glabrous; anther thecae offset but overlapping, somewhat oblique, lower theca 0.9–1 mm long, base with short pale tail; upper theca 0.75–0.9 mm long. Ovary pubescent at apex, elsewhere glabrous; style pubescent in distal half. Capsule 10.5–14 mm long, eglandular-pubescent with few interspersed short glandular hairs. Seeds brown, c.1.75 mm wide, tuberculate.

Zambia. W: Mwinilunga Dist., Lisombo R., fl. 13.vi.1963, *Loveridge* 950 (K, LISC, NDO, SRGH); Mwinilunga Dist., Zambezi rapids, fl. 18.v.1969, *Mutimushi* 3281 (K, NDO).

Also from Sierra Leone to Uganda, NW Tanzania, D.R. Congo and Angola. Riverine forest (mushitu) and riverine margins; c.1400 m.

Conservation notes: Widespread and common in Guineo-Congolian forest; Least Concern.

Plants in NW Zambia are very slender and small-flowered, closely resembling populations from parts of West Africa, e.g. Sierra Leone; the above description covers only these small-flowered forms. Populations from the central Congo Basin through to Uganda and Tanzania are notably larger in all parts and lack glandular hairs on the inflorescence axes. Whilst these two extremes appear quite distinct, variation appears clinal.

Previously, Darbyshire & Harris (Kew Bull. **61**: 416, 2006) maintained *Rhinacanthus minimus* as distinct mainly because the type specimen had not been studied in depth. On further study and on consultation with Dominique Champluvier (BR), I have concluded that it is best treated as a small-flowered variant of *R. virens*.

5. **Rhinacanthus submontanus** T. Harris & I. Darbysh. in Kew Bull. **61**: 411, fig.2 (2006). —Darbyshire in F.T.E.A., Acanthaceae **2**: 610 (2010). Type: Malawi, Chitipa Dist., Musitu Forest, 21 km SE of Chisenga, fl.& fr. 11.vii.1970, *Brummitt* 11987 (K holotype, LISC, MAL, SRGH).

Decumbent perennial herb, 15–50(90) cm tall, base usually trailing and rooting at nodes; stems 6-angular, shortly pubescent mainly on angles, hairs mostly antrorse but sometimes retrorse or spreading. Leaves (ovate-)elliptic to narrowly so, 3.5–14 × 1–3.7 cm, base cuneate or attenuate, becoming ± rounded in uppermost leaves, apex acute to shortly attenuate, upper surface with sparse long, multicellular hairs, veins beneath and margin shortly antrorse-pubescent; lateral veins 4–7 pairs; petiole 2.5–15(30) mm. Inflorescences axillary and terminal, the latter extending beyond uppermost leaves, laxly paniculate, 3–18.5 × (1)2.5–8.5 cm, few to several-branched; flowers subsessile along axis, solitary or paired at each inflorescence node, few flowers open at any given time; axis shortly antrorse-pubescent towards inflorescence base, with short spreading blunt-tipped hairs and increasing numbers of capitate stalked glands distally; bracts lanceolate, 1.5–2.5 mm long, bracteoles somewhat shorter. Calyx lobes 2–3.5 mm long, with dense spreading blunt-tipped hairs and numerous capitate stalked glands externally. Corolla (9)12–15(17) mm long, white with pink to purple markings on lower lip, spreading-pubescent, with scattered glandular hairs externally; tube (6)7.5–9(10) mm long, straight; upper lip ovate, 3–4.5 mm long, apex notched; lower lip 3.5–7 mm long, lobes (oblong-)rounded, 1.5–3.5 mm long. Staminal filaments 1.8–2.5 mm long, glabrous; thecae highly oblique and separated by 0.3–0.6 mm, lower theca 0.9–1.2 mm long, parallel to filament, upper theca 0.6–0.8 mm long, held almost patent to filament. Ovary glabrous; style pubescent towards base or glabrous. Capsule 12–14 mm long, sparsely pubescent and with short stalked-glands towards apex. Seeds black, 2–2.7 mm wide, tuberculate.

Zambia. E: Isoka Dist., Makutu Mts, fr. 26.x.1972, *Fanshawe* 11526 (K, NDO). **Malawi**. N: Nkhata Bay Dist., Viphya Plateau, 48 km SW of Mzuzu, Mpalampala, fl. 9.x.1976, *Pawek* 11890 (K, MAL, MO, SRGH). C: Nkhotakota Dist., Nchisi Mt, fl. 30.vii.1946, *Brass* 17036 (BM, BR, K, SRGH, NY). **Mozambique**. MS: Gorongosa Dist., Serra da Gorongosa, slopes of Mt Nhandore, fl.& fr. 22.x.1965, *Torre & Pereira* 12552 (BR, LISC, LMU, WAG).

Also in Tanzania. Submontane and montane wet forest; 1200–2000 m.

Conservation notes: Assessed as Vulnerable under IUCN criterion B by Harris & Darbyshire (2006) when it was thought to be restricted to N Malawi and adjacent Zambia. But as isolated populations have since been recorded in the Pare Mts of N Tanzania and Mt Gorongosa in Mozambique, the Extent of Occurrence is greatly extended, although its Area of Occupancy remains small; probably Vulnerable.

An immature specimen from dense deciduous forest at Inhamitanga in C Mozambique (*Simão* 710, LISC) has flower buds which appear to be similar in size to this species and the related *Rhinacanthus selousensis* I. Darbysh. from lowland Tanzania. However, the young stamens have parallel, immediately superposed thecae. More mature material is required to determine whether this is a different species.

Excluded species.

Rhinacanthus gracilis Klotzsch in Peters, Naturw. Reise Mossamb. **6**: 218 (1861). Type: Mozambique, Goa Is., bud & fr., *Peters* s.n. (B† holotype); neotype: Mozambique, 4.8 km E of Mandimba, 25.v.1961, *Leach & Rutherford-Smith* 11013 (K, SRGH), chosen by Darbyshire & Harris in Kew Bull. **61**: 408 (2006), but rejected by Darbyshire in Kew Bull **67**: 768 (2012).

The holotype of *R. gracilis* was destroyed in the bombing of the Berlin Herbarium during WW II. Although the name has been applied widely in the past to the Zambesian and East African species of *Rhinacanthus*, the description in the protologue does not match any currently known African *Rhinacanthus*, and probably belongs in a different genus. For further discussion, see Darbyshire (2012).

36. ANISOTES Nees [14]

Anisotes Nees in De Candolle, Prodr. **11**: 424 (1847). —Baden in Nordic J. Bot. **1**: 623–664 (1981).

Himantochilus T. Anderson in Bentham & Hooker, Gen. Pl. **2**: 1117 (1876).
Macrorungia C.B. Clarke in F.T.A. **5**: 254 (1900); in Fl. Cap. **5**: 89 (1901). —Baden in Nordic J. Bot. **1**: 143–153 (1981).
Metarungia Baden in Kew Bull. **39**: 638 (1984).

Shrubby herbs, shrubs or small trees; cystoliths conspicuous or not. Leaves opposite, equal, entire to crenate. Plants often flowering when leafless, flowers in dorsiventral racemoid cymes with strobilate bracts or in 1–12-flowered axillary dichasia or short cymes with minute bracts;

[14] By Kaj Vollesen

bracts persistent, from large and showy to minute; bracteoles small. Calyx divided into 5 equal segments. Corolla glandular puberulous, sometimes also hairy (rarely glabrous) on the outside; basal part of tube short, cylindrical, widening slightly upwards; throat short, very indistinct; limb very distinctly 2-lipped with lips usually more than twice as long as tube and throat combined; upper lip hooded; lower lip usually pendent and coiled at anthesis, without rugula. Stamens 2, no staminodes, inserted at top of tube; anthers bithecous, thecae subequal, one a little below others, parallel, both mucronate or lower with minute appendage at base. Style filiform, glabrous or hairy; stigma minutely bifid with 2 ellipsoid lobes. Capsule 2–4-seeded, clavate with a solid basal stalk, placentae solid and not splitting from capsule wall or splitting from base or wall splitting longitudinally. Seeds spheroid to discoid, compressed or not, smooth to rugose or reticulate-tuberculate.

A genus of 24 species in tropical Africa, Arabia and Madagascar. The centre of diversity is from S Ethiopia and S Somalia through E Kenya to Tanzania, and to a lesser degree onwards through Mozambique to E Zimbabwe and N South Africa.

1. Flowers in elongated racemoid cymes; bracts large and conspicuous, (3)8–30 mm long 2
   - Flowers in 1-flowered cymules; bracts 2–5 mm long 6
2. Bracts with broad pale red to dark red margins; capsule with placentae rising elastically from capsule wall at maturity . . . . **4.** *pubinervis*
   - Bracts without differently coloured margins or with pale green margins; placentae not rising elastically at maturity (except in *A. longistrobus*) 3
3. Cymes terminal on main branches or also terminal on short lateral branches, or some axillary; peduncle to 1 cm long; corolla 4.3–5.8 cm long 4
   - Cymes all axillary; peduncle 1.5–4 cm long; corolla 2.2–3.5 cm long. 5
4. Corolla 4.3–4.8 cm long; bracts with 5 distinct rib-like longitudinal veins; largest leaf 3–7 cm long . . . . **1.** *formosissimus*
   - Corolla 4.9–5.8 cm long; bracts without rib-like longitudinal veins; largest leaf 8.5–20 cm long . . . . **5.** *longistrobus*
5. Corolla greenish yellow to orange or brownish yellow, lower lip with 3 short lobes; bracts pale green or yellowish green with darker venation and pale crinkly margins . . . . **2.** *bracteatus*
   - Corolla purple, lower lip divided almost to base; bracts without darker venation or pale crinkly margins. . . . . **3.** *nyassae*
6. Mature leaves 3–5.5(7.5) × 1.5–2.5(3.7) cm, developing with flowers; bracts (2.5)3–5 mm long; bracteoles 3–4.5 mm long; calyx 3–5 mm long . . . . .**7.** *rogersii*
   - Mature leaves leaves 6–15 × 3–6 cm, developing after flowers; bracts 2–3.5 mm long; bracteoles 2–3.5 mm long; calyx 2.5–4 mm long . . . . **6.** *sessiliflorus*

1. **Anisotes formosissimus** (Klotzsch) Milne-Redh. in Mem. N.Y. Bot. Gard. **9**: 25 (1954). Type: Mozambique, Caia Dist., Rios de Sena, n.d., *Peters 8* (K lectotype, BM), lectotypified by Baden (1981).

    Adhatoda formosissima Klotzsch in Peters, Naturw. Reise Mossamb. **6**: 215 (1862).
    Macrorungia formosissima (Klotzsch) C.B. Clarke in F.T.A. **5**: 255 (1900). —Binns, Checklist Herb. Fl. Malawi: 15 (1968).

Erect or scrambling shrub to 2.5(?4) m tall; young branches green, 4-angled with longitudinal grooves, glabrous to finely antrorsely sericeous, soon glabrescent. Leaves glabrous or finely antrorsely sericeous along major veins; petiole to 2.5 cm long; lamina ovate to elliptic, largest 3–7 × 2–4 cm; apex acute to obtuse, base cuneate. Flowers solitary, sessile, in axillary 4-sided (all bracts supporting flowers) (0.5)2–11 cm long racemoid cymes, terminal on main stems or on short side branches towards stem apex; peduncle to 5 mm long, puberulous or sericeous-puberulous; rachis with similar indumentum; bracts slightly imbricate or not, green usually with indistinct pale yellow margins, with 5 distinct longitudinal rib-like veins, narrowly elliptic,

0.8–1.5(2) × 0.3–0.5 cm, subglabrous to sparsely antrorsely sericeous on veins and ciliate, with numerous minute subsessile capitate glands, apex acute, base cuneate; bracteoles similar in colour and indumentum, narrowly elliptic, 6–11 mm long. Calyx 7–10 mm long, divided to c.2 mm from base, sparsely sericeous-puberulous and with numerous minute capitate glands; lobes narrowly ovate-triangular, acuminate. Corolla red to dark red, dull red, claret or purplish red (rarely yellow), 4.3–4.8 cm long, puberulous with mixture of capitate glands and hairs; tube 1.2–1.8 cm long, 3–4 mm wide; upper lip 3–3.3 cm long, lower lip 2.5–3 cm long, with 6–8 mm long lobes. Filaments 2.5–3 cm long; thecae 2.5–3 mm long. Capsule 4-seeded, 2.7–3 cm long, sparsely puberulous with mixture of hairs and capitate glands. Seeds (very immature) reticulate-tuberculate.

Zimbabwe. E: Buhera Dist., Save (Sabi) R., Birchenough Bridge, fl. 18.v.1964, *Chase* 8160 (BM, EA, K, SRGH). **Malawi**. S: Chikwawa Dist., Chikwawa, fl. 3.vii.1955, *Jackson* 1711 (K, MAL). **Mozambique**. Z: Morrumbala Dist., Morire (Murire) to Chirombe, fl. 27.vii.1949, *Andrada* 1795 (LISC). T: Changara Dist., Boroma, Sisitso, N bank of Zambezi R., fl. 17.vii.1950, *Chase* 2701 (BM). MS: Gorongosa Dist., Gorongosa Nat. Park, Urema Flats, fl. 14.vii.1957, *Chase* 6624 (K, SRGH). GI: Guijá Dist., Guijá to Mabalane, Lake Mafucuè, fl. 3.vi.1959, *Barbosa & Lemos* 8590 (K, LISC). M: Magude Dist., Magude, 325 m, fl. 10.vii.1948, *Gerstner* 6614 (K, PRE).

Also in South Africa. In dry low altitude alluvial *Acacia* and mopane bushland and grassland, dry thicket and forest patches on alluvium, termite mounds, riverine thicket, streamsides; 50–600 m.

Conservation notes: Widespread; not threatened.

2. **Anisotes bracteatus** Milne-Redh. in Hooker's Icon. Pl: t.3268 (1935). —White, For. Fl. N. Rhod.: 381 (1962). —Vollesen in F.T.E.A., Acanthaceae 2: 653 (2010). Type: Tanzania, Mpwapwa Dist., Gulwe, 19.viii.1930, *Greenway* 2407 (EA lectotype, K), lectotypified by Baden (1981).

Shrub to 3 m tall; young branches pale yellowish, with 4 ridges running down from petiole base, densely and finely whitish antrorsely sericeous, soon glabrescent. Leaves glabrous when mature, finely sericeous when young, densely so on petiole; petiole ill-defined, to 3 cm long; lamina ovate or broadly so, largest 8–12 × 4–6 cm; apex acute, base attenuate, decurrent. Flowers solitary or in 3-flowered cymules (or 1 + 2 aborted buds) towards base of cyme, in axillary dorsiventral 2.5–5 cm long racemoid cymes from upper part of branches, often aggregated to a large pseudo-panicle; peduncle 1.5–2.5 cm long, densely and finely antrorsely sericeous or puberulous; rachis with similar indumentum; bracts imbricate, pale green or yellowish green with conspicuously darker green venation and a well-defined paler margin, ovate or broadly so, 2.5–3 × 1.3–1.8 cm, sparsely sericeous-puberulous, densest on veins, apex acuminate to cuspidate, base cordate basally in cyme to truncate or cuneate towards tip, margin distinctly crinkly; bracteoles similar in colour and indumentum, ovate, 2–2.5 × 0.7–1.2 cm. Calyx 6–7 mm long, divided to 2–3 mm from base, finely uniformly antrorsely sericeous or sericeous-puberulous; lobes narrowly triangular, acute. Corolla greenish yellow to dirty orange yellow or brownish yellow, 2.7–3.1 cm long, retrorsely sericeous with sharply bent hairs and stalked capitate glands; tube 0.9–1.2 cm long, 3–4 mm wide; upper lip 1.8–2 cm long, lower lip 1.5–1.8 cm long, lobes 2–3 mm long. Filaments 1.8–2 cm long; thecae c.3 mm long. Capsule 2–4-seeded, c.2.3 cm long, densely glandular puberulous. Seeds c.6.5 × 6.5 mm, discoid, smooth and glabrous, with weak ridge on inside.

Zambia. C/E: ?Mpika Dist., Luangwa R., fl. 5.vi.1958, *Fanshawe* 4543 (K). **Zimbabwe**. N: Guruve Dist., Angwa R., Mana Pools, fl. 3.vi.1965, *Bingham* 1550 (K, LISC, SRGH). S: Masvingo Dist., Mushandike Nat. Park, Mushandike Dam, fl.& fr. 26.iv.1971, *Wright* T313 (K, LISC, SRGH).

Also in Tanzania. In riverine scrub, alluvial *Acacia* bushland and grassland, and on rocky slopes; 600–1200 m.

Conservation notes: Fairly widespread; not threatened.

The limited material from the Flora area has consistently larger corollas than material from Tanzania (2.7–3.1 vs. 2.2–2.5 cm long); more collections are needed to assess whether there is a real difference. If so, a case might be made for recognising two separate subspecies.

3. **Anisotes nyassae** C. Baden in Nordic J. Bot. **1**: 36 (1981). —White *et al.*, Evergr. For. Fl. Malawi: 113 (2001). —Vollesen in F.T.E.A., Acanthaceae **2**: 656 (2010). Type: Malawi, Misuku Hills, Wilindi Forest, n.d., *Müller* 1675 (K holotype, SRGH).

   Anisotes sp. 1 sensu White *et al.*, Evergr. For. Fl. Malawi: 113 (2001).

Shrub or small tree, sometimes scrambling, to 5 m tall; young branches 4-angled, antrorsely sericeous or sparsely so or puberulous, soon glabrescent. Leaves sparsely antrorsely sericeous or puberulous along major veins; petiole to 3.5 cm long; lamina narrowly elliptic to elliptic (rarely slightly obovate), largest 6.5–22.5 × 2.5–6.5 cm; apex subacuminate, base attenuate, decurrent. Flowers solitary or 2 towards base of cyme, in axillary dorsiventral 1–4 cm long racemoid cymes from near branch apex; peduncle 2–4 cm long, antrorsely sericeous to puberulous; rachis with similar indumentum; bracts imbricate or not, uniformly green, narrowly ovate-elliptic, (0.3)1.1–1.7 × (0.1)0.3–0.7 cm (see note), sparsely antrorsely sericeous on veins, ciliate, apex subacuminate, base cuneate to attenuate; bracteoles similar in colour and indumentum, lanceolate to narrowly triangular, 2–7 × 1–2 mm. Calyx (4)5–8 mm long, divided to c.2 mm from base, whitish puberulous or sparsely so, with sessile gland dots; lobes narrowly ovate to narrowly triangular, acuminate. Corolla purple, (2.2)3–3.5 cm long, glandular puberulous; tube (0.8)1–1.5 cm long, 2 (base) to 4 (apex) mm in diameter; upper lip (1.4)2–2.5 cm long, lower lip 1.5–2 cm long, divided almost to base. Filaments (1)1.5–2 cm long; thecae 1.5–2 mm long. Capsule 4-seeded, 2–2.3 cm long, densely pubescent with non-glandular hairs and scattered short capitate glands. Seeds discoid, reticulate-tuberculate, with very weak ridge on inside, c.5 × 4 mm.

   Zambia. E: Isoka Dist., Makutu Mts, fl.& fr. 28.x.1972, *Fanshawe* 11623 (K). **Malawi**. N: Chitipa Dist., Misuku Hills, Wilindi Forest, 1850 m, fl. 5.x.1982, *Dowsett-Lemaire* 427 (K).

Also in Tanzania. In wet evergreen montane forest; 1700–2000 m.

Conservation notes: Local montane distribution; possibly Near Threatened.

Some inflorescences from *Dowsett-Lemaire* 840 (Malawi, S Viphya, Ntungwa Forest) have reduced bracts c.3 mm long and very small corollas (minimum dimensions in description), although other branches from the same collection have unopened buds c.2.5 cm long and would have had opened flowers of normal size. The anthers are normally developed. I am treating this merely as an abnormal, possibly diseased, specimen.

4. **Anisotes pubinervis** (T. Anderson) Heine in Fl. Gabon **13**: 189 (1966). —Vollesen in F.T.E.A., Acanthaceae **2**: 654 (2010). Type: Malawi, Mt Tohiradzovu, 3.x.1859, *Kirk* s.n. (K holotype). FIGURE 8.6.**63**.

   Rungia pubinervia T. Anderson in J. Linn. Soc., Bot. **7**: 46 (1863).

   Himantochilus marginatus Lindau in Bot. Jahrb. Syst. **20**: 60 (1894). Type: Tanzania, Lushoto Dist., Usambara Mts, Kwa Mshuza, viii.1893, *Holst* 9063 (B† holotype, BM, K).

   Himantochilus pubinervius (T. Anderson) Lindau in Engler, Pflanzenw. Ost-Afr. **C**: 373 (1895).

   Macrorungia pubinervia (T. Anderson) C.B. Clarke in F.T.A. **5**: 255 (1900). —Binns, Checklist Herb. Fl. Malawi: 15 (1968). —Baden in Nordic J. Bot. **1**: 148 (1981). —Mapura & Timberlake, Checklist Zimb. Vasc. Pl.: 14 (2004). —Phiri, Checklist Zamb. Vasc. Pl.: 19 (2005).

   Metarungia pubinervia (T. Anderson) C. Baden in Kew Bull. **39**: 638 (1984). —White *et al.*, Evergr. For. Fl. Malawi: 116 (2001).

Shrub or small tree to 4(7) m; young branches ± 4-angled, dark green, finely antrorsely whitish sericeous or densely so (rarely tomentellous), soon glabrescent, older branches brownish. Mature leaves subglabrous to densely puberulous or sericeous, densest on midrib and large veins; petiole to 5 cm long; lamina ovate to elliptic or narrowly so, largest 11.5–25 × 4–9 cm; apex

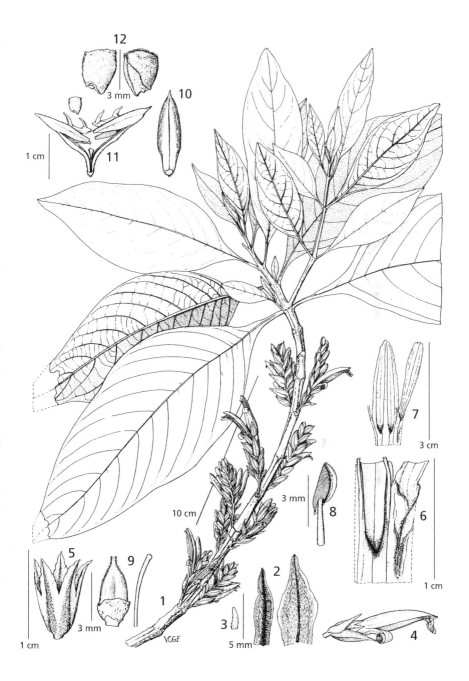

Fig. 8.6.**63**. ANISOTES PUBINERVIS. 1, habit; 2, bracts; 3, bracteole; 4, corolla with bracts and calyx; 5, calyx; 6, corolla tube, opened up; 7, corolla limb; 8, anther; 9, ovary, style and stigma; 10, capsule; 11, dehisced capsule; 12, seeds. 1, 4 & 6–8 from *Perdue & Kibuwa* 9475, 2, 3 & 5 from *Whellan* 261, 9–12 from *Fanshawe* 7081. Drawn by Victoria Gordon-Friis. Reproduced with permission from Nordic Journal of Botany (1981).

acuminate to cuspidate (rarely acute or obtuse), base attenuate, decurrent. Flowers solitary, subsessile, in axillary dorsiventral 1–4(5.5) cm long racemoid cymes, or from old branches below leaves; peduncle 0.3–1.5 cm long (to lowermost bract with red margin), with a number of small sterile bracts, sparsely to densely sericeous; rachis with similar indumentum; fertile bracts imbricate or not, green with broad pale red to claret margin, ovate to elliptic or narrowly so, 0.7–1.7(2) × 0.3–0.7 cm, apex acute or subacute, base truncate or cuneate, subglabrous to sericeous; bracteoles absent, rarely present, filiform to 2 mm long. Calyx green with broad pale red to claret margins, 8–13 mm long, fused to ± middle but sometimes to within 2–3 mm of apex, subglabrous to sparsely puberulous or sericeous, densest along broad midrib, ciliate; lobes triangular, acute. Corolla dark claret to crimson (very rarely white), 3–4.2 cm long, glabrous; tube 1–1.4 cm long, 2–4 mm wide; upper lip 2–3 cm long, lower lip 2–3 cm long, with 3 minute lobes c.1 mm long. Filaments 2–3 cm long; thecae 3–4 mm long. Capsule with placentae rising elastically from base at maturity and splitting from outer wall, each valve usually splitting longitudinally from base, 4-seeded, 1.6–1.9 cm long, glabrous or sparsely puberulous near apex. Seeds densely reticulate-verrucose when young and with transverse fold, ± smooth and without fold at maturity, triangular-conical in shape, c.4 × 3 mm.

Zambia. W: Mpongwe Dist., fl.& fr. 11.x.1962, *Fanshawe* 7086 (K, NDO). E: Chama Dist., Nyika Plateau, Kasoma Forest, 2000 m, fl.& fr. 2.ix.1981, *Dowsett-Lemaire* 259 (K). **Zimbabwe**. W: Matobo Dist., Matopos Nat. Park, Maleme Dam, fl. 30.iv.1967, *Biegel* 2097 (K, SRGH). E: Chipinge Dist., Chirinda, 1175 m, fl. 23.iv.1947, *Wild* 1899 (K, SRGH). S: Masvingo Dist., Zimbabwe Ruins, fl. 5.iii.1970, *Mavi* 1073 (K, SRGH). **Malawi**. N: Chitipa Dist., Misuku Hills, Mugesse Forest, 1800 m, fl. 14.vii.1970, *Brummitt* 12100 (K, MAL). C: Nkotakota Dist., Chipata Hill, fr. 5.viii.1963, *Ball* 64 (K, SRGH). S: Machinga Dist., Liwonde Forest Reserve, Mt Chikala, Mposa Village, 1450 m, fl. 20.vii.1982, *Chapman & Patel* 6358 (K). **Mozambique**. MS: Cheringoma Dist., Inhamitanga, 11 km from Marromeu turnoff on Lacerdonia road, 200 m, fl. 12.v.1998, *Burrows* 6282 (K). GI: Xai Xai Dist., Chipenhe, Chirindzene Forest, fl. 9.vi.1960, *Lemos & Balsinhas* 47 (BM, K).

Also in Nigeria, Sudan, Ethiopia, D.R. Congo, Burundi, Uganda, Kenya and Tanzania. In intermediate and montane (rarely lowland) evergreen forest and riverine forest, often a dominant understorey shrub; 20–2000 m.

Conservation notes: Widespread; not threatened.

The genus *Metarungia* is normally separated from *Anisotes* by the placentae rising elastically from the base at maturity and splitting from the outer capsule wall. It is the only character separating the two genera and in recent years this has been found to be unreliable in other pairs of genera (*Justicia/Rungia, Dicliptera/Peristrophe*).

In some collections, e.g. *Fanshawe* 7086, the placentae do not rise elastically or only partly so in some capsules, although the lateral capsule walls are thin and break irregularly.

5. **Anisotes longistrobus** (C.B. Clarke) Vollesen, comb. nov. Type: South Africa, Mpumalanga, near Barberton, Avoca, 1890, *Galpin* 888 (K holotype).

> *Macrorungia longistrobus* C.B. Clarke in Fl. Cap. 5: 89 (1901). —Baden in Nordic J. Bot. 1: 150, fig.5 (1981).
>
> *Metarungia longistrobus* (C.B. Clarke) Baden in Kew Bull. **39**: 638 (1984). —da Silva *et al.*, Prelim. Checklist Vasc. Pl. Mozamb.: 19 (2004).

Shrubby herb or shrub to 1.5 m; young branches ± 4-angled, green, antrorsely sericeous to tomentellous, soon glabrescent, older branches brownish. Mature leaves subglabrous to puberulous or sericeous, densest on midrib and large veins; petiole to 1.2 cm long; lamina elliptic or narrowly so, largest 8.5–15(20) × 2.5–4(5) cm; apex subacuminate to obtuse, base attenuate, decurrent. Flowers solitary, subsessile, in (2)3–7.5(10) cm long dorsiventral racemoid terminal cymes, or on small axillary branches, or some axillary; peduncle absent or 0.3–1 cm long in axillary cymes, with a single pair of sterile bracts; fertile bracts imbricate, green with broad paler hyaline margin, ovate, (1.2)1.6–2.6 × (0.5)0.8–1.5 cm, subglabrous to sericeous along veins, apex acute or subacute, base truncate or cuneate; bracteoles narrowly triangular, 1–3 mm long. Calyx green with broad paler hyaline margins, 7–11 mm long, fused to ± middle but sometimes to within 2–3 mm of apex; sparsely puberulous or sericeous, densest along broad midrib, ciliate; lobes triangular, acute. Corolla salmon to brick red, 4.9–5.8 cm long, glabrous or sparsely puberulous; tube 1.9–2.5 cm long, c.3 mm wide; upper lip 3–3.3 cm long, lower lip 2.5–3.1 cm long, with 3 minute lobes 1–2 mm long. Filaments 2.4–3.1 cm long; thecae 4–5 mm long. Capsule with placentae rising elastically from base at maturity and splitting from outer wall, each valve usually splitting longitudinally from base, 4-seeded, 1.8–2.2 cm long, puberulous. Seeds smooth, triangular-conical in shape, c.4 × 3 mm.

Mozambique. M: Namaacha Dist., Goba Estate, fl. 20.i.1980, *de Koning* 7958 (BM). Also in South Africa (Limpopo, Mpumulanga). In coastal thicket and bushland; c.100 m.

Conservation notes: Fairly widespread but local in the Flora area; possibly Near Threatened.

6. **Anisotes sessiliflorus** (T. Anderson) C.B. Clarke in F.T.A. **5**: 226 (1900). — Vollesen in F.T.E.A., Acanthaceae **2**: 662 (2010). Type: Malawi, Shire R., Shibisa (Chikwawa), viii-ix.1861, *Meller* s.n. (K holotype, K).

Himantochilus sessiliflorus T. Anderson in Bentham & Hooker, Gen. Pl. **2**: 1117 (1876). *Anisotes sessiliflorus* subsp. *iringensis* Baden in Nordic J. Bot. **1**: 36 (1981); in Nordic J. Bot. **1**: 660 (1981). Type: Tanzania, Kilosa Dist., Ruaha R. Gorge, 16.v.1971, *Mhoro* 1236 (UPS holotype, EA, K).

Shrub to 3(?4) m tall; young branches pale yellowish brown, ± 4-angled and longitudinally ridged, sparsely antrorsely sericeous on two sides, very quickly glabrescent, older branches grey to dark grey or purplish grey. Leaves sparsely antrorsely sericeous along midrib and larger veins, sessile or with ill-defined petiole to 2 mm long; lamina ovate to elliptic or broadly so, largest 6–15 × 3–6 cm; apex subacuminate to obtuse; base attenuate, decurrent to stem. Flowers in a 1-flowered cymule enclosed by pairs of bracts and bracteoles, sometimes 2(3) cymules per axil; peduncle absent or to 1 mm long; bracts uniformly green, ovate to triangular, 2–3.5 mm long, subacute to obtuse, densely minutely antrorsely sericeous; bracteoles triangular or ovate-triangular, 2–3.5 mm long, with similar indumentum. Calyx 2.5–4 mm long, divided to c.1 mm from base, with similar indumentum; lobes narrowly triangular, subacute, with faint white edges and conspicuous rib-like midrib. Corolla tube yellow, limb bright orange, orange-red or bright red, 4–5.2 cm long, glandular puberulous and with long pilose hairs, densest at base; tube 1.4–1.6 cm long, 2.5–4 mm wide; upper lip 2.6–3.6 cm long, lower lip 2.5–3.5 cm long, lobes 2–3 mm long. Filaments 2.5–3.5 cm long; thecae 2.5–3.5 mm long. Capsule and seeds not seen (see note).

Zimbabwe. S: Chiredzi Dist., Chitsa's Camp to Chibilia Falls, near Save R., fl. 15.vi.1950, *Chase* 2521 (BM, SRGH). **Malawi.** S: Chikwawa Dist., Chikwawa, 175 m, fl. 19.v.1949, *Gerstner* 7060 (EA, K, PRE). **Mozambique.** Z: Morrumbala Dist., Chire, fl. 4.iv.1972, *Bowbrich* T220 (LISC). T: Mágoè Dist., Chitengo, 4 km towards Sangrassa camp, fl. 4.v.1978, *Diniz et al.* 189 (LMU). MS: Chemba Dist., Chiramba, fl. vi.1926, *Surcouf* 129 (K, P).

Also in Tanzania. In mopane woodland, bushland and dry riverine scrub; 100–300 m.

Conservation notes: Widespread; not threatened.

I do not see any justification for maintaining subsp. *iringensis*. Including recent material, the overlap in corolla size is larger than indicated by Baden (1981) and there is also an overlap in calyx size. The differences are smaller than the ones between the northern and southern populations of *Anisotes bracteatus*, a species which has a very similar distribution but was kept as one taxon by Baden.

There are c.3 cm long mis-shapen (?galled) capsules on *Hall-Martin* 1235 from Malawi; they are sparsely antrorsely sericeous with no glandular hairs.

7. **Anisotes rogersii** S. Moore in J. Bot. **57**: 91 (1919). Type: South Africa, Messina, Soutpansberg, n.d., *Rogers* 19349 (BM holotype, K).

Anisotes sessiliflorus sensu Codd, Trees Shrubs Kruger Nat. Park: 168 (1951), non (T. Anderson) C.B. Clarke.

Shrub to 1.5(2) m tall; young branches pale yellowish brown, ± 4-angled and longitudinally ridged, finely antrorsely sericeous to (more commonly) puberulous, soon glabrescent, older branches dark purplish grey. Leaves sparsely antrorsely sericeous or puberulous along midrib and larger veins only, or also on lamina, sessile or with ill-defined petiole to 2 mm long; lamina ovate to elliptic, largest 3–5.5(7.5) × 1.5–2.5(3.7) cm; apex subacuminate to obtuse, base attenuate, decurrent to stem. Flowers in a 1-flowered cymule enclosed by pairs of bracts and bracteoles, sometimes 2 cymules per axil; peduncle absent; bracts uniformly green, ovate or narrowly triangular, (2.5)3–5 mm long, subacute, finely antrorsely sericeous or densely so; bracteoles triangular, 3–4.5 mm long, indumentum similar. Calyx 3–5 mm long, divided to c.1 mm from base, indumentum similar; lobes narrowly triangular, acute, with faint white edges and conspicuous rib-like midrib. Corolla tube yellow, limb bright orange-red or bright red, 3.8–4.8 cm long, sparsely glandular puberulous and with dense long pilose hairs; tube (1)1.2–1.6 cm long, 2.5–4 mm wide; upper lip 2.8–3.3 cm long, lower lip 2.5–3.2 cm long, lobes 2–3.5 mm long. Filaments 2.5–3.2 cm long; thecae 3–3.5 mm long. Capsule (from Baden) 2.5–3 cm long, 'sparsely strigose'. Seeds c.6 × 5 mm, reticulate-tuberculate.

Zimbabwe. S: Beitbridge Dist., Nulli Range, fl. 26.ii.1961, *Wild* 5431 (K, SRGH); Chiredzi Dist., Gonarezhou Nat. Park (Game Reserve), Swimuwina Camp, 4.vi.1971, *Grosvenor* 620 (K, LISC, PRE, SRGH).

Also in South Africa. In mopane woodland and bushland, and rocky outcrops with *Commiphora–Grewia* thicket; 250–700 m.

Conservation notes: Localised in Flora area, but probably not threatened.

The species is very closely related to *Anisotes sessiliflorus* and perhaps should be considered no more than a subspecies of that, although in general appearance they are very different and easily recognisable. *A. rogersii* has small leaves and generally flowers with the developing leaves, while *A. sessiliflorus* has bigger leaves and flowers before the leaves develop or with old last season leaves still on the plant. Bracts, bracteoles and calyces are also generally larger in *A. rogersii*.

The material of *A. sessiliflorus* is sparse and of poor quality and more and better collections are needed to solve this problem.

37. ECBOLIUM Kurz[15]

Ecbolium Kurz in J. Asiat. Soc. Bengal, 2, Nat. Hist. **40**: 75 (1871). —Vollesen in Kew Bull. **44**: 638–680 (1989).

Justicia Kuntze, Rev. Gen.: 491 (1891), non *Justicia* L. (1753).

Erect perennial or shrubby herbs or shrubs; stems articulated with transverse lines at nodes, usually swollen above nodes and with numerous longitudinal ribs. Leaves opposite, with short often inconspicuous cystoliths, margin entire, usually recurved. Flowers usually sessile, solitary or 3(7) per bract, in dense strobilate terminal spiciform cymes, axes often flattened; bracts papery, usually caducous, pale to yellowish green (rarely purplish), with straight non-pungent tip, usually with inconspicuous to ± invisible venation; bracteoles 2, subulate to narrowly triangular. Calyx deeply divided into 5 equal lanceolate to narrowly triangular acuminate to cuspidate lobes. Corolla livid yellowish to bluish green or turquoise green (rarely white to lilac), upper part of tube usually pubescent outside; tube long, cylindric, slightly widening upwards, straight; lower lip with 3 horizontal narrowly elliptic to circular rounded lobes, middle one widest; upper

---

[15] By Kaj Vollesen

lip with one linear-lanceolate 2-veined bifurcate lobe held erect or curved back. Stamens 2, held erect and parallel under upper lip, inserted near base of lower lip, usually 1–3 mm long, glabrous; anthers usually 2–3 mm long, bithecous, medifixed, held parallel with filament, thecae oblong, subequal, curved, rounded at both ends. Style filiform, glabrous or hairy near base (rarely over whole length); stigma lobes equal, broadly elliptic, rounded, erect; ovary with 2 ovules per locule. Capsule 2–4-seeded, club-shaped, apiculate; retinaculae strong. Seed discoid, cordiform in outline, from smooth to densely tuberculate, usually with a broad raised rim with entire to jagged edge.

A genus of 22 species, 12 in eastern and southern Africa (10 endemic), 7 on Madagascar and the Comoro Is. (all endemic), 3 in southern Arabia (1 endemic), 3 in India (2 endemic), one of which extends to Malaysia (probably introduced).

1. Corolla white to lilac, tube c.0.7 cm long, lobes c.1 cm long; filaments c.7 mm long; stem 4-angular with 4 longitudinal furrows on edges . . . . **1.** *hastatum*
–   Corolla livid bluish green or turquoise green, tube 2–3.5 cm long, lobes shorter than tube; filaments 1–3 mm long; stem rounded, with numerous longitudinal ribs 2
2. Inflorescence axis glabrous or with scattered capitate glands; bracts glabrous; calyx glabrous except for finely ciliate edges; capsule glabrous . . . . . **4.** *glabratum*
–   Inflorescence axis puberulous or sericeous, often also with capitate glands; bracts uniformly puberulous, with capitate glands; calyx uniformly puberulous; capsule puberulous 3
3. Seeds smooth or very slightly tuberculate; calyx 4–5(7) mm long; corolla tube 2–2.7 cm long . . . . **3.** *clarkei*
–   Seeds densely tuberculate; calyx (4)5–11 mm long; corolla tube 2.5–3.5 cm long . . . . **2.** *amplexicaule*

1. **Ecbolium hastatum** Vollesen in Kew Bull. **44**: 643 (1989). Type: Mozambique, Vilanculos, viii.1936, *Gomes e Sousa* 1882 (K holotype).

Shrubby herb to 2 m tall; young branches 4-angular with 4 longitudinal furrows on edges, glabrous, glaucous. Leaves glabrous, fleshy; petiole 1–4 mm long; lamina elliptic to obovate, largest 4–7.5 × 1.5–2.2 cm; apex acute to rounded, finely apiculate; base slightly to distinctly hastate. Inflorescence single, 2–5 cm long; peduncle 2–5 mm long; axis finely glandular puberulous with short non-capitate glands and subsessile capitate glands; bracts broadly elliptic to orbicular, broadly rounded with mucro to 1 mm, 6–10 × 5–8 mm, upper finely glandular puberulous with non-capitate and capitate glands; bracteoles 1–2(4) mm long; flowers solitary. Calyx 4–7 mm long, finely glandular puberulous with non-capitate glands and scattered capitate glands. Corolla white to lilac, tube c.7 mm long; lobes in lower lip c.10 mm long, middle c.3 mm wide. Filaments c.7 mm long. Capsule (immature) glabrous. Seeds (immature) densely tuberculate.

Mozambique. GI: Xai-Xai Dist., Xai-Xai (João Belo), Praia Sepúlveda, fl. 9.x.1968, *Balsinhas* 1359 (LISC, LMA, PRE). M: Maputo Dist., Inhaca Is., Ponta Ponduini, fl. 3.vi.1970, *Correia & Marques* 1612 (LMU).

Not known elsewhere. Coastal forest and bushland on old sand dunes; 0–25 m.

Conservation notes: Endemic to coastal southern Mozambique, where it is possibly Vulnerable.

Known only from three collections, this species shows some characters reminiscent of *Megalochlamys* – corolla lobes longer than tube, long filaments, stem 4-angular with 4 longitudinal furrows. But the caducous bracts, curved anthers which are held erect and parallel with the parallel filaments, and erect stigma lobes indicate that it is better placed in *Ecbolium*. The pollen is exactly like in other species of *Ecbolium* (see Furness in Kew Bull. **44**: 688, 1989).

2. **Ecbolium amplexicaule** S. Moore in J. Bot. **32**: 136 (1894). —Clarke in F.T.A. **5**: 237 (1900), for type. —Vollesen in F.T.E.A., Acanthaceae **2**: 667 (2010). Type: Kenya, Kilifi, Galana (Sabaki) R., 11.viii.1893, *Gregory* s.n. (BM holotype).

Ecbolium auriculatum C.B. Clarke in F.T.A. **5**: 237 (1900), invalid name, non *E. auriculatum* (Nees) Kuntze. Type: Tanzania, no locality, n.d., *Hannington* s.n. (K lectotype); Maramo, n.d. *Stuhlmann* 6987 (B† syntype); Usambara, n.d. *Holst* 608a (B† syntype), lectotypified by Vollesen (1989).

Perennial or shrubby herb to 2 m tall; young branches subglabrous to sericeous-pubescent. Leaves subglabrous to puberulous, sessile (rarely with petiole to 5(15) mm); lamina pandurate to elliptic-pandurate, largest 8–22 × 2–8.5 cm; apex acuminate to subacute; base auriculate, clasping stem. Inflorescence 3–22(33) cm long; peduncle 4–25 mm long; axis sericeous-puberulous, with few to many stalked capitate glands; bracts ovate (lower) to elliptic, acute to acuminate (or lowermost rounded) with straight mucro 1–2 mm, 13–30 × 7–22 mm, upper puberulous or densely so, usually with dense stalked capitate glands, indistinctly ciliate; bracteoles 1.5–3.5(6.5) mm long; flowers solitary or in 3s at lower nodes. Calyx (4)5–9(11) mm long, puberulous, with stalked capitate glands. Corolla tube 2.5–3.5 cm long; middle lobe in lower lip 13–18 × 7–11 mm, lateral lobes to 17 × 7 mm, upper lip 10–14 × 1–2 mm. Filaments pale green, anthers dark blue. Capsule 1.7–2.3 cm long, puberulous. Seeds 7.5–10 × 6.5–9 mm, both surfaces densely tuberculate with tubercles denser towards edge, rim prominent.

Mozambique. N: Pemba Dist., 3 km S of Pemba (Porto Amelia), fl.& fr. 11.iii.1960, *Gomes e Sousa* 4529 (K, PRE, SRGH); Mossuril Dist., Matibane, 20.ii.1984, *Groenendijk et al.* 1212 (K, LMU).

Also in Kenya and Tanzania. Coastal bushland; 10–50 m.

Conservation notes: Restricted distribution within the Flora area, where it is possibly threatened.

3. **Ecbolium clarkei** Hiern in Cat. Afr. Pl. Welw. **1**(4): 1032 (1900). —Clarke in F.T.A. **5**: 514 (1900). —Meyer in Merxmüller, Prodr. Fl. SW Afr. **130**: 29 (1968). Type: Angola, Luanda, n.d., *Welwitsch* 5124 (BM lectotype, K), lectotypified by Vollesen (1989). FIGURE 8.6.64.

Ecbolium amplexicaule sensu Clarke in F.T.A. **5**: 237 (1900), excl. type. —da Silva *et al.*, Prelim. Checklist Vasc. Pl. Mozambique: 18 (2004), non S. Moore.

Perennial herb to 1.25 m tall, rarely shrubby, usually with several stems from a woody rootstock; young stems glabrous (rarely slightly puberulous on uppermost node) or uniformly puberulous, often glaucous. Leaves glabrous (hairy along midrib) or uniformly puberulous, often glaucous, sessile or petiole to 3 mm long; lamina ovate-elliptic to pandurate, largest 5–15 × 2.7–6 cm; apex acute or subacute; base auriculate to cordate, clasping stem. Inflorescence 3–21 cm long; peduncle 4–12 mm long; axis puberulous, with few to many stalked capitate glands (rarely without); bracts ovate to elliptic, acuminate with straight mucro 1–3 mm, 12–22 × 7–15 mm, upper puberulous or sparsely so, with usually dense stalked capitate glands (rarely without), indistinctly ciliate; bracteoles 2–3.5 mm long; flowers solitary or in 3s at lower nodes. Calyx 4–5(7) mm long, puberulous, with few to many stalked capitate glands. Corolla tube 2–2.7 cm long; middle lobe in lower lip 12–15 × 6–8 mm, lateral lobes to 13 × 5 mm, upper lip 8–12 × 1–2 mm. Filaments pale green, anthers dark blue. Capsule 1.7–2.2 cm long, puberulous. Seeds 7.5–9 × 6.5–8.5 mm, both surfaces smooth or very slightly tuberculate.

a) Var. **clarkei**.

Young stems glabrous, rarely slightly puberulous on uppermost node. Leaves glabrous apart from hairy midrib.

Malawi. S: Machinga Dist., Liwonde Nat. Park, fl.& fr. 17.iv.1980, *Blackmore et al.* 1256 (BM, C, K).

Also in Angola and Namibia. Mopane woodland, often on rocky slopes; c.500 m.

Conservation notes: Wide distribution; not threatened.

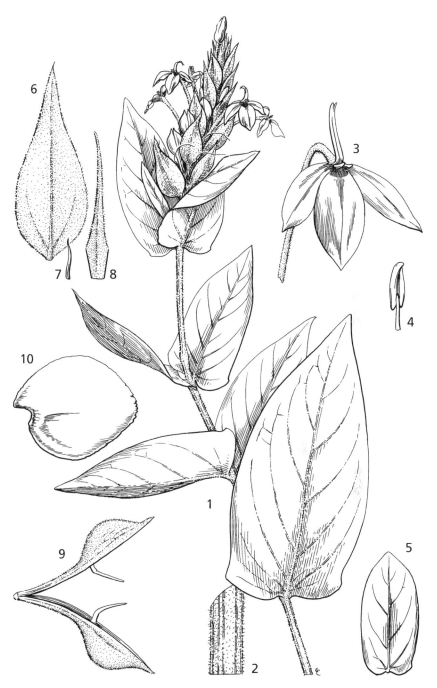

Fig. 8.6.**64**. ECBOLIUM CLARKEI var. PUBERULUM. 1, flowering branch (× ²/₃); 2, detail of stem (× 6); 3, corolla (× 2); 4, upper part of filament and anther (× 6). ECBOLIUM CLARKEI var. CLARKEI. 5, leaf (× ²/₃); 6, bract (× 2); 7, bracteole (× 6); 8, calyx lobe (× 6); 9, capsule (× 2); 10, seed (× 5). 1–4 from *Merxmüller& Giess* 30540, 5,9,10 from *Merxmüller & Giess* 1411, 6–8 from *Merxmüller & Giess* 35010. Drawn by Eleanor Catherine. Reproduced from Kew Bulletin (1989).

b) Var. **puberulum** Vollesen in Kew Bull. **44**: 668 (1989). Type: Zimbabwe, Mwenezi R., Buffalo Bend, 28.iv.1961, *Drummond & Rutherford-Smith* 7576 (K holotype, LISC, PRE, SRGH).

Young stems uniformly puberulous. Leaves uniformly puberulous on both sides.

Zimbabwe. S: Chiredzi Dist., near Fishans, 25.iv.1962, *Drummond* 7744 (K, SRGH). **Mozambique**. GI: Guijá Dist., Caniçado, 7.2 km from Massingir to Macuma, 14.xi.1970, *Correia* 1960 (LMU).

Also in Angola and Namibia. Mopane woodland, rocky slopes, river banks and riverine forest; 50–300 m.

Conservation notes: Wide distribution; not threatened.

4. **Ecbolium glabratum** Vollesen in Kew Bull. **44**: 669 (1989). Type: South Africa, Mpumalanga, Komati Poort, 17.xii.1897, *Schlechter* 11792 (K holotype, BM, E, G, HBG, P, S, Z).

Ecbolium amplexicaule sensu Ross, Fl. Natal: 326 (1972). —Compton in J. S. Afr. Bot. (suppl.) **11**: 561 (1976). —Gibbs Russell in Mem. Bot. Surv. S. Afr. **48**: 118 (1984). —da Silva *et al.*, Prelim. Checklist Vasc. Pl. Mozambique: 18 (2004), non S. Moore.

Perennial or shrubby herb to 1 m tall with several stems from a woody rootstock; young stems glabrous and glaucous. Leaves glabrous (rarely a few hairs on midrib), glaucous, sessile (rarely petiole to 1 mm long); lamina ovate-elliptic or narrowly so or slightly pandurate, largest 3–7(9.5) × 1.7–3(4.2) cm; apex subacute to broadly rounded; base truncate to slightly auriculate, then clasping stem. Inflorescence 2–11 cm long; peduncle 4–15(25) mm long; glabrous or with scattered stalked capitate glands; bracts elliptic, rounded (basal) to acute with straight mucro less than 0.5 mm, 15–30(40) × 7–18 mm, glabrous; bracteoles 1–3 mm long; flowers solitary. Calyx 5–8 mm long, glabrous apart from finely ciliate edges. Corolla tube 2.3–3.3 cm long; middle lobe in lower lip 13–17 × 7–10 mm, lateral lobes to 15 × 7 mm, upper lip 10–13 × 1–2 mm. Filaments pale green, anthers dark blue. Capsule 1.8–2.5 cm long, glabrous. Seeds 8–10 × 6.5–8.5 mm, both surfaces tuberculate, tubercles often merging into crests towards edges.

Mozambique. GI: Chibuto Dist., on road to Chaimite, fl.& fr. 11.ii.1942, *Torre* 3965 (BR, COI, LISC, LMA, LMU). M: Magude Dist., Chobela, fl. 15.i.1964, *Balsinhas* 688 (B, K); Matola Dist., Matola to Umbeluzi, fl. 2.xii.1947, *Barbosa* 623 (BR, LISC, LMU, WAG).

Also in Swaziland and South Africa. Coastal and lowland *Acacia* bushland on clay; 10–50 m.

Conservation notes: Restricted distribution, but probably not threatened.

38. MEGALOCHLAMYS Lindau[16]

Megalochlamys Lindau in Bot. Jahrb. Syst. **26**: 345 (1899). —Vollesen in Kew Bull. **44**: 605 (1989).

Ecbolium Kurz subgen. *Choananthus* C.B. Clarke in F.T.A. **5**: 236 (1900), excl. *E. striatum* and *E. barlerioides*.

Erect (rarely scrambling) shrubby herbs or shrubs; stems 4-angular with 4 longitudinal furrows on edges. Leaves opposite, usually with conspicuous cystoliths, sometimes with dark sessile glands, entire. Flowers solitary, sessile, in dense strobilate terminal spiciform cymes; bracts persistent, usually glossy, usually with a straight or recurved pungent tip and with conspicuously raised venation; bracteoles 2, subulate. Calyx deeply divided into 5 equal linear to lanceolate or narrowly triangular lobes. Corolla pale blue to bright sky-blue, puberulous outside; tube cylindric, not or very slightly widening upwards, straight or very slightly curved; lower lip with

---

[16] By Kaj Vollesen

3 horizontal narrowly elliptic to elliptic subequal lobes; upper lip with one narrowly elliptic 2-veined entire (rarely retuse) lobe. Stamens 2, spreading straight out between lower and upper lip and diverging, inserted near base of lower lip, usually with short downwardly directed hairs along their whole length; anthers usually 2–3 mm long, bithecous, medifixed, held at right angle to filament, thecae oblong, subequal, straight, rounded at both ends. Style filiform, hairy in lower part or in whole length; stigma lobes equal, oblong-elliptic, flat, rounded, usually recurved; ovary with 2 ovules per locule. Capsule 2-seeded, club-shaped, apiculate; retinaculae strong. Seeds discoid, broadly elliptic to orbicular in outline with a slightly raised usually entire rim, rugose with dense glandular-glochidiate hairs.

A genus of 10 species in eastern and southern Africa, two extending to the southern part of the Arabian Peninsula.

Hairs on young branches appressed; corolla lobes 8–14 mm long; filaments 5–8 mm long . . . . **1.** *revoluta*
– Hairs on young branches spreading, bent down near apex, rarely straight; corolla lobes 13–18 mm long; filaments 7–10 mm long . . . . **2.** *hamata*

1. **Megalochlamys revoluta** (Lindau) Vollesen in Kew Bull. **44**: 616 (1989); in F.T.E.A., Acanthaceae 2: 671 (2010). Type: Tanzania, Lushoto Dist., W of Pare Mts, n.d., *von Höhnel* 8 (B† holotype); Tanzania, Lushoto Dist., Lake Manka, v.1967, *Procter* 3652 (K neotype, EA, FT), neotypified by Vollesen (1989).

   Schwabea revoluta Lindau in Bot. Jahrb. Syst. **20**: 59 (1894).
   Ecbolium revolutum (Lindau) C.B. Clarke in F.T.A. **5**: 239 (1900).

   Shrubby herb or shrub to 1.25 m tall; young branches, petioles and peduncles pubescent or finely sericeous. Leaves pubescent or sparsely so, often densest on veins and margins; petiole 1–8(15) mm; lamina narrowly ovate to ovate or narrowly elliptic to elliptic, largest 2–8(12) × 0.5–2.3(5.2) cm; apex acute to retuse, apiculate, base attenuate to truncate. Inflorescence 1–8 cm long; peduncle 1–8 mm; axis finely sericeous to puberulous with non-glandular hairs and non-capitate glandular hairs, scattered capitate glands sometimes present; bracts green or with purplish veins, ovate or broadly so or elliptic, with a sharp recurved acumen 1–6 mm long, 7–20 × 3–14 mm, subglabrous to puberulous or pubescent with non-glandular hairs or non-capitate glandular hairs, short (less than 0.5 mm long) capitate glands present or not, ciliate or not; bracteoles 1–4 mm long. Calyx 5–8 mm long, puberulous to pubescent with non-capitate glands, usually also with stalked capitate glands. Corolla bright sky-blue; tube 6–13 mm long; lobes 8–14 mm long, middle lobe in lower lip 2–4.5 mm wide. Filaments 5–8 mm long. Capsule 9–14 mm long, puberulous, mainly along sutures with central part of valves usually glabrous. Seeds 4.5–6 × 3.5–5 mm, greyish, rugose with numerous glandular-glochidiate hairs, rim entire.

   Subsp. **cognata** (N.E. Br.) Vollesen in Kew Bull. **44**: 619 (1989). Type: Botswana, Central Dist., Chukutsa salt pan, ii.1897, *Lugard* 223 (K holotype).

   Ecbolium schlechteri Lindau in Bot. Jahrb. Syst. **38**: 72 (1905). Type: Mozambique, near Beira, iv.1895, *Schlechter* s.n. (B† holotype).
   Ecbolium cognatum N.E. Br. in Bull. Misc. Inform., Kew **1909**: 130 (1909). —Grignon & Johnsen, Check-list Vasc. Pl. Botswana: 4 (1986).
   Ecbolium hamatum sensu Clarke in F.T.A. **5**: 239 (1900) for *Lugard* 223, non sensu stricto.

   Young branches finely sericeous or densely so. Leaves narrowly ovate to ovate or narrowly elliptic to elliptic. Inflorescence axis glandular puberulous with short non-capitate and longer capitate glands, sometimes also with scattered non-glandular hairs. Corolla tube 9–13 mm long. Capsule entirely puberulous or sparsely so or only apically. Seeds 5–6 × 4–5 mm.

   Botswana. N: Ngamiland, Kwebe Hills, 1050 m, 22.ix.2009, *Gwafila et al.* MSB 621 (K). SE: Central Dist., 75 km W of Lethlakane (Lothlekane), between Lake Xau (Dow) and Ntshukuksa (Chukutsa) Pan, fl. 23.iii.1965, *Wild & Drummond* 7242 (K, PRE, SRGH). **Zimbabwe**. S: Beitbridge Dist., Nottingham Ranch, 550 m, fl.& fr. 15.xii.1956,

Davies 2287 (K, LISC, PRE, SRGH). **Mozambique**. MS: Beira, iv.1895, *Schlechter* s.n. (not seen). GI: Chigubo Dist., Banhine, fr. x.1973, *Tinley* 2997 (LISC, M, PRE, SRGH). M: Moamba Dist., Mangulane, fl. iv.1931, *Gomes e Sousa* 530 (K).

Also in South Africa. *Acacia* bushland on alluvial flats or rocky hills, alkaline pans with *Sesamothamnus* bushland and riverine scrub; 20–1100 m.

Conservation notes: Widespread; not threatened.

There are two further subspecies in Kenya and Tanzania – subsp. *revoluta* with spreading indumentum and subsp. *nyanzae* Vollesen with appressed indumentum but with generally shorter (7–10 mm) corolla tube, glabrous (rarely hairy apically) capsule and smaller (4–5 × 3.5–4 mm) seeds.

2. **Megalochlamys hamata** (Klotzsch) Vollesen in Kew Bull. **44**: 622 (1989). Type: Mozambique, Sena, n.d., *Peters* s.n. (B† holotype); Zimbabwe, Chipinge Dist., Birchenough Bridge, 15.iv.1963, *Chase* 7991 (K neotype, LISC, M, PRE, SRGH), neotypified by Vollesen (1989). FIGURE 8.6.**65**.

   Blechum hamatum Klotzsch in Peters, Naturw. Reise Mossamb. **6**: 220 (1861).
   Ecbolium hamatum (Klotzsch) C.B. Clarke in F.T.A. **5**: 239 (1900).
   Megalochlamys strobilifera C.B. Clarke in F.T.A. **5**: 240 (1900). —Grignon & Johnsen, Checklist Vasc. Pl. Botswana: 4 (1986). Type: Mozambique, Tete, 1.ii.1860, *Kirk* s.n. (K holotype).
   Ecbolium lugardae N.E. Br. in Bull. Misc. Inform., Kew **1909**: 130 (1909). —Grignon & Johnsen, Check-list Vasc. Pl. Botswana: 4 (1986). Type: Botswana, Ngamiland, Kwebe Hills, 7.iii.1898, *Lugard* 212 (K holotype, GRA, Z).

Erect or scrambling shrubby herb or shrub to 1(1.5) m tall; young branches, petioles and peduncles pubescent or densely so, hairs usually bent down apically, rarely straight. Leaves pubescent or sparsely so, densest on veins, lamina often glabrous; petiole (3)5–13 mm; lamina narrowly ovate to ovate or elliptic, largest (4.5)5.5–9.5(11) × 1.5–3.8(4.5) cm; apex acuminate to acute, apiculate, base attenuate, decurrent on petiole. Inflorescence 1.5–5.5(8) cm long; peduncle 1–5 mm; axis pubescent, rarely also with stalked capitate glands; bracts pale green or green, sometimes tinged purplish towards apex, ovate or elliptic, 10–19 × 6–13 mm, with a sharp recurved acumen 1–3 mm long, subglabrous to sparsely puberulous, with (rarely without) subsessile or short-stalked capitate glands, usually conspicuously ciliate; bracteoles 3–5 mm long. Calyx (5)6–8 mm long, glabrous to sparsely puberulous, with or without stalked capitate glands; lobes conspicuously ciliate. Corolla pale blue to bright sky-blue; tube 10–13 mm long; lobes 13–18 mm long, lower lip 2.5–4 mm wide. Filaments 7–10 mm long. Capsule 9–13 mm long, sparsely puberulous along sutures and laterally on valves, central part of valves glabrous. Seeds 4.5–5.5 × 3.5–4 mm, greyish, rugose or slightly so with numerous glandular-glochidiate hairs, rim entire.

Botswana. N: Ngamiland Dist., Ngwanalekau Hills, fl.& fr. 17.iv.1968, *Cudmore* K3 (K). **Zambia**. C: Luangwa Dist., Katondwe, fl. 23.ii.1965, *Fanshawe* 9214 (K, LISC). S: Siavonga Dist., Siavonga area, Lukwechele, 500 m, fl. 14.iii.1997, *Zimba et al.* 1121 (K, MO). **Zimbabwe**. N: Kariba Dist., 20 km ESE of Chirundu, 400 m, fl. 15.iii.1966, *Simon* 698 (K, LISC, SRGH). W: Hwange Dist., Victoria Falls Nat. Park, fl. 6.ii.1980, *Kabisa* 8 (K, SRGH). E: ?Mutare Dist., Odzi R., fl. ii.1931, *Myres* 672 (K). S: Chipinge Dist., Devuli Project, fl. i.1957, *Davies* 2415 (K, LISC, SRGH). **Malawi**. S: Mangochi Dist., 2 km SW of Monkey Bay, 525 m, fl. 20.ii.1982, *Brummitt et al.* 16007 (K, MAL, SRGH). **Mozambique**. T: Mágoè Dist., Chicoa to Mágoè, 350 m, fl. 13.ii.1970, *Torre & Correia* 17940 (BR, LD, LISC, LMA, LMU, MO, PRE, WAG).

Also in South Africa. In *Acacia, Combretum–Terminalia* and mopane woodland and bushland, riverine forest and thicket, on rocky slopes or hills or on black clay soils; 200–1200 m.

Conservation notes: Widespread; not threatened.

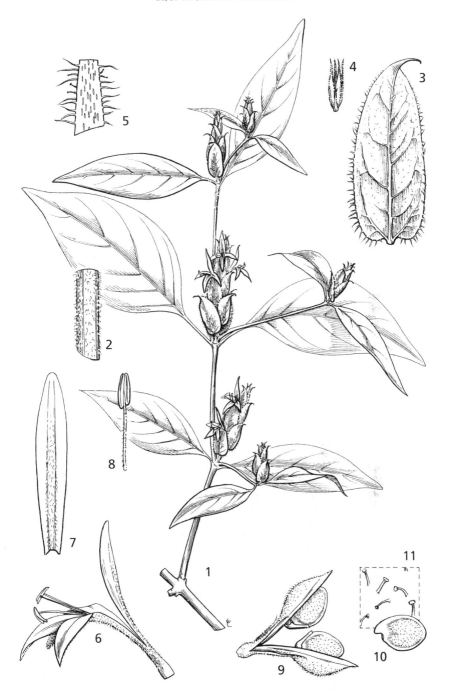

Fig. 8.6.**65**. MEGALOCHLAMYS HAMATA. 1, flowering branch (× 1); 2, detail of stem (× 3); 3, bract (× 3); 4, bracteole and calyx (× 2); 5, part of calyx lobe (× 24); 6, corolla (× 2); 7, upper lip of corolla (× 3); 8, upper of part of filament and anther (× 4); 9, capsule (× 3); 10, seed (× 3); 11, detail of seed surface (× 30). 1–8 from *Biegel* 2942, 9–11 from *Bruce* 59. Drawn by Eleanor Catherine. Reproduced from Kew Bulletin (1989).

39. **CEPHALOPHIS** Vollesen[17]

Cephalophis Vollesen in F.T.E.A., Acanthaceae **2**: 676 (2010).

Herbs. Leaves opposite, equal, crenate, with conspicuous cystoliths. Flowers solitary or in 3-flowered cymules in narrow racemiform panicle; bracts often foliaceous at basal nodes, upwards bract-like persistent; bracteoles present, large. Calyx deeply divided into 5 subequal often recurved segments. Corolla outside with long curly glossy hairs all over; basal tube linear, long, widening into a short indistinct throat; limb distinctly 2-lipped, upper lip strongly falcate and laterally flattened, with 2 minute apical teeth, lower lip subcircular in outline, longitudinally plicate with 2 lateral parts folded over and ± covering central part, with 3 minute lobes, without conspicuous pattern of transverse differently coloured lines ('herring-bones'). Stamens 2, attached just inside tube; filaments strongly flattened, strap-shaped, enclosed in upper lip; anthers bithecous, medifixed; thecae equal or one very slightly longer, held at same height and parallel, rounded. Style filiform, held between stamens; stigma with 2 erect ovoid rounded lobes. Capsule 4-seeded, clavate, with sterile basal stalk about equal in length to fertile part; retinaculae strong, curved. Seeds discoid, circular in outline, without raised rim, densely covered all over with retrorsely barbed glochidiae.

A single species in Kenya and Mozambique.

Cephalophis lukei Vollesen in F.T.E.A., Acanthaceae **2**: 678 (2010). Type: Kenya, Kwale Dist., Gongoni Forest Reserve, 10.xi.1992, *Harvey & Vollesen* 41 (K holotype, CAS, DSM, EA, NHT). FIGURE 8.6.**66**.

Perennial herb; basal part of stem creeping and rooting, apical part erect, to 1 m tall; young stems finely puberulous or sparsely so with antrorsely curved hairs, sometimes also with scattered stalked capitate glands. Leaves puberulous along midrib and lateral veins on both sides, glabrous on lamina; petiole to 5.5 cm; lamina ovate to elliptic or narrowly so, largest 10–20 × 3.5–9 cm; apex subacute to shortly acuminate; base cuneate to truncate, slightly to distinctly unequal-sided; margin slightly crenate. Inflorescence 15–30 cm long, sometimes with short lateral inflorescences from lowermost pair of nodes; peduncle 3–5 cm; peduncle and rachis puberulous and with stalked capitate glands. Flowers solitary or in 3-flowered cymules towards base of inflorescence (rarely 2 cymules together), solitary towards apex; main axis bracts foliaceous at basal nodes, upwards narrowly oblong to narrowly obovate, 7–15 × 1.5–3 mm, with purplish apical part, puberulous, with scattered to dense stalked capitate glands (rarely glandular hairs only); cymule bracts and bracteoles linear-lanceolate, 0.5–1 cm long, with similar indumentum. Calyx 7–11 mm long, divided to c.1 mm from base, puberulous with glossy capitate glands and minute non-glandular hairs; lobes purple towards apex, linear, acute, often recurved. Corolla 2.2–2.8 cm long along upper lip, basal part of tube greenish, upper part of tube whitish, lower and upper lips purple with long (c.1 mm) curly glossy hairs all over, lower lip with white area along midrib and weakly rugulate with transverse white bands; basal tube cylindrical, 0.8–1.1 cm long; throat indistinct, 2–3 mm long; upper lip strongly curved and laterally flattened, 1.2–1.6 cm long, with 2 minute apical teeth; lower lip 7–9 × 6–8 mm, with 3 minute lobes c.1 mm long. Filaments 1.1–1.5 cm long, glabrous; thecae 1.5–2 mm long, linear-oblong. Style 2.1–2.7 cm long, glabrous. Capsule 1–1.3 cm long, finely puberulous all over. Seeds 3–4 mm wide.

Mozambique. MS: Marromeu Dist., between Inhamitanga and Lacerdónia, fl.& fr. 7.v.1942, *Torre* 4088 (LISC, LMA, MO, PRE, WAG); Nhamatanda Dist., Serra do Chiluvo, Nhamatanda (Vila Machado), fl. 14.iv.1948, *Mendonça* 3929 (BR, COI, LISC, LMU). Also in Kenya. In moist evergreen lowland forest; 50–150 m.

Conservation notes: Local distribution in the Flora area in a much-threatened habitat; probably Vulnerable.

---

[17] By Kaj Vollesen

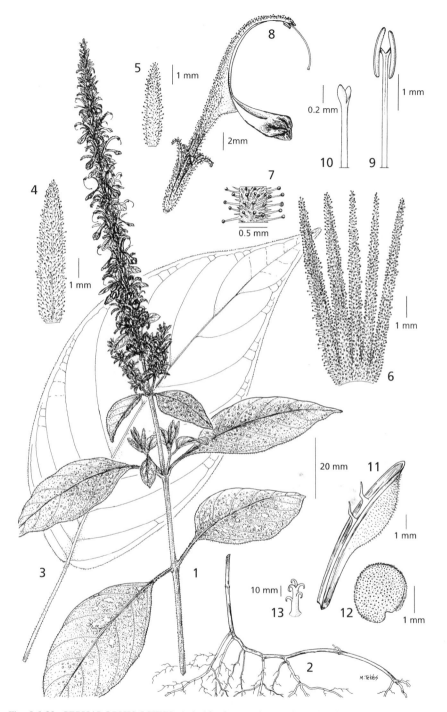

Fig. 8.6.**66**. CEPHALOPHIS LUKEI. 1, habit; 2, creeping and rooting basal part of stem; 3, large leaf; 4, bract; 5, bracteole; 6, calyx opened up; 7, detail of calyx indumentum; 8, calyx and corolla; 9, stamen and anther; 10, stigma; 11, capsule; 12, seed; 13, glochidia from seed surface. 1 from *Torre* 4088, 2 from *Harvey & Vollesen* 41, 3–10 from *Mendonça* 3929, 11–13 from *Luke* 3377. Drawn by Margaret Tebbs. Reproduced from Flora of Tropical East Africa (2010).

40. CHORISOCHORA Vollesen[18]

Chorisochora Vollesen in Kew Bull. **49**: 474–477 (1993).

Shrubs or shrubby herbs; stems indistinctly articulated, rounded and longitudinally striate or indistinctly 4-angled. Leaves opposite, with short often inconspicuous cystoliths, margin entire. Flowers pedicellate, solitary or paired or 3-flowered cymules, in erect racemiform cymes or these aggregated into panicles; bracts papery, caducous, imbricate or not; bracteoles 2, subulate to lanceolate. Calyx deeply divided into 5 subequal linear lobes. Corolla distinctly 2-lipped, puberulous on outside; tube cylindric, not or slightly widening upwards, straight or slightly curved; lower lip deeply 3-lobed, lobes held horizontally, recurved and with recoiled tips; upper lip with one linear-lanceolate 2-veined bifurcate lobe, recurved and with recoiled entire or notched apex. Stamens 2, parallel, held ± equidistant between upper and lower lips, inserted near base of lower lip, with short downwardly directed hairs; anthers bithecous, medifixed, held parallel with filament, thecae oblong, subequal, slightly curved, rounded at both ends. Style filiform, glabrous or hairy near base; stigma lobes equal, narrowly oblong-elliptic, rounded, erect; ovary with 2 ovules per locule. Capsule 2–3-seeded, club-shaped, apiculate; retinaculae strong. Seeds discoid, broadly elliptic to circular in outline, with smooth to verrucose or spinulose sides and distinctly raised almost smooth to spinulose rim.

A genus of 4 species. One in southern Africa (Botswana and South Africa), one in Somalia and two on Socotra.

Chorisochora transvaalensis (Meeuse) Vollesen in Kew Bull. **49**: 476 (1993). Type: South Africa, Limpopo, Waterberg, 22 km N of Hermanusdoorns, 27.iv.1949, *Codd & Erens* 5454 (PRE holotype, K). FIGURE 8.6.**67**.

Angkalanthus transvaalensis Meeuse, Fl. Pl. Afr. **31**: pl. 1227 (1956). —Dyer *et al.*, Wild Fl. Transvaal: 317, pl.159 (1962).

Shrub to 2 m tall; young branches whitish sericeous-puberulous or densely so, indistinctly 4-angled. Leaves crisped puberulous or sparsely so; petiole 0.4–1.7 cm long; lamina ovate to elliptic or broadly so, largest 2–5.5 × 1.3–4 cm; apex obtuse to broadly rounded or retuse; base cuneate to truncate, often slightly unequal. Inflorescence a simple racemiform cyme when young, eventually becoming a branched panicle to 10 cm long; flowers solitary or in 3-flowered cymules towards base of inflorescence; peduncle to 2 cm long, peduncle, axes and bracts whitish sericeous-puberulous or densely so; bracts green or tinged purplish, lanceolate to narrowly triangular-ovate, acute, 2–5 × 1–2 mm; pedicels 1–2 mm long; bracteoles 2–4 mm long, lanceolate. Calyx 6–8(10 in fruit) mm long, green or tinged purplish, finely puberulous and with or without scattered (rarely dense) short-stalked orange capitate glands. Corolla deep mauve, paler and more bluish on outside, tube white to yellowish; tube 7–14 mm long; lobes linear to oblong, subacute, 15–25 mm long, in lower lip 2–4 mm wide, in upper lip c.2 mm wide; filaments white to pale mauve, held horizontally, 15–20 mm long, anthers deep mauve, c.2 mm long; style glabrous. Capsule 15–22 mm long, glabrous to minutely puberulous. Seeds 5–6 × 4–5 mm, both surfaces scatteredly verrucose, rim finely spinulose.

Botswana. SE: Central Dist., Tswapong Hills, S slope of Moeng valley, 1 km E of Moeng College, 1000 m, fl. i.v.1996, *Rossiter* 132 (K).
Also in South Africa. In wooded grassland on sandy soil; c.1000 m.
Conservation notes: Local distribution in the Flora area but probably not threatened.

---

[18] By Kaj Vollesen

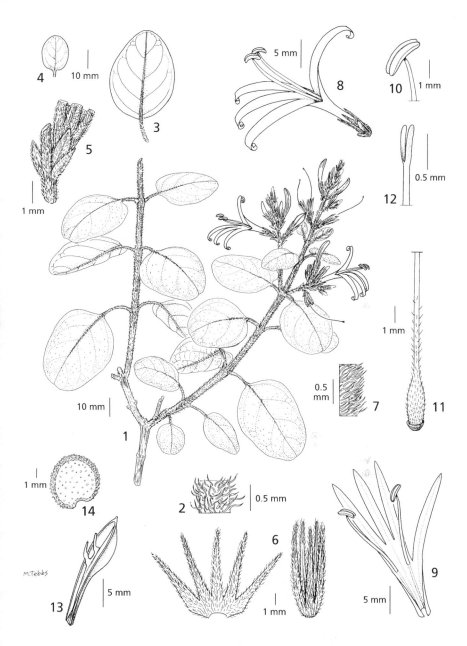

Fig. 8.6.**67**. CHORISOCHORA TRANSVAALENSIS. 1, habit; 2, detail of stem indumentum; 3, large leaf; 4, small leaf; 5 bract and bracteole; 6, calyx, whole and opened up; 7, detail of calyx indumentum; 8, corolla; 9, corolla, opened up; 10, apical part of filament and anther; 11, basal part of style and ovary; 12, apical part of style and stigma; 13, capsule; 14, seed. 1–3 & 10 from *Codd* 4008, 4 & 11–12 from *Rossiter* 132, 5 from *Daniel* 9379, 6–7 & 13–14 from *Meeuse* s.n., 8–9 from *Codd* 7741. Drawn by Margaret Tebbs.

41. DICLIPTERA Juss.[19]

Dicliptera Juss. in Ann. Mus. Natl. Hist. Nat. **9**: 267 (1807), conserved name.
—K. Balkwill, Getliffe Norris & M.-J. Balkwill in Kew Bull. **51**: 1–61 (1996).
—Darbyshire & Vollesen in Kew Bull. **62**: 119–128 (2007). —Darbyshire in
Kew Bull. **63**: 361–383 (2009).
Diapedium K.D. Koenig in Ann. Bot. **2**: 189 (1805). —Steudel, Nomencl. Bot., ed.2 **1**: 501, 504
(1840), in synonymy, name rejected.
Peristrophe Nees in Wallich, Pl. Asiat. Rar. **3**: 112 (1832). —Clarke in F.T.A. **5**: 242 (1900).
—K. Balkwill in Bothalia **26**: 83–93 (1996).

Annual or perennial herbs or subshrubs, sometimes suffruticose and pyrophytic; cystoliths linear, numerous. Stems 6-angular, often ridged. Leaves evergreen or deciduous, petiolate or subsessile, pairs equal or unequal; blade margins entire or uneven. Inflorescences axillary and/ or terminal, comprising a series of monochasial cymules, usually umbellately arranged, more rarely solitary, umbels solitary to several in each axil, sometimes compounded and paniculate or contracted into dense terminal heads, rarely arranged in a spike; main axis bracts paired, pairs equal to somewhat unequal, free, often much-reduced, rarely foliaceous; cymule bracts paired, pairs subequal to highly unequal, free but ± adpressed, usually conspicuous, sometimes with pale 'windows' of sclerids between principal veins in basal portion (Fig. 8.6.**69**,3); bracteoles narrower than cymule bracts and often much-reduced. Calyx shortly tubular, 5-lobed, lobes subequal or one lobe longer, linear-lanceolate. Corolla bilabiate, tube usually white, limb variously white to pink or purple, lip held in upper position usually with darker guidelines on a paler palate; corolla resupinate, tube twisted through 180°, cylindrical, somewhat widened towards mouth, rarely strongly so; lip held in upper position ± recurved distally, apex shortly 3-lobed; lip held in lower position unlobed, apex obtuse or shallowly emarginate. Stamens 2; filaments arising from mouth of corolla tube, rarely from within tube; anthers exserted, bithecous, thecae elliptic, offset, superposed or slightly overlapping. Ovary (oblong-)ovoid, bilocular, 2 ovules per locule; style filiform; stigma bifid, exserted. Capsule short- to long-stipitate, placental base elastic, placental base and membranous capsule walls then tearing from thickened flanks at dehiscence, or inelastic. Seeds 4 per capsule, rarely 2 by abortion, held on retinacula, lenticular, discoid or compressed-ellipsoid, with a hilar excavation, smooth or tuberculate.

A genus of c.175 species with a pantropical and subtropical distribution. Species delimitation in tropical Africa is very challenging with a number of widespread complexes including, in our region, the *D. heterostegia* group (species 1–6) and the *D. maculata–clinopodia* group (species 9–14, with species 15–18 also related). A full revision is desirable, with molecular analyses essential to test species limits and to make further inroads into the more difficult groups. As in the F.T.E.A. region, there are a number of potentially distinct taxa known only from one or two specimens which require further collection before they can be fully assessed.

Identification of *Dicliptera* taxa is most easily achieved by comparison to named material. Strong spot characters for separation of species are few, it usually being the combination of a suite of characters that enables species delimitation. Such differences are difficult to convey in a key, which is therefore necessarily descriptive in places, and some species key out more than once due to the considerable infraspecific variation. Geography and ecology can sometimes be helpful.

1. Cymule bracts broadly ovate or broadly elliptic(-obovate); length to width ratio of larger bract in each pair <2(2.3):1, green or grey-green even when dry, without 'windows', often at least the smaller of each pair turning pale-scarious in fruit; inflorescences axillary 2

---

[19] By Iain Darbyshire

- Cymule bracts variously shaped but usually narrower; if length to width ratio <2.5: 1 then with conspicuous pale or coloured 'windows' (Fig. **69**) and/or drying blackish- or brownish-green, not turning scarious in fruit and/or inflorescences crowded into a terminal head or spike 7
2. Corolla 8–8.5 mm long; larger cymule bract of each pair 7–9.5 mm long; cymules held in lax umbels (Mozambique MS) . . . . **6.** *sp. B*
- Corolla 10–27 mm long; larger cymule bract of each pair 9–25 mm long; umbels variously lax to contracted 3
3. Cymule bracts 3-veined from base, elliptic or smaller of each pair obovate; plants sericeous; calyx lobes with long wispy hairs along margin . . . .**5.** *zambeziensis*
- Cymule bracts 5–9-veined from base, larger of each pair ovate or ovate-elliptic; if bracts 3-veined from base and elliptic (*D. extenta*) then plants not sericeous and calyx lobes lacking long wispy hairs. 4
4. Corolla 20–27 mm long including 10–16 mm long lip held in upper position; bracteoles 7–10.5 mm long, pale green with pale hyaline margin; larger cymule bract of each pair elliptic or somewhat obovate; cymules held in ± lax umbels; plants never sericeous. . . . .**1.** *extenta*
- Corolla 11–18.5 mm long including 4–10.5 mm long lip held in upper position; bracteoles 1.5–5.5 mm long, pale-hyaline throughout; if corolla and bracteoles larger then umbels contracted and plants sericeous 5
5. Umbels solitary at each fertile node, each with (1)2(4) cymules (Malawi S). . . . .
. . . . **2.** *sp. A*
- At least some fertile nodes with 2–7 umbels of (2)3–7 cymules forming dense axillary fascicles 6
6. Leaves sometimes absent at flowering and fruiting, ovate or elliptic, 2–7.5 cm wide, length to width ratio <3:1, margin entire, lateral veins conspicuous . . . . . .
. . . . **3.** *heterostegia*
- Leaves always present at flowering and fruiting, linear or lanceolate, 0.6–1.6 cm wide, length to width ratio 4–14.5:1, margin shallowly repand, lateral veins inconspicuous. . . . . **4.** *swynnertonii*
7. Placental base of capsule rising elastically at dehiscence, capsule walls much thinner than thickened flanks (Figs. **68,69**); cymule bracts either with only midrib prominent or often prominently 3-veined from base, usually paler towards base and sometimes conspicuously 'windowed' (Fig. **69**) 8
- Placental base of capsule not or only very slightly rising elastically at dehiscence, capsule walls not clearly thinner than flanks (Fig. **71**); cymule bracts either with only midrib prominent or with prominent pinnate-reticulate venation particularly on inner surface, unicolourous except for narrow pale-hyaline margin, not paler towards base 27
8. Cymules several-flowered, subcapitate or shortly spiciform, each flower subtended by 1 bract and 2 bracteoles (Fig. **70**); pyrophytic suffrutex usually flowering on short, largely leafless stems after burning. . . . . **22.** *pumila*
- Cymules 1 to few-flowered enclosed within a pair of cymule bracts, each flower subtended by pair of bracteoles; habit various 9
9. Inflorescences axillary, together sometimes forming loose panicles or thyrses on largely leafless upper portion of stem but not compounded into dense terminal heads or spikes 10
- Inflorescences compounded into dense terminal heads or spikes, these capitate, globose, conical, cylindrical or becoming verticillate in lower portion (often also with cymes in upper leafy axils). 19

10. Corolla of chasmogamous flowers 4–6 mm long, limb barely protruding from between cymule bracts (in our area many specimens apparently cleistogamous, with only flower buds present); capsule 3–4.5 mm long; annual herb with inflorescences often developing even at lowermost nodes, often compounded into dense axillary fascicles . . . . **8.** *verticillata*
- Corolla over 10 mm long, limb clearly protruding from between cymule bracts; capsule ≥5.5 mm long; perennial herb or if annual (*D. carvalhoi*) then inflorescences restricted to upper portion of branches and not forming dense fascicles 11
11. Stamens long-exserted, anthers extending well beyond corolla lips, filaments often curved upwards; cymule bracts subulate, narrowly oblanceolate or narrowly lanceolate, 1–2.7 mm wide. . . . . **18.** *carvalhoi*
- Stamens with anthers not or barely extending beyond corolla lips, filaments ± straight; cymule bracts variously shaped, 2–10.5 mm wide 12
12. Cymule bracts with conspicuous long spreading eglandular hairs along margin, often also on midrib and main veins; surfaces with short antrorse-eglandular and/or short spreading glandular hairs 13
- Cymule bracts lacking long spreading eglandular hairs along margin; surfaces either glabrous or eglandular- and/or glandular-puberulous. 15
13. Cymule bracts not membranous, often at least smaller bract of each pair with conspicuous pale 'windows'; leaves ovate, if length to width ratio <2:1 then mature stem with long white sericeous hairs . . . . **9.** *maculata*
- Cymule bracts thin and rather membranous, turning somewhat scarious in fruit, 'windows' usually inconspicuous; mature stem lacking sericeous hairs 14
14. Leaves broadly ovate, length to width ratio <2:1; at least some fertile axils with 2–3 umbels of 3–6 cymules; more robust erect, decumbent or spreading herb of coastal sand forest and woodland (Mozambique M) . . . . **12.** *quintasii*
- Leaves more narrowly ovate, length to width ratio >2:1; fertile axils with 1 umbel of 2–3 cymules; slender trailing and weakly decumbent herb of upland forest (Malawi S) . . . . **13.** *sp. D*
15. Capsule with eglandular hairs only 16
- Capsule with predominantly glandular hairs 17
16. Corolla (excluding guidelines) white or cream-coloured; at least some fertile nodes with 2–3 umbels, together forming rather dense axillary fascicles; leaves ovate or elliptic, length to width ratio >2:1; seeds with minute hooked tubercles most numerous towards rim; plants of submontane forest . . . . **7.** *laxata*
- Corolla pink or magenta; fertile nodes with solitary or rarely 2 lax umbels, not forming dense fascicles; leaves broadly ovate, length to width ratio <2:1; seeds with long slender hooked tubercles numerous throughout; plants of drier habitats . . . . **16.** *syringifolia*
17. Mature leaves over 4 cm long; cymule bracts with ± conspicuous pale 'windows'; plants erect, decumbent, scrambling or scandent; plants of wet upland or riverine forest . . . . **9.** *maculata* subsp. *maculata*
- Mature leaves (sometimes fallen by fruiting) ≤4 cm long; cymule bracts either not conspicuously 'windowed' or if so then branches often becoming procumbent, trailing or weakly scrambling; plants usually of drier habitats. 18
18. Corolla 14–20 mm long; umbels mainly on short peduncles 1–7 mm long, appearing contracted into axils, those at lowermost fertile nodes sometimes to 14(25) mm long but then held at acute angle to stems; branches at first short and erect but often becoming procumbent, trailing or weakly scrambling. . **14.** *minor*
- Corolla 21.5–33 mm long; umbels mainly on long peduncles 11–70 mm long, ± widely spreading from stems; branches straggling or erect, sometimes forming untidy masses, not procumbent or trailing. . . . . .15. *monroi*

19. Seeds smooth except for minute reticulation, discoid with a keeled membranous rim, 2.2–3 mm wide; at least smaller cymule bract of each pair prominently 3-veined 20
– Seeds tuberculate or if smooth (*D. carvalhoi* subsp. *carvalhoi*) then only subflattened, lacking a keeled rim, to 1.7 mm wide; cymule bracts not prominently 3-veined 22
20. Pyrophytic suffrutex, producing few to numerous short flowering shoots from a woody base and rootstock . . . . **19.** *melleri*
– Slender annual herbs 21
21. Inflorescence distinctly elongate, cylindrical or conical, (1.5)2.5–5.5 cm long; cymule bract pairs usually strongly dimorphic, smaller bract obovate with obtuse or rounded apex, sometimes with a short attenuate tip; capsule (in our region) shortly pubescent towards apex . . . . **20.** *betonicoides*
– Inflorescence hemispheric or capitate, 0.5–2.5 cm wide; cymule bract pairs not clearly dimorphic, smaller bract lanceolate or elliptic with acute or attenuate apex; capsule glabrous . . . .**21.** *capitata*
22. Flowering stems soon becoming procumbent or prostrate (if erect then typically less than 10 cm tall); cymule bracts largely glabrous or eglandular- and glandular-puberulent . . . .**14.** *minor* subsp. *minor*
– Flowering stems erect, decument or scrambling; cymule bracts more conspicuously hairy. 23
23. Stamens long-exserted, anthers extending well beyond corolla lips, filaments often curved upwards; cymule bracts subulate, narrowly oblanceolate or narrowly lanceolate, length to width ratio (3.15)3.5–11:1, darker towards apex but not conspicuously 'windowed' below; wiry perennial or annual herb. . . .**18.** *carvalhoi*
– Stamens with anthers not or barely extending beyond corolla lips, filaments ± straight; cymule bracts variously ovate, elliptic, obovate, or if lanceolate or oblanceolate and with length to width ratio >3.5:1 then ± conspicuously 'windowed' below; usually more robust perennial herbs (except *sp. C*) 24
24. Leaves broadly ovate, length:width ratio <2:1; cymule bracts rather membranous, usually pale olive-green with inconspicuous 'windows', larger bract of each pair (oblong-)obovate, apex obtuse, rounded or shortly attenuate with a mucro; plant of coastal sand forest and woodland (Mozambique M) . . . . **12.** *quintasii*
– Either leaves more narrowly ovate or lanceolate, with length to width ratio >2:1, or if broadly ovate then cymule bracts very different to above, not membranous, ovate, lanceolate, elliptic or somewhat oblanceolate with apex acute-mucronulate to caudate, often conspicuously 'windowed'; not in coastal sand forest. 25
25. Corolla 14.5–19.5 mm long; cymule bracts somewhat paler towards base but not 'windowed', apex usually obtuse or rounded below mucro. . . . . **17.** *nyangana*
– Corolla usually larger (to 36 mm long), or if less than 20 mm then cymule bracts with conspicuous pale 'windows' in basal portion and apex acute to caudate below mucro 26
26. Seeds with verruculiform tubercles without hooks; corolla 17.5–36 mm long, if <20 mm then cymule bracts with large and conspicuous pale 'windows'; perennial herbs, usually not branching from a woody base . . . . **10.** *clinopodia*
– Seeds with more conical or elongate tubercles, some minutely hooked; corolla to 19 mm long; cymule bracts with only narrow 'windows'; suffruticose perennial branching from a woody base . . . . **11.** *sp. C*
27. Inflorescence (except sometimes those at lowermost fertile axils) a subsessile umbel; cymule bracts narrowly elliptic, larger bract of each pair 2.5–5.5 mm wide (length to width ratio 3–4.4:1), green at first but soon turning pink-brown or purple . . . . **25.** *colorata*

– Inflorescence (except sometimes those at uppermost fertile axils) clearly pedunculate; cymule bracts subulate, narrowly lanceolate or oblanceolate, larger bract of each pair 0.7–3(3.5) mm wide (length to width ratio usually over 5:1), colour variable but never purple 28

28. Plant drying a rather vivid yellow-green; corolla 13–14 mm long; suffruticose perennial, restricted to serpentine on Great Dyke of Zimbabwe . . **27.** *serpenticola*
– Plant drying green or blackish-green, not yellow-green; corolla over 15 mm long, if shorter (*D. paniculata*) then an annual herb with a simple taproot. 29

29. Stem rather densely hairy at least when young, with soft to somewhat hispid spreading or retrorse hairs (usually) not restricted to ridges 30
– Stem glabrous or coarsely long-hispid only along ridges. 32

30. Leaves linear-lanceolate or narrowly oblong (length to width ratio 4.8–10:1); cymule bracts with conspicuous short capitate glandular hairs externally; corolla with mixed eglandular and glandular hairs externally . . . . **26.** *gillilandiorum*
– Leaves ovate or lanceolate (length to width ratio 1.4–6.6:1); cymule bracts either with glandular hairs absent or only with minute and inconspicuous glandular hairs; corolla with only eglandular hairs externally 31

31. Leaves, if present at flowering (some plants pyrophytic, flowering before leaves develop), ovate to broadly so, length to width ratio 1.4–2.6:1, base rounded to cordate; cymules potentially several-flowered, capitate or shortly spiciform; capsule 7.5–9.5 mm long . . . . **23.** *brevispicata*
– Leaves lanceolate or narrowly ovate, length to width ratio 2.7–6.6:1, base cuneate or attenuate; cymules 1–2-flowered, never spiciform; capsule 9.5–11 mm long . .
. . . . **24.** *transvaalensis*

32. Annual herb with a simple taproot; corolla 9–14 mm long including lip held in upper position 4.5–8 mm long . . . .**28.** *paniculata*
– Suffruticose perennial or subshrub, branching from a woody base and rootstock; corolla over 15 mm long including lip held in upper position ≥9 mm long . . . 33

33. Capsule glabrous; cymule bracts glabrous or only with eglandular hairs; woody portion of stems with white or pale-grey flaking bark 34
– Capsule at least sparsely hairy with mixed short eglandular and glandular hairs towards apex, though sometimes glabrescent; cymule bracts with few to numerous short-stalked or subsessile capitate glands; woody portion of stems (if present) without pale flaking bark. 35

34. Corolla 15.5–20 mm long including lip held in upper position 9–11.5 mm long; cymule bracts green, 1–1.7 mm wide, external surface with antrorse eglandular hairs at least along midrib . . . . **30.** *decorticans*
– Corolla 24–25 mm long including lip held in upper position 15–16.5 mm long; cymule bracts grey-green, 1.7–2.1 mm wide, external surface glabrous except for few inconspicuous hairs along margin . . . . **31.** *sp. E*

35. Leaves (if present at flowering) ovate, length to width ratio 1.8–2.9:1; inflorescence rather densely glandular- and eglandular-puberulous throughout. . . . .**29.** *cernua*
– Leaves (if present at flowering) lanceolate or linear-lanceolate, length to width ratio 3.2–18:1; inflorescence not puberulous, short-stalked or subsessile glands restricted to cymule bracts and bracteoles . . . . **32.** *aculeata*

1. **Dicliptera extenta** S. Moore in J. Linn. Soc., Bot. **40**: 162 (1911). —K. Balkwill in Kew Bull. **51**: 21, fig.1 (1996). Type: Central Mozambique, Mt Maruma, fl. 14.ix.1906, *Swynnerton* 1937 (BM holotype).

   Dicliptera zeylanica sensu auct., non Nees; —Clarke in Fl. Cap. 5: 91 (1901). —Gledhill, Eastern Cape Wildfl.: 213 (1969).

Perennial herb or subshrub, scrambling or trailing, to 150 cm tall; stems 6-angular, ± prominently pale-ridged, shortly antrorse-pubescent to sparsely so, later glabrescent. Leaves ovate to lanceolate, 3–12 × 0.8–5.7 cm, base cuneate or attenuate, margin often shallowly repand, apex attenuate or acuminate, upper surface sparsely pilose, veins beneath and margin antrorse-pubescent or leaves largely glabrous; lateral veins (4)5–7(8) pairs; petiole 12–53 mm long. Inflorescence axillary, 1–2(4) umbels of (1)2–5 cymules per axil, rarely partially compounded; umbel peduncle (6)10–58 mm long, antrorse- or spreading-pubescent; main axis bracts usually linear-lanceolate, 3–8 × 0.5–1 mm, occasionally more leafy, then lanceolate or elliptic, to 19.5 × 2.5 mm; longest cymule peduncle in each umbel 3–20(32) mm long; cymule bract pairs unequal, ratio 1.2–1.55(1.7) : 1, green, smaller of each pair somewhat paler towards base and later turning scarious, larger bract elliptic or somewhat obovate, 13–19 × 6.5–14.5 mm, base cuneate, apex acute or attenuate, mucronulate, palmately 3–7-veined but often only midrib prominent, surface largely glabrous or antrorse-pubescent and/or spreading-pilose on midrib, margin, and sometimes exposed portion of interior surface, with or without short glandular hairs towards margin or scattered throughout; bracteoles pale green with pale-hyaline margin, lanceolate, 7–10.5 mm long, apex attenuate. Calyx lobes 4–5.5 mm long, glandular-puberulent, margin strigulose. Corolla 20–27 mm long, white, pink or mauve, eglandular-pubescent externally, limb with interspersed glandular hairs; tube 8–11 mm long; lip held in upper position oblong, 10–16 × 3.5–5.5 mm, palate puberulous; lip held in lower position elliptic, 9.5–15 × 5–7.5 mm. Staminal filaments 7.5–11.5 mm long, hairy mainly beneath; anther thecae 0.8–1.2 mm long, superposed and sometimes becoming separated. Style sparsely strigulose. Capsule 7–9.5 mm long, eglandular-pubescent, with interspersed glandular hairs towards apex; placental base elastic. Seeds discoid, 2.2–2.7 mm in diameter, with minute hooked tubercles mainly towards rim.

Zimbabwe. E: Mutare Dist., Tshakwe Mt, Burma Farm, fl. 21.viii.1955, *Chase 5724* (BM, K, LMA, SRGH). S: Mberengwa Dist., N aspect of Mt Buchwa, fl. 1.v.1973, *Simon et al.* 2411 (BR, LISC, PRE, SRGH). **Mozambique**. MS: Manica Dist., Penhalonga Forest, fl. 27.viii.1950, *Chase* 2910 (BM, BR, K, SRGH).

Also in Swaziland and South Africa (Limpopo, Mpumalanga, KwaZulu-Natal, Eastern Cape). Moist forest, under closed canopy and along margins and roadsides; 800–1900 m.

Conservation notes: In the Flora area largely found in submontane forest, but in South Africa it extends to coastal forest. Reasonably widespread and known historically from numerous sites so unlikely to be globally threatened; Least Concern.

Müller 3042 (K, SRGH) from Bunga Forest, Vumba Mts and *Mavi* 1780 (SRGH) from Mutare Dist. are unusual in having lanceolate leaves (length: width ratio 4–8.8 : 1, versus 1.8–4 : 1 in typical *D. extenta*) which are somewhat reminiscent of those of *Dicliptera swynnertonii* but the inflorescence is a good match for *D. extenta*. These are both rather small, delicate specimens and it may be that they are merely young plants with immature foliage that have flowered early, or they may represent a distinct variety.

Balkwill (1996) remarked on the similarity of this species to *D. grandiflora* Gilli from Tanzania and *D. elegans* W.W. Sm. from China. The former is easily separated by its (presumably) bird-pollinated flowers which have an extended cylindrical throat, the stamens being attached well below the corolla mouth (see Darbyshire in Kew Bull. **63**: 373, 2008). On cursory inspection the latter is remarkably similar to *D. extenta* but, based on brief observations, it differs in the anther thecae overlapping for ± half of their length (not fully superposed) and in having finer and more appressed hairs along the cymule bract margins.

2. **Dicliptera sp. A** (= *Brummitt* 15133).

Weakly decumbent ?perennial herb, branches to 50 cm long, rooting along nodes of trailing portion; stems subangular, shortly pubescent with crisped antrorse and retrorse hairs, later glabrescent. Leaves ovate, 1.8–3.4 × 0.7–1.8 cm, base obtuse or shortly and broadly attenuate, apex attenuate to obtuse, upper surface, margin and veins beneath sparsely pubescent, with shorter antrorse hairs on midrib above; lateral veins 4–5 pairs; petiole 3–8.5 mm long. Inflorescence axillary, 1 umbel of (1)2(4) cymules per axil; umbel peduncle 1–3 mm long; main axis bracts linear-lanceolate, 2.5–4.5 × 0.5 mm; longest cymule peduncle in each umbel 3.5–5.5 mm long, pilose; cymule bract pairs unequal, ratio 1.25–1.4:1, green, smaller of each pair later turning scarious, larger bract broadly ovate, 9–12 × 7–9.5 mm, base rounded or obtuse, apex obtuse, mucronulate, inconspicuously 5–7-veined from base, margin pale-pilose, sparsely so on main veins externally and on exposed internal surface of larger bract, sparsely puberulent externally with occasional gland-tipped hairs; bracteoles pale green with pale-hyaline margin, lanceolate, 4–5.5 mm long. Calyx lobes 3.5–4 mm long, mixed glandular- and eglandular-puberulent, with occasional long spreading eglandular hairs. Corolla 11–12.5 mm long, purple, eglandular-pubescent externally; tube 6–6.5 mm long; lip held in upper position oblong-elliptic, 5–6.5 × 2.5 mm, palate sparsely puberulous; lip held in lower position elliptic, 4.5–5.5 × 3.5 mm. Staminal filaments 3.5–4 mm long, sparsely hairy; anther thecae 0.6–0.75 mm long, superposed and oblique. Ovary glabrous. Capsule and seeds not seen.

Malawi. S: Mangochi Dist., Mangochi Mt, 23 km E of Mangochi, fl. 20.xi.1977, *Brummitt* 15133 (K, MAL).

Not known elsewhere. Evergreen montane forest; c.1700 m.

Conservation notes: Possibly endemic to Mangochi Mt. and known only from a single collection in a small and threatened forest patch; probably Endangered.

Brummitt 15133 is allied to *Dicliptera sp. A* of F.T.E.A. (= *Lovett et al.* 2309) from the Mufindi Escarpment of S Tanzania, and it may be conspecific. However, it differs in having ovate, not (obovate-)elliptic cymule bracts and smaller corollas (11–12.5 vs. 14.5–16 mm), although more material is desirable to assess their significance. The broadly ovate bracts of the Mangochi plant are similar to those of *D. heterostegia* but it differs in the fewer inflorescences per axil, the more slender trailing habit and the less hairy bracts, as well as the difference in habitat.

3. **Dicliptera heterostegia** Nees in De Candolle, Prodr. **11**: 478 (1847). —Clarke in Fl. Cap. **5**: 90 (1901). —Balkwill *et al.* in Kew Bull. **51**: 24 (1996). —Darbyshire in Kew Bull. **63**: 364 (2009); in F.T.E.A., Acanthaceae **2**: 690, fig.90 (2010). Type: South Africa, between Omsamwubo (Umzimvubu) and Omsamcaba (Umsikaba), fl. 1837, *Drège* s.n. (B† holotype, G-DC, K). FIGURE 8.6.68.

Justicia heterostegia E. Mey. in Drège, Cat. Pl. Afr. Austral. **1**: 3 (1837); in Drège, Zwei Pflanzengeogr. Doc.: 152, 159, 195 (1843), invalidly published.

Dicliptera mossambicensis Klotzsch in Peters, Naturw. Reise Mossamb. **6**: 220 (1861). —Clarke in F.T.A. **5**: 258 (1900). —Binns, Checklist Herb. Fl. Malawi: 13 (1968). Type: Mozambique Is. and mainland, 1842–1846, *Peters* s.n. (B† holotype), cited by Clarke (1900) as "Mozambique, Klotzsch" in error.

Dicliptera insignis Mildbr. in Notizbl. Bot. Gart. Berlin-Dahlem **12**: 521 (1935). Type: Tanzania, Lindi Dist., Lake Lutamba, c.40 km W of Lindi, fl. 3.ix.1934, *Schlieben* 5240 (B† holotype, BM, BR, G, HBG, LISC, M, MA, PRE, S).

Erect or scrambling perennial (rarely annual) herb or subshrub, 25–200 cm tall; stems 6-angular or weakly so, base often woody, antrorse-pubescent at least when young, often also ± densely sericeous, sometimes glabrescent. Leaves sometimes absent at flowering, ovate or elliptic, 3.5–16 × 2–7.5 cm, base (cuneate-)attenuate, apex attenuate or acuminate, upper surface and veins beneath sparsely pilose or largely glabrous, margin and midrib above pubescent; lateral veins 5–8 pairs; petiole 1–6.5 cm long. Inflorescence axillary, (1)2–5(7) umbels of (2)3–5(7) cymules per axil forming dense fascicles; primary peduncle of innermost umbel 2–14(55) mm long but outer umbels usually subsessile, peduncles antrorse-pubescent and/or sericeous; main axis bracts linear-lanceolate, 2.5–6(7.5) mm long; longest cymule peduncle in each umbel 3–9(12.5) mm long; cymule bracts often unequal in size within each umbel, those of central cymule sometimes markedly larger, green, often turning scarious in fruit, pairs slightly to strongly unequal, ratio 1.2–1.9(2.3):1, larger bract broadly ovate or ovate-elliptic, 9–20(25) × 6–12.5(14.5) mm, base rounded or obtuse, margin sometimes revolute,

Fig. 8.6.**68**. DICLIPTERA HETEROSTEGIA. 1, habit, fertile portion (× ²/₃); 2, habit sketch, full plant (× ¹/₆); 3, detail of stem indumentum (× 4); 4, mature leaf at fertile node (× ²/₃); 5, cymule bract pair and bracteole, external faces (× 3); 6, calyx and bracteoles (× 8); 7, corolla with stamens and ovary (× 4); 8, detail of stamens and stigma (× 6); 9, capsule (× 5); 10, seed (× 10). 1 & 6–8 from *Chase* 5047; 2, 3 & 9 from *Brass* 17945; 4 from *T. Harris* s.n.; 5 from *Brummitt & Harrison* 12402; 10 from *Fanshawe* 2585. Drawn by Juliet Williamson.

apex acute or attenuate, mucronate, smaller bract often more elliptic or suborbicular with apex (acute) obtuse or rounded, surfaces and particularly margin white-pilose to sericeous, also glandular- and eglandular-puberulent at least towards apex of larger bract within, surface palmately 5–9-nerved; bracteoles pale throughout, linear-lanceolate, 1.8–5(7.5) mm long. Calyx lobes 2.5–4(5.5) mm long, glandular-puberulent, sometimes with long eglandular hairs towards margin. Corolla (10)13–18.5(26) mm long, pink, purple or rarely white, eglandular-pubescent externally; tube (6)8–11.5 mm long; lip held in upper position oblong-obovate, (4)6–10.5(15) × 3–5 mm, palate sparsely puberulous towards mouth; lip held in lower position broadly elliptic-obovate, (4)6–8.5(14.5) × 4–6 mm. Staminal filaments (4)4.5–8.5(17.5) mm long, glabrous or with sparse hairs towards base; anther thecae 1–1.2 mm long, superposed and somewhat oblique. Style sparsely strigulose distally. Capsule 5–7.5 mm long, puberulous to pubescent, with short glandular hairs and few to numerous, sometimes longer eglandular hairs; placental base elastic. Seeds 1.5–2 mm in diameter, with minute hooked tubercles mainly towards rim, sometimes sparse at maturity.

Zambia. W: Masaiti Dist., Katanino, fl.& fr. 3.xi.1955, *Fanshawe* 2585 (BR, K, NDO). **Zimbabwe.** N: Mazowe Dist., tributary of Mazoe R., Mazoe Citrus Estates, fl.& fr. 9.ix.1957, *Drummond* 5313 (BR, K, LISC, SRGH). E: Mutare Dist., Mutare commonage, fl.& fr. 2.viii.1953, *Chase* 5047 (BM, K, SRGH). S: Chiredzi Dist., Save–Runde (Sabi-Lundi) Junction, Chitsa's Kraal, fl. 4.vi.1950, *Wild* 3333 (K, LISC, SRGH). **Malawi.** C: Lilongwe Dist., Nature Sanctuary Forest Zone A, Lingazi R., fl.& fr. 20.ix.1985, *Patel & Kwatha* 2699 (K, MAL). S: Blantyre, Sunnyside suburb, by Mudi R., fl.& fr. 3.viii.1970, *Brummitt & Harrison* 12402 (K, MAL, SRGH). **Mozambique.** N: Mecúfi Dist., Mecúfi, road to Pemba (Porto Amélia), fl.& fr. 27.x.1942, *Mendonça* 1094 (BR, LISC, LMA, MO). Z: Lugela Dist., Mabu Mt, fl.& fr. 14.x.2008, *Mphamba* 62 (K, LMA). MS: Gorongosa Dist., Chitengo, near to Sangarassa Forest, fl.& fr. 4.v.1978, *Diniz et al.* 190 (BR, LMU). GI: Govuro Dist., 37 km from Tessolo to Jofane, fl. 11.ix.1973, *Correia & Marques* 3334 (LMU). M: Maputo (Lourenço Marques), fl.& fr. 6.vi.1920, *Borle* 1102 (K, PRE).

Also in eastern D.R. Congo, Kenya, Tanzania, Swaziland and South Africa (Limpopo, Mpumalanga, KwaZulu-Natal, Eastern Cape). Riverine forest and thickets, mid-altitude forest, dry coastal forest and woodland, and rocky hillslopes; 20–1350 m.

Conservation notes: Widespread and fairly common in a variety of shaded habitats; Least Concern.

A widespread and polymorphic species, with a number of inter-related regional forms. Several taxa may be recognizable in time, but would require a full revision of this and related species in Africa, Arabia, Madagascar and the Indian subcontinent (*Dicliptera zeylanica* Nees and allies). At present, suffice to note the more significant variation within our region: (i) the most widespread form has a ± dense sericeous stem indumentum and plants are usually leafless at flowering and fruiting; particularly hairy plants ('*D. insignis*') are found in NE Mozambique and SE Tanzania; (ii) plants from the Rovuma River catchment on the Mozambique–Tanzania border, also sericeous and largely leafless at flowering, can have unusually large corollas (bracketed upper measurements), long bracteoles and the calyx lobes can have long wispy eglandular hairs; (iii) plants from S Mozambique lack the sericeous indumentum and are often leafy at flowering, they tend to have less dense inflorescences at each axil; this form matches typical *D. heterostegia* from South Africa; (iv) a robust, large-leaved form with largely glabrous stems and dense inflorescences with ± large bracts is recorded from mid-altitude forests of mountains in S Malawi (e.g. *J.D. & E.G. Chapman* 9268 from Mulanje) and adjacent Mozambique – the single collection from Mt Mabu (*Mphamba* 62) is particularly striking in having long-pedunculate umbels. This form would seem a strong a candidate for separation but it can be confusingly similar to the South African plants and also to forms from lowland Kenya.

4. **Dicliptera swynnertonii** S. Moore in J. Linn. Soc., Bot. **40**: 163 (1911). —K. Balkwill in Kew Bull. **51**: 24 (1996). Type: Zimbabwe, Chipinge Dist., Chirinda Forest outskirts, fl. 26.v.1906, *Swynnerton* 528 (BM lectotype), lectotypified by Balkwill (1996), see note.

Perennial herb up to 100 cm tall; stems 6-angular, shortly antrorse-pubescent. Leaves somewhat unequal, lanceolate or linear, 2.8–12.3 × 0.6–1.6 cm, base cuneate to attenuate, margin shallowly repand, apex acute to obtuse, upper surface sparsely pilose or glabrous, veins beneath and margin antrorse- or spreading-pubescent; lateral veins inconspicuous; petiole 4–24 mm long. Inflorescence axillary, (1)2–4 umbels of 3–5 cymules per axil; peduncle of innermost umbel at each axil 2–9 mm long but outermost umbels usually subsessile, peduncles pilose; main axis bracts lanceolate or linear-lanceolate, 4–8 × 0.3–2 mm, apex ± cuspidate; longest cymule peduncle in each umbel 2–6 mm long; cymule bracts often unequal in size within each umbel, those of central cymule sometimes markedly larger, pairs unequal, ratio 1.3–1.55:1, green, smaller of each pair somewhat paler, larger bract broadly ovate, 11–18 × 7–11 mm, base rounded, apex attenuate into a prominent mucro (smaller bract usually obtuse with a short mucro), palmately 5–7-veined, outer surface finely pubescent, with longer hairs on veins, margin and exposed inner surface of larger bract with dense long silky hairs; bracteoles pale-hyaline, linear-lanceolate, 1.5–3 mm long. Calyx lobes 2–3 mm long, glandular-puberulent. Corolla 12–15 mm long, pink, eglandular-pubescent externally; tube 7–9 mm long; lip held in upper position oblong-elliptic, 5–6.3 × 2–3.2 mm, palate puberulous; lip held in lower position (obovate-)elliptic, 5–5.8 × 2.3–3.8 mm. Staminal filaments c.5 mm long, shortly hairy; anther thecae 0.6–0.8 mm long, superposed and oblique. Style sparsely strigulose or glabrous. Capsule 6 mm long, shortly glandular-pubescent with occasional long eglandular hairs; placental base elastic. Seeds discoid, c.2 mm in diameter, with minute hooked tubercles.

Zimbabwe. E: Chipinge Dist., Chirinda Forest, fl. 17.vi.1906, *Swynnerton* 528b (BM, K). **Mozambique**. MS: Manica Dist., Bandula, fl.& fr. 8.ix.1956, *Chase* 6190 (K, LISC, SRGH).

Also in South Africa (KwaZulu-Natal). Moist forest and forested ravines; 1000–1250 m.

Conservation notes: Very scarce and known from only 3 sites. Recorded as common in Chirinda by Swynnerton but not collected there since 1906; the status of the Mozambique site is not known; possibly threatened.

The inflorescence of *Dicliptera swynnertonii* is very close to *D. heterostegia*, of which it could easily be considered an extreme variant, but it is maintained as distinct here pending a full revision of this group since the foliage is strikingly different. Balkwill (1996) records the cymule bracts as having short glandular hairs on both surfaces but in the material I have seen these are absent – an additional difference to *D. heterostegia*. The single specimen from KwaZulu-Natal cited by Balkwill (*Strey & Moll* 3895), collected from lowland swamp forest, looks rather different to plants from our region, being more delicate with a trailing habit and having a solitary few-flowered umbel at each node with small cymule bracts. This collection is rather scant and it would be desirable to see more material from KwaZulu-Natal to assess whether the two are truly conspecific.

In designating *Swynnerton* 528 at BM as the lectotype, Balkwill also listed an isolectotype at K. However, the K sheet labelled 528 has a different date which matches the BM specimen *Swynnerton* 528b, one of the other original syntypes; the K sheet is therefore not an isolectotype.

5. **Dicliptera zambeziensis** I. Darbysh., sp. nov. Similar to sericeous forms of *D. heterostegia* but differing in having fewer and less crowded inflorescences per axil, in the elliptic (larger) or obovate (smaller) 3-veined cymule bracts (larger bract ovate or ovate-elliptic and 5–9-veined in *D. heterostegia*), and in the calyx lobes having long wispy hairs along the margin (these usually absent or sparse in *D. heterostegia*). Type: Zambia, banks of Zambezi R., Katambora, fl.& fr. 23.viii.1947, *Greenway & Brenan* 7957 (K holotype, BR).

Straggling or procumbent perennial herb, stems 30–90 cm long, 6-angular, prominently ridged, densely long spreading-sericeous throughout or tardily glabrescent on old stems, uppermost internodes also with short spreading glandular hairs. Leaves rather sparse or absent on flowering stems, ovate-elliptic, 2.4–5 × 1.2–2.8 cm (though probably can get considerably larger), base cuneate or attenuate, apex acute to acuminate, surfaces at first sericeous particularly on margin, veins beneath and midrib above, later glabrescent; lateral veins 4 pairs; petiole to 23 mm long. Inflorescence axillary, 1–2 umbels of (1)2–4 cymules per axil; peduncle of longest umbel at each axil 2–11(16.5) mm long, sericeous, with short spreading glandular hairs; main axis bracts linear-lanceolate to narrowly elliptic-lanceolate, 1.8–6 × 0.5–1 mm, rarely foliaceous and to 17 × 7.5 mm; longest cymule peduncle in each umbel 1.5–5.5 mm long; cymule bract pairs unequal, ratio 1.25–1.7:1, green, smaller of each pair paler towards base, larger bract elliptic to broadly so, 9–14.5 × 4.5–7.3 mm (smaller bract often obovate), base cuneate, apex acute, mucronulate, 3-veined from base, prominently so on smaller bract, external surface and exposed portion of internal surface of larger bract densely sericeous particularly on margin and main veins, with ± numerous short spreading glandular hairs; bracteoles pale green with a pale-hyaline margin, linear-lanceolate, 6–6.5 × 0.7 mm. Calyx lobes 3–3.3 mm long, glandular-puberulent, with long wispy eglandular hairs along margin towards apex. Corolla 15–15.5 mm long, pink or purple, eglandular-pubescent externally; tube 8–8.7 mm long; lip held in upper position oblong-elliptic, 6.5–7.2 × 3.3–3.6 mm, palate sparsely puberulous; lip held in lower position elliptic, 6.3–6.8 × 3.5–3.7 mm. Staminal filaments c.6 mm long, shortly hairy mainly beneath; anther thecae 0.7–0.8 mm long, superposed and somewhat oblique. Style sparsely strigulose or glabrous. Capsule 5.7 mm long, shortly glandular-pubescent with occasional long eglandular hairs; placental base elastic. Only immature seeds seen, c.1.8 mm in diameter, with minute hooked tubercles.

Zambia. B: Senanga Dist., 11 km SW of Senanga, fl. 5.viii.1952, *Codd* 7403 (BM, K, PRE). S: Livingstone Dist., banks of Zambezi R., Katambora, fl.& fr. 23.viii.1947, *Greenway & Brenan* 7957 (BR, K).

Not known elsewhere. Amongst bushy clumps on open grassy plains, in open riparian forest of *Syzygium, Phoenix* and *Trichilia emetica;* 900–1050 m.

Conservation notes: Very localised in the upper Zambezi and known from only two sites; it may prove to be more widespread in SW Zambia and perhaps into Angola. Still extant at the type locality where it has recently been photographed by B. Wursten (pers. comm.). Threats are unknown so it is best considered Data Deficient.

Closely allied to *Dicliptera heterostegia* but readily distinguished by the diagnostic characters given in the diagnosis.

6. **Dicliptera sp. B** (= *Simão* 711).

Perennial(?) herb, erect to trailing, stems to 50 cm long, 6-angular, shortly antrorse-pubescent, with or without numerous longer spreading or somewhat retrorse hairs. Leaves ovate-elliptic, 3.2–6 × 1.3–2.4 cm, base attenuate, apex bluntly attenuate, upper surface and margin sparsely pilose or largely glabrous, veins beneath sparsely antrorse-pubescent; lateral veins 4–6 pairs; petiole 8–18 mm long. Inflorescence axillary, (1)2–3 lax umbels of (1)2–5 cymules per axil, sometimes partially compounded; peduncle of longest umbel at each axil (3)5–24(35) mm long, antrorse- or spreading-pubescent; main axis bracts usually linear-lanceolate, 1.7–4 × 0.5 mm, occasionally more leafy and up to 11 × 4.5 mm; longest cymule peduncle in each umbel (3)4.5–25(62) mm long; cymule bracts rather membranous, pairs unequal, ratio 1.1–1.5:1, green, larger bract broadly ovate, 7–9.5 × 4.5–7 mm, base obtuse or subrounded, apex acute or attenuate, mucronate or shortly so, surface palmately 5–7-veined, antrorse-pubescent mainly on midrib and margin, margin sometimes with few more spreading hairs, these sometimes also sparse on exposed inner surface of larger bract; bracteoles pale green or pale-hyaline throughout, linear-lanceolate, 2.5–3.3 × 0.2–0.3 mm, apex acuminate. Calyx lobes 2.2–2.8 mm long, glandular-puberulent and with occasional longer eglandular hairs. Corolla 8–8.5 mm long, lilac, eglandular-pubescent externally; tube 5–5.5 mm long; lip held in upper position oblong, 3.2–3.6 × 1.8–2 mm, palate puberulous with short conical hairs; lip held in lower position ± broadly obovate, 3.3–3.4 × 2.5–2.8 mm. Staminal filaments 3–3.5 mm long, largely glabrous; anther thecae 0.55–0.65 mm long, superposed and strongly oblique. Style glabrous. Capsule not seen.

Mozambique. MS: Cheringoma Dist., Inhamitanga, fl. 4.vii.1946, *Simão* 711 (LMA); Marromeu Dist.?, between Inhamitanga and Lacerdónia, imm.fl. 7.v.1942, *Torre* 4081 (LISC).

Not known elsewhere. Dense lowland forest; 50–250 m.

Conservation notes: Known from only two collections; recorded as very frequent by Simão. These lowland forest areas are threatened; possibly Endangered.

This species falls within the *Dicliptera heterostegia* complex and appears close to *D. inconspicua* I. Darbysh. from coastal Kenya and Tanzania and *D. elliotii* C.B. Clarke from the Guineo-Congolian forests, with which it shares the slender, weak-stemmed habit, the small broad bracts and small flowers. It differs from both most notably in the more lax mature umbels. *Torre* 4081, which has only immature inflorescences, appears very similar to *D. inconspicua*, but the mature Simão collection is clearly different in having larger corollas (5.5–6.5 mm in *D. inconspicua*) with the lip held in the upper position being obovate and longer than wide (vs. broadly trullate and wider than long in *D. inconspicua*), in having more numerous inflorescences per axil (1–2 umbels of (1)2–3(4) cymules in *D. inconspicua*), and in having more ovate cymule bracts. *Dicliptera elliotii* has considerably more contracted inflorescences than *D. sp. B* and additionally differs in having more clearly ovate bracts with a more rounded or subcordate base, and usually with a more hairy surface. The Simão collection additionally differs in having densely hairy stems, but in the Torre specimen these are much more sparsely hairy and so similar to the other species. Further specimens, including fruiting material, are required to confirm the status of this taxon. Similar looking plants from Madagascar should also be investigated.

7. **Dicliptera laxata** C.B. Clarke in F.T.A. **5**: 261, 515 (1900). —Mildbraed in Notizbl. Bot. Gart. Berlin-Dahlem **9**: 507 (1926). —Heine in F.W.T.A., ed.2 **2**: 425 (1963). —Darbyshire in F.T.E.A., Acanthaceae **2**: 694, fig.90.6 (2010). Type: Kenya, Kavirondo, Samia, fr. 1893–1894, *Scott-Elliot* 7098 (K lectotype, BM), lectotypified here.

Dicliptera humbertii Mildbr. in Bull. Jard. Bot. État **14**: 356 (1937). Type: D.R. Congo, between Lubero and Libongo, fl. 1929, *Humbert* 8740 (BR lectotype, P), lectotypified here.

Hypoestes phaylopsoides sensu Clarke in F.T.A. **5**: 248 (1900) in part for *Whyte* s.n. from Masuka Plateau, Malawi.

Erect or straggling perennial herb or subshrub, (10)30–120(200) cm tall; stems (sub)6-angular, shortly antrorse-pubescent mainly on the angles, soon glabrescent. Leaves dark green, soon drying green-black, ovate or elliptic, (3)7–12.5(16) × 1.5–5.5(7) cm, base attenuate or cuneate, apex acuminate, largely glabrous or upper surface with sparse hairs, midrib above antrorse-pubescent; lateral veins (5)6–7(8) pairs, prominent; petiole 5–55 mm long. Inflorescence axillary, 1–3 umbels of (3)4–5(7) cymules per axil, often forming dense fascicles; innermost umbel at each axil pedunculate for 1.5–14(24) mm, but subsequent umbels often subsessile, peduncles antrorse-pubescent; main axis bracts linear-lanceolate, (1)2–4.5 mm long; longest cymule peduncle in each umbel 2–8 mm long, occasionally with one cymule on a greatly extended peduncle to 26 mm long; cymule bract pairs slightly unequal, ratio 1.1–1.3:1, green or turning brown, paler towards base and often faintly 'windowed', larger bract elliptic or obovate-elliptic, 9–14.5 × 4.5–10.5 mm, apex acute, shortly attenuate or obtuse, minutely apiculate, surface inconspicuously 3–5-veined from base, largely glabrous except for sparse antrorse eglandular hairs at base, with or without minute subsessile glands towards apex; bracteoles linear-lanceolate, 5–10 mm long, margin broadly hyaline. Calyx lobes 3–5 mm long, glandular-puberulent. Corolla 10.5–15.5 mm long, white or cream with purple guidelines on both lips, eglandular-pubescent, with interspersed glandular hairs externally; tube 4–5.5 mm long; lip held in upper position oblong, 6–10 × 2–3.5 mm, palate puberulous towards mouth; lip held in lower position elliptic, 6.5–9.5 × 3–4.5 mm. Staminal filaments (3.5)5.5–8 mm long, pubescent above; anther thecae 0.6–0.85 mm long, superposed. Style sparsely strigulose. Capsule 5.5–7 mm long, ± densely eglandular-pubescent; placental base elastic. Seeds 2–3 mm in diameter, with minute hooked tubercles more dense towards rim.

Malawi. N: Chitipa Dist., Mugesse Forest, fl. 13.ix.1977, *Pawek* 13017 (K, LISC, MAL, MO, SRGH, UC).

Also in Nigeria, Bioko, Cameroon, D.R. Congo, South Sudan, Ethiopia, Uganda, Kenya and Tanzania. Montane and submontane evergreen forest and forest margins; 1750–2100 m.

Conservation notes: A widespread and often common species of Afromontane forest; Least Concern.

The above description covers only the populations from Malawi and from the Southern Highlands of Tanzania which tend to have longer bracteoles and calyces and broader, more markedly 'windowed' cymule bracts than plants from elsewhere in its range, though the differences are not sufficiently consistent to recognise infraspecific taxa.

da Silva *et al.* (Prelim. Checklist Vasc. Pl. Mozamb.: 18, 2004) record this species from Gaza Province in S Mozambique; this is almost certainly incorrect.

8. **Dicliptera verticillata** (Forssk.) C. Chr. in Dansk Bot. Ark. **4**(3): 11 (1922). —Raynal in Adansonia, sér.2 **7**: 304 (1967). —Wood *et al.* in Kew Bull. **38**: 450 (1983). — Darbyshire in F.T.E.A., Acanthaceae **2**: 697 (2010). Type: Yemen, fl.& fr. 1763, *Forsskål* 393 (C microfiche 39: I.3–4 lectotype, BM, K photo), lectotypified by Wood *et al.* (1983).

   Dianthera verticillata Forssk., Fl. Aegypt.-Arab.: 9 (1775).

   Justicia cuspidata Vahl, Symb. Bot. **2**: 9, 16 (1791), illegitimate name. Type as for *D. verticillata*.

   Justicia umbellata Vahl, Enum. Pl. **1**: 111 (1804). Type: Senegal, fl.& fr., n.d., *Jussieu* s.n. [28?] (P-JU holotype).

   Dicliptera umbellata (Vahl) Juss. in Ann. Mus. Natl. Hist. Nat. **9**: 268 (1807). —Clarke in F.T.A. **5**: 258 (1900), in part for *Whyte* 28 from Malawi.

   Dicliptera micranthes Nees in Wallich, Pl. Asiat. Rar. **3**: 112 (1832). —Clarke in F.T.A. **5**: 258 (1900). —Binns, Checklist Herb. Fl. Malawi: 13 (1968). Type: Yemen, fr. 1763, *Forsskål* 392 (C microfiche 39: I.1–2 lectotype, K photo), lectotypified by Wood *et al.* (1983).

   Dicliptera spinulosa K. Balkwill in Kew Bull. **51**: 53, fig.14a (1996). —Mapaura & Timberlake, Checklist Zimb. Vasc. Pl.: 13 (2004). —Setshogo, Prelim. Checklist Pl. Botswana: 18 (2005). —Welman in Germishuizen *et al.*, Checklist S. Afr. Pl.: 83 (2006). —Makholela in Figueiredo & Smith, Pl. Angola, Strelitzia **22**: 22 (2008). Type: Sudan, 'Cordofanum Milbeo', fl.& fr. 29.xi.1839, *Kotschy* 277 (G-DC holotype, K, M, P, WAG).

Erect or decumbent annual herb, 5–50(90) cm tall; stems 6-angular with pale ridges, largely glabrous or sparsely retrorse-pubescent. Leaves often immature or absent at flowering; blade ovate or elliptic, 1.5–7(10.5) × 1–4(5.5) cm, base attenuate, apex (sub)attenuate, apiculate, surfaces sparsely antrorse-pubescent, soon glabrescent except for short hairs along the margin; lateral veins 4–6 pairs, conspicuous beneath; petiole 5–25(50) mm long. Inflorescence axillary, developing even at lowermost nodes, (1)2–3 umbels of 3–5 cymules per axil, umbels often becoming compounded and forming dense axillary fascicles; primary peduncle 1–5(20) mm long, shortly pubescent; main axis bracts linear-lanceolate, 2.5–7.5 mm long; cymules subsessile, bract pairs slightly to strongly unequal, ratio 1.2–1.5:1, green with a pale-hyaline margin towards base, larger bract elliptic, obovate, oblanceolate or linear, 5–8.5(10) × 1–2.7(4) mm, apex attenuate into a curved mucro, margin often densely pilose-ciliate below, hairs shorter towards apex, surface with hairs of variable length, longest on prominent midrib; bracteoles linear-lanceolate, 4–6 mm long, green with hyaline margin, indumentum as cymule bracts or with additional short glandular hairs. Calyx lobes 2.5–3.5 mm long, sparsely pubescent and with minute glandular hairs externally. Corolla 4–6 mm long, pink to purple (rarely ?white), pubescent externally, in our region often found only in bud and then probably cleistogamous; chasmogamous flowers with tube 3–4 mm long; lip held in upper position oblong, 1.5–2.5 × c.1 mm; lip held in lower position broadly flabellate, 1.2–1.8 × 1.5–3 mm, curved around stamens. Staminal filaments 0.8–1.6 mm long, pubescent above; anther thecae 0.2–0.4 mm long, slightly overlapping, highly oblique. Style glabrous. Capsule 3–4.5 mm long, puberulous, sometimes with short glandular hairs at apex; placental base elastic. Seeds 0.75–1 mm in diameter, with minute slender hooked tubercles.

Botswana. N: Chobe Dist., road to Kasane, fl.& fr. 3.viii.1967, *Lambrecht* 277 (K, SRGH). SE: Central Dist., Boteti Delta area, NE of Mopipi, fl. bud 22.iv.1973, *Thornton* 1 (K, SRGH). **Zambia.** N: Kaputa Dist., R. Kangele, 14.5 km from Lake Chishi on track to Nsama, fr. 20.vii.1962, *Tyrer* 51 (BM, SRGH). W: Kitwe, fl.& fr. 21.v.1958, *Fanshawe* 4450 (K, NDO). C: Lusaka, fl.& fr. 26.v.1955, *Best* 102 (K, LISC, LMA, SRGH). E: Petauke Dist., Luangwa Bridge, Great East Road, fl.& fr. 7.x.1958, *Robson* 6 (BM, K, LISC, SRGH). S: Choma Dist., Mapanza Mission, fl.& fr. 9.v.1953, *Robinson* 202 (K). **Zimbabwe.** N: Hurungwe Dist., Hurungwe Safari Area, c.4.5 km from road to Zambia, near Nyangombe Is., 100 m from Rifa campsite, fr. 25.vi.1997, *Mapaura* 86 (SRGH). W: Nyamandhlovu Dist., no locality, fr. 17.vi.1943, *Hopkins* in SRGH 13394 (SRGH). C: Chegutu Dist., Poole Farm, fl. 14.iv.1953, *Hornby* 3331 (K, LISC, SRGH). E: Chipinge Dist., Save (Sabi Valley) Expt. Station, fl.& fr. v.1960, *Soane* 303 (K, SRGH). S: Chiredzi Dist., Save–Runde (Sabi-Lundi) junction, Chitsa's Kraal, fl.& fr. 6.vi.1950, *Wild* 3382 (K, LISC, SRGH). **Malawi.** N: Mzimba Dist., Great North Road, between two Mzuzu junctions, fl.& fr. 13.vi.1971, *Pawek* 4900 (K, MAL). C: Lilongwe Dist., Lilongwe suburbs, house of Postmaster General, fr. 13.vii.1994, *Brummitt et al.* 19049 (CAS, K, LISC, MAL, MO, PRE, SRGH, WAG). S: Mangochi Dist., Namwera, Sr. Martha Hospital, fr. 23.viii.1976, *Pawek* 11667 (K, MAL, MO). **Mozambique.** T: Angónia Dist., near Ulónguè, fl.& fr. 18.x.1982, *Nuvunga* 973 (LMU, SRGH). MS: "Meringua Dist., Sabi R.", fl.& fr. 27.vi.1950, *Chase* 2581 (BM, BR, K, SRGH).

Widespread in tropical Africa from Cape Verde and Senegal to Somalia, south to Namibia and South Africa; also in Madagascar, tropical Arabia and India. Found in a variety of open and partially shaded habitats, including dry woodland, grassland and riverbanks, often weedy on disturbed ground; 150–1400 m.

Conservation notes: Widespread and common; Least Concern.

9. **Dicliptera maculata** Nees in De Candolle, Prodr. **11**: 485 (1847), excl. var. *b senegambica.* —Clarke in F.T.A. **5**: 257 (1900). —Burrows & Willis, Pl. Nyika Plateau: 48, 50 (2004). —Darbyshire in Kew Bull. **63**: 374, table 2 (2009); in F.T.E.A., Acanthaceae **2**: 699, fig.92 (2010). Type: Ethiopia, Tigray, Djeladjeranne (Dscheladscheranne), fl. 3.xi.1838, *Schimper* II.701 (K lectotype, BM, BR, M, MPU, P, W, WAG), lectotypified by Darbyshire (2010).

Perennial herb or subshrub, erect, decumbent, scrambling or trailing, 40–400 cm tall; stems 6-angular, sometimes ridged, indumentum variable (see subspecies). Leaves sometimes immature or absent at anthesis; ovate, 2–12.5 × 1–6 cm, base rounded, obtuse or shortly attenuate, apex (sub)acuminate, surfaces largely glabrous or sparsely pubescent on margin and main veins beneath; lateral veins (3)4–8(10) pairs; petiole 5–78 mm long. Inflorescence axillary, 1–2(3) umbels of (1)2–5(6) cymules per axil, together sometimes forming loose panicles on largely leafless branches; umbel peduncle 2–30(73) mm long; main axis bracts linear-lanceolate, 2.5–7.5 × 0.3–1 mm, those at lower nodes rarely more foliaceous and lanceolate, to 20 × 7 mm; longest cymule peduncle in each umbel 2.5–28 mm long, shortly glandular and/or eglandular pubescent, with or without long spreading eglandular hairs; cymule bract pairs unequal to highly so, ratio 1.2–2.1:1, green(-brown) or purple, ± conspicuously pale-'windowed', larger bract variously ovate, lanceolate, elliptic, oblong, obovate or oblanceolate, 7–18 × 2.3–8.5 mm, base cuneate, apex acute to rounded or attenuate, mucronate, surface prominently 3-veined, often shortly glandular-pubescent, margin and midrib often with long spreading eglandular hairs and midrib with short antrorse eglandular hairs, rarely largely glabrous; bracteoles green or brown along midrib and apex, with broad pale-hyaline margin, lanceolate, 5.5–10 mm long, apex acuminate. Calyx lobes 3–5.3 mm long, glandular-puberulent, margin strigulose, sometimes also with few long spreading eglandular hairs. Corolla 10.5–23 mm long, limb bright pink, purple or white, eglandular-pubescent externally, limb with interspersed shorter glandular hairs; tube 6–10 mm long; lip held in upper position narrowly oblong, 4.5–14.5 × 2–4 mm, palate puberulous towards mouth; lip held in lower position (ovate-)elliptic to broadly so, 4.5–14 × 3–8 mm. Staminal filaments 4–14.5 mm long, sparsely hairy beneath or largely glabrous; anther thecae 0.6–1.2 mm long, superposed or slightly overlapping, oblique. Style sparsely strigulose or glabrous. Capsule 5.5–8.5 mm long, glandular-puberulous; placental base elastic. Seeds lenticular, 1.2–2 mm in diameter, with short verruciform or conical tubercles with or without minute hooks.

A widespread and polymorphic species. The above description refers only to variation occurring within our region, where three principal forms are easily recognisable. These are here separated as subspecies, though it is debatable as to whether they will be maintained once the *Dicliptera maculata* complex is revised across its full geographic range.

Key to infraspecific taxa:

1. Mature stems glabrescent; young stems usually only with short eglandular hairs, rarely also with glandular hairs; plants of montane to mid-altitude and riverine forest . . . . a) subsp. *maculata*
   – Mature stems usually sericeous; young stems with ± dense short glandular indumentum; plants of low altitude woodland, riverine thicket and dry riverine forest 2
2. Corolla white, cream or mauve-tinged, (13.5)18–23 mm long; smaller cymule bract of each pair ± elliptic with acute apex; seeds with unhooked verruculiform or conical tubercles . . . . b) subsp. *pallidiflora*
   – Corolla bright pink or purple, 10.5–15.5 mm long; smaller cymule bract of each pair usually obovate with obtuse or rounded apex; seeds with minutely hooked conical tubercles . . . . c) subsp. *hirta*

a) Subsp. **maculata**. Darbyshire in Kew Bull. **63**: 374 (2009); in F.T.E.A., Acanthaceae, **2**: 701 (2010) in part, excl. fig.92.

> *Dicliptera maculata* Nees var. *a panicularis* Nees in De Candolle, Prodr. **11**: 485 (1847), illegitimate name. Type as for subsp. *maculata*.
> *Dicliptera lingulata* C.B. Clarke in F.T.A. **5**: 257 (1900). —Binns, Checklist Herb. Fl. Malawi: 13 (1968). —Phiri, Checklist Zamb. Vasc. Pl.: 18 (2005). —Strugnall, Checklist Sperm. Mt. Mulanje: 33 (2006). Type: Malawi, Nyika Plateau, fl.& fr. 1896, *Whyte* 192a (K holotype).
> *Dicliptera longipedunculata* Mildbr. in Bull. Jard. Bot. État **14**: 357 (1937). Type: D.R. Congo, Ile Idjwi in Lake Kivu, fl. v.1929, *Humbert* 8374 (BR lectotype, P), lectotypified here.
> *Dicliptera wittei* Mildbr. in Bull. Jard. Bot. État Brux. **17**: 89 (1943). Type: D.R. Congo, Lake Magera, fl. 1.iii.1934, *de Witte* 1435 (BR holotype).
> *Dicliptera* sp. [=*Brass* 17161] sensu Milne-Redhead in Mem. N.Y. Bot. Gard. **9**: 28 (1954).
> ?*Dicliptera umbellata* sensu Binns, Checklist Herb. Fl. Malawi: 13 (1968), non (Vahl) Juss.

Mature stems glabrescent; young stems usually only with short antrorse and/or retrorse eglandular hairs, rarely also with short glandular hairs; plants leafy at anthesis. Cymule bracts elliptic, narrowly oblong-elliptic, lanceolate or obovate-elliptic, ratio of larger to smaller bract length 1.2–1.55:1, surfaces often shortly glandular-pubescent, with antrorse eglandular hairs along midrib, with or without long spreading eglandular hairs along margin, or bracts sometimes largely glabrous. Corolla bright pink or purple, 14–21.5 m long; staminal filaments 6.5–12 mm long. Seeds with short unhooked conical or verruciform tubercles.

Zambia. N: Isoka Dist., Mafinga Mts, fl. 24.v.1973, *Chisumpa* 69 (K, NDO). E: Chama Dist., Nyika, fl. 26.vi.1966, *Fanshawe* 9740 (K, NDO, SRGH). **Malawi**. N: Rumphi Dist., Nyika Plateau, Kafwimba Forest, fl. 24.viii.1977, *Pawek* 12960 (K, MAL, MO, SRGH, UC). S: Mulanje Dist., Mchese Mt, Namphende Valley, fl. 28.vii.1988, *J.D. & E.G. Chapman* 9229 (BR, K, MO, SRGH).

Also in Cameroon, Gabon, Central African Republic, South Sudan, Eritrea, Ethiopia, Uganda, Kenya, Tanzania, D.R. Congo, Burundi and Angola. Montane to mid-altitude moist forest and fringing riverine forest, including margins and clearings; 1200–2300 m.

Conservation notes: Widespread and often common; Least Concern.

Plants from the Flora region generally have less hairy inflorescences than those in E & NE Africa – the bracts often lack the long spreading eglandular cilia and can sometimes be almost glabrous. However, there is considerable variation with some notable differences between the three main population centres (Misuku Hills, Nyika Plateau and Mulanje massif), e.g. the Misuku Hills plants have the most hairy and proportionally broadest bracts and also have more contracted umbels than elsewhere.

b) Subsp. **pallidiflora** I. Darbysh., subsp. nov. Differs from subsp. *maculata* in having the combination of pale corollas, a densely glandular indumentum on the young stems and usually densely sericeous lower stems. Differing from subsp. *hirta* in the larger and paler corollas and in having seeds with verruculiform, unhooked tubercles. Type: Tanzania, Selous Game Reserve, Matandu Camp, fl.& fr. 19.ix.1977, *Vollesen* in MRC 4702 (C sheet 1 holotype, C sheet 2, DSM, EA). FIGURE 8.6.**69**, 1–6.

> *Dicliptera* sp. nov. aff. *umbellata* (= *Haerdi* 620/0) sensu Vollesen in Opera Bot. **59**: 80 (1980).
>
> *Dicliptera hirta* K. Balkwill in Kew Bull. **51**: 50 (1996) in part for *Torre* 483 from Mozambique.
>
> *Dicliptera maculata* Nees subsp. *maculata* sensu Darbyshire in F.T.E.A., Acanthaceae 2: 701 (2010) in part, incl. fig.92.

Mature stems often white-sericeous, hairs usually also present but less dense on younger stems; young stems also ± densely glandular-pubescent; leaves often immature or absent at anthesis. Cymule bracts oblong-elliptic, oblanceolate or larger of pair somewhat ovate, smaller bract with acute or attenuate apex, ratio of larger to smaller bract length 1.35–1.85:1, surfaces shortly glandular-pubescent, with antrorse eglandular hairs along midrib and with few to numerous long spreading eglandular hairs along margin. Corolla white, cream or mauve-tinged, 18–23 mm long; staminal filaments 8–14.5 mm long. Seeds with short unhooked conical or verruciform tubercles.

Mozambique. N: Cuamba Dist., Cuamba, road to Mandimba, fl. 3.viii.1934, *Torre* 483 (COI, LISC).

Also in SE Tanzania. Riverine woodland, dry sandy or rocky streambeds and open dry woodland; up to c.700 m.

Conservation notes: A localised subspecies that may have declined due to loss of riparian habitat; possibly Near Threatened.

In F.T.E.A., this taxon was treated as a lowland, drier country form of subsp. *maculata* (see note, p. 701), but it is usually easy to separate from the forest plants by the combination of pale flowers and differing stem indumentum. It is here separated as a geographic and ecological subspecies.

le Testu 813 from Nhandoa (Mozambique T) appears intermediate between this and subsp. *hirta* – the flowers are white but small (c.13.5 mm long; filaments 6.5 mm). Unfortunately the specimen lacks fruits so the seed sculpturing is unknown.

c) Subsp. **hirta** (K. Balkwill) I. Darbysh., comb. & stat. nov. Type: Zimbabwe, Lomagundi, Manyame (Hunyani) R., fl.& fr. 22.vii.1921, *Eyles* 3139 (SRGH holotype, BOL, K, PRE, SAM). FIGURE 8.6.**69**, 7.

> *Dicliptera hirta* K. Balkwill in Kew Bull. **51**: 50, fig.14b (1996), excl. *Torre* 483 from Mozambique. —Mapaura & Timberlake, Checklist Zimb. Vasc. Pl.: 13 (2004). —Setshogo, Prelim. Checklist Pl. Botswana: 18 (2005). —Darbyshire in F.T.E.A., Acanthaceae 2: fig.92.7 (2010).
>
> *Dicliptera maculata* sensu Phiri, Checklist Zamb. Vasc. Pl.: 18 (2005), non Nees sensu stricto.

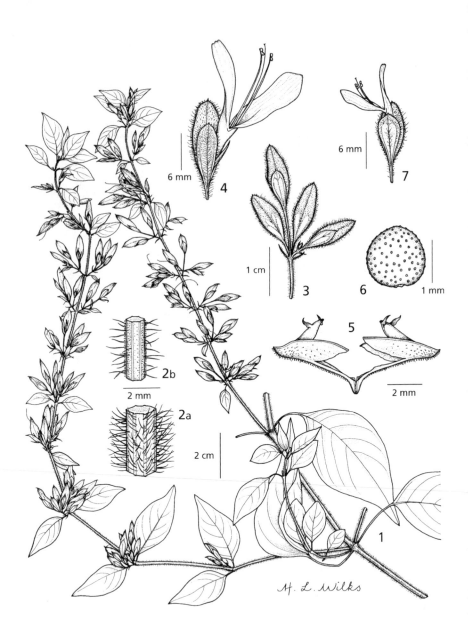

Fig. 8.6.**69**. DICLIPTERA MACULATA subsp. PALLIDIFLORA. 1, habit; 2, variation in stem indumentum: a, lower stem, b, upper stem between fertile nodes; 3, inflorescence; 4, cymule with flower; 5, dehisced capsule; 6, mature seed. DICLIPTERA MACULATA subsp. HIRTA. 7, cymule with flower. 1, 2, 5 and 6 from *Vollesen* in MRC 4702; 3 from *Haerdi* 620/0; 4 from *Flock* 91; 7 from *Eyles* 3139. Drawn by Hazel Wilks. Reproduced from Flora of Tropical East Africa (2010).

Mature stems densely white-sericeous, hairs usually also present but less dense on younger stems; young stems ± densely glandular-pubescent; leaves often immature or absent at anthesis. Larger cymule bract of each pair elliptic, ovate, oblong or obovate, smaller bract always obovate with ± rounded apex, ratio of larger to smaller bract length (1.2)1.35–2.1:1, surfaces shortly glandular-pubescent, with ± numerous long spreading eglandular hairs along margin and midrib. Corolla bright pink or purple, 10.5–15.5 mm long; staminal filaments 4–7 mm long. Seeds with minutely hooked conical tubercles.

Botswana. N: Okavango swamps, Karo Is., c.16 km S of Seronga, fl.& fr. 28.ix.1954, *Story* 4767 (K, PRE). **Zambia**. W: Kitwe, fl.& fr. 4.vii.1963, *Mutimushi* 340 (K, NDO). E: Lundazi Dist., Lukusuzi Nat. Park, near Kalindi, fl.& fr. 4.vii.1971, *Sayer* 1228 (SRGH). S: Choma Dist., Mapanza, fl. 24.v.1957, *Robinson* 2206 (K, SRGH). **Zimbabwe**. N: Makonde Dist. (Lomagundi), Manyame (Hunyani) R. between Mhangura (Mangula) and Raffingora, fl.& fr. 21.ix.1969, *Jacobsen* 3951 (K, PRE). S: Chiredzi Dist., Save–Runde (Sabi-Lundi) junction, Danga, Chitsa's Kraal, fl.& fr. 5.vi.1950, *Chase* 2310 (BM, BR, K, LISC, SRGH).

Not known elsewhere. In dry riverine forest, riverine thicket and woodland; 250–1250 m.

Conservation notes: Endemic to the Flora area. Widespread but apparently rather local; probably Least Concern.

In the protologue to *Dicliptera hirta*, Balkwill included in synonymy "*D. maculata* sensu C.B. Clarke in F.T.A.", stating that Clarke's description better matches plants of *D. hirta* than the type of *D. maculata*, particularly in relation to stem indumentum. However, Clarke's description was based on material from Ethiopia and Sudan, including Nees' type, whilst Balkwill records *D. hirta* as being from southern Africa only. Stem indumentum is, in fact, quite variable in *D. maculata* sensu stricto, particularly in the northern part of its range, with some specimens having the long sericeous hairs as in *D. hirta*. The southern African plants differ somewhat from true subsp. *maculata* in the minutely hooked seed tubercles and in often having more strikingly uneven cymule bract pairs, so, I maintain it as a subspecies here.

10. **Dicliptera clinopodia** Nees in De Candolle, Prodr. **11**: 483 (1847). —Anderson in J. Linn. Soc., Bot. **7**: 47 (1864). —Clarke in Fl. Cap. **5**: 91 (1901). —Balkwill *et al.* in Kew Bull. **51**: 29, map 4 (1996). —Bandeira *et al.*, Fl. Nat. Sul Moçamb.: 161, 194 (2007). Type: South Africa, KwaZulu-Natal "In Promont. Bonae Spei" in 'valley of Umlazi R.', fl. *Drège* s.n. (K 000378960 lectotype), lectotypified here (see note).

Justicia clinopodia E. Mey. in Drège, Zwei Pflanzengeogr. Docum.: 158, 195 (1843–1844), invalid name.

Diapedium clinopodium (Nees) Kuntze, Revis. Gen. Pl. **2**: 485 (1891).

Diapedium clinopodium (Nees) Kuntze var. *minor* S. Moore in J. Bot. **40**: 385 (1902). Type: South Africa, KwaZulu-Natal, Isandhlwana, fl. 1901, *Pateshall Thomas* s.n. (BM holotype).

Perennial herb, erect or scrambling, 30–300 cm tall; stems (sub)6-angular, sometimes with prominent ridges, shortly pale-pubescent with retrorse and/or antrorse hairs or soon glabrescent. Leaves ovate or lanceolate, 3.3–15 × 0.8–5 cm, base rounded or obtuse to cuneate or attenuate, apex acute to acuminate, apiculate, surfaces finely antrorse-pubescent, particularly on veins beneath, or largely glabrous; lateral veins 3–7 pairs; petiole 3–63 mm long or uppermost leaves sessile. Inflorescence of ± numerous umbellately arranged cymules aggregated into dense conical, cylindrical, subcapitate or verticillate spikes 1.5–10(25) × 2–4 cm, often interrupted by reduced, bract-like leaves; umbels subsessile or peduncles to 8.5 mm long; main axis bracts linear-lanceolate to ovate-acuminate or obovate-caudate, 5–21 × 0.5–7 mm, green, green-brown or purplish, sometimes with pale 'windows'; cymules sessile or peduncules to 5 mm, antrorse- or retrorse-pubescent; cymule bract pairs subequal to unequal, ratio 1–1.6:1, green, green-brown or purple-brown, with ± conspicuous pale or purple 'windows', larger bract lanceolate, narrowly ovate, elliptic or somewhat oblanceolate, 9–21.5 × 2–8 mm, base cuneate, apex acute-mucronulate to caudate, surface prominently 3-veined, often with few to numerous short glandular- and eglandular-hairs, main veins and margin antrorse-pubescent, margin and midrib often also with few to numerous long spreading hairs; bracteoles green, green-brown or purple-brown with broad pale-hyaline margin, lanceolate, 5–16.5 × 1–2.5 mm, apex acuminate to caudate. Calyx lobes 3–6 mm long, glandular- and eglandular-puberulent and shortly ciliate, sometimes with scattered longer ascending or spreading eglandular hairs. Corolla 17.5–34(36) mm long, pink, mauve, purple or rarely white, with purple guidelines, pubescent externally, limb with interspersed shorter glandular hairs; tube 7.5–14.5 mm long; lip held in upper position narrowly oblong to ligulate, 10–20(23) × 2.5–4.3 mm, palate glabrous or sparsely puberulous at mouth; lip held in lower position oblong-elliptic or ovate-elliptic, 10.5–21(26) × 4–10 mm. Staminal filaments 9.5–22 mm long, sparsely hairy beneath towards base; anther thecae 0.8–1.2 mm long, superposed or slightly overlapping, somewhat oblique. Style sparsely strigulose. Capsule 5–9 mm long, glandular-puberulous towards apex, often with interspersed short eglandular hairs or these along flanks; placental base elastic. Seeds lenticular, 1.2–2.3 mm in diameter, with verruciform tubercles.

I have taken a pragmatic approach here in maintaining *Dicliptera clinopodia* as distinct from *D. maculata* since the two are so easily separable in the Flora region. However, there are forms of *D. maculata* from E and NE Africa with more contracted inflorescences forming terminal heads very similar to those in *D. clinopodia* sensu lato, and the two could easily be considered conspecific. That said, I am reluctant to sink *D. clinopodia* until the *D. maculata* complex has been fully revised.

Three geographically and ecologically allopatric forms of "*D. clinopodia*" are easily separable in our region and it is tempting to consider them as separate species, but this is complicated by the picture in South Africa. The first two correspond to Balkwill's (1996) *D. clinopodia* forms A and B, both of which extend into eastern South Africa and Swaziland. He noted that although these are readily distinguished by differences in shape and colouring of the main axis bracts in the northern part of their range (see key below), they appear to converge and become virtually indistinguishable in the southernmost part of the range in the Eastern Cape and southern KwaZulu-Natal (see his map 4, p.33).

There are two Drège collections at Kew bearing the name *Justicia clinopodia*, both of which were annotated by C.B. Clarke as being from the valley of the Umlazi River, KwaZulu-Natal. The sheet seen by Balkwill (K000378960) and annotated as *D. clinopodia* by Nees, is one of these intermediate forms – it has the much-reduced main axis bracts of form A, but small, rather narrow leaves and a mainly retrorse stem indumentum as in form B. The second sheet (K000378962) appears rather different and could be placed in form B, though both the main axis and cymule bracts are admittedly rather narrower and less prominently 'windowed' than usual in that form. However, this sheet does not appear to have been annotated by Nees. It therefore seems best to select K000378962 as the lectotype, but this leaves some difficulty in applying the name *D. clinopodia* sensu stricto if forms A and B are to be formally recognized at some rank. In view of this problem, together with the uncertainty over the status of *D. clinopodia* in relation to *D. maculata*, I have chosen not to give the three forms formal names at present.

Key to infraspecific forms:

1. Main axis bracts ovate- or lanceolate-acuminate or obovate-caudate, 2.5–7 mm wide, with conspicuous pale 'windows', ± similar to or broader than cymule bracts; petioles short, 3–15 mm long. . . . .*form B*
- Main axis bracts linear-lanceolate or rarely lanceolate, 0.5–2.5 mm wide, without pale 'windows', much narrower than cymule bracts; petioles often over 15 mm long 2

2. Leaves usually lanceolate or narrowly ovate, length to width ratio 2.6–8.3:1; cymule bracts with large and conspicuous pale 'windows', larger bract of each pair with a prominent flexuose mucro; corolla 17.5–27 mm long . . . . *form C*
- Leaves ovate, length to width ratio 1.7–2.4:1; cymule bracts usually with smaller and less conspicuous 'windows', whole bract often purple (drying purple-brown), larger bract of each pair with a short mucro; corolla (23)26–34 mm long . . *form A*

Form **A**. —Balkwill *et al.* in Kew Bull. **51**: 32, map 4 (1996).

Dicliptera nobilis S. Moore in J. Linn. Soc., Bot. **40**: 164 (1911). Type: Zimbabwe, near Chirinda, fl. 27.vi.1906, *Swynnerton* 1936 (BM holotype, K, SRGH).

Stem hairs antrorse, more rarely spreading. Leaves ovate, (4.3)6–10.5 × (2.3)2.6–5 cm, length to width ratio 1.7–2.4:1; petiole (12)22–63 mm long. Main axis bracts linear-lanceolate or lanceolate, 5–11(14.5) × 0.5–2.5 mm, without pale 'windows'; cymule bract pairs unequal, ratio 1.3–1.6:1, green, green-brown or purple-brown, 'windowed' but windows inconspicuous when bracts purple, larger bract elliptic(-lanceolate) or narrowly ovate, 12–21.5 × 3.5–8 mm, apex acute to attenuate, mucronulate; bracteoles 5–9 mm long. Corolla (23)26–34 mm long; lip held in upper position ligulate, (12)14–20 mm long, palate glabrous. Capsule 6.5–9 mm long. Seeds 1.7–2.3 mm in diameter.

Zimbabwe. E: Mutare Dist., Vumba Mts, Mopaff Estate (Rhodesia Wattle Company), R bank of Nzombe R. at coffee factory, fl. 12.iv.1965, *Chase* 8300 (BR, K, LISC, SRGH). S: Bikita Dist., c.8 km SE of Silveira Mission, fl. 7.v.1969, *Biegel* 3048 (K, LISC, SRGH). **Mozambique**. MS: Manica Dist., Manica (Vila de Manica) to Vumba, fl. 22.vi.1949, *Pedro & Pedrógão* 6805 (LMA).

Also in Swaziland and South Africa (Limpopo, Mpumalanga, KwaZulu-Natal, Eastern Cape). Dense bushland and scrub at forest margins and along streams, road verges and on rocky ground; 700–1500 m.

Conservation notes: Locally common and likely to tolerate some disturbance; not threatened.

The material is rather uniform in our area and has previously been separated as *Dicliptera nobilis*; largely identical plants are frequent in NE South Africa. Further south, Balkwill's form A becomes more variable and the plants often have a smaller stature than in our region, although the variation appears clinal.

Form **B**. —Balkwill *et al.* in Kew Bull. **51**: 34, map 4 (1996).

Stem hairs retrorse or sometimes with interspersed antrorse hairs. Leaves ovate or lanceolate, 3.3–8.5 × 1.3–3.3 cm, length to width ratio (1.75)2–3.6:1; petiole to 15 mm long or uppermost leaves subsessile. Main axis bracts ovate- or lanceolate-acuminate to obovate-caudate, 9–21 × 2.5–7 mm, green-brown or brown with conspicuous pale 'windows'; cymule bract pairs subequal or somewhat unequal, ratio 1–1.35:1, green-brown or purple-brown with conspicuous pale 'windows', larger bract lanceolate to somewhat oblanceolate, 10–21 × 2–4(6) mm, apex caudate; bracteoles (7.5)9.5–16.5 mm long. Corolla 21–32(36) mm long; lip held in upper position ligulate, 11–18.5(23) mm long, palate glabrous. Capsule 6.5–8.5 mm long. Seeds 1.8–2.2 mm in diameter.

Mozambique. M: Namaacha Dist., Goba Fronteira, fl.& fr. 28.vi.1961, *Balsinhas* 491 (K, LISC, LMA, LMU, PRE); Namaacha, 11 km along road to Matutuine (Matianine), fl. 16.vi.1974, *Marques* 2477 (LISC, LMU, SRGH).

Also in Swaziland and South Africa (Mpumalanga, KwaZulu-Natal, Eastern Cape). In open woodland and dry forest margins on rocky slopes, riverine thicket and disturbed ground; 10–650 m.

Conservation notes: This form is fairly widespread in the KwaZulu-Natal lowlands extending into the Lebombo Mts, where it is fairly common; Least Concern.

Most specimens from our region have large corollas 27–32 mm long, but occasional smaller-flowered specimens are noted, for example *Torre* 7675 from Rio Maputo, Salamanga (LISC).

Form C (not form C of Balkwill in Kew Bull. **51**: 34, 1996).

Dicliptera leonotis sensu auct., non C.B. Clarke. —Clarke in F.T.A. **5**: 260 (1900). —Binns, Checklist Herb. Fl. Malawi: 13 (1968). —Moriarty, Wild Fl. Malawi: 84, pl.42(3) (1975). — Strugnall, Checklist Sperm. Mt. Mulanje: 33 (2006).

Stem hairs retrorse and/or antrorse. Leaves lanceolate or narrowly ovate, rarely more broadly ovate, 6.5–15 × 0.8–4.8 cm, length to width ratio 2.6–8.3:1; petiole to 4–40 mm long. Main axis bracts linear-lanceolate, 6–15 × 0.5–1.5 mm, without pale 'windows'; cymule bract pairs unequal, ratio (1.1)1.2–1.4(1.5):1, green or purple-brown with large and conspicuous pale 'windows', larger bract lanceolate to narrowly elliptic, 9–17.5 × 2.5–4 mm, apex acute or attenuate with a conspicuous mucro or sometimes caudate-mucronate; bracteoles 6.5–9.5 mm long. Corolla 17.5–27 mm long; lip held in upper position narrowly oblong(-lanceolate), 10–16.5 mm long, palate sparsely puberulous at mouth. Capsule 5–7 mm long. Seeds 1.2–1.6 mm in diameter.

Malawi. C: Ntchisi Dist., Chipata Mt, fl. 2.v.1980, *Brummitt, Banda & Patel* 15569 (K, MAL). S: Zomba Dist., Zomba Mt forestry road from Old Naisi, fl. 30.iv.1980, *Blackmore et al.* 1363 (BM, BR, K, MAL). **Mozambique**. N: Mandimba Dist., Mandimba, fl.& fr. 31.vii.1934, *Torre* 473 (LISC). Z: Morrumbala Dist., Metolola, fl. 23.v.1943, *Torre* 5358 (LISC, MO, PRE). T: Moatize Dist., Zóbuè, fl. 17.vi.1947, *Hornby* 2762 (K, SRGH).

Not known elsewhere. Miombo and *Combretum–Acacia* woodland, sometimes on rocky slopes, riparian thicket and woodland, damp shaded gullies, amongst tall grass in partial shade and persisting in pine plantations; 600–1500 m.

Conservation notes: Apparently mostly confined to the hills around the S end of Lake Malawi and the Shire Highlands. Locally common in a variety of habitats, it appears tolerant of disturbance; Least Concern.

Clarke includes this taxon within the Indian *Dicliptera leonotis*, but the two are quite clearly different – true *D. leonotis* has much smaller bracts which lack the conspicuous 'windows'.

This form comes particularly close to the occasional specimen of *D. maculata* from NE Africa with contracted terminal heads, such as plants from Mt Debasien (Kadam) in N Uganda (e.g. *Eggeling* 2662), differing from those mainly in the narrower leaves. However, it is very different to sympatric forms of *D. maculata* in S Malawi.

Leach & Rutherford-Smith 10974 from 59.5 km W of Nampula in N Mozambique has exceptionally large flowers for this form, similar to those of forms A and B.

11. **Dicliptera sp. C** (= *Faulkner* PRE 196).

Spreading suffruticose perennial to 50 cm tall, much-branched from a woody base; stems 6-angular, furrowed between ridges when dry, at first sparsely antrorse-pubescent but soon glabrescent. Leaves lanceolate, 3–6 × 0.7–1.8 cm, base cuneate, decurrent onto petiole, apex acute or obtuse, glabrous except for few inconspicuous hairs along margin; lateral veins 4–5 pairs; petiole 3–10 mm long. Inflorescences axillary and terminal, (1)2–3 umbels of 3–5 cymules per axil, together forming a verticillate terminal thyrse; umbel peduncle 1–2.5 mm long; main axis bracts linear-lanceolate, 4–10.5 × 0.5–0.7 mm; longest cymule peduncle in each umbel 1–2.5 mm long, glandular- and eglandular-puberulous; cymule bracts in dry state green-brown or pink-tinged, with narrow pale-brownish or pink-tinged 'windows', pairs somewhat unequal, ratio 1.1–1.3:1, larger bract (elliptic-)lanceolate, 8–9.5 × 2–3 mm, apex acute, mucronate, surface 3-veined, densely glandular- and eglandular-puberuolous, with occasional long spreading eglandular hairs on margin and towards base; bracteoles with dark midrib and apex and broad pale-hyaline margin, lanceolate, 6–6.5 × 1.2 mm, apex acuminate. Calyx lobes 3.3 mm long, glandular- and eglandular-puberulent. Corolla 18.5–19 mm long, colour not recorded, eglandular-pubescent externally with few interspersed glandular hairs; tube 7.5–8 mm long; lip held in upper position oblong, 10–11 × 2.5–3.2 mm, palate puberulous towards mouth; lip held in lower position elliptic, 10–11 × 4.7–5.2 mm. Staminal filaments c.10 mm long, sparsely hairy beneath in proximal half; anther thecae c.0.8 mm long, superposed and oblique. Style glabrous. Capsule c.5.3 mm long, with mixed short glandular and eglandular hairs; placental base elastic. Only immature seeds seen, tuberculate, tubercles conical but those towards margin more elongate and minutely hooked.

Mozambique. Z: Lugela Dist., Namagoa, 200 km inland from Quelimane, fl.& fr. ix.1943, *Faulkner* PRE 196 (BR, EA, K, LMA, PRE).

Not known elsewhere. Habitat information on the two Kew sheets is contradictory – "common in one or two places close to a stream, but not in damp ground" vs. "growing in dampish ground, and along drain sides. Common in our area but probably also in other damp situations"; c.150 m.

Conservation notes: Known only from the single specimen cited which records this species as locally common; Data Deficient.

This specimen has previously been identified as "*Dicliptera leonotis*" (= *D. clinopodia* form C) but differs in the suffruticose habit, the less congested compound inflorescence, the smaller 'windows' on the bracts, the denser glandular indumentum on peduncles and bracts, and the more elongate, sometimes minutely hooked seed tubercles. Some of these characters approach *D. carvalhoi* and it is possible that this specimen represents a hybrid population, though it could equally be of a distinct and highly localised species.

12. **Dicliptera quintasii** Lindau in Bot. Jahrb. Syst. **22**: 121 (1897). —Clarke in Fl. Cap. **5**: 92 (1901). —Balkwill *et al.* in Kew Bull. **51**: 44, fig.13 (1996). Type: Mozambique, Moamba Dist., Pessene, fl. v.1893, *Quintas* 85 (B† holotype, C, COI).

Erect, decumbent or spreading perennial herb to 60 cm tall; stems 6-angular, shortly retrorse-pubescent. Leaves papery, ± broadly ovate, (2.3)3–9.5 × (1.4)2–5.2 cm, base broadly attenuate to obtuse or rounded, apex acute or attenuate, tip blunt and apiculate, surfaces shortly antrorse-pubescent, sparse at maturity; lateral veins 3–5 pairs; petiole (5)9–40 mm long. Inflorescences axillary and terminal, (1)2–3 umbels of 3–6 cymules per axil, sometimes crowded in upper axils and forming a verticillate spike interrupted by reduced leaves, sometimes with a cylindrical terminal 'head' to 3.5 cm long; umbel peduncle 1–7.5 mm long, retrorse-pubescent; main axis bracts oblanceolate or elliptic, more rarely linear-lanceolate, (3)5–11 × (0.5)1–3 mm, apex often acuminate, margin long-ciliate; longest cymule peduncle in each umbel 0.8–3.5 mm long; cymule bracts membranous, at first olive-green with usually inconspicuous pale 'windows' towards base, later turning somewhat scarious, rarely flushed purple towards apex, pairs unequal, ratio 1.15–1.55:1, larger bract (oblong-)obovate, 8–14.5 × 3.2–7 mm, apex obtuse, rounded or shortly attenuate, mucronate, surface 3-veined from base, margin and main veins long-pilose in basal half, antrorse-pubescent elsewhere externally, short glandular hairs sparse towards apex and on exposed internal surface of larger bract, rarely denser throughout; bracteoles pale-hyaline with pale green midrib and apex, lanceolate, 5.7–10.5 × 0.8–1.4 mm, apex caudate. Calyx lobes 3.5–6.7 mm long, glandular-and eglandular-puberulent, midrib of each lobe with longer eglandular hairs, margin long-ciliate. Corolla 16–24 mm long, limb purple or bright pink, eglandular-pubescent externally; tube 7.5–10.5 mm long; lip held in upper position oblong, 8–13.5 × 2.5–3 mm; lip held in lower position (oblong-)elliptic, 7.5–13.5 × 3.5–5.5 mm. Staminal filaments 7–11 mm long, sparsely hairy; anther thecae c.0.7 mm long, superposed. Style strigulose. Capsule 6–7.5 mm long, shortly pubescent with glandular and few eglandular hairs; placental base elastic. Only immature seeds seen, with verruciform or conical tubercules.

Mozambique. M: Matutuine Dist., Maputo, 6.4 km from ferry on R. Maputo in direction of Zitundo, fl. 21.iv.1969, *Correia & Marques* 671 (LMU); Marracuene Dist., between R. Incomati and Marracuene, fl. 20.v.1981, *de Koning & Boane* 8762 (K, LMU, SRGH).

Also in South Africa (KwaZulu-Natal). Dry dense forest, secondary forest and degraded woodland on sand; 0–100 m.

Conservation notes: Species restricted to the sand forests of the Maputaland Centre of Endemism, known from 2 localities in South Africa and 7 in Mozambique. Some sites are protected but the species is likely to have suffered from loss of habitat; possibly threatened.

There is no doubt that *Dicliptera quintasii* is closely allied to both *D. clinopodia* and *D. maculata*. The rather broad, long-petiolate, papery leaves and the often pale green (washed-out) and membranous cymule bracts help to distinguish it. Although these may seem rather 'soft' characters, *D. quintasii* is easily identified by comparison to named material.

Balkwill (1996) records the seeds of this species as having tall, hooked tubercles, but under the microscope they appear rather squat and conical or verruculose and the hooks are not clearly visible.

13. **Dicliptera sp. D** (= *Brass* 17817).

   Dicliptera sp. (=*Brass* 17817) sensu Milne-Redhead in Mem. N.Y. Bot. Gard. **9**: 28 (1954).

   Slender herb, trailing towards base then ascending to 20–30 cm high; stems 6-angular, furrowed between ridges when dry, at first sparsely antrorse-pubescent on ridges but soon glabrescent. Leaves papery, ovate, pairs rather unequal, larger 2.8–5.8 × 1–2 cm, base obtuse or rounded, apex acuminate, upper surface pilose, lower surface glabrous or with sparse antrorse hairs on main veins and margin; lateral veins 3–5 pairs; petiole 6–20 mm long. Inflorescence axillary, 1 umbel of (1)2(3) cymules per axil; umbel peduncle 0.5–2 mm long; main axis bracts linear-lanceolate, 2–3.5 × c.0.5 mm; longest cymule peduncle in each umbel 2–4.5 mm long, antrorse-pubescent; cymule bracts membranous, pale green with inconspicuous paler 'windows' towards base of smaller bract, pairs unequal, ratio 1.3–1.45:1, larger bract elliptic or obovate, 8–12.5 × 4–6 mm, apex acute to obtuse, prominently mucronate, smaller bract obovate with a more obtuse or rounded apex, both triplinerved but 2 lateral veins more prominent on smaller bract, margin and midrib sparsely long-pilose and antrorse-pubescent, long hairs sometimes also present on exposed portion of larger bract within, short glandular hairs sparse either along margins only or scattered on exposed surfaces; bracteoles pale-hyaline with pale green midrib and apex, lanceolate, 5.3–8.5 × 1 mm, apex caudate. Calyx lobes 2.7–3.8 mm long, long-pilose or sparsely so. Corolla pale pink with darker pink guidelines, c.15 mm long, eglandular-pubescent externally with few interspersed glandular hairs; tube c.7 mm long; lips 7–8 mm long, lip held in upper position oblong, lip held in lower postion elliptic. Staminal filaments c.4.5 mm long; anther thecae 0.7 mm long, superposed and oblique. Style glabrous. Capsule c.6.5 mm long, glandular-pubescent, placental base elastic. Seeds with verruciform tubercules.

   Malawi. S: Thyolo Dist., Thyolo (Cholo) Mt, fl.& fr. 26.ix.1946, *Brass* 17817 (K, NY). Not known elsewhere. Moist forest undergrowth; c.1350 m.

   Conservation notes: Known only from the single specimen cited. As there is very little forest remaining on Thyolo Mt, this species may be Critically Endangered or Extinct.

   The membranous pale green cymule bracts are reminiscent of *Dicliptera quintasii*, but the Malawi specimen clearly differs in being more slender with narrower leaves and in having solitary, few-flowered/fruited axillary cymes.

14. **Dicliptera minor** C.B. Clarke in Fl. Cap. **5**: 92 (1901), excl. *Burchell* 2147 from Namibia. —Balkwill *et al.* in Kew Bull. **51**: 37, fig.12a (1996). Type: Botswana, Kweneng Dist. (Bakwena Territory), fl.& fr., n.d., *Holub* s.n. (K lectotype), lectotypified by Balkwill *et al.* (1996).

   Perennial herb or subshrub, erect, scrambling, procumbent or trailing, to 60 cm tall, branching from a woody rootstock, sometimes rooting along trailing branches; stems 6-angular,

± prominently ridged, glabrous or with sparse short appressed hairs when young, nodal line often pubescent. Leaves ovate or narrowly so, 0.7–4 × 0.5–1.5 cm, base rounded to shortly attenuate, apex rounded, acute or attenuate, apiculate, surfaces largely glabrous; lateral veins 2–5 pairs; petiole 0–9 mm long. Inflorescences axillary or crowded into short spikes on short lateral branches, 1(2) umbels of (1)3–4 cymules per axil; umbel peduncle 1–7 mm long, those at lower axils sometimes to 14(25) mm long, held close to stem, ± glabrous; main axis bracts green, linear-lanceolate, then 3.5–8 × 0.7–1.2 mm, those of umbels at lower axils more leafy and lanceolate, to 14(20) × 4(7) mm; longest cymule peduncle in each umbel 1–5.5(8.5) mm long, puberulous, antrorse-pubescent or glabrous; cymule bract pairs ± unequal, ratio 1.1–1.45:1, green or green-brown, with pale or pinkish-brown 'windows' towards base sometimes inconspicuous, larger bract narrowly elliptic(-lanceolate) to obovate, 6.5–13.5 × 2.2–5.5 mm, apex acute, acuminate or rounded, mucronate, prominently 3-veined from base, with short antrorse hairs on midrib and margin, sometimes also glandular- and eglandular-puberulent at least on exposed inner surface of larger bract; bracteoles green with broad pale-hyaline margin, lanceolate, (4)5.5–9 × 1–2.2 mm, acuminate. Calyx lobes 2.5–4 mm long, glandular-puberulent or sparsely so, margin strigulose. Corolla 14–20 mm long, pink, mauve, purple-red or white, pubescent externally, limb with few interspersed short glandular hairs; tube 6–8.5 mm long; lip held in upper position narrowly oblong, 7.8–11.5 × 2.5–4 mm, palate puberulous towards mouth; lip held in lower position elliptic to broadly ovate, 7.8–12.5 × 4.3–10 mm. Staminal filaments 6–9 mm long, shortly hairy below; anther thecae 0.7–1.1 mm long, superposed but slightly overlapping and oblique. Style glabrous or sparsely strigulose. Capsule 5.5–7 mm long, glandular-puberulous towards apex, often also with few eglandular hairs; placental base elastic. Seeds lenticular, 1.7–1.8 mm in diameter, tuberculate.

Key to infraspecific taxa:

1. Cymule bracts largely glabrous or sparsely glandular- and eglandular-puberulent; seeds with squat, conical tubercles . . . . a) subsp. *minor*
– Cymule bracts densely glandular- and eglandular-puberulent; seeds with more slender tubercles. 2
2. Corolla pink, mauve or pale purple; branches often erect, decumbent or scrambling, rarely procumbent . . . . b) subsp. *divaricata*
– Corolla white or tinged pink; branches at first erect but soon becoming procumbent or prostrate [not recorded in our region] . . . . . subsp. *pratis-manna*

a) Subsp. **minor**. Balkwill *et al.* in Kew Bull. **51**: 37 (1996).

Perennial herb; branches at first erect, typically when less than 10 cm tall, but soon becoming procumbent, trailing or weakly scrambling, sometimes rooting along trailing nodes. Leaf apex rounded, acute or attenuate. Umbels at lower fertile nodes sometimes on peduncles to 14(25) mm long; main axis bracts either linear-lanceolate and 3.5–8 × 0.7–1.2 mm or more leafy and lanceolate to 14(20) × 4(7) mm. Cymule bracts often glabrous except for short antrorse hairs along midrib and margin, sometimes also sparsely (rarely densely) glandular- and eglandular-puberulent; bracteoles (4)5.5–7 mm long. Corolla deep pink, mauve or reddish-purple. Seeds with squat, conical tubercles which are minutely hooked.

Botswana. SE: Kgatleng Dist., Mochudi, fl. iii.1914, *Rogers* 6358 (BM, K); Gaborone, near Motwane Sewage Pond, fl.& fr. 10.xii.1981, *Barnes* 265 (PRE).

Also in South Africa (Limpopo, North West). Riverbanks in woodland and *Acacia* bushland including heavily grazed areas, damp soils on floodplains and in pans; 750–1100 m.

Conservation notes: Restricted to a small area in SE Botswana and NE South Africa. Unlikely to be threatened since it tolerates some disturbance; Least Concern.

The type material of *Dicliptera betschuanica* Lindau at Berlin has been destroyed and I have been unable to locate any duplicates. From the description and locality it is quite possible that it belongs here – see note under Excluded Species.

b) Subsp. **divaricata** (Compton) I. Darbysh., comb. & stat. nov. Type: Swaziland, Ingwavuma Poort, fl. 6.iii.1959, *Compton* 28614 (NBG holotype, PRE).

Dicliptera divaricata Compton in J. S. Afr. Bot. **41**: 49 (1975). —Compton, Fl. Swaziland: 558 (1976). —Balkwill *et al.* in Kew Bull. **51**: 35 (1996). —Welman in Germishuizen *et al.*, Checklist S. Afr. Pl.: 83 (2006).

Dicliptera minor C.B. Clarke subsp. *pratis-manna* K. Balkwill in Kew Bull. **51**: 38 (1996), in part for *Kirk* 112 (South Africa).

Perennial, sometimes shrubby herb, branches erect, decumbent, scrambling or more rarely procumbent. Leaf apex acute or attenuate. Umbel peduncles to 6.5(10.5) mm long; main axis bracts linear-lanceolate or lanceolate, 3.5–13.5 × 0.7–2.5(4) mm. Cymule bracts densely glandular- and eglandular-puberulent in addition to short antrorse hairs on midrib and margin; bracteoles 6–9 mm long. Corolla pink, mauve or pale purple. Seeds with more slender small tubercles without hooks.

Mozambique. M: Moamba Dist., Ressano Garcia, fl.& fr. 24.xii.1897, *Schlechter* 11910 (BM, BR, K, PRE).

Also in Swaziland and South Africa (Mpumalanga, KwaZulu-Natal). Habitat not recorded in our region; elsewhere in dry bushland, open wooded grassland including heavily grazed areas and bare soil; c.300 m.

Conservation notes: A very local subspecies but not threatened as it occurs in open areas and appears tolerant of disturbance and grazing; Least Concern.

In their revision of South African *Dicliptera*, Balkwill *et al.* (1996) recognised two subspecies in *D. minor* – subsp. *minor* with dark reddish-purple corollas, largely glabrous cymule bracts and seeds with mainly squat, minutely hooked tubercles, and subsp. *pratis-manna* with white corollas, densely glandular cymule bracts and seeds with more slender and tall, sparsely hooked tubercles. They maintained *D. divaricata* as distinct from both, citing the differing growth habit, different flower colour (lilac or pink), and the more ovate corolla lip in lower position which is ruffled in bud. Flower colour is recorded as quite variable in both species and seems insignificant in view of the differing colours of the two subspecies of *D. minor*. Moreover, the shape of the lip in lower position is somewhat variable in both species with significant overlap. In its densely glandular bracts and slender seed tubercles, subsp. *pratis-manna* in fact more closely resembles *D. divaricata* than subsp. *minor*. Indeed, I consider one specimen cited by Balkwill *et al.* in the protologue of subsp. *pratis-manna* (*Kirk* 112) to be referable to subsp. *divaricata*. In light of this evidence, it seems more appropriate to consider *divaricata* as a third subspecies of *D. minor*.

15. **Dicliptera monroi** S. Moore in J. Bot. **49**: 189 (1911). Type: Zimbabwe, Masvingo (Fort Victoria), fl. 1909, *Monro* 1039 (BM holotype).

Dicliptera eenii sensu auct., non S. Moore. —Balkwill *et al.* in Kew Bull. **51**: 46 (1996) in part. —Mapaura & Timberlake, Checklist Zimb. Vasc. Pl.: 13 (2004).

Straggling or erect perennial herb, 30–90 cm tall, much-branched and sometimes forming untidy masses; stems brittle, 6-angular, ± prominently pale-ridged, with white appressed to (spreading-)retrorse hairs mainly on ridges or more widespread. Leaves often absent at fruiting; ovate, 2–4 × 0.9–2.3 cm, base rounded, obtuse or shortly attenuate, apex acute or attenuate, mucronulate, surfaces pale antrorse-pubescent, more sparse above, sometimes restricted to veins beneath; lateral veins 3–4 pairs; petiole 4–18 mm long. Inflorescence axillary, 1(2) lax umbels of (1)2–4 cymules per axil; umbel peduncle 11–70 mm long, indumentum as young stems, sometimes with interspersed short spreading glandular hairs; main axis bracts linear-lanceolate to narrowly elliptic, 2.5–7 × 0.5–2 mm; longest cymule peduncle in each umbel 2–37 mm long; cymule bract pairs unequal, ratio 1.3–1.5(1.65):1, green, later turning brownish, with inconspicuous 'windows' and a narrow pale hyaline margin to lower half, larger bract (elliptic-) lanceolate, 8.5–18 × 2–4 mm, apex attenuate to acuminate, 3-veined from base but only midrib prominent, surface antrorse-pubescent sometimes restricted to midrib, usually also sparsely to more densely glandular- and eglandular-puberulous; bracteoles green with pale-hyaline margin, lanceolate, 5–9 × 1 mm, acuminate, glandular- and eglandular-puberulous. Calyx lobes 4–6 mm long, glandular-puberulent, margin strigulose. Corolla 21.5–33 mm long, pink, mauve or red-violet with darker guidelines, eglandular-pubescent externally, with or without interspersed shorter glandular hairs; tube 11–14 mm long; lip held in upper position narrowly oblong, 10.5–19 × 2.5–5 mm, palate glabrous; lip held in lower position elliptic, 10–17.5 × 3–7 mm. Staminal filaments 10.5–17.5 mm long, sparsely hairy or largely glabrous; anther thecae 1–1.3 mm long, superposed. Style strigulose in lower half. Capsule 8–9 mm long, puberulous towards apex, hairs mainly glandular; placental base elastic. Seeds lenticular, c.1.7 mm in diameter, tuberculate, tubercles without conspicuous hooks.

Zimbabwe. W: Matobo Dist., Besna Kobila, fl.& fr. ix.1953, *Miller* 1908 (K, SRGH). C: Shurugwi Dist., Wanderer Valley, on road below Wanderer Mine, fl.& fr. 28.iv.1968, *Biegel* 2620 (K, SRGH). S: Masvingo Dist., 6.4 km E of Great Zimbabwe, fl. 1.vii.1930, *Hutchinson & Gillett* 3361 (K).

Also in South Africa (Limpopo). Damp ground on stream banks, ravine forest and in miombo woodland; 1000–1500 m.

Conservation notes: A very local and scarce species, having been collected on only a few occasions despite its showy flowers; recorded as common on the summit ridge of Mt Buchwa (M. Hyde, pers. comm.); potentially threatened.

Balkwill *et al.* (1996) included some of the specimens of this species from Zimbabwe in *Dicliptera eenii* S. Moore, a species described from Namibia. Whilst the two are certainly close, the Namibian plants (and similar populations from North West Province and Gauteng in South Africa) have broader, elliptic or rhombic cymule bract (length to width ratio 2–3:1 vs. (2.5)3.5–6.5:1 in *D. monroi*) with a finer indumentum, and somewhat smaller flowers, 16–19 mm long. Whilst the variation may eventually prove to be clinal, I prefer to maintain *D. monroi* as distinct at present. It should also be noted that *D. fruticosa* Balkwill from South Africa is closely allied to *D. monroi*; indeed, in some characters such as flower size, habit and cymule bract shape, the two are more similar than they are to *D. eenii*. *D. fruticosa* usually has considerably smaller cymule bracts than *D. monroi* but there is some overlap at the extremes.

This species seems to shed its leaves during or after flowering and the glandular indumentum on the bracts develops during this time so that plants at the onset of flowering can appear quite different to those at fruiting. Intermediate specimens however indicate that a single taxon is involved.

16. **Dicliptera syringifolia** Merxm. in Mitt. Bot. Staatssamml. München 1: 204 (1953). Type: Zimbabwe, Rusape, fl.& fr. viii.1952, *Dehn* R.34 (M holotype, K, SRGH), see note.

Small straggling or decumbent perennial herb to 35 cm tall, rooting at lower nodes; stems slender, 6-angular, antrorse-puberulous, with or without interspersed longer more spreading pale hairs. Leaves papery, broadly ovate, 3–5 × 1.8–3 cm, base rounded or truncate to broadly cuneate, apex attenuate, apiculate, upper surface and veins beneath with few to numerous conspicuous pale multicellular hairs, midrib on both surfaces antrorse-puberulous; lateral veins 4–7 pairs; petiole 8–33 mm long. Inflorescence axillary, 1(2) lax umbels of (1)2–3(5) cymules per axil, sometimes partially compounded; umbel peduncle 8–50 mm long, antrorse-puberulous; main axis bracts green, lanceolate, those of umbels at lower axils ovate, 2–6.5(10.5) × 0.8–2.5(7) mm, often recurved; cymule peduncles very unequal within each umbel, longest 3–32 mm long; cymule bract pairs unequal, ratio 1.15–1.5:1, green or brown-green, sometimes paler towards base but not 'windowed', with a narrow pale hyaline margin in lower half, larger bract obovate, elliptic or narrowly oblong-obovate, 8–11.5 × 2.8–5 mm, apex rounded, obtuse or rarely subacute, apiculate, 3-veined from base but only midrib prominent, surface minutely and ± sparsely antrorse-puberulous, hairs on midrib and margin somewhat strigulose, sometimes with interspersed short glandular hairs; bracteoles green with paler margin, lanceolate, 5.5–8 × 0.8–1.5 mm, apex acute or attenuate. Calyx lobes 4–5 mm long, glandular- and eglandular-puberulous and shortly ciliate. Corolla 17–18.5 mm long, pink or magenta with darker guidelines, eglandular-pubescent externally, limb with interspersed glandular hairs; tube 7–8.5 mm long; lip held in upper position narrowly oblong, 9.5–10.5 × 3.5–4 mm, palate glabrous; lip held in lower position elliptic to broadly so, 9.5–10.5 × 5.5–6.7 mm. Staminal filaments 8.5–9 mm long, sparsely hairy or largely glabrous; anther thecae 0.65–0.9 mm long, superposed and becoming separated at anthesis. Style glabrous. Capsule 6–7 mm long, eglandular-pubescent; placental base elastic. Only immature seeds seen, with slender elongate hooked tubercles.

Zimbabwe. E: Mutare Dist., Bonda, near Nyanga, fl.& fr. iv.1963, *Garley* 677 (K, SRGH); Nyanga Dist., site of Rhodes' Outspan, Rusape–Nyanga road, fl.& fr. 26.ii.1966, *Chase* 8376 (K, LISC, SRGH).

Not known elsewhere. Amongst rocks and "amongst grass and other herbs"; 1300–1850 m.

Conservation notes: A very local species known from only 3 collections, endemic to the Rusape–Nyanga area of E Zimbabwe; potentially threatened.

On the M holotype the date is given as August 1952 and the locality simply as Rusape (which is in Zimbabwe C); on the duplicates at both K and SRGH the date is given as July 1952 and the locality as Rusape Dist., Valhalla (Zimbabwe E).

17. **Dicliptera nyangana** I. Darbysh., sp. nov. Differs from *D. carvalhoi* in having broader, elliptic or obovate cymule bracts (length to width ratio of larger bract 1.55–3.05:1 vs. bracts subulate or oblanceolate, ratio (3.15)3.5–9(11):1) and in the stamens being shorter in proportion to the corolla, not extending beyond the limb. Differing from *D. quintasii* in the narrower leaves (length to width ratio over 2.5:1 vs. less than 2:1), the finer and more appressed stem indumentum, the more clearly defined and contracted terminal synflorescence, the thicker and brighter green or green-brown (not membranous and olive-green) cymule bracts with a shorter indumentum, and the mixed eglandular and glandular hairs on the corolla (vs. eglandular hairs only). Type: Zimbabwe, Nyanga Dist., above Mare Dam, fl.& fr. 19.iv.1966, *Biegel* 1128 (K holotype, SRGH).

Erect, spreading or scrambling perennial herb, 5–90 cm tall or more; stems 6-angular, furrowed when young, with fine short appressed-antrorse or -retrorse hairs mainly on the angles, later glabrescent. Leaves narrowly ovate or lanceolate, 2–4 × 0.5–1.5 cm, base obtuse or acute, apex acute or attenuate, upper surface and veins beneath shortly antrorse-pubescent, sparse at maturity; lateral veins 3–4 pairs; petiole 1.5–4.5 mm long. Inflorescence of numerous umbellately arranged cymules aggregated into dense conical or cylindrical heads 1.8–3.7 × 1.5–2 cm, subcapitate in small plants, often becoming verticillate towards base and interrupted by reduced leaves, usually also with umbels in upper leafy axils; umbels subsessile or those at lower axils on peduncles to 3.5–9 mm long; main axis bracts linear-lanceolate, 3–8 mm long, acuminate; cymule bracts either green throughout or green-brown in upper portion, paler towards the base but not 'windowed', pairs unequal, ratio 1.2–1.5:1, larger bract elliptic or obovate, 7.5–12.5 × 2.8–7 mm, apex obtuse, rounded or rarely acute, mucronate, surface 3-veined, antrorse-pubescent on margin and main veins, margin also long-pilose, internal surface of larger bract with numerous mixed short eglandular and glandular hairs on exposed portion, sparse on smaller bract externally; bracteoles lanceolate, 5–9 × 0.6–1.1 mm, attenuate. Calyx lobes 4.3–4.7 mm long, glandular-puberulent, margin long-ciliate. Corolla 14.5–19.5 mm long, magenta-pink, purple or dark wine-red, mixed eglandular- and glandular-pubescent externally; tube 6.5–8.5 mm long; lip held in upper position oblong(-ovate), 8–11.5 × 3.3–4.5 mm, palate puberulous; lip held in lower position ± broadly ovate to elliptic, 7.5–10.5 × 3.5–7 mm. Staminal filaments 6–8 mm long, hairy above; anther thecae 0.6–0.9 mm long, superposed but slightly overlapping, oblique. Style strigulose. Capsule 6–6.5 mm long, mixed glandular- and eglandular-pubescent; placental base elastic. Only immature seeds seen, tuberculate, tubercles not hooked.

Zimbabwe. E: Nyanga Dist., E of Rhodes Nyanga (Inyanga Mountains) Hotel, fl. 11.ii.1965, *Chase* 8266 (K, SRGH); Marora R. between Rhodes Nyanga Orchards and Hotel, fl. 22.iii.1966, *Simon* 750 (K, PRE, SRGH).

Not known elsewhere. Rocky grassland, riverbanks, pine plantations, edges of firebreaks and roadsides; 1700–2250 m.

Conservation notes: Restricted and endemic to the Nyanga National Park area in E Zimbabwe (14 collections), but not threatened since it occurs in a protected area and also in disturbed habitats.

In many respects, this newly described species appears intermediate between *Dicliptera carvalhoi* and members of the *D. maculata–clinopodia* complex (particularly *D. quintasii*); it may possibly have originated as a hybrid but is now clearly established as a good species. Populations of *D. carvalhoi* from E Zimbabwe are of subsp. *carvalhoi* and so have smooth seeds, whilst those of *D. nyangana* are tuberculate, a further diagnostic character.

The only sympatric species within the *D. maculata–clinopodia* complex is *D. clinopodia* form A which is very different to *D. nyangana*, being a more robust plan with much broader leaves, larger bracts which are 'windowed' and often purple, much larger corollas and seeds with verruciform tubercles.

18. **Dicliptera carvalhoi** Lindau in Engler, Pflanzenw. Ost-Afrikas **C**: 371 (1895). — Clarke in F.T.A. **5**: 257 (1900). —Darbyshire in Kew Bull. **63**: 376, table 3 (2009). —Darbyshire in F.T.E.A., Acanthaceae **2**: 709 (2010). Type: Mozambique, between lower and middle Zambezi, fl.& fr. 1884–1885, *de Carvalho* s.n. (COI lectotype), lectotypified by Darbyshire (2009).

Dicliptera angustifolia Gilli in Ann. Naturhist. Mus. Wien **77**: 48 (1973). Type: Tanzania, Lumbila, fl. 8.viii.1958, *Gilli* 520 (W holotype, K).

Annual or perennial herb or subshrub, erect, scrambling or decumbent, 15–150(200) cm tall; stems often wiry and brittle, 6-angular, sulcate when young, with or without pale ridges, indumentum variable (see subspecies). Leaves sometimes absent or immature at flowering, ovate to linear-lanceolate, 1–8(14.5) × 0.2–2.8 cm, base rounded to attenuate, apex acute or shortly attenuate, apiculate, surfaces antrorse-pubescent mainly on midrib and margins or pilose throughout, upper surface sometimes hispid, rarely glabrous; lateral veins 3–6 pairs; petiole 1–12(22) mm long. Inflorescence either of widely spaced solitary or umbellate cymules in largely bare upper axils or umbels aggregated into a verticillate or dense globose, cylindrical or subcapitate heads; main axis bracts linear-lanceolate, pairs unequal, larger bract 2–8(12.5) mm long, sometimes caducous; cymule bracts darker and often tinged purplish towards apex or sometimes green throughout, pairs subequal to unequal, ratio 1.05–1.8:1 larger bract subulate, (oblong-)oblanceolate or lanceolate, 5.5–16 × 1–3(4) mm, apex either attenuate or abruptly narrowed, mucronate, with sparse to dense capitate glandular hairs towards apex particularly on internal surface, margin and often also midrib densely pilose, outer surface also with antrorse to spreading eglandular hairs; bracteoles linear-lanceolate, 4.5–8 mm long, apex attenuate, margin hyaline towards base. Calyx lobes 2.5–5 mm long, glandular-puberulent, sometimes with few eglandular hairs along margin. Corolla (9.5)13–20 mm long, limb pink to purple or white, with purple guidelines, pubescent externally; tube (5)6.5–10.5 mm long; lip held in upper position oblong, (5.5)6.5–11.5 × 2–4.5 mm; lip held in lower position obovate or elliptic, (5)6.5–10.5 × 3.5–7.5 mm. Stamens long-exserted; filaments (5)7–13(16) mm long, sparsely hairy above; anther thecae 0.5–1 mm long, superposed. Style glabrous. Capsule 4.5–8 mm long, eglandular-puberulous and/or with short glandular hairs towards apex; placental base elastic. Seeds only subflattened, 1–1.7 mm in diameter, smooth or tuberculate, tubercles with or without minute hooks.

A complex species with five currently recognised subspecies, four of which occur in our region. Further additions or modifications to the infraspecific taxa are likely to result from targeted collecting of fruiting material across the species' range (see note to subsp. *laxiflora*).

Key to infraspecific taxa:

1. Cymules aggregated into globose, cylindrical or subcapitate heads, sometimes interrupted by reduced leaves and/or becoming verticillate in lower portion (often with additional subsessile umbels in uppermost leafy axils) 2
 – Cymules held in pedunculate or sessile axillary umbels in largely bare upper portion of branches, not aggregated into heads 3
2. Seeds smooth except for microscopic verrucae . . . . a) subsp. *carvalhoi*
 – Seeds tuberculate . . . . .b) subsp. *nemorum*
3. Stems sparsely antrorse- or retrorse-pubescent or sparsely pilose mainly in upper portion, or stems largely glabrous . . . . .c) subsp. *laxiflora*
 – Stems densely pale spreading-pilose throughout . . . . d) subsp. *petraea*

a) Subsp. **carvalhoi**. Darbyshire in Kew Bull. **63**: 377 (2009) in part, excl. *Gomes e Sousa* 880 from Mozambique; circumscription modified here.

> *Dicliptera rogersii* Turrill in Bull. Misc. Inform., Kew **1911**: 314 (1911). —Champluvier in Fl. Rwanda, Spermatophytes **3**: 452 (1985). —Phiri, Checklist Zamb. Vasc. Pl.: 18 (2005). Type: Zambia, Kalomo, fl.& fr. 9.v.1909, *Rogers* 8249 (K 000378996 lectotype), lectotypified here (see note).
>
> *Dicliptera cephalantha* S. Moore in J. Linn. Soc., Bot. **40**: 162 (1911). Type: Zimbabwe, near Chirinda, fl.& fr. 27.v.1906, *Swynnerton* 514 (BM holotype, K).
>
> *Dicliptera olitoria* Mildbr. in Notizbl. Bot. Gart. Berlin-Dahlem **11**: 1085 (1934). —Vollesen in Opera Bot. **59**: 80 (1980). Type: Tanzania, Mahenge, Ngombe, fl.& fr. 9.vi.1932, *Schlieben* 2295 [collector and number not recorded in protologue] (BM lectotype, BR, G, HBG, K, LISC, M, MA, PRE, S), lectotypified here.

Wiry annual or perennial herb or subshrub; stems shortly antrorse- or retrorse pubescent, sometimes also hispid or pilose. Leaves lanceolate, linear-lanceolate or rarely ovate. Inflorescences compounded into terminal, subglobose, cylindrical or subcapitate heads, usually comprising numerous subsessile umbellate cymes (reduced to 1–2 cymes in small plants), sometimes verticillate towards base of synflorescence, often with additional subsessile umbels in upper leafy axils. Larger cymule bract of each pair (6)9–13.5(16) mm long. Capsule 4.5–5.5 mm long. Seeds smooth except for microscopic verrucae.

Zambia. C: Kafue Dist., Iolanda, Lusaka Waterworks, Muchuto R. gorge, fl.& fr. 17.v.1998, *Bingham & Fichtl* 11668 (K). E: Petauke Dist., Kacholola, fl.& fr. 21.x.1967, *Mutimushi* 2185 (K, NDO). S: Choma Dist., Siamambo Forest Reserve, fl.& fr. 20.vi.1952, *White* 2947 (BM, BR, FHO, K). **Zimbabwe**. C: Harare Dist., Christon Bank, Mazoe headwaters, fl.& fr. 22.v.1966, *Loveridge* 1469 (K, SRGH). E: Nyanga Dist., mountain slope beyond Nyanga village, fl.& fr. 24.iv.1953, *Chase* 4941 (BM, BR, K, SRGH). **Malawi**. N: Rumphi Dist., Livingstonia escarpment below bend 11, fl.& fr. 3.vii.1970, *Pawek* 3560 (K, MO). **Mozambique**. T: Mutarara? Dist., "between lower and middle Zambesi", fl.& fr. 1884–1885, *de Carvalho* s.n. (COI). MS: Sussundenga Dist., Chimanimani Mts, c.68 km WSW of Chimoio, fl.& fr. 15.vi.2012, *Bester* 11188 (K, PRE).

Also in Rwanda, Burundi, eastern D.R. Congo and Tanzania. Miombo woodland, rough grassland, rocky hillslopes, margins and rocky beds of streams and fallows; 100–1850 m.

Conservation notes: Widespread and fairly common; Least Concern.

The type of *Dicliptera rogersii* is listed as *Rogers* 8249 but two collections bear that number. The first K sheet (K000378996) matches the locality given by Turrill in the protologue, Kalomo, and was collected in May 1909, so is chosen as the lectotype. The second K sheet and the duplicate at GRA were collected from Pemba in June 1909 and so are not considered to be isolectotypes.

b) Subsp. **nemorum** (Milne-Redh.) I. Darbysh. in Kew Bull. **63**: 379, fig.6c-d (2009); in F.T.E.A., Acanthaceae **2**: 710 (2010). Type: Zambia, Solwezi Dist., Mbulungu stream, fl.& fr. 15.vii.1931, *Milne-Redhead* 712 (K sheet 1 lectotype, BR, K sheet 2), lectotypified by Darbyshire (2009).

> *Dicliptera nemorum* Milne-Redh. in Bull. Misc. Inform. Kew **1937**: 429 (1937). —Phiri, Checklist Zamb. Vasc. Pl.: 18 (2005).
>
> *Dicliptera carvalhoi* Lindau subsp. *carvalhoi* sensu Darbyshire in Kew Bull. **63**: 377 (2009), in part for *Gomes e Sousa* 880 from Mozambique.

Wiry perennial herb or subshrub, rarely annual; stems shortly antrorse- or retrorse-pubescent. Leaves ovate, lanceolate or rarely linear-lanceolate. Inflorescences compounded into a terminal, subglobose to cylindrical head comprising numerous umbellate subsessile cymes, often with additional subsessile umbels in upper leafy axils. Larger cymule bract of each pair (6.5)9–14.5 mm long. Capsule 4.5–6.5 mm long. Seeds tuberculate, tubercles with or without minute hooks.

Zambia. N: Kawambwa, fl.& fr. 25.viii.1957, *Fanshawe* 3620 (K, NDO). W: Luanshya Dist., Sunken Lake, 'Swahili Native Reserve', fl.& fr. 12.vii.1953, *Fanshawe* 141 (BR, EA, K, NDO). C: Serenje Dist., along Chisomo–Serenje road, 41.2 km from Chisomo village at bridge over Fukwe R., fl.& fr. 7.v.1994, *Harder et al.* 3062 (K, MO). E: Lukusuzi Dist., Lukusuzi Nat. Park (Game Reserve), Mburuzi R., fl.& fr. 30.vii.1970, *Sayer* 671 (SRGH). **Mozambique**. N: Malema Dist., near Malema, fl.& fr. xi.1931, *Gomes e Sousa* 880 (K).

Also in SW Tanzania and southern D.R. Congo. Riparian woodland and thicket, miombo woodland, tall grassland and grassy roadsides; 600–1500 m.

Conservation notes: Widespread but under-recorded; Least Concern.

The discovery of two specimens of this subspecies from N Mozambique and one from E Zambia shows that it is more widespread than previously thought and may well occur in Malawi. However, all fruiting Malawian material of *Dicliptera carvalhoi* seen to date has been of subsp. *carvalhoi*.

c) Subsp. **laxiflora** I. Darbysh. in Kew Bull. **63**: 380 (2009); in F.T.E.A., Acanthaceae **2**: 711 (2010). Type: Zambia, 16 km on Mupulungu–Mbala (Abercorn) road, fl. 21.v.1963, *Boaler* 951 (K sheet 1 holotype, EA, K).

Wiry annual or perennial herb or spindly subshrub; young stems sparsely antrorse-pubescent, sometimes also spreading-pilose, later glabrescent. Leaves narrowly ovate, lanceolate or rarely linear. Inflorescence axillary, 1–2 umbels of 2–3(4) cymules in largely leafless upper axils, umbels in lower portion of stems often pedunculate, peduncles 1.5–16 mm long. Larger cymule bract of each pair 6–10(13) mm long. Capsule 5.5–6.5 mm long. Seeds smooth or tuberculate, tubercles usually with minute hooks (see note).

Zambia. B: Sesheke Dist., Kataba, fl.& fr. 11.vi.1963, *Fanshawe* 7825 (K, NDO). N: Mbala Dist., Kalambo R., Kalambo Farm, Saisi valley, fl. 21.v.1952, *Richards* 1775 (BR, K). S: Namwala Dist., 13 km N of Ngoma, Kafue Nat. Park, fl. 12.v.1962, *Mitchell* 14/73 (SRGH). **Mozambique**. Z: Morrumbala Dist., Metolola, fl.& fr. 24.v.1943, *Torre* 5386 (LISC, PRE, WAG).

Also in Tanzania and ?Angola. Amongst long grass, miombo and *Baikiaea* woodland, margins of riverine forest and thickets, amongst boulders on rocky hillslopes and on termitaria; 400–1800 m.

Conservation notes: Widespread; assessed as Least Concern in Darbyshire (2009).

The current circumscription of this subspecies is rather unsatisfactory since it contains specimens with both smooth and tuberculate seeds. It is likely that more than one taxon is actually involved (perhaps lax-flowered varieties of both subsp. *carvalhoi* and subsp. *nemorum*), but it is difficult to make firm decisions based on the few specimens currently available. The single specimens from isolated populations in Barotseland and Mozambique both have tuberculate seeds, whilst in NE Zambia both seed types occur. Indeed, in *Mutimushi* 904 from Lake Chila, Mbala Dist. both types are present on the K sheet, albeit from two plants with a rather different appearance. Whilst the type specimen unfortunately lacks fruits, an almost identical specimen from nearby (*Nash* 127, BM) has smooth seeds and so it could be taken that subsp. *laxiflora* sensu stricto has this seed type if this taxon were to be split up in the future.

d) Subsp. **petraea** I. Darbysh., subsp. nov. Differs from subsp. *laxiflora* in stems being densely pale spreading-pilose throughout. Type: Zambia, Mpika, fl. 13.viii.1965, *Fanshawe* 9252 (K holotype, NDO).

Perennial herb from a woody base and rootstock, ± many branched; stems densely pale spreading-pilose throughout. Leaves absent or immature at flowering, ovate, 1–1.5 × 0.4–0.8 cm (possibly considerably larger at maturity). Inflorescence axillary, ± widely spaced on largely bare branches, cymules solitary or in umbels of 2–3, most inflorescences sessile but some on peduncles 1.5–3.5 mm long. Larger cymule bract of each pair 5.5–8.5 mm long. Capsule 6.5 mm long. Seeds tuberculate, tubercles without hooks.

Zambia. N: Mpika Dist., Koloswe, near Kapoko, fl. 16.vii.1930, *Hutchinson & Gillett* 3755 (BM, K); Mpika Dist., Lavushi Manda Nat. Park, fl.& fr. 9.xi.2010, *Byng & Johnson* 126 (K).

Not known elsewhere. Miombo woodland on rocky hillslopes and quartzite kopjes; 1300–1600 m.

Conservation notes: Apparently endemic to a small area on the watershed of north-central Zambia (3 collections), part of which is protected; possibly threatened.

This distinctive subspecies was previously misidentified as "*Peristrophe* sp." at Kew and so overlooked during an earlier review of *Dicliptera carvalhoi*. The dense stem indumentum and the differing habit, being more densely branched from the base and having less wiry stems than most specimens of *D. carvalhoi*, give it a distinct appearance, but it is otherwise close to subsp. *laxiflora*.

19. **Dicliptera melleri** Rolfe in Oates, Matabeleland Victoria Falls, ed.2 app.5: 405 (1889).—Clarke in F.T.A. **5**: 261, 515 (1900).—Darbyshire in F.T.E.A., Acanthaceae **2**: 707 (2010). Types: Malawi, Manganja Hills, Mt Chiradzulu, fl. ix.1861, *Meller* s.n. (K syntype); Zimbabwe, Matabeleland, fl.& fr. ii.1888, *Oates* s.n. (K syntype).

Diapedium melleri (Rolfe) S. Moore in J. Bot. **38**: 205 (1900).

Peristrophe mellerioides Merxm. in Proc. Trans. Rhod. Sci. Assoc. **43**: 123 (1951). Type: Zimbabwe, Marondera (Marandellas), fl. 23.vii–31.x.1942, *Dehn* 676 & 678 (M syntypes).

Decumbent or procumbent pyrophytic perennial, producing few to numerous stems 5–30(60) cm long from a woody base and rootstock; stems 6-angular, furrowed when dry, sparsely (rarely more densely) pubescent or glabrous. Leaves ?immature at flowering; oblong-elliptic, oblanceolate or rarely linear, 1.5–5(8) × 0.25–1.3 cm, base cuneate to obtuse, apex acute or obtuse, apiculate, surfaces largely glabrous; lateral veins 3–4 pairs, inconspicuous; petiole 0–3 mm long. Inflorescence of numerous umbellately arranged cymules aggregated into dense conical or cylindrical terminal heads, (1.5)2.5–5(7) × 1–2.5 cm; umbels subsessile or those towards base of spike on peduncles 2–6.5(15) mm long, antrorse-pubescent; main axis bracts linear-lanceolate, 5–7.5 mm long, ciliate, with short hairs along midrib; cymules subsessile or peduncle to 3 mm long; bract pairs unequal, and dimorphic, ratio (1.05)1.2–1.5(1.8):1, larger bract (pale) green with pale hyaline margin, sometimes pale-'windowed' between main veins, linear-oblanceolate or narrowly subpandurate, (6)7–10.5(13) × 0.75–2 mm, apex acuminate, smaller bract more conspicuously 3-veined and usually pale-'windowed', obovate or oblanceolate, 5–8 × 1.7–3 mm, apex rounded-apiculate, both bracts with conspicuous (rarely sparse) long-pilose margin, surfaces glabrous or usually shortly pubescent at least on main veins, with short-stalked glands towards apex; bracteoles lanceolate, 6–7 × 1 mm, acuminate, hyaline except for prominent green midrib. Calyx lobes 4–5 mm long, ciliate. Corolla 12.5–16 mm long, white or greenish-cream, guidelines variously purple to green or brown, eglandular-pubescent externally;

tube 6.5–9 mm long; lip held in upper position oblong-ovate, 6–8 × 3–5 mm, palate pubescent towards mouth; lip held in lower position sub-rounded to broadly flabelliform, 5–6 × 5.5–7.5 mm, margin irregular. Staminal filaments 3.5–4 mm long, shortly hairy mainly above; anther thecae 0.8–1.1 mm long, superposed but slightly overlapping. Style sparsely strigulose in lower half. Capsule 6.5–7.5 mm long, eglandular-puberulous towards apex; placental base elastic. Seeds discoid with keeled membranous rim, 2.2–3 mm in diameter, smooth except for minute reticulation.

Zambia. N: Mbala Dist., edge of Nkali dambo, fl.& fr. 14.ix.1963, *Richards* 18192 (K). B: Mankoya Dist., near Mankoya resthouse, fl.& fr. 20.xi.1959, *Drummond & Cookson* 6659 (K, SRGH). W: Mufulira, fl. 17.x.1948, *Dell* 399 (BR, K). C: Lusaka Dist., Lusaka SE, c.20 km along Chifwema road, fl. 14.x.1999, *Bingham & Nefdt* 12037 (K). E: Chipata Dist., Chipata (Fort Jameson), fl.& fr. 13.x.1967, *Mutimushi* 2161 (K, LMA, NDO). **Zimbabwe**. N: Hurungwe Dist., Zwipani, fl. 12.x.1957, *Phipps* 784 (K, SRGH). W: Insiza Dist., 10 km W of Shangani on main road to Bulawayo, fl. 13.xii.1989, *Adams* 899 (SRGH). C: Harare Dist., Borrowdale, fl.& fr. 8.xi.1953, *Wild* 4148 (BR, K, LISC, SRGH). E: Mutare Dist., E of Morningside, fl. 2.xi.1961, *Chase* 7553 (K, LISC, SRGH). S: Masvingo Dist., Masvingo (Fort Victoria), fl., n.d., *?Brown* 474 (SRGH). **Malawi**. N: Mzimba Dist., 4.8 km W of Mzuzu at Katato, fl. 24.x.1975, *Pawek* 10326 (BR, K, MAL, MO, SRGH, UC). C: Dedza Dist., Chongoni Forest Reserve, fl. 18.ix.1967, *Salubeni* 836 (K, SRGH). S: Chiradzulu Dist., Manganja Hills, Mt Chiradzulu, fl. ix.1861, *Meller* s.n. (K). **Mozambique**. T: Macanga Dist., Furancungo, road to Angónia, fl. 29.ix.1942, *Mendonça* 517 (BR, LISC, WAG).

Also in Tanzania and D.R. Congo (Katanga). In burnt grassland and miombo woodland, on sandy soil or over laterite or dolerite, sometimes along dambo margins; 750–1600 m.

Conservation notes: Common in periodically burnt miombo; Least Concern.

Mutimushi 2161 from Chipata, E Zambia, differs from typical *Dicliptera melleri* in having longer flowering stems with long linear leaves and flowers at the larger end of the species' size range. It looks close to the type of *D. katangensis* De Wild. from D.R. Congo which Champluvier maintains as a good species in the forthcoming Flore d'Afrique Centrale (pers. comm.), but I consider the Mutimushi specimen to be a variant of *D. melleri*.

20. **Dicliptera betonicoides** S. Moore in J. Bot. **49**: 312 (1911). —Darbyshire in Kew Bull. **63**: 373 (2009). Type: Angola, Kassuango–Kuiriri, fl.& fr. iii.1906, *Gossweiler* 3680 (BM lectotype), lectotypified by Darbyshire (2009).

 Dicliptera arenaria Milne-Redh. in Bull. Misc. Inform., Kew. **1937**: 427 (1937). Type: Zambia, Mwinilunga Dist., near R. Wamibobo, fl.& fr. 6.viii.1930, *Milne-Redhead* 842 (K holotype, BR).

Erect or decumbent annual herb, 20–60 cm tall, few to many-branched; stems wiry, 6-angular, appressed pale-pubescent on ridges, rarely also spreading-pilose. Leaves sometimes absent at fruiting, blades lengthening up stem, upper stem leaves oblong-lanceolate to linear, 3–8.3 × 0.25–2.3 cm, base shallowly cordate or rounded, sometimes subauriculate, apex acute or obtuse, upper surface sparsely pilose, margin, veins beneath and midrib above with shorter antrorse hairs; lateral veins (3)4–6 pairs, prominent beneath; blade sessile or petiole to 2 mm long. Inflorescence of numerous umbellately arranged cymules aggregated into cylindrical or conical heads, (1.5)2.5–5.5 × 0.8–1.8 cm; umbels (sub)sessile; main axis bracts linear-lanceolate, 3.7–6(9.5) × 0.5–1.5 mm, acuminate, ciliate, hyaline except for green midrib; cymule bracts unequal, ratio 1.3–1.8:1 and ± dimorphic, larger bract green with a broad pale yellowish margin in lower half, sometimes faintly 'windowed', oblong-lanceolate, oblanceolate or narrowly pandurate, 5.7–10 × 1.2–2.5 mm, apex acute or obtuse, often only midrib prominent, margin pale-pilose, midrib and sometimes upper portion antrorse-pubescent, smaller bract with veins and apex green, with conspicuous pale yellowish 'windows' and margin, usually obovate, 4–5.5(8) × 1.5–3.2 mm, apex obtuse to rounded, sometimes with a short attenuate tip, surface prominently 3-veined, veins sparsely to densely white-pubescent, with few short-stalked glands between veins, margin pale-pilose; bracteoles lanceolate, 4.5–6.5 × 1 mm, acuminate, hyaline except for prominent midrib. Calyx hyaline, lobes 3.5–4.5 mm long, ciliate. Corolla 11–13.5 mm long, white, pink or mauve, with purple guidelines on lip in lower position and sometimes with greenish guidelines on lip in upper position, eglandular-pubescent externally; tube 6–7.7 mm long; lip held in upper position ovate(-elliptic), 5–6 × 3.5–4.2 mm; lip held in lower position flabellate, 4–5 × 4.5–6 mm, margin irregular. Staminal filaments 2.7–4.5 mm long, shortly hairy mainly above; anther thecae 0.7–1 mm long, superposed but slightly overlapping. Style sparsely strigulose. Capsule 4.5–7.5 mm long, with short eglandular hairs towards apex or glabrous; placental base elastic. Seeds discoid with keeled membranous rim, 2.5–3 mm in diameter, smooth except for minute reticulation.

Zambia. B: Mongu, fl.& fr. 11.iv.1966, *Robinson* 6933 (K, LMA, SRGH). N: Kaputa Dist., Lake Mweru Wantipa, 20 km N of Nsama, fl. 16.iv.1989, *Goyder et al.* 3053 (K, MAL, NDO). W: Mwinilunga Dist., West Lunga, near R. Wamibobo, fl.& fr. 6.viii.1930, *Milne-Redhead* 842 (BR, K). **Zimbabwe**. N: Makonde Dist., Trelawney, Tobacco Research Station, fl. 27.iii.1943, *Jack* 171 (SRGH), see note.

Also in D.R. Congo (Katanga) and Angola. *Brachystegia* and *Cryptosepalum* woodland on sand and sandy roadsides; 900 m.

Conservation notes: Known from few collections but fairly widespread; probably not threatened.

The isolated specimen from Lake Mweru Wantipa differs slightly from other material in having very dense pubescence on the veins of the smaller cymule bract, less conspicuous long cilia on the bract margins, and in the corolla having green speckling on the lip held in the upper position. More material and further field observations of flower colour and markings are desirable to clarify whether two taxa are involved.

The presence of this species in central Zimbabwe is most unusual since this is well beyond its usual geographic and ecological range. The single specimen seen has less dimorphic cymule bract pairs than in most other material seen, but is otherwise a close match. It is possibly an accidental introduction as it has not been found in Zimbabwe since.

21. **Dicliptera capitata** Milne-Redh. in Bull. Misc. Inform., Kew **1937**: 428 (1937). — Darbyshire in F.T.E.A., Acanthaceae **2**: 707 (2010). Type: Zambia, Solwezi Dist., Solwezi Boma, fl.& fr. 13.vi.1930, *Milne-Redhead* 493 (K holotype).

Erect or decumbent annual herb, 5–90 cm tall, unbranched or laxly branched; stems wiry, 6-angular, ridged, appressed pale-pubescent mainly on ridges. Leaves lengthening up stem, upper stem leaves oblong-lanceolate or linear-lanceolate, (1.5)3–7.8 × 0.4–1.5 cm, base rounded to subcordate, apex acute, upper and sometimes lower surface ± sparsely pubescent, veins beneath and midrib above with shorter antrorse hairs; lateral veins (3)5–7 pairs, prominent beneath; blade sessile or petiole to 3 mm long. Inflorescence of few to numerous umbellately arranged cymules aggregated into sessile hemispheric or capitate terminal heads, 0.5–2.5 cm in diameter, immediately subtended by pair or pseudowhorl of 4 leaves; main axis bracts lanceolate, 5–9.5 mm long, acuminate, hyaline except for green midrib; cymule bracts (greyish-)green, often faintly 'windowed', pairs somewhat unequal, ratio (1)1.15–1.3(1.4) : 1, larger bract lanceolate, 7.5–16 × 1.5–3.5 mm, apex acute to attenuate, often only midrib prominent, surface pubescent to pilose, with inconspicuous short-stalked glands, margin densely pale-pilose, smaller bract lanceolate or elliptic, prominently 3-veined; bracteoles lanceolate, 5–8.5 mm long, acuminate, hyaline except for prominent midrib. Calyx hyaline, lobes 3.5–4 mm long, ciliate. Corolla 10.5–15 mm long, pink, purple or rarely white, with purple or blackish guidelines, eglandular-pubescent externally; tube 5–9 mm long; lip held in upper position oblong, 5–7.5 × 2–3 mm; lip held in lower position ovate-rhombic or flabellate, 4–7.5 mm long and wide, margin irregular. Staminal filaments 3–5.5

mm long, shortly hairy mainly above; anther thecae 0.5–0.7 mm long, superposed. Style glabrous or sparsely strigulose. Capsule 5–6 mm long, glabrous; placental base elastic. Seeds discoid with keeled membranous rim, 2.5–3 mm in diameter, smooth except for minute reticulation.

Zambia. N: Mporokoso Dist., Lumangwe Falls on Kalungwishi R., 45 km NE of Kawambwa, fl. 14.iv.1989, *Goyder et al.* 3024 (K, NDO). W: Mwinilunga Dist., Kalene Hill, fl.& fr. 9.vi.1974, *Chisumpa* 153 (K, NDO). C: Lusaka Dist., Lusaka East Forest Reserve, Trotover, fl. 26.iii.2000, *Bingham* 12169 (K). S: Mazabuka Dist., Muvuma Hills, 22 km S of Kafue Town on road to Mazabuka, fl. 25.iii.1972, *Kornaś* 1447 (K). **Mozambique**. N: Malema Dist., Mutuáli, Estação Experimental do CICA, near R. Nàlume, fl. 28.iv.1961, *Balsinhas & Marrime* 463 (K, LISC, LMA).

Also in Burundi, D.R. Congo, Tanzania and Angola. Open miombo woodland and wooded grassland on sandy soils, sandy roadsides; 600–1400 m.

Conservation notes: Locally common in miombo woodland; assessed as Least Concern by Darbyshire (2010).

The single collection seen from Mozambique has white flowers and more clearly dimorphic cymule bracts than typical, the smaller bract being elliptic. It is matched by *Milne-Redhead & Taylor* 10489 from SE Tanzania (see further discussion in Darbyshire, Kew Bull. **63**: 373, 2009). This may be a distinct regional variant but more material is required.

22. **Dicliptera pumila** (Lindau) Dandy in Mem. N.Y. Bot. Gard. **9**: 27 (1954). — Darbyshire in F.T.E.A., Acanthaceae 2: 713 (2010). Type: Malawi, Shire Highlands, fl. 1891, *Buchanan* 1474 (K lectotype, BM), lectotypified by Darbyshire (2010). FIGURE 8.6.70.

Duvernoia pumila Lindau in Bot. Jahrb. Syst. **20**: 44 (1894).

Peristrophe usta C.B. Clarke in F.T.A. 5: 244 (1900). —Binns, Checklist Herb. Fl. Malawi: 15 (1968). Types: Malawi, top of Zomba Mt, Shire Highlands, fl. n.d, *Buchanan* 127 (K syntype); Zomba Mt, fl. n.d., *Whyte* s.n. (K syntype); Tanganyika Plateau, *Carson* s.n. (not seen).

Peristrophe pumila (Lindau) Lindau in Wiss. Ergebn. Schwed. Rhod.-Kongo-Exped. **1**: 307 (1916). —as *P. pumila* (Lindau) Gilli in Ann. Naturhist. Mus. Wien **77**: 53 (1973). —Burrows & Willis, Pl. Nyika Plateau: 54 (2004).

Pyrophytic suffrutex, usually with many ± erect flowering shoots, 5–15 cm tall, later leafy shoots straggling or decumbent, to 30–40 cm long, rarely bearing flowers; stems 6-angular, furrowed, flowering shoots puberulous, sometimes also pilose (rarely densely so), with short capitate glandular hairs; leafy stems glabrous or sparsely puberulous. Flowering shoots often largely leafless; mature leaves elliptic, obovate or oblanceolate, 2–7 × 0.7–2 cm, base cuneate, apex acute, shortly ciliate and with short hairs on veins beneath; lateral veins 4–5 pairs; petiole to 5 mm long. Inflorescences axillary and terminal, 1(2) per axil, cymules usually solitary but appearing umbellately arranged at apex of branches; primary peduncle (2)7–35(95) mm long, puberulous, usually with interspersed capitate glandular hairs, sometimes also sparsely to densely pilose; main axis bracts absent; cymules several- to many-flowered, subcapitate or spiciform, to 1–5.5(9) cm long; flowers sessile, each subtended by one bract and a pair of bracteoles, except terminal flower where bracts paired; bracts green towards base, often darker blackish- or brown-green towards apex, linear-lanceolate, subulate or oblanceolate, bracts subequal throughout head/spike, 4–11 × 0.8–3.2 mm, apex abruptly or more gradually attenuate, apiculate, surface puberulous and with few to numerous capitate glandular hairs, margin and midrib often finely long-pilose, or whole surface rarely densely so, margin narrowly hyaline towards base; bracteoles linear, 3.5–10.5 × 0.6–1 mm. Calyx lobes 2–5 mm long, puberulent and with sparse glandular hairs, ciliate. Corolla 10–22 mm long, white, pale pink, mauve or blue, with pink to purple guidelines, eglandular-pubescent externally; tube 5–11 mm long; lip held in upper position oblong(-ovate), 5–12.5 × 3–5.5 mm; lip held in lower position elliptic or obovate, 4.5–11.5 × 3–6.5 mm. Staminal filaments 4.5–13 mm long, sparsely hairy; anther thecae 0.75–1 mm long, superposed and becoming separated. Style glabrous. Capsule 5.5–8.5 mm long, puberulous, with interspersed short glandular hairs; placental base elastic. Seeds 1.5–2 mm in diameter, tuberculate.

Zambia. N: Mbala Dist., Nkali dambo, fl. 21.viii.1956, *Richards* 5881 (K). W: Mwinilunga Dist., West Lunga, Mwanamitowa R., fl. 12.viii.1930, *Milne-Redhead* 886 (K). C: Kabwe Dist., c.9 km SW of Kabwe (Broken Hill), fl. 14.viii.1966, *Gillett* 17456 (EA, K, SRGH). E: Nyika Plateau, by main road, 4 km SW from Rest House, fl.& fr. 22.x.1958, *Robson* 236 (BM, K, LISC, SRGH). **Zimbabwe**. N: Kariba Dist., road from Zwipani over Sanyati (Kanyati) R. via Karabazu Gate to upper Gatshe Gatshe (Gashegashe) valley, fl.& fr. 19.x.1957, *Phipps* 812 (BR, K, LISC, LMA, SRGH). C: Harare Dist., University site, fl. 23.ix.1957, *Seagrief* 3067 (BM, BR, K, LMA, SRGH). **Malawi**. N: Nyika Plateau, Chosi, fl. 5.xii.1975, *Phillips* 468 (K). C: Nkhotakota Dist., Chintembwe, fl. 9.ix.1946, *Brass* 17582 (K, NY, SRGH). S: Zomba Dist., Zomba Plateau, Chingwe's Hole, fl. 24.ix.1985, *Banda et al.* 2488 (K, MAL). **Mozambique**. N: Lago Dist., Maniamba, fl. 20.xi.1934, *Torre* 490 (LISC).

Also in South Sudan, Ethiopia, Uganda, D.R. Congo, Burundi, Tanzania and Angola. Recently burnt montane grassland, short grassland, periodically burnt miombo woodland, and bare ground by roadsides; 600–2350 m.

Conservation notes: Common constituent of periodically burnt grassland and open woodland; Least Concern.

A rather variable species in terms of indumentum, bract shape and corolla size, but the inflorescence form is very distinctive and quite different to most species of *Dicliptera* (Fig. 8.6.**70**, 2,3), although care should be taken in separating plants of *D. brevispicata* with spiciform inflorescences.

23. **Dicliptera brevispicata** I. Darbysh. in Kew Bull. **62**: 123, fig.1 (2007); in F.T.E.A., Acanthaceae 2: 714, fig.93 (2010). Type: Tanzania, Ufipa Dist., 2 km NW of Kalaela on Matai–Mwimbi road, fl.& fr. 23.xi.1994, *Goyder et al.* 3798 (K holotype, C, CAS, DSM, EA, WAG).

> *Peristrophe brevispicata* (I. Darbysh.) Y.F. Deng & Z.P. Hao in J. Trop. Subtrop. Bot. **19**: 327 (2011).

Scrambling, trailing or decumbent suffruticose perennial, often pyrophytic, few to many-branched from a woody rootstock, branches 5–90 cm long; stems 6-angular, ± ridged, with pale, somewhat hispid spreading and retrorse hairs, young stems often with minute glandular and eglandular hairs. Leaves sometimes immature or absent at flowering (pyrophytic plants), ovate to broadly so, 1.5–8.5 × 0.8–5 cm, base rounded to shallowly cordate, apex acute or subacuminate, upper surface and veins beneath hispid; lateral veins 4–7 pairs; petiole 3–17 mm long. Inflorescences axillary and terminal, 1(2) per axil, cymules often solitary or sometimes in lax umbel of 2(3) cymules; primary peduncle 4–45(60) mm long, indumentum as young stems but hairs often antrorse; main axis bracts (if present) linear-lanceolate, (3.5)5–7(9) mm long, margin and midrib antrorse-pubescent; cymules dense, capitate or shortly spiciform, flowers sessile, each subtended by one bract and a pair of bracteoles; cymule bracts initially green, tinged brown or pinkish with age, pairs equal or usually somewhat unequal, ratio 1–1.35:1, larger bract subulate, (5.5)8–13(16) × 1.5–2(3) mm, the outermost (cymule) bracts subequal to inner bracts, apex acute to subattenuate, outer surface pale antrorse-pubescent particularly on midrib, with inconspicuous minute hairs, venation reticulate, prominent within, margin pale-hyaline; bracteoles as bracts but 7–10.5(12) × 0.7–1.7 mm. Calyx lobes 4.5–8 mm long, glandular-puberulent, midrib with occasional eglandular hairs, margin hyaline, ciliate. Corolla 17.5–20.5 mm long, pale pink to mauve or rarely white, sometimes with pink or purple guidelines, eglandular-pubescent externally; tube 7–9 mm long; lip held in upper position oblong, 9–12 × 3–4.5 mm, palate puberulous towards mouth; lip held in lower position ovate-elliptic, 8.5–11.5 × 6.5–8 mm. Staminal filaments 4.5–7 mm long, pubescent; anther thecae c.1 mm long, superposed and becoming slightly separated. Style sparsely strigulose. Capsule 7.5–9.5 mm long, densely eglandular-pubescent; placental base inelastic or rising slightly at base. Seeds 1.9–2.3 mm in diameter, sparsely tuberculate, tubercles shortly clavate.

Fig. 8.6.**70**. DICLIPTERA PUMILA. 1, habit (× 1); 2, inflorescence, spiciform type (× 1½); 3, inflorescences, capitate type (× 1.5); 4, cymule bract, external face (× 6). 5, calyx (× 6) with detail of indumentum; 6, corolla with stamens and pistil (× 4.5); 7, detail of anthers (× 10); 8, capsule (× 6); 9, seed (× 12). 1 from *Banda et al.* 2488; 2 & 7 from *Richards* 1893B; 3 from *Brummitt* 11795; 4–6 from *Phipps* 812, 8 & 9 from *Pawek* 10225. Drawn by Juliet Williamson.

Zambia. N: Mbala Dist., Vomo Gap by Mwandwizi R., fl. 7.x.1966, *Richards* 21505 (BR, K, SRGH).

Also in SW Tanzania. Miombo woodland, particularly in disturbed or recently burnt areas, and along roadsides; c.1500 m.

Conservation notes: Known from only a single collection in the Flora region, but locally common in miombo woodland in Tanzania; assessed as Least Concern by Darbyshire (2007).

This species can flower both precociously as a pyrophyte (as in the single Zambian specimen seen) and at the onset of the dry season on the mature leafy stems.

24. **Dicliptera transvaalensis** C.B. Clarke in Fl. Cap. **5**: 92 (1901). Type: South Africa, former Transvaal, no locality, fl. n.d., *Holub* s.n. (K holotype).

Peristrophe transvaalensis (C.B. Clarke) K. Balkwill in S. Afr. J. Bot. **51**: 489 (1985). —K. Balkwill in Bothalia **26**: 89 (1996). —Setshogo, Prelim. Checklist Pl. Botswana: 19 (2005). —Welman in Germishuizen *et al.*, Checklist S. Afr. Pl.: 87 (2006).

Scrambling, trailing or decumbent suffruticose perennial, several-branched from a woody rootstock, to 1 m long, sometimes rooting at nodes of trailing portion; stems 6-angular, pale-ridged, with rather dense pale spreading or retrorse eglandular hairs. Leaves sometimes immature at flowering, lanceolate or narrowly ovate, 2.3–5 × 0.4–1.3 cm, base cuneate or shortly attenuate, apex acute, apiculate, surfaces pale-pubescent, particularly on veins beneath and midrib above; lateral veins 4–5 pairs; petiole 2–7 mm long. Inflorescence axillary, 1(2) umbels of (1)2–4(5) cymules per axil, umbels often partially compounded; umbel peduncle 4–32 mm long, umbels in uppermost axils rarely sessile; main axis bracts linear-lanceolate, 4–9 × 0.5–1 mm, those of lowermost umbels on stems sometimes more leafy and lanceolate, to 13 × 3 mm; cymules pedunculate, usually with peduncle of one cymule much longer than others, (2)5–32 mm long; cymule bracts green or smaller of pair turning brown, pairs ± highly unequal, ratio 1.25–1.9:1, larger bract subulate or slightly oblanceolate, (9)12–18 × 1.5–3(3.5) mm, apex mucronulate, external surface pale spreading-pubescent throughout, only midrib prominent, margin pale-hyaline in lower half; bracteoles as cymule bracts but lanceolate, 8–10.5 × 1.5 mm. Calyx lobes 5–6 mm long, with or without few minute glandular hairs, margin hyaline, ciliate. Corolla 17–20 mm long, pink to mauve with purple guidelines, pubescent externally; tube 7–9 mm long; lip held in upper position narrowly oblong, 8–11 × 2.5–3.5 mm, palate puberulous towards mouth; lip held in lower position ovate or elliptic, 8–11 × 5.5–8.5 mm. Staminal filaments 4.5–7 mm long, pubescent; anther thecae 0.7–1 mm long, superposed. Style sparsely strigulose or glabrous. Capsule 9.5–11 mm long, pale eglandular-pubescent; placental base inelastic. Seeds c.2 mm in diameter, sparsely tuberculate mainly around rim, tubercles shortly clavate.

Botswana. SE: Southern Dist., 72 km from Lobatse to Ghanzi, fl. ix.1967, *Lambrecht* 344 (K, SRGH); Central Dist., c.20 km NW of Serowe, fl.& fr. 23.iv.2005, *Darbyshire* 456 (GAB, K).

Also in South Africa (Limpopo). Open bushland and woodland, e.g. *Terminalia, Combretum, Croton*, on rocky hillslopes and river valleys; 1100–1400 m.

Conservation notes: Very local and apparently scarce, known from less than 10 localities, but locally frequent in the Serowe area; probably not threatened.

25. **Dicliptera colorata** C.B. Clarke in F.T.A. **5**: 260 (1900), in part excl. *Scott-Elliot* 6766. —Darbyshire in Kew Bull. **63**: 381 (2009). Type: Malawi, Nyika Plateau, fl. vi–vii.1896, *Whyte* s.n. (K sheet 1 lectotype, K), lectotypified by Darbyshire (2009).

Erect or decumbent suffruticose perennial, 10–55 cm tall, several-branched from a small woody rootstock; stems 6-angular, pale-ridged, pale-hispid along ridges, two opposite sides with inconspicuous minute antrorse hairs. Leaves ovate, 3.7–8.5 × 1.5–3.2 cm, base rounded or subcordate to obtuse or shortly attenuate, apex acute or attenuate, surfaces pale-hispid; lateral veins 4–6 pairs; petiole 4–26 mm long. Inflorescences axillary and terminal, together forming a ± dense verticillate thyrse when mature; 1–3 (or more) umbels of (1)2–4 cymules per axil; umbels subsessile or those at lowermost fertile axils with primary peduncle up to 11 mm long; main axis bracts lanceolate to elliptic, 6–13.5 × 1–4 mm; cymules subsessile, cymule bracts initially green but soon turning pink-brown or purple, pairs equal or somewhat unequal, ratio 1–1.35 : 1, larger bract narrowly elliptic, 10–17.5 × 2.5–5.5 mm, base cuneate, apex narrowed to a mucro, external surface with numerous pale, somewhat hispid spreading hairs, venation pinnate-reticulate, prominent within, margin pale-hyaline in basal half; bracteoles as cymule bracts but (elliptic-)lanceolate, 9–12 × 1.5–3 mm. Calyx lobes 4.5–7 mm long, shortly retrorse-pubescent and minutely glandular-puberulent or sparsely so, margin hyaline. Corolla 15–18 mm long, white, magenta or purple, retrorse eglandular-pubescent externally; tube 8–9 mm long; lip held in upper position oblong, 7.5–10 × 2.5–3.5 mm, palate pubescent towards mouth; lip held in lower position elliptic, 7.5–10.5 × 3.5–6.5 mm. Staminal filaments 5–6 mm long, pubescent; anther thecae 0.7–0.9 mm long, superposed and slightly oblique. Style sparsely hairy towards base or largely glabrous. Capsule not seen.

Malawi. N: Mzimba Dist., Great North Road, between two Mzuzu junctions, fl. 13.vi.1971, *Pawek* 4894 (K, MAL).

Not known elsewhere. Dry miombo woodland and woodland margins; 1200–2200 m.

Conservation notes: Rare and highly range-restricted species; assessed as Endangered under IUCN criterion B by Darbyshire (2009).

Dicliptera sp. F of F.T.E.A. (p.716) from S Tanzania is allied to this species, but differs in having clearly pedunculate cymes and cymules, longer calyx lobes (8.5–9.5 mm), larger corollas (23–25 mm), and in the main axis bracts being much-reduced in comparison to the large, rhombic cymule bracts.

26. **Dicliptera gillilandiorum** (K. Balkwill) I. Darbysh. in Kew Bull. **62**: 123 (2007). Type: South Africa, Limpopo Prov., Dongola, Schroda, fl. 15.iii.1948, *Bruce* 58 (PRE holotype, K).

Peristrophe gillilandiorum K. Balkwill in S. Afr. J. Bot. **51**: 488, fig.2 (1985). —K. Balkwill in Bothalia **26**: 90 (1996). —Mapaura & Timberlake, Checklist Zimb. Vasc. Pl.: 14 (2004). — Welman in Germishuizen *et al.*, Checklist S. Afr. Pl.: 87 (2006).

Suffruticose perennial herb or shrub, 45–100 cm tall; young stems slender, 6-angular, ± prominently ridged, densely and finely white retrorse-pubescent; older woody stems subangular, glabrescent. Leaves linear-lanceolate or narrowly oblong, 2–7 × 0.2–0.9 cm, base cuneate or attenuate, apex acute, surfaces at first finely white antrorse-puberulous, sparse at maturity; lateral veins 3–4 pairs, inconspicuous; petiole 0–4.5 mm long. Inflorescence axillary, one umbel of (1)2(3) cymules per axil; umbel peduncle 0.5–3.5(7.5) mm long; main axis bracts linear-lanceolate or narrowly oblanceolate, 2–9.5 × 0.5–1.5 mm; cymules shortly pedunculate or one of a pair sessile, longest cymule peduncle in each umbel 0.5–4.5 mm long, indumentum as young stem; cymule bracts green or dark brown, pairs unequal, ratio 1.1–1.6:1, larger bract subulate, 4.5–10 × 1.2–1.6 mm, apex acute, external surface pale spreading-pubescent, also puberulent with mixed eglandular and capitate-glandular hairs, only midrib prominent, margin sometimes narrowly pale-hyaline; bracteoles linear-lanceolate, 4.5–6.5 mm long. Calyx lobes 2.3–3.3 mm long, ciliate, surface glandular-puberulent, sometimes a few long eglandular hairs towards apex. Corolla (13.5)18.5–24 mm long, pale purple or pale lilac to white, with dark purple guidelines, shortly eglandular and glandular-pubescent externally; tube (6)8.5–11 mm long; lip held in upper position oblong, (7)10–14 × 2–3.5 mm, palate puberulous towards mouth; lip held in lower position ovate or elliptic, (7.5)10–13 × (2.5)4.5–5 mm. Staminal filaments c.7.5 mm long, largely glabrous; anther thecae 1–1.2 mm long, superposed. Style sparsely strigulose towards base. Capsule 10–10.5 mm long, shortly pubescent with mixed eglandular and glandular hairs; placental base inelastic. Seeds c.2.2 mm in diameter, tuberculate, tubercles minutely hooked.

Zimbabwe. S: Beitbridge Dist., Sentinel Ranch, c.6.4 km NE of Pazhi–Limpopo confluence, fl.& fr. 24.iii.1959, *Drummond* 5973 (K, PRE, SRGH); Sentinel Ranch, c.48 km W of Beitbridge, fl. 2.ii.1973, *Grosvenor* 820 (SRGH).

Also in South Africa (Limpopo). Rocky sandstone hillslopes; in South Africa also in *Hyphaene* woodland on clay soils; c.500 m.

Conservation notes: A very local species, endemic to a small area of the Limpopo valley on Forest Sandstone. In the Flora area, known only from two collections on Sentinel Ranch. Balkwill (1985) recorded it as locally common in the Dongola area of South Africa, but later (1996) noted that the *Hyphaene* woodland locality was being destroyed by cotton planting. With only three locations known and with at least one threatened, it would qualify as Endangered under IUCN criterion B.

Corolla measurements from Balkwill's original description have here been added to my own measurements since he has seen extra material with more variation, notably the bracketed lower corolla size range.

Peristrophe cliffordii K. Balkwill (combination not yet made in *Dicliptera*), also restricted to the Limpopo valley but as yet known only from the South African side, is closely allied to this species. It was separated primarily on having very short cymule bracts up to 5 mm long and with conspicuous rust-coloured glandular hairs. However, there is some overlap in these characters with the Zimbabwe specimens of *D. gillilandiorum*. *P. cliffordii* nevertheless appears to be a good species, separated by its broader (ovate or oblong-ovate) leaves, its less clearly ridged stems which have a denser and finer indumentum, its more lax inflorescence, and by the peduncles having mixed eglandular and capitate-glandular hairs.

27. **Dicliptera serpenticola** (K. Balkwill & Campb.-Young) I. Darbysh. in Kew Bull. **62**: 123 (2007). Type: Zimbabwe, NW of Harare, on Great Dyke, Vanad Pass, fl. 15.v.1998, *Balkwill, Cron & Coates Palgrave* 10544 (J holotype, B, C, E, K, M, MO, NU, PRE, RSA, SRGH).

Peristrophe serpenticola K. Balkwill & Campb.-Young in S. Afr. J. Sci. **97**: 551 (2001). — Mapaura & Timberlake, Checklist Zimb. Vasc. Pl.: 14 (2004).

Suffruticose perennial with several decumbent branches to 15–100 cm tall from a woody base and rootstock, often rooting along trailing basal portions; vegetative parts a rather vivid yellow-green; leafy stems slender and rather brittle, furrowed, glabrous or finely pale subappressed-puberulous, hairs both retrorse and antrorse, nodal line strigulose. Leaves rather thick, elliptic to narrowly ovate-elliptic, 2–5 × 0.6–2 cm, base cuneate, shortly attenuate or obtuse, apex acute, apiculate, surfaces glabrous or with minute curved hairs mainly on veins beneath; lateral veins 3–4 pairs; petiole 0–2.5 mm long. Inflorescence axillary, of (1)2–3 cymules, monochasially arranged (scorpioid); primary peduncle 3–12.5 mm long; main axis bracts often caducous, linear-lanceolate or those on inflorescences of lower stem axils narrowly elliptic or ovate, 2–9 × 0.5–1.5 mm; cymules pedunculate, longest peduncle 1–11.5 mm long, glabrous to rather densely retrorse-puberulous; cymule bracts yellow-green, pairs somewhat unequal, ratio 1.05–1.35:1, larger bract narrowly lanceolate or subulate, 7–10 × 1–1.5 mm, apex attenuate, outer surface glabrous to densely antrorse-puberulous, sometimes with few interspersed short capitate glandular hairs, only midrib prominent, margin narrowly pale-hyaline in basal half; bracteoles linear-lanceolate, 5.5–8 mm long, surface often with more numerous glandular hairs than bracts, margin hyaline. Calyx pale green-brown, lobes 4.5–5.5 mm long, ciliate, surface glandular-puberulent or sparsely so. Corolla 13–14 mm long, limb white or pale lilac, shortly eglandular-pubescent externally; tube 6.5–7 mm long; lip held in upper position oblong, 6.5–7 × 3.5–4 mm, palate puberulous towards mouth; lip held in lower position broadly ovate or elliptic, 5–6.8 × 4.5–5.5 mm. Staminal filaments c.4 mm long, with few long spreading hairs; anther thecae 0.75–0.85 mm long, superposed but touching. Style sparsely strigulose. Capsule 9–10.5 mm long, shortly pubescent with mixed eglandular and glandular hairs; placental base inelastic. Seeds c.1.5 mm in diameter, tuberculate, tubercles prominent, not hooked.

Zimbabwe. N: Guruve Dist., Nyamunyeche Estate, on Great Dyke, fl.& fr. 7.viii.1978, *Nyariri* 284 (LMA, SRGH).

Not known elsewhere. Large termite mounds on rocky serpentine soils; 1300–1500 m.

Conservation notes: Endemic to the northern Great Dyke of Zimbabwe where it is known from only three locations. With an Extent of Occurrence of 100 km², and with some of its habitat being degraded through mining, it is likely to qualify as Endangered under IUCN criterion B.

28. **Dicliptera paniculata** (Forssk.) I. Darbysh. in Kew Bull. **62**: 122 (2007); in F.T.E.A., Acanthaceae **2**: 716, fig.91:11 (2010). Type: Yemen, no locality, 1763, *Forsskål* 385 (C microfiche 38: III.3-4 lectotype, K photo), lectotypified by Wood *et al.* (Kew Bull. **38**: 451, 1983). FIGURE 8.6.**71**.

   Dianthera paniculata Forssk., Fl. Aegypt.-Arab.: 9 (1775).
   Peristrophe bicalyculata (Retz.) Nees in Wallich, Pl. Asiat. Rar. **3**: 113 (1832); in De Candolle, Prodr. **11**: 496 (1847). —Clarke in F.T.A. **5**: 242 (1900), excl. *Holub* s.n. & *Scott Elliot* 8247; in Fl. Cap. **5**: 85 (1901). —Heine in F.W.T.A., ed.2 **2**: 424 (1963). —Binns, Checklist Herb. Fl. Malawi: 15 (1968). —Meyer in Merxmüller, Prodr. Fl. SW Afr. **130**: 49 (1968). Type: India, Malabar, n.d. *Koenig* s.n. (UPS-LINN holotype).
   Peristrophe paniculata (Forssk.) Brummitt in Kew Bull. **38**: 451 (1983). —Balkwill in Bothalia **26**: 87, fig.5 (1996). —da Silva *et al.*, Prelim. Checklist Vasc. Pl. Mozamb.: 19 (2004). —Mapaura & Timberlake, Checklist Zimb. Vasc. Pl.: 14 (2004). —Phiri, Checklist Zamb. Vasc. Pl.: 19 (2005) in part. —Setshogo, Prelim. Checklist Pl. Botswana: 19 (2005).

Erect to straggling annual herb with a short taproot, stems 60–180(260) cm tall, much-branched, 6-angular with prominent pale ridges, ridges pale-hispid or rarely glabrous, young stems often with short eglandular and capitate glandular hairs. Leaves often falling before or during flowering/fruiting, ovate or lanceolate, 1.5–10 × 0.8–4.5 cm, base rounded to shortly attenuate, apex acute to acuminate, apiculate, surfaces hispid particularly margin and veins beneath, rarely largely glabrous; lateral veins (3)5–7 pairs; petiole 2–22 mm long. Inflorescences axillary and terminal, compounded into a very lax 20(40) cm pseudopanicle on largely bare upper branches; peduncles wiry, with short eglandular (often aculeolate) hairs and sparse to dense capitate glandular hairs, sometimes also pilose; main axis bracts linear-lanceolate, 2–4.5(5.5) mm long; cymule peduncles (3.5)8–22(38) mm long, apex often abruptly bent; cymule bracts green, pairs highly unequal, ratio 1.2–2.5:1, larger bract subulate or linear-lanceolate, 7–17.5 × 0.7–1.7 mm, apiculate, margin narrowly hyaline towards base, external surface with or (when young) without minute eglandular and glandular hairs, often also with capitate glandular hairs, only midrib prominent, midrib usually aculeolate, margin sparsely ciliate; bracteoles as bracts but 3.5–7 mm long, margin hyaline. Calyx lobes 2–3.5 mm long, ciliate, with occasional minute glandular hairs externally, margin hyaline. Corolla 9–14 mm long, pink or mauve, rarely white, with purple guidelines, eglandular-pubescent externally; tube 4–6 mm long; lip held in upper position oblong, 4.5–8 × 1.5–3 mm, palate puberulous at mouth; lip held in lower position elliptic, 4.5–7 × 2.5–4 mm. Staminal filaments 3.5–7 mm long, sparsely hairy; anther thecae 0.5–0.75 mm long, superposed and somewhat oblique. Ovary largely glabrous. Capsule 8–12.5 mm long, eglandular-pubescent, with scattered short-stalked glands; placental base inelastic. Seeds 1.8–2.5 mm in diameter, tuberculate particularly towards rim, tubercles short, minutely hooked.

Botswana. N: Ngamiland Dist., Sennonore, Maun, fl.& fr. 15.iv.1994, *P.A. Smith* 5697 (K, PRE). SW: Ghanzi Dist., Ghanzi Pan Farm 59, fl. bud 11.iii.1970, *Brown* 8801 (K, PRE, SRGH). SE: Central Dist., Boteti Delta area, NE of Mopipi, fr. 22.iv.1973, *Thornton* 4 (SRGH). **Zambia.** B: Masese, fl.& fr. 18.v.1962, *Fanshawe* 6829 (K, NDO, SRGH). N: Mpulungu Dist., Kumbula (Nmbulu) Is., Lake Tanganyika, fl. 11.iv.1955, *Richards* 5390 (K). S: Livingstone, fl. bud viii.1921, *Borle* 314 (SRGH). **Zimbabwe.** N: Darwin Dist., Mkumvura R., Mukumbura (Mukumbeira), fl.& fr. 7.vii.1950, *Chase* 2602 (BM, K, SRGH). W: Hwange Dist., Victoria Falls, S bank of Zambezi, fl.& fr. v.1915, *Rogers* 13179 (BR, K). S: Mwenezi Dist., NW Mateke Hills, fl. 4.v.1958, *Drummond*

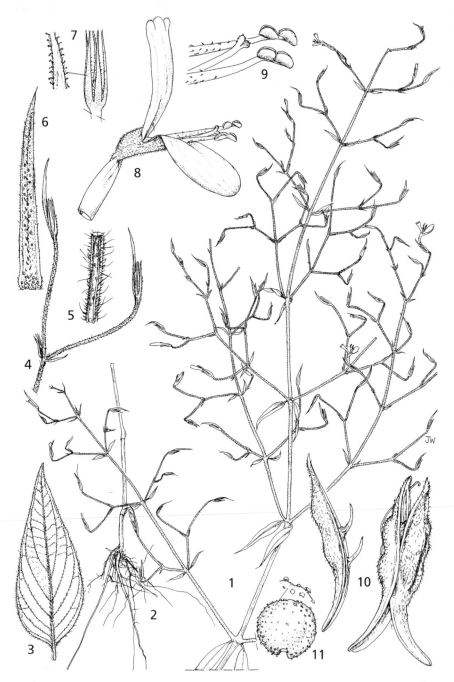

Fig. 8.6.**71**. DICLIPTERA PANICULATA. 1, habit (× ²/₃); 2, rootstock (× ²/₃); 3, mature leaf, abaxial surface (×1); 4, partial inflorescence, 2 cymules and main axis bracts (× 2); 5, detail of peduncle indumentum (× 12); 6, longer cymule bract, external surface (× 6); 7, calyx with detail of indumentum (× 6); 8, corolla with stamens and stigma (× 5); 9, detail of anthers and stigma (× 10); 10, capsule and single valve (× 5); 11, seed with sculpturing detail (× 8). 1 from *Biegel et al.* 5088, 2, 4 & 5 from *Fanshawe* 6829, 3 from *Smith* 5697, 6–9 from *Lambrecht* 125, 10 & 11 from *Brummitt* 11423. Drawn by Juliet Williamson.

5560 (BR, K, LMA, SRGH). **Malawi**. N: Karonga Dist., 27 km N of Chilumba at Ngara (Ngala), fl.& fr. 4.vii.1970, *Pawek* 3564 (K, MAL). C: Nkhotakota Dist., Nkhotakota, fl.& fr. 15.vi.1970, *Brummitt* 11423 (K, MAL, SRGH). **Mozambique**. T: Zambezi Valley, 'Boroma', near Magoé (Magwe), fl. 23.vii.1950, *Chase* 2689 (BM, K, SRGH).

Widespread in tropical Africa from Cape Verde and Senegal to Somalia, south to Namibia and South Africa, and likely to occur in the Caprivi Strip; also in Arabia, India and Thailand. Open to dense woodland, thicket and wooded grassland, often in riverine fringes or on lakeshores, frequently on disturbed ground; 250–1100 m.

Conservation notes: Widespread and often common; Least Concern.

Balkwill (1996, p.89) notes that the first flowers produced are cleistogamous, ensuring rapid replenishment of the seed bank; chasmogamous flowers develop later in the growing season.

Burrows in Pl. Nyika Plateau: 54 (2004) lists this species from the Nyika Plateau based upon an earlier record by Mill. This is almost certainly a misapplied name since I have seen no material of *Dicliptera paniculata* from this region and the habitat and altitude do not appear suitable – it probably refers *to D. aculeata*.

29. **Dicliptera cernua** (Nees) I. Darbysh., comb. nov. Type: South Africa, Eastern Cape, "inter frutices in campis ad flumen Zwartkopsrivier prope praedium Pauli Maré, alt. I (Uitenhage)", fl. x.[year not known] *Ecklon* Un It. 556 (GZU holotype, K?, STE).

Peristrophe cernua Nees in Linnaea 15: 374 (1841); in De Candolle, Prodr. 11: 498 (1847). —Clarke in Fl. Cap. 5: 85 (1901). —Balkwill in Bothalia 26: 91 (1996). —Welman in Germishuizen *et al.*, Checklist S. Afr. Pl.: 87 (2006).

Peristrophe oblonga Nees in Linnaea 15: 375 (1841) in part. Type: South Africa, Eastern Cape, "prope sedem Chali, regis Caffrorum in montibus Chumi, alt. III (Terrae Caffrorum)" [on label: Kafferland, Schumiberg, südöstlich vom Katberg], fl. vi.[year not known], *Ecklon* s.n. (S ?holotype), mixed collection, see note.

Peristrophe krebsii C. Presl in Abh. Königl. Böhm. Ges. Wiss. ser.5 3: 524 (1845); in Bot. Bemerk.: 94 (1846). Type: South Africa, Eastern Cape, no locality, *Krebs* [pl. cap. exs. n.] 251 (BR).

Peristrophe caulopsila Nees in De Candolle., Prodr. 11: 498 (1847). —Clarke in Fl. Cap. 5: 84 (1901). Type: South Africa, Eastern Cape, Somerset Division, between Zuurberg Range and Kleinbruintjieshoogte, fl. 1837, *Drège* s.n. (K lectotype), lectotypified by Balkwill (1996).

Rhinacanthus oblongus (Nees) Nees in De Candolle., Prodr. 11: 444 (1847) in part, see note.

Peristrophe natalensis T. Anderson in J. Proc. Linn. Soc., Bot. 7: 48 (1863). —Clarke in Fl. Cap. 5: 85 (1901). Type: South Africa, Durban (Port Natal), fl.& fr., n.d., *Gueinzius* s.n. (K lectotype), lectotypified by Balkwill (1996).

Peristrophe hensii sensu Clarke in Fl. Cap. 5: 85 (1901), non C.B. Clarke in F.T.A.

Erect, straggling or trailing perennial herb or subshrub, 80–200 cm tall; stems 6-angular, ± pale-ridged, glabrous, nodal line pale-hispid. Leaves sometimes absent at flowering, ovate, 1.5–4.5 × 0.5–2 cm, base obtuse or rounded, apex acute or attenuate, mucronulate, surfaces glabrous except for few conspicuous pale hispid hairs on margin towards base and sometimes on midrib beneath; lateral veins 4–5 pairs; petiole 2.5–6 mm long. Inflorescence axillary, 1–2 umbels of 2–4 cymules per axil, umbels sometimes partially compounded, together sometimes forming rather slender terminal pseudopanicles on largely leafless branches; umbel peduncles 4–17 mm long, rather densely puberulous with mixed eglandular and capitate-glandular hairs; main axis bracts often caducous, linear-lanceolate, 1.5–3.5 × 0.5 mm; cymules pedunculate, longest peduncle in each umbel 7–14 mm long, lateral peduncles often abruptly bent at apex; cymule bracts green, pairs unequal, ratio 1.1–1.65:1, larger bract subulate, 5–11 × 0.9–1.3 mm, apex shortly attenuate, outer surface puberulous with mixed eglandular and capitate-glandular hairs, only midrib prominent, margin narrowly pale-hyaline; bracteoles linear-lanceolate, 4.5–7 mm long. Calyx pale green-brown, lobes 2.7–3.3 mm long, ciliate, surface glandular-puberulent or sparsely so. Corolla 19–22.5 mm long, pale pink or white tinged purple, eglandular-pubescent and with few short glandular hairs externally; tube 7–9.5 mm long; lip held in upper position narrowly oblong, 9.5–13.5 × 3.3–3.8 mm, palate sparsely puberulous at mouth; lip held in lower position oblong-elliptic to narrowly ovate-elliptic, 11–13.5 × 5.5–6.5 mm. Stamens markedly unequal, the longer with filament 10.5–12.5 mm long, largely glabrous; anther thecae 0.9–1.1 mm long, superposed but touching. Style glabrous. Capsule 10–15 mm long, puberulous on fertile portion with mixed eglandular and glandular hairs, rather sparse at maturity; placental base inelastic. Seeds 1.7–2.2 mm in diameter, tuberculate, tubercles minutely hooked.

Mozambique. GI: Massingir Dist., 23 km from Massingir towards R. Singuédzi, fl.& fr. 19.vii.1969, *Correia & Marques* 958 (LMU); Massingir Dist., near Mavodze village, fl.& fr. 28.vii.1982, *Matos* 5127 (LISC).

Also in South Africa (Eastern Cape, KwaZulu-Natal, Mpumalanga). Dry riverine forest with *Trichilia*, *Ficus sycomorus* and *Acacia xanthophloea;* also in open mopane and Acacia woodland; c.150 m.

Conservation notes: Widespread and often common in South Africa, but in the Flora area only known from three collections within the Massingir area of S Mozambique; possibly threatened in Mozambique.

Dicliptera cernua is here recorded for the first time from outside South Africa. Within its South African range, it varies considerably, though the variation in e.g. flower and bract size is reported as clinal (Balkwill in Bothalia **26**: 92, 1996). The outlier populations in Mozambique display some minor differences to the South African plants, most notably in having the most densely hairy inflorescences. They also differ in vegetative indumentum from KwaZulu-Natal populations where the stems have fine white antrorse or retrorse hairs on the ridges and the leaves are antrorsely hairy above and on the margin and veins beneath. However, some populations from the Eastern Cape more closely match the Mozambique plants in vegetative indumentum. Only the Mozambican plants are covered in the above description.

The type material of *Peristrophe oblonga* Nees held at S is a mixed collection, with one piece (S-G-4715) referable to *Dicliptera cernua* and the second (S08-564) to *Rhinacanthus latilabiatus* (K. Balkwill) I. Darbysh. Although originally described in *Peristrophe*, Nees compared this species to the Indian *Peristrophe pubigera* Nees (= *Justicia pubigera* (Nees) C.B. Clarke) and noted that it was also near *Rhinacanthus*. *J. pubigera* is similar to *R. latilabiatus* in its inflorescence form, and it is clear from the protologue that Nees based his description of *P. oblonga* largely on S08-564. Indeed, he later made the new combination *Rhinacanthus oblongus* (Nees) Nees. It may, therefore, be preferable to select S08-564 as the lectotype of *P. oblongus*, in which case *Rhinacanthus oblongus* would become an accepted name with nomenclatural priority over *R. latilabiatus*. On the other hand, K. Balkwill (F.S.A. **30**: 14, 1995) took the stance that, since Nees' *oblonga* was originally described in *Peristrophe*, it should be regarded as a synonym of *Peristrophe cernua*, although he did not lectotypify the name accordingly in his subsequent revision of southern African *Peristrophe*.

30. **Dicliptera decorticans** (K. Balkwill) I. Darbysh. in Kew Bull. **62**: 123 (2007). Type: South Africa, Louis Trichardt Dist., beside N1 at gate to Plaas Marius, N of Wylliespoort, fl.& fr. 12.v.1983, *K. Balkwill* 801 (J holotype, E, K, NU, PRE).

Peristrophe decorticans K. Balkwill in S. Afr. J. Bot. **55**: 254, fig.1 (1989). —Balkwill in Bothalia **26**: 90 (1996). —Mapaura & Timberlake, Checklist Zimb. Vasc. Pl.: 14 (2004). —Setshogo, Prelim. Checklist Pl. Botswana: 19 (2005). —Welman in Germishuizen *et al.*, Checklist S. Afr. Pl.: 87 (2006).

Peristrophe bicalyculata sensu Clarke in F.T.A. **5**: 242 (1900), in part for *Holub* s.n., non (Retz.) Nees.

Peristrophe kotschyana sensu K. Balkwill *et al.* in S. Afr. J. Bot. **51**: 488 (1985); in S. Afr. J. Bot. **52**: 514 (1986), non Nees.

Suffruticose perennial, ± much-branched from a woody rootstock, branches often scrambling or sprawling, 20–100 cm tall; herbaceous stems 6-angular, prominently ridged, at first pale-hispid on ridges, later glabrescent but with swollen hair bases persisting, rarely glabrous throughout except for hispid nodal line; woody stems with pale grey flaking bark. Leaves sometimes immature at flowering, ovate or lanceolate, 2–6 × 0.5–2.3 cm, base obtuse to shortly attenuate, apex acute or subattenuate, mucronulate, surfaces pale (hispid-)pilose or hairs restricted to margin and main veins beneath; lateral veins (3)4–5 pairs; petiole 1–7 mm long. Inflorescence axillary, 1(2) umbels of (1)2–3(4) cymules per axil, umbels often partially compounded, together sometimes forming loose slender terminal pseudopanicles on largely leafless branches; umbel peduncle (1)3–17(27) mm long, indumentum as stems; main axis bracts usually linear-lanceolate, 2–7 × 0.5–1 mm, more rarely leafy and elliptic-lanceolate, up to 16 × 5.5 mm; longest cymule peduncle in each umbel (2.5)4–23 mm long, lateral peduncles sometimes abruptly bent at apex; cymule bracts green, tardily turning (pinkish-)brown, pairs unequal, ratio (1.1)1.25–1.5(1.7):1, larger bract linear-lanceolate to subulate, 7–15.5 × 1–1.7 mm, apex shortly attenuate, external surface rather stiffly antrorse-pubescent, hairs sometimes restricted to prominent midrib, margin narrowly pale-hyaline; bracteoles linear-lanceolate, 5.5–10 mm long, external surface as bracts or sometimes also sparsely puberulent. Calyx lobes 3.3–5.5 mm long, margin hyaline, ciliate, external surface glabrous. Corolla 15.5–20 mm long, pale pink, mauve, purple or white, with purple guidelines, eglandular-pubescent externally; tube 6.5–9 mm long; lip held in upper position narrowly oblong, 9–11.5 × 3–3.5 mm, palate puberulous towards mouth; lip held in lower position broadly ovate-elliptic, 8–11 × 5.5–8 mm. Staminal filaments 6–7 mm long, shortly pubescent above; anther thecae 0.75–1 mm long, superposed but touching. Style glabrous. Capsule 8.5–11 mm long, glabrous; placental base inelastic. Seeds 1.5–2 mm in diameter, tuberculate, tubercles minutely hooked when immature.

Botswana. N: NE Dist., Francistown, fl. ii.1926, *Rand* 77 (BM). SE: Central Dist., Mahalapye, Expt. Station, fl.& fr. 15.ii.1961, *Yalala* 125 (K, SRGH). **Zimbabwe**. S: Gwanda Dist., Pye R. weir, Special Nature Area G, fl. 17.xii.1956, *Davies* 2342 (K, SRGH).

Also in South Africa (NorthWest, Limpopo, Mpumalanga). *Acacia* bushland and mopane *Terminalia–Commiphora* woodland on sandy or alluvial soils, riverine bushland, base of sandstone outcrops; 500–1000 m.

Conservation notes: Fairly common in South Africa but local in the Flora area; Least Concern.

Balkwill records the fruits as sometimes being hairy, but they are glabrous in all the material I have seen. He also records corollas down to only 12 mm long, but the material in the Flora region falls within the measurements given here.

31. **Dicliptera sp. E** (= *Mendonça* 4310).

Perennial herb; herbaceous stems 6-angular, narrowly ridged, pale-hispid on ridges; woody stems with white flaking bark. Leaves ovate, 4.7–6.5 × 2–2.5 cm, base obtuse, rounded or broadly cuneate, apex subattenuate, margin and sometimes midrib above hispid towards leaf base, elsewhere very sparsely pubescent or glabrous; lateral veins 4–5 pairs; petiole 0–2 mm long. Inflorescence axillary, 1 umbel of 2–3 cymules per axil, or umbels often compounded, those of lateral branches together forming lax terminal pseudopanicles; umbel peduncle 14–38 mm long, glabrous or sparsely hispid; main axis bracts linear-lanceolate, 3.5–5(8) × c.0.5 mm; longest cymule peduncle in each umbel 8–17 mm long, lateral peduncles gradually curved to more abruptly bent at apex; cymule bracts grey-green or smaller of pair turning pinkish-brown, pairs highly unequal, ratio 1.4–1.65:1, larger bract subulate, 12–16 × 1.7–2.1 mm, apex shortly attenuate, external surface glabrous, margin sparsely and shortly ciliate, narrowly pale-hyaline in basal half; bracteoles linear-lanceolate, 9–10.5 mm long. Calyx lobes 6–6.5 mm long, margin hyaline, ciliate, external surface glandular-puberulent. Corolla 24–25 mm long, pink-purple, eglandular-pubescent externally; tube 8.5–9 mm long; lip held in upper position narrowly oblong, 15–16.5 × 3.5 mm, palate puberulous towards mouth; lip held in lower position oblong-elliptic, 14.5–17 × 5.5–6 mm. Stamens unequal in length, longer filament 11–12.5 mm long, shortly pubescent; anther thecae 1–1.25 mm long, superposed and slightly separated, oblique. Style glabrous. Capsule c.10 mm long, glabrous; placental base inelastic. Seeds c.2 mm in diameter, sparsely tuberculate, tubercles clavate.

Mozambique. T: Mutarara Dist., Tete, 148 km along railway line, fl.& fr. 17.v.1948, *Mendonça* 4310 (LISC).

Not known elsewhere. Clearings in dry forest; c.300 m.

Conservation notes: Known only from the type; probably threatened.

This distinctive specimen differs from *Dicliptera aculeata* in the proportionally broader, glabrous, grey-green bracts, the glabrous capsules, the broader, ovate leaves and the presence of white flaking bark on the woody basal stems. *Dicliptera decorticans* also has pale flaking bark, but sp. E differs in the larger corollas and the broader cymule bracts that lack the distinctive eglandular hairs of *D. decorticans* and are grey-green (not green) in colour. It also bears a striking resemblance to *Peristrophe teklei* Ensermu from Yemen, Eritrea and Ethiopia. More material is required to confirm its status.

32. **Dicliptera aculeata** C.B. Clarke in F.T.A. **5**: 257 (1900). —Darbyshire in F.T.E.A., Acanthaceae **2**: 718 (2010). Type: Malawi, Nyika Mts, fl. 1896, *Whyte* s.n. (K lectotype), lectotypified by Brummitt (1991).

    Peristrophe aculeata (C.B. Clarke) Brummitt in Kew Bull. **46**: 290 (1991). —Burrows & Willis, Pl. Nyika Plateau: 51, 53 (2004).

    Peristrophe bicalyculata sensu Clarke in F.T.A. **5**: 242 (1900) in part for *Scott Elliot* 8247, non (Retz.) Nees.

    Peristrophe hensii sensu da Silva *et al.*, Prelim. Checklist Vasc. Pl. Mozamb.: 19 (2004), non C.B. Clarke.

Erect or straggling suffruticose perennial, (15)45–200 cm tall, few- to much-branched from a woody rootstock; stems 6-angular, with or without prominent pale ridges, hispid on ridges and nodal line or glabrous. Leaves often immature or absent at flowering, lanceolate or linear-lanceolate, (2)4.5–15 × 0.3–3 cm, base shortly attenuate, apex acute to acuminate, apiculate, shortly hispid particularly on margin and veins beneath, or largely glabrous; lateral veins 3–7 pairs, pale and prominent beneath; petiole 1–8 mm long. Inflorescences axillary and terminal, lax or rather congested, 1–2 umbels of (2)3–4 cymules per axil, often compounded into terminal pseudopanicles on largely leafless branches; umbel peduncles wiry, (1)4–55 mm long, glabrous or sparsely antrorse-hispid; main axis bracts linear-lanceolate, (2)3.5–8(14) mm long; cymules subsessile or peduncles to 25(40) mm long, sometimes with scattered subsessile glands; cymule bracts green, pairs ± unequal or rarely subequal, ratio (1.05)1.2–1.55(1.85):1, larger bract linear-lanceolate, 5.5–24 × 0.8–2 mm, apex attenuate to acuminate, apiculate, surface usually with scattered short-stalked or subsessile glands, sometimes also sparsely eglandular-puberulous, midrib prominent, sometimes antrorsely hispid, margin hyaline towards base; bracteoles as bracts but 4.5–15 mm long, margin hyaline. Calyx lobes 3–8 mm long, ciliate and with occasional minute glandular hairs externally, margin hyaline. Corolla 15.5–30 mm long, pink to purple or rarely white, with purple guidelines, eglandular-pubescent and often with scattered short glandular hairs externally; tube 7–11 mm long; lip held in upper position oblong, 8.5–19 × 3–5 mm, puberulous towards mouth; lip held in lower position (oblong-)elliptic, 8–20 × 4–7.5 mm. Staminal filaments 6.5–14.5 mm long, with short hairs particularly above; anther thecae 0.8–1 mm long, superposed and somewhat oblique. Style sparsely strigulose. Capsule 9.5–14(16) mm long, with sparse mixed short eglandular and glandular hairs at least towards apex, sometimes glabrescent; placental base inelastic. Seeds 1.5–2.5 mm in diameter, tuberculate, tubercles short, somewhat clavate.

Zambia. N: Isoka Dist., road from Nakondi, bank of Nafungu R., fl.& fr. 8.xi.1967, *Richards* 22414 (K). C: Mkushi Dist., fl. 2.v.1957, *Fanshawe* 3256 (BR, K, NDO). E: Katete Dist., Katete R., mile 6 Katete–Chadiza, fl.& fr. 8.x.1958, *Robson* 12 (BM, BR, K, SRGH). **Malawi**. N: Mzimba Dist., Mzimba, fl. 30.vii.1960, *Leach & Brunton* 10371 (K, SRGH). C: Dowa Dist., by Ngara (Ngala) Mt, fl.& fr. 28.iv.1970, *Brummitt* 10231 (K, MAL, SRGH). S: Mangochi Dist., Namwera, fl. 13.vi.1971, *Binns* 847 (SRGH).

Mozambique. N: Cuamba Dist., 45 km NW of Cuamba (Nova Freixo), fl. 25.v.1961, *Leach & Rutherford-Smith* 11004 (K, LISC, SRGH). Z: Gurue Dist., between Molumbo and Lioma, fl. 8.vii.1949, *Andrada* 1973 (LISC). T: Angónia Dist., R. Namanza, Furancungo road, fl. 1.vi.1980, *Stefanesco & Nyongani* 527 (LISC, SRGH).

Also in Tanzania. Miombo woodland on sandy soils, often in seasonally damp areas such as dambos, secondary woodland, riverbanks amongst scrub or grassland, and roadsides; 450–1800 m.

Conservation notes: Widespread and locally common; Least Concern.

The type material represents a particularly striking form restricted to N Malawi and parts of C & S Tanzania where the inflorescence is contracted on short lateral branches and forms rather dense pseudopanicles. In this form, the larger cymule bract of each pair is generally in the range 13–24 mm long, the bracteoles are correspondingly long and the calyx lobes are up to 8 mm long. Elsewhere, the inflorescence units are all pedunculate, forming (very) lax pseudopanicles, the larger cymule bracts are generally 5.5–14.5 mm long with correspondingly short bracteoles, and the calyx lobes are up to 5 mm long. The first form tends to have more prominently ridged and hispid stems, though these can also be present in lax forms from Malawi. However, whilst the two extremes look quite different, the variation appears to be largely clinal, and so I have chosen not to recognize two taxa here.

Excluded species:

Dicliptera betschuanica Lindau in Bot. Jahrb. Syst. **30**: 112 (1901). Types: Botswana, N bank of Limpopo ("des nördlichen Limpopoufers"), fl. vii.1896, *Passarge* 29 & 68 (B† syntypes).

From the type locality and from the general description, this taxon is most likely to be conspecific with *Dicliptera minor*. The name *D. betschuanica* (March 1901) predates *D. minor* (June 1901) and so would have nomenclatural priority. However, there are some minor discrepancies in the description, most notably in the fact that Lindau records the cymule bracts as having a ciliate margin which does not seem to fit with *D. minor*. Since the types have been destroyed, this uncertainty remains and so I consider it best to maintain the name *D. minor* at present. It would, in any case, be preferable to conserve the name *D. minor* over *D. betschuanica* in view of the fact that Balkwill used the former in his regional revision of the genus.

42. HYPOESTES R. Br.[20]

Hypoestes R. Br., Prodr. Fl. Nov. Holl.: 474 (1810). —Nees in De Candolle, Prodr. **11**: 501 (1847). —Benoist in Notul. Syst. (Paris) **10**: 241–248 (1942). —Balkwill & Getliffe-Norris in S. Afr. J. Bot. **51**: 133–144 (1984).

Annual or perennial herbs or subshrubs; cystoliths present; stems sub-angular, sulcate or ridged. Leaves opposite-decussate, petiolate, pairs equal or anisophyllous, usually evergreen. Inflorescences axillary and/or terminal, comprising a series of monochasial cymules, arrangement of cymules variously fasciculate, umbellate, spiciform, strobiliform or cylindrical, sometimes compounded into verticillate or paniculate thyrses; main axis bracts paired, equal to highly unequal, free or basally fused, sometimes foliaceous; cymule bracts paired, equal to unequal, free or fused for up to ²/₃ their length; bracteoles paired, equal to ± unequal, free, narrower than cymule bracts. Calyx divided almost to base or with a distinct tube, 4- or

[20] By Iain Darbyshire

5-lobed, lobes subequal to ± unequal in length, often hyaline. Flowers protandrous. Corolla bilabiate, variously white, pink or purple, usually with darker guidelines; limb resupinate, tube twisted through 180°, cylindrical below twist, weakly to strongly expanded from above twist to mouth; lip held in upper position recurved distally, 3-lobed or with only median lobe well-developed, palate longer than lobes; lip held in lower position unlobed, apex acute, obtuse or shallowly emarginate, recurved distally. Stamens 2; filaments attached at corolla mouth, exserted, initially straight, becoming divergently reflexed and bending out of corolla in female phase; anthers monothecous, thecae ellipsoid; staminodes absent. Disk cupular with V-shaped slit, 2-awned. Ovary oblong-ovoid, glabrous, bilocular, 2 ovules per locule; style filiform; stigma bifid, exserted from corolla tube, held below anthers. Capsule clavate, shortly stipitate, placental base inelastic. Seeds held on retinacula, lenticular or compressed-ellipsoid, with a hilar excavation, smooth or tuberculate.

Genus confined to the Old World tropics and subtropics from Africa to Australia, with the highest diversity on Madagascar where a revision is sorely needed. The number of accepted species varies from 40 to 150. In continental Africa, c.10 species are recognised. The taxonomy is complicated by the presence of three widespread and polymorphic species: *H. triflora*, *H. aristata* and *H. forskaolii*, the delimitations of which have not been fully resolved.

1. Cymule bract pairs free but clasping, broadly obovate to narrowly oblong-elliptic; cystoliths arcuate, prominent on leaf blade; inflorescence umbellate or a solitary pedunculate cymule, umbels of 2–4(6) cymules, sometimes partially compounded and laxly verticillate; staminal filaments glabrous; seeds tuberculate . . . . **1.** *triflora*
   – Cymule bract pairs partially fused, lanceolate or linear, with or without a long arista; cystoliths elliptic or linear, often inconspicuous; inflorescence fasciculate or a unilateral spike, often compounded into a dense verticillate or paniculate thyrse; cymules 5 to many per inflorescence; staminal filaments usually hairy; seeds smooth or rarely tuberculate 2
2. Cymule bract pairs fused for ½ to ²/₃ their length, bract apex not aristate; calyx lobes 5; corolla lip in lower position trullate-caudate; capsule puberulous towards apex or (rarely) largely glabrous . . . . **3.** *forskaolii*
   – Cymule bract pairs fused for up to ¹/₃ their length (often considerably less), bract apex with long flexuose arista; calyx lobes 4; corolla lip in lower position narrowly oblong-elliptic or lanceolate; capsule glabrous 3
3. Corolla tube rapidly expanded above twist into a markedly enlarged throat 5–6.5 mm wide at mouth; lip in upper position hooded with reflexed lobes, widest at base when flattened; staminal filaments 5–6 mm long . . . . . **2b.** sp. aff. *aristata B*
   – Corolla tube more gradually expanded above twist, throat usually much narrower or rarely up to 5.5 mm wide at mouth; lip in upper position not hooded, oblong or oblong-obovate, narrowed towards base; staminal filaments 6–14 mm long . . 4
4. Seeds smooth; stem leaves ovate, to 19.5 × 9.5 cm, usually much smaller, larger ones with prominent acumen; cymule bracts usually with few to numerous spreading glandular hairs on arista . . . . **2.** *aristata*
   – Seeds tuberculate; stem leaves elliptic, 20–24.5 × 9.5–12 mm, apex shortly attenuate; cymule bracts with eglandular hairs only . . . . **2a.** sp. aff. *aristata A*

1. **Hypoestes triflora** (Forssk.) Roem. & Schult., Syst. Veg. **1**: 141 (1817). —Clarke in F.T.A. **5**: 247 (1900); in Fl. Cap. **5**: 87 (1901). —Darbyshire in F.T.E.A., Acanthaceae **2**: 720 (2010). Types: Yemen, fl. 1763, *Forsskål* s.n. (C microfiche 60: I.3–4 & 5–6 syntypes, LD, K photo).

   Justicia triflora Forssk., Fl. Aegypt.-Arab.: 4 (1775).

   Hypoestes phaylopsoides S. Moore in Trans. Linn. Soc., Bot. **4**: 34 (1894). —Clarke in F.T.A. **5**: 248 (1900) in part, excl. *Whyte* s.n. from Masuka Plateau; in Fl. Cap. **5**: 88 (1901). —Binns,

Checklist Herb. Fl. Malawi: 14 (1968). Type: Malawi, Mt Mulanje, fl.& fr. x.1891, *Whyte* 126 (BM holotype).

Hypoestes kilimandscharica Lindau in Bot. Jahrb. Syst. **19**: 47 (1894). Type: Tanzania, Kilimanjaro, Kiboscho, W of Sina's Boma, fl. x.1893, *Volkens* 1663 (B† holotype, BM).

Hypoestes consanguinea Lindau in Bot. Jahrb. Syst. **20**: 50 (1894). Types: Togo, Bismarckburg, fl. 23.x.1890, *Büttner* 315 (B† syntype); Togo, Bismarckburg, fl. 16.xi.1889, *Kling* 189 (B† syntype); Cameroon, W of Buea, fl. 17.i.1891, *Preuss* 599 (B† syntype, K).

Hypoestes rosea sensu Clarke in F.T.A. **5**: 248 (1900) in part for *Büttner* 315, *Kling* 189 & *Preuss* 599, non P. Beauv.

Creeping, scrambling or more rarely erect herb, annual or usually perennial, 10–300 cm tall, often rooting at lower nodes; stems pubescent to pilose when young, sometimes also glandular-pubescent, ± soon-glabrescent except for long hairs along nodal line. Leaf pairs ± unequal, ovate, elliptic or lanceolate, 1.5–14 × 0.8–7 cm, base ± asymmetric, rounded to attenuate, margin crenate-serrate or subentire, apex acuminate, upper surface and veins beneath pubescent; lateral veins 4–7(10) pairs; cystoliths arcuate, conspicuous; petiole 0.7–6(10) cm. Inflorescences axillary and terminal, cymules solitary or usually in umbels of 2–4(6), sometimes partially compounded to form verticils at stem apex; primary peduncle 3–70 mm long, sparsely to densely pubescent, with or without spreading glandular hairs; main axis bracts foliaceous, ovate to oblanceolate, to 20 mm long, one pair per umbel; cymules subsessile or peduncles to 33 mm long, one cymule peduncle in umbel often much longer than others; cymule bracts free but clasping, broadly obovate to narrowly oblong-elliptic, unequal, larger bract 7.5–18.5 × 2.5–10.5 mm, apex rounded, acute or acuminate, variously short-pubescent to long-pilose, sometimes only along margin, surface with or without short spreading glandular hairs; bracteoles unequal, linear-lanceolate, the longer 6–11 mm long, margin hyaline. Calyx divided almost to base, lobes 5, subequal, linear-lanceolate, 2–8 mm long, glabrous or ciliate. Corolla 17–32 mm long, white to mauve, with (blackish-)purple markings on palate of lip held in upper position, pubescent externally; tube 10–13.5 mm long; lip held in upper position oblong, 7–18.5 mm long, apex shortly 3-lobed; lip held in lower position obovate or oblong-elliptic, 7–16 mm long, apex rounded or emarginate. Staminal filaments 6–14 mm long, glabrous; anthers 0.6–2 mm long. Style pubescent towards apex. Capsule 10.5–15.5 mm long, puberulous towards apex with mixed eglandular and glandular hairs, sometimes sparse. Seeds lenticular, (1)1.5–2.5 mm wide, tuberculate.

Zambia. N: Mbala Dist., Mwengo, fl.& fr. 10.vi.1951, *Bullock* 3958 (K). W: Mufumbwe Dist., Mufumbwe R., fl.& fr. 25.vi.1953, *Fanshawe* 123 (K, NDO). C: Serenje Dist., Kundalila Falls, 13 km SE of Kanona, fl.& fr. 15.x.1967, *Simon & Williamson* 1020 (K, LISC, SRGH). E: Isoka Dist., Makutu Hills, fl. 27.x.1972, *Fanshawe* s.n. (K, NDO). **Malawi**. N: Chitipa Dist., Misuku Hills, Mughesse Forest, fl. 6.vii.1973, *Pawek* 7018 (CAH, K, MAL, MO, UC). C: Dedza Dist., Dedza Mt, fl. 16.viii.1976, *Pawek* 11602 (K, MAL, MO, SRGH, UC). S: Zomba Dist., Chingwe's Hole Nature Trail, Zomba Plateau, fl. 24.ix.1985, *Banda et al.* 2511 (K, MAL). **Mozambique**. Z: Milange Dist., Milange, at side of mountain, fl.& fr. 10.ix.1941, *Torre* 3405 (LISC, MO, PRE).

Widespread in tropical Africa from Sierra Leone to Eritrea, south to Angola, Swaziland and South Africa, also in southern Arabia, India to China and Thailand. In montane forest, often in clearings, along margins or in dense scrub along streams; at lower altitudes usually in wet riverine forest (mushitu), also in roadside scrub; often common or dominant in the herb layer; 1100–2200 m.

Conservation notes: Widespread and common; Least Concern.

The extremes of variation in our region are represented by populations from the Misuku Hills and the Nyika Plateau, both in N Malawi. In the former, the cymules are subsessile in well-defined umbels, the cymule bracts are broadly obovate and sparsely hairy with only eglandular hairs, and corollas are at the lower end of the size range recorded above. On the Nyika, the cymules are often held on extended peduncles, the umbels being lax and less well defined, cymule bracts are rather narrowly oblong-elliptic or obovate and have numerous glandular hairs, and corollas are larger.

However, populations from elsewhere have different combinations of these characters and are more variable, such that recognition of infraspecific taxa appears impossible. The apparent absence of this species from the Eastern Highlands of Zimbabwe and most of the montane regions in Mozambique is surprising.

2. **Hypoestes aristata** (Vahl) Roem. & Schult., Syst. Veg. 1: 140 (1817). —Clarke in F.T.A. **5**: 245 (1900); in Fl. Cap. **5**: 86 (1901), both as *H. aristata* R. Br. —Burrows & Willis, Pl. Nyika Plateau: 48 (2005). —Bandeira *et al.*, Fl. Nat. Sul Moçamb.: 161, 194 (2007). —Darbyshire in F.T.E.A., Acanthaceae **2**: 724 (2010). Type: South Africa, Cape of Good Hope (Caput Bonae Spei), *Bülow* s.n. (C holotype).

Justicia aristata Vahl, Symb. Bot. Upsal. **2**: 2 (1794).

Justicia verticillaris L.f., Suppl. Pl.: 85 (1782). Type: South Africa, Cape of Good Hope, *Herb. Thunberg* 427 (UPS lectotype), lectotypified by Brummitt *et al.* in Taxon **32**: 658 (1983); name rejected by Brummitt *et al.* in Taxon **32**: 658–659 (1983) & Taxon **36**: 432 (1987).

Hypoestes verticillaris (L.f.) Roem. & Schult., Syst. Veg. 1: 140 (1817).

Hypoestes aristata (Vahl) Roem. & Schult. var. *macrophylla* Nees in De Candolle, Prodr. **11**: 510 (1847). —Benoist in Notul. Syst. (Paris) **10**: 243 (1942). Types: South Africa, Cape, fl. n.d., *Drège* s.n. (B†, GZU, K syntypes).

Hypoestes insularis T. Anderson in J. Linn. Soc., Bot. **7**: 49 (1862). —Clarke in F.T.A. **5**: 246 (1900). Type: Bioko (Fernando Po), fl. i.1860, *Mann* 179 (K holotype), number not cited in protologue.

Hypoestes antennifera S. Moore in J. Bot. **18**: 41 (1880). —Clarke in F.T.A. **5**: 245 (1900); in Fl. Cap. **5**: 87 (1901). —Binns, Checklist Herb. Fl. Malawi: 14 (1968). Type: Kenya, N'di (Taita), fl. ii.1877, *Hildebrandt* 2563 (BM holotype, CORD, K).

Hypoestes staudtii Lindau in Bot. Jahrb. Syst. **22**: 122 (1897). —Clarke in F.T.A. **5**: 246 (1900). Type: Cameroon, Yaoundé (Jaundestation), fl.& fr. x.1893, *Zenker & Staudt* 36 (B† holotype, BM, K).

Hypoestes aristata var. *insularis* (T. Anderson) Benoist in Notul. Syst. (Paris) **10**: 244 (1942).

Hypoestes aristata var. *staudtii* (Lindau) Benoist in Notul. Syst. (Paris) **10**: 244 (1942).

Hypoestes aristata var. *alba* K. Balkwill in S. Afr. J. Bot. **51**: 141 (1985). —Mapaura & Timberlake, Checklist Zimb. Vasc. Pl.: 14 (2004). —Burrows & Willis, Pl. Nyika Plateau: 48 (2005). Type: South Africa, Mpumulanga, Wonderkloof Nature Reserve, fl.& fr. 23.v.1978, *Kluge* 1317 (PRE holotype).

Erect or scrambling perennial herb or subshrub 40–300 cm tall; stems sparsely to densely puberulous or pubescent, hairs spreading, antrorse or retrorse, older stems often glabrescent except for persistent hairs along nodal line. Leaves broadly to narrowly ovate, 1.5–19.5 × 1.2–9.5 cm, base attenuate to rounded or shallowly cordate, margin entire or obscurely undulate, apex acuminate or rarely obtuse, surfaces sparsely to densely puberulous or pubescent, particularly along veins beneath; lateral veins 6–9 pairs; cystoliths minute, elliptic; petiole to 75 mm, uppermost leaves often subsessile. Inflorescences verticillate in upper axils and terminal, cymes densely fasciculate, sessile; fascicles subtended by leafy bracts; main axis bracts inconspicuous, lanceolate, 1.5–4(10) mm long; cymules numerous, each 1-flowered, subsessile or rarely on 5 mm peduncles; cymule bracts subequal to ± unequal, lanceolate-aristate, 5–21.5 mm long including flexuose arista, 0.8–2 mm wide towards base, fused in basal 1–2.5 mm, antrorse-puberulous to pilose or sparsely so, usually with few to numerous spreading glandular hairs on arista, margin hyaline towards base; bracteoles linear-caudate, 4–8.5 mm long, margin broadly hyaline. Calyx tube 1–2(4) mm long; lobes 4, subequal, lanceolate or acuminate, 2.8–5 mm long, sparsely ciliate. Corolla 15–32 mm long, white, pink, purple or red, usually with darker markings on palate of lip held in upper position, retrorse-pubescent and with short spreading glandular hairs externally; tube 7.5–19 mm long; lip held in upper position oblong-obovate, 8–16 × 3–9.5 mm, conspicuously 3-lobed, ciliate; lip held in lower position narrowly oblong-elliptic to lanceolate, 7–14.5 × 2–5 mm, apex obtuse or acute. Staminal filaments 6–14 mm long, puberulous to pilose or largely glabrous; anthers 1.2–2 mm long. Style sparsely pubescent. Capsule 10–14.5 mm long, glabrous. Seeds subellipsoid, 2–3 mm wide, smooth.

Zambia. N: Samfya Dist., c.5 km W of Samfya, Kasamba mushitu, fl. 21.iv.1989, *Pope et al.* 2205 (BR, K, LISC, NDO, SRGH). W: Ndola, fl. 6.iii.1954, *Fanshawe* 937 (K, NDO). E: Isoka Dist., Makutu Hills (Isoka Dist. Nat. Park), between Lupita and Chipata, , fl.& fr. 20.viii.1965, *Lawton* 1268 (K). **Zimbabwe**. C: Wedza Dist., Wedza Mt, fl. 22.v.1968, *Mavi* 744 (K, LISC, SRGH). E: Chimanimani Dist., Pasture Res. Station, fl. 8.vii.1949, *F. Williams* 20 (K, SRGH). S: Masvingo Dist., Zimbabwe Ruins, fl.& fr. 1.vii.1930, *Hutchinson & Gillett* 3314 (BM, K). **Malawi**. N: Mzimba Dist., Mtangatanga Forest, fl. 14.viii.1971, *Salubeni* 1697 (K, LISC, SRGH). C: Dedza Dist., Chencherere hill, N of Dedza on Linthipe road, fl. 30.iv.1989, *Pope et al.* 2252 (BR, K, LISC, MAL). S: Zomba Dist., Chiradzulu peak, Zomba Plateau, fl. 24.vii.1986, *Salubeni & Mussa* 4636 (K, MAL). **Mozambique**. Z: Gurué Dist., Namuli peaks, W face, fl.& fr. 26.vii.1962, *Leach & Schelpe* 11474 (K, LISC, SRGH). MS: Manica Dist., 23.4 km from Chimoio (Vila Pery) towards Garuso, by bridge, fl. 21.v.1949, *Pedro & Pedrógão* 5772 (K, LMA, LMU). M: Namaacha Dist., c.20 km from Namaacha, Ferradura, fl.& fr. 24.iv.1947, *Pedro & Pedrógão* 688 (LMA, LMU).

Nigeria to Ethiopia south to South Africa. Mid-altitude to montane forest, particularly forest margins and clearings where it is often dominant, sometimes extending into miombo woodland and wooded grassland; at lower altitudes in swamp and wet riverine forest (mushitu), bushland or amongst boulders on rocky hillslopes, riverine woodland and thicket and in ditches; 400–2400 m.

Conservation notes: Widespread and common, often dominant in suitable habitat; Least Concern.

A polymorphic species with much regional variation in, for example, vegetative indumentum, leaf shape and size, inflorescence density, bract length and flower size and colour. However, these characters vary independently such that a matrix of different forms are found making it difficult to apply any meaningful infraspecific classification. Whilst the subdivision into three varieties proposed by Balkwill & Getliffe-Norris (1985) works well in South Africa, var. *alba* is less clearly defined in our region and so is not upheld here.

The above description covers only the range of variation from the Flora region. It excludes two particularly distinctive 'variants' from N Mozambique which are provisionally separated below; whilst falling within the *Hypoestes aristata* complex, they are worthy of particular consideration if and when a full revision of the genus is undertaken.

2a. **Hypoestes sp. aff. aristata A** (= *Torre* 5595).

Robust erect perennial herb 100–200 cm tall, differing from *H. aristata* in having elliptic leaves (ovate-elliptic towards stem apex), 20–24.5 × 9.5–12 cm, apex only shortly attenuate, lateral veins 8–11 pairs. Main axis bracts more conspicuous, 6–13 × 2–3 mm; cymule bracts lacking glandular hairs on the aristae; bracteoles 10–12 mm long. Calyx tube 5–7 mm long, lobes 4, triangular, 2.3–3.6 mm long; corolla 36–39 mm long, magenta or purple; tube 21.5–22.5 mm long, rather broadly expanded in throat. Anthers 2.4–2.9 mm long. Capsule 16–18 mm long, glabrous; seeds tuberculate, tubercles verruciform with reticulate micro-sculpturing.

Mozambique. Z: Gurué Dist., Gurué (Vila Junqueiro), edge of mountain, fl. 28.vi.1943, *Torre* 5595 (LISC); W of Morrece peak, NE of Namuli peaks, fl. 27.vii.1962, *Leach & Schelpe* 11491 (K, LISC, SRGH); Lugela Dist., Mt Mabu, fr. 14.x.2008, *Mphamba* 59 (K, LMA).

Not known elsewhere. By streams and rivers in forest and deep shade; 700–1000 m.

Conservation notes: Known from only four collections from mountains in NC Mozambique where it can be locally abundant in undergrowth. Potentially threatened by mid-altitude forest clearance.

Originally I planned to describe this as a distinct species, with the tuberculate seeds as the key diagnostic character. However, a review of specimens currently included within *Hypoestes aristata* sensu lato from West Africa reveals several specimens with tuberculate seeds, e.g. *Morton* K1203 from Nigeria (K). Although the Gurué plants are easily separated from *H. aristata* in the Flora region by a number of additional characters (large leaves, flowers and fruits, eglandular bracts, calyx with a relatively long tube and short lobes), none of these are wholly diagnostic across the full geographic range of *H. aristata* when taken in isolation. It is, therefore, only the combination of characters that is unique and this is perhaps insufficient to warrant species status.

2b. **Hypoestes sp. aff. aristata B** (= *Leach & Schelpe* 11407).

Densely pubescent erect or decumbent herb 90–150 cm tall, differing from *H. aristata* in the verticillate inflorescence being congested with very short internodes, together forming a ± well defined cylindrical head to 5 cm long. Corolla 30–32 mm long, lilac or pale grey-mauve; tube 20.5–22 mm long, with pronounced expanded throat 12–13 × 5–6.5 mm wide at mouth; lip held in upper position somewhat hooded, broadly ovate when flattened, 8–10.5 × 9–11 mm. Staminal filaments 5–6 mm long.

Mozambique. N: Ribáuè Dist., S face of Ribáuè Mt, fl. 19.vii.1962, *Leach & Schelpe* 11407 (K, LISC); Murrupula Dist., Serra Chinga, between Chinga 1 and 2, fl. 27.v.1968, *Macedo & Macuácua* 3287 (K, LMA).

Not known elsewhere. Woodland on steep slopes and in rocky valleys; 900–1100 m.

Conservation notes: Known only from the localities cited with no known threats; Data Deficient.

These two collections look close to hairy forms of *Hypoestes aristata* from woodland habitats elsewhere in the Flora region, but the corolla form in the Leach & Schelpe collection is strikingly different, appearing more similar to flowers of some (presumably) bird-pollinated species on Madagascar. Both specimens also have a more congested compound inflorescence than in typical *H. aristata*, forming a cylindrical head. It may well be a distinct species from *H. aristata*, but I cannot rule out the possibility that corollas are aberrant in the Leach & Schelpe collection, the lips not having separated normally from the corolla tube. Unfortunately the Macedo collection is badly insect-damaged such that the corolla form is not discernable. Further material is required before reaching any firm conclusion.

3. **Hypoestes forskaolii** (Vahl) R. Br. in Salt, Voy. Abyss., app.: 63 (1814) as *forskalii*. — Clarke in F.T.A. **5**: 249 (1900); in Fl. Cap. **5**: 89 (1901), both as *forskalei*. —Burrows & Willis, Pl. Nyika Plateau: 48 (2005). —Darbyshire in F.T.E.A., Acanthaceae 2: 726, fig.94 (2010). Type: Yemen, Jebel Melhan, fr. 1763, *Forsskål* 387 (C lectotype, microfiche 60: I.1–2, K photo, LUND), lectotypified by Wood, Hillcoat & Brummitt (Kew Bull. **38**: 455, 1983).

Justicia forskaolii Vahl, Symb. Bot. **1**: 2 (1790) as *forskoalei*.

Justicia paniculata Forssk., Fl. Aegypt.-Arab.: 4 (1775), illegitimate name, non *J. paniculata* Burm. f. Type as for *H. forskaolii*.

Hypoestes latifolia Nees in De Candolle, Prodr. **11**: 509 (1847). Type: Sudan, "in valle ad montem Cordofanum Turra locis umbrosis", fl. 5.xii.1839, *Kotschy* 296 (GZU [000250762] lectotype, BM, HAL, HBG, K, LD, M, MPU, P, S, STU, TUB), lectotypified here.

Hypoestes latifolia Nees var. *integrifolia* Nees in De Candolle, Prodr. **11**: 509 (1847). Types: Senegal, Casamance, fl. 1837 [1838 on K sheet], *Heudelot* 552 (GZU [herb. no. 6], P, K syntypes); Sudan, Sennaar, *Acerbi* s.n. (G-DC syntype).

Hypoestes mollis T. Anderson in J. Linn. Soc., Bot. **7**: 49 (1864). Type: D.R. Congo, fl.& fr., n.d., *Smith* s.n. (K holotype, P).

Hypoestes verticillaris R. Br. var. *glabra* S. Moore in J. Bot. **18**: 363 (1880). Type: Angola, "inter Lag. de Ivantala et Quilongues juxta rip. flum. Caculuvar", fl. ii.1860, *Welwitsch* 5059 (BM holotype, K).

Hypoestes preussii Lindau in Bot. Jahrb. Syst. **20**: 48 (1894). —Clarke in F.T.A. **5**: 251 (1900). Type: Cameroon, Buea, fl.& fr. 6.ii.1891, *Preuss* 755 (B† holotype, K).

Hypoestes echioides Lindau in Bot. Jahrb. Syst. **20**: 52 (1894). Type: Tanzania, Mpwapwa, fl. 16.vii.1890, *Stuhlmann* 286 (B† holotype).

Hypoestes depauperata Lindau in Bot. Jahrb. Syst. **20**: 52 (1894). Type: South Africa, N Cape, "Betschuanaland, Kuruman zwischen Steinen" [on PRE sheet: Ebene nr. Kuruman], fl. ii.1886, *Marloth* 1120 (B† holotype, PRE).

Hypoestes violaceotincta Lindau in Bot. Jahrb. Syst. **24**: 323 (1897). —Clarke in F.T.A. **5**: 251 (1900). Type: Togo, Misahöhe, fl. 15.iv.1893 [1895 on K sheet], *Baumann* 476 (B† holotype, K).

Hypoestes mlanjensis C.B. Clarke in F.T.A. **5**: 250 (1900). Types: Kenya, Ukambani, fl. vii.1893, *Gregory* s.n. (BM syntype); Malawi, Mt Mulanje, fl.& fr. 1891, *Whyte* 151 (BM, K syntypes).

Hypoestes tanganyikensis C.B. Clarke in F.T.A. **5**: 252 (1900). Types: Malawi, Tanganyika Plateau, fl. vii.1896, *Whyte* s.n. (K syntype); Nyika Plateau, fl. vii.1896, *Whyte* 192 (K syntype).

Hypoestes verticillaris R. Br. var. *forskaolii* (Vahl) Benoist in Notul. Syst. (Paris) **10**: 246 (1942).

Hypoestes verticillaris var. *mollis* (T. Anderson) Benoist in Notul. Syst. (Paris) **10**: 246 (1942).

Hypoestes verticillaris var. *latifolia* (Nees) Benoist in Notul. Syst. (Paris) **10**: 247 (1942).

Hypoestes verticillaris var. *violaceotincta* (Lindau) Benoist in Notul. Syst. (Paris) **10**: 247 (1942).

Hypoestes verticillaris sensu auct., non (L.f.) Roem. & Schult. —Nees in De Candolle, Prodr. **11**: 507 (1847). —Clarke in F.T.A. **5**: 250 (1900); in Fl. Cap. **5**: 88 (1901). —Binns, Checklist Herb. Fl. Malawi: 14 (1968).

Erect, scrambling or trailing perennial herb, subshrub or suffrutex 5–100(200) cm tall, sometimes pyrophytic; stems solitary or often much-branched from a woody base, sometimes procumbent and rooting along lower nodes, glabrous to densely pubescent or pilose or whole plant appressed-canescent. Leaves ovate, elliptic or lanceolate, 2.2–18 × 0.5–8 cm, base attenuate, acute, rounded or subcordate, margin entire to shallowly undulate, apex shortly acuminate, surfaces sparsely to densely pubescent or pilose particularly along veins, or largely glabrous; lateral veins 5–8(10) pairs; cystoliths elliptic to linear, ± inconspicuous; petiole 2–25(50) mm. Inflorescences terminal and axillary, cymes unilateral, spiciform or fasciculate, to 4 cm long, sometimes compounded into a dense and verticillate or lax and paniculate thyrse; main axis bracts inconspicuous, triangular to lanceolate, 1.5–4(11) mm long, pairs basally fused; cymules (2)5 to many per cyme, subsessile; cymule bracts equal or ± unequal, lanceolate, linear or acuminate, 5–11(15) mm long, fused in lower ½ to ²/₃, united portion often hyaline, surfaces sparsely to densely puberulous, often also densely pilose with numerous gland-tipped hairs on lobes; bracteoles elliptic-lanceolate, 5–9.5 mm long, margin broadly hyaline. Calyx tube 2.5–4.5 mm long; lobes 5, subequal, lanceolate, 1.2–3.5 mm long, with minute spreading hairs externally. Corolla 13–20(25) mm long, white to purple with purple or pink markings on paler palate of lip held in upper position, pubescent and with ± prominent glandular hairs externally; tube 8–14 mm long; lip held in upper position oblong-obovate, 5.5–8(11) × 3.5–5(6) mm, apex shortly 3-lobed; lip held in lower position trullate-caudate, 5–7(9.5) × 2–4.5 mm, apex of acumen rounded to emarginate. Staminal filaments 4–8 mm long, pilose above, hairs inconspicuously glandular; anthers 1.1–1.6 mm long. Style glabrous or sparsely pubescent. Capsule 6.5–8(9) mm long, pubescent in upper half or rarely largely glabrous. Seeds broadly ellipsoid, 1.3–2 mm wide, smooth.

Subsp. **forskaolii**. FIGURE 8.6.**72**.

Vegetative indumentum highly variable – largely glabrous to densely pubescent and/or pilose, hairs spreading, antrorse or retrorse, but never appressed-canescent.

Caprivi. E of Kwando (Cuanda) R., ?1930, *Curson* 1083 (PRE). **Botswana**. N: Chobe Dist., Tsetse Fly Camp 16 km W of Parakarungu on Chobe R., fl. 5.viii.1967, *Lambrecht* 271 (K, SRGH). SE: Kgatleng Dist., Mochudi, fl.& fr. i–iv.1914, *Harbor* in *Herb. Rogers*

6611 (K). **Zambia**. B: Sesheke Dist., Masese, fl. 18.vi.1960, *Fanshawe* 5746 (K, NDO).
N: Mpika Dist., North Luangwa Nat. Park, Elephants Playground, c.14 km SE of
Mano Camp, fl. 14.v.1994, *Schmidt et al.* 1347 (K, MO). W: Solwezi Dist., NE corner of
Lualaba P.F.A., fl. 10.vi.1962, *Holmes* 1487 (K, NDO). C: Serenje Dist., Serenje, fl.& fr.
2.vi.1965, *Robinson* 6677 (K). E: Chipata Dist., Chipata (Fort Jameson), fl. 1.vi.1958,
Fanshawe 4488 (K, NDO). S: Choma Dist., Mapanza Mission, fl. 10.v.1953, *Robinson* 204
(K). **Zimbabwe**. N: Hurungwe Dist., Hurungwe communal land (Urungwe Special
Native Area), L.D.O.'s camp, fl. vi.1957, *Davies* 2443 (K, SRGH). W: Bubi Dist., Bubi R.
on Goodwood Farm, fl. v.1956, *Goldsmith* 100a/56 (K, LISC, SRGH). C: Darwin Dist.,
Msengesi Camp, fl. 8.v.1955, *Whellan* 858 (K, SRGH). E: Mutare Dist., E Vumba Mts,
fl. 19.v.1957, *Chase* 6493 (K, LISC, SRGH). S: Mwenezi Dist., Bubi (Bubye) R., near
Bubye Ranch homestead, fl.& fr. 8.v.1958, *Drummond* 5689 (K, LISC, SRGH). **Malawi**.
N: Nkhata Bay Dist., 3.2 km S of Chikangawa, fl. 21.viii.1978, *Phillips* 3809 (K, MO).
C: Lilongwe Dist., Lilongwe Nature Sanctuary, fl. 25.vi.1987, *Salubeni & Tawakali* 4955
(K, MAL). S: Nsanje Dist., 5 km NW of Nsanje, near Nyamadzere Rest House, fl.& fr.
28.v.1970, *Brummitt* 11132 (K, MAL). **Mozambique**. N: Meluco Dist., R. Muaguide,
near bridge on Pemba (Porto Amelia) road, fl. 25.x.1960, *Gomes e Sousa* 4580 (K,
LISC, LMA). Z: Lugela Dist., Namagoa, 200 km inland from Quelimane, fl. iv.1943,
Faulkner PRE 314 (K sheet 1, LMA, PRE). T: Moatize (Zóbuè) Dist., fl. 17.vi.1947,
Hornby 2763 (K, SRGH). MS: Susundenga Dist., c.30 km N of Dombe, fl.& fr. 4.vi.1971,
Biegel & Pope 3534 (K, LISC, LMA, SRGH). GI: Guijá Dist., R margin of R. Limpopo,
fl. 9.vi.1947, *Pedrógão* 272 (K, LMA). M: Marracuene Dist., Arricata settlement, 23 km
on road from Maputo to Marracuene, fl. 30.iv.1964, *Balsinhas* 712 (K, LISC, LMA).

Widespread in tropical Africa from Senegal to Somalia south to Namibia and South
Africa; also extending to the Saharan highlands, Arabia and Madagascar. In a wide
variety of habitats including miombo woodland on sandy soils or rocky slopes, dry
mixed woodland and thicket, termitaria, grassland, riverine fringes, dry lowland to
wet montane forest, particularly in clearings and along margins, conifer plantations,
roadsides and other disturbed areas; sometimes in recently burnt woodland and
grassland; 20–2300 m.

Conservation notes: Widespread and abundant, tolerant of disturbance; Least
Concern.

Given its large geographic and ecological range it is not surprising that the species
is very variable, as reflected in the sizable synonymy. With the exception of the largely
discrete subsp. *hildebrandtii* (Lindau) I. Darbysh. from NE Africa, which has an
appressed-canescent indumentum, previously recognized taxa merely represent the
extremes of a continuum of variation.

In the Flora area, the most common form, particularly abundant in miombo
and other types of woodland, has a dense inflorescence, white flowers with purple
markings, usually densely hairy bracts and small to medium-sized, ovate or lanceolate
leaves. In some areas (notably Malawi C, Zambia C, E & S), pyrophytic populations
occur, flowering on short erect, largely leafless shoots shortly after fire (e.g. *Brass*
17584, *Angus* 181), but are otherwise close to the miombo form. Populations
previously separated as *Hypoestes tanganyikensis* have more slender, lax flowering spikes
than others; this form was described from N Malawi but similar plants are also found
in W and E Zimbabwe (e.g. *Miller* 2747 and *Seine* 1290, respectively).

Forest forms (including *H. violaceotincta* Lindau and *H. preussii* Lindau) usually have
larger, more papery, long-petiolate leaves, with a more paniculate inflorescence and
smaller, often only shortly hairy bracts, though this is variable with some forest forms
having densely long-hairy bracts. A particularly striking form was previously separated
as *H. latifolia* Nees, being found in our region in riverine thickets and woodland. It
has broadly ovate, long-petiolate leaves with an abruptly narrowed base, a densely

Fig. 8.6.**72**. HYPOESTES FORSKAOLII subsp. FORSKAOLII. 1, habit (× ²/₃); 2 & 3, mature leaves, extreme large and small variants (× ²/₃); 4, paired main axis bracts and cymule bracts, lateral view (× 6); 5, bracteoles and calyx (× 6); 6, flower, lateral view (× 4); 7, detail of corolla lip in lower position (× 4); 8, detail of stamens (× 8); 9, stamens, recurved during female phase of protandry (× 12); 10, capsule with seeds (× 5); 11, mature seed (× 12). 1 from *Kayombo* 2392, 2 from *Bidgood et al.* 919, 3, 6–8 from *Greenway* 13902, 4–5, 10–11 from *Drummond & Hemsley* 2913, 9 from *Kayombo* 576. Drawn by Juliet Williamson. Reproduced from Flora of Tropical East Africa (2010).

pilose or sericeous stem indumentum, narrow cymule bracts and mauve to purple corollas with darker purple markings on a pale palate to the lip in upper position. It is scattered through the Flora area (e.g. Malawi: *Brummitt & Seyani* 18669; Zambia: *Bingham* 11096; Zimbabwe: *Wild* 5762). It is perhaps the strongest candidate for varietal recognition, but I refrain from recognising varieties until a full Africa-wide revision is undertaken.

INDEX TO BOTANICAL NAMES

FAMILIES OF VASCULAR PLANTS REPRESENTED IN THE FLORA ZAMBESIACA AREA

PTERIDOPHYTA
(Flora Zambesiaca families and family number. Published 1970)

Actiniopteridaceae		Gleicheniaceae	9	Parkeriaceae	
see Adiantaceae	18	Grammitidaceae	20	see Adiantaceae	18
Adiantaceae	18	Hymenophyllaceae	15	Polypodiaceae	21
Aspidiaceae	27	Isoetaceae	4	Psilotaceae	1
Aspleniaceae	23	Lindsaeaceae	19	Pteridaceae	
Athyriaceae	25	Lomariopsidaceae	26	see Adiantaceae	18
Azollaceae	13	Lycopodiaceae	2	Salviniaceae	12
Blechnaceae	28	Marattiaceae	7	Schizaeaceae	10
Cyatheaceae	14	Marsileaceae	11	Selaginellaceae	3
Davalliaceae	22	Oleandraceae		Thelypteridaceae	24
Dennstaedtiaceae	16	see Davalliaceae	22	Vittariaceae	17
Dryopteridaceae		Ophioglossaceae	6	Woodsiaceae	
see Aspidiaceae	27	Osmundaceae	8	see Athyriaceae	25
Equisetaceae	5				

GYMNOSPERMAE
(Flora Zambesiaca families and family number. Volume 1(1) 1960)

Cupressaceae	3	Cycadaceae	1	Podocarpaceae	2

ANGIOSPERMAE
(Flora Zambesiaca families, volume and part number and year of publication)

Acanthaceae			Asphodelaceae	12(3)	2001
tribes 1–5	8(5)	2013	Avicenniaceae	8(7)	2005
tribes 6–7	8(6)	2014	Balanitaceae	2(1)	1963
Agapanthaceae	13(1)	2008	Balanophoraceae	9(3)	2006
Agavaceae	13(1)	2008	Balsaminaceae	2(1)	1963
Aizoaceae	4	1978	Barringtoniaceae	4	1978
Alangiaceae	4	1978	Basellaceae	9(1)	1988
Alismataceae	12(2)	2009	Begoniaceae	4	1978
Alliaceae	13(1)	2008	Behniaceae	13(1)	2008
Aloaceae	12(3)	2001	Berberidaceae	1(1)	1960
Amaranthaceae	9(1)	1988	Bignoniaceae	8(3)	1988
Amaryllidaceae	13(1)	2008	Bixaceae	1(1)	1960
Anacardiaceae	2(2)	1966	Bombacaceae	1(2)	1961
Anisophylleaceae			Boraginaceae	7(4)	1990
see Rhizophoraceae	4	1978	Brexiaceae	4	1978
Annonaceae	1(1)	1960	Bromeliaceae	13(2)	2010
Anthericaceae	13(1)	2008	Buddlejaceae		
Apocynaceae	7(2)	1985	see Loganiaceae	7(1)	1983
Aponogetonaceae	12(2)	2009	Burmanniaceae	12(2)	2009
Aquifoliaceae	2(2)	1966	Burseraceae	2(1)	1963
Araceae	12(1)	2012	Buxaceae	9(3)	2006
Araliaceae	4	1978	Cabombaceae	1(1)	1960
Arecaceae			Cactaceae	4	1978
see Palmae	13(2)	2010	Caesalpinioideae		
Aristolochiaceae	9(2)	1997	see Leguminosae	3(2)	2006
Asclepiadaceae	-	-	Campanulaceae	7(1)	1983
Asparagaceae	13(1)	2008	Canellaceae	7(4)	1990

Family		
Cannabaceae	9(6)	1991
Cannaceae	13(4)	2010
Capparaceae	1(1)	1960
Caricaceae	4	1978
Caryophyllaceae	1(2)	1961
Casuarinaceae	9(6)	1991
Cecropiaceae	9(6)	1991
Celastraceae	2(2)	1966
Ceratophyllaceae	9(6)	1991
Chenopodiaceae	9(1)	1988
Chrysobalanaceae	4	1978
Colchicaceae	12(2)	2009
Combretaceae	4	1978
Commelinaceae	-	-
Compositae		
tribes 1–5	6(1)	1992
tribes 6–12	-	-
Connaraceae	2(2)	1966
Convolvulaceae	8(1)	1987
Cornaceae	4	1978
Costaceae	13(4)	2010
Crassulaceae	7(1)	1983
Cruciferae	1(1)	1960
Cucurbitaceae	4	1978
Cuscutaceae	8(1)	1987
Cymodoceaceae	12(2)	2009
Cyperaceae	-	-
Dichapetalaceae	2(1)	1963
Dilleniaceae	1(1)	1960
Dioscoreaceae	12(2)	2009
Dipsacaceae	7(1)	1983
Dipterocarpaceae	1(2)	1961
Dracaenaceae	13(2)	2010
Droseraceae	4	1978
Ebenaceae	7(1)	1983
Elatinaceae	1(2)	1961
Ericaceae	7(1)	1983
Eriocaulaceae	13(4)	2010
Eriospermaceae	13(2)	2010
Erythroxylaceae	2(1)	1963
Escalloniaceae	7(1)	1983
Euphorbiaceae	9(4)	1996
Euphorbiaceae	9(5)	2001
Flacourtiaceae	1(1)	1960
Flagellariaceae	13(4)	2010
Fumariaceae	1(1)	1960
Gentianaceae	7(4)	1990
Geraniaceae	2(1)	1963
Gesneriaceae	8(3)	1988
Gisekiaceae		
see Molluginaceae	4	1978
Goodeniaceae	7(1)	1983
Gramineae		
tribes 1–18	10(1)	1971
tribes 19–22	10(2)	1999
tribes 24–26	10(3)	1989
tribe 27	10(4)	2002
Guttiferae	1(2)	1961
Haloragaceae	4	1978
Hamamelidaceae	4	1978
Hemerocallidaceae	12(3)	2001
Hernandiaceae	9(2)	1997
Heteropyxidaceae	4	1978
Hyacinthaceae	-	-
Hydnoraceae	9(2)	1997
Hydrocharitaceae	12(2)	2009
Hydrophyllaceae	7(4)	1990
Hydrostachyaceae	9(2)	1997
Hypericaceae		
see Guttiferae	1(2)	1961
Hypoxidaceae	12(3)	2001
Icacinaceae	2(1)	1963
Illecebraceae	1(2)	1961
Iridaceae	12(4)	1993
Irvingiaceae	2(1)	1963
Ixonanthaceae	2(1)	1963
Juncaceae	13(4)	2010
Juncaginaceae	12(2)	2009
Labiatae		
see Lamiaceae & Verbenacaeae		
Lamiaceae		
Viticoideae, Pingoideae	8(7)	2005
Lamiaceae		
Scutellaroideae-		
Nepetoideae	8(8)	2013
Lauraceae	9(2)	1997
Lecythidaceae		
see Barringtoniaceae	4	1978
Leeaceae	2(2)	1966
Leguminosae,		
Caesalpinioideae	3(2)	2007
Mimosoideae	3(1)	1970
Papilionoideae	3(3)	2007
Papilionoideae	3(4)	2012
Papilionoideae	3(5)	2001
Papilionoideae	3(6)	2000
Papilionoideae	3(7)	2002
Lemnaceae		
see Araceae	12(1)	2012
Lentibulariaceae	8(3)	1988
Liliaceae sensu stricto	12(2)	2009
Limnocharitaceae	12(2)	2009
Linaceae	2(1)	1963
Lobeliaceae	7(1)	1983
Loganiaceae	7(1)	1983
Loranthaceae	9(3)	2006
Lythraceae	4	1978
Malpighiaceae	2(1)	1963
Malvaceae	1(2)	1961
Marantaceae	13(4)	2010
Mayacaceae	13(2)	2010
Melastomataceae	4	1978
Meliaceae	2(1)	1963
Melianthaceae	2(2)	1966

Menispermaceae	1(1)	1960	Ptaeroxylaceae	2(2)	1966
Menyanthaceae	7(4)	1990	Rafflesiaceae	9(2)	1997
Mesembryanthemaceae	4	1978	Ranunculaceae	1(1)	1960
Mimosoideae			Resedaceae	1(1)	1960
see Leguminosae	3(1)	1970	Restionaceae	13(4)	2010
Molluginaceae	4	1978	Rhamnaceae	2(2)	1966
Monimiaceae	9(2)	1997	Rhizophoraceae	4	1978
Montiniaceae	4	1978	Rosaceae	4	1978
Moraceae	9(6)	1991	Rubiaceae		
Musaceae	13(4)	2010	subfam. Rubioideae	5(1)	1989
Myristicaceae	9(2)	1997	tribe Vanguerieae	5(2)	1998
Myricaceae	9(3)	2006	subfam. Cinchonoideae	5(3)	2003
Myrothamnaceae	4	1978	Rutaceae	2(1)	1963
Myrsinaceae	7(1)	1983	Salicaceae	9(6)	1991
Myrtaceae	4	1978	Salvadoraceae	7(1)	1983
Najadaceae	12(2)	2009	Santalaceae	9(3)	2006
Nesogenaceae	8(7)	2005	Sapindaceae	2(2)	1966
Nyctaginaceae	9(1)	1988	Sapotaceae	7(1)	1983
Nymphaeaceae	1(1)	1960	Scrophulariaceae	8(2)	1990
Ochnaceae	2(1)	1963	Selaginaceae		
Olacaceae	2(1)	1963	see Scrophulariaceae	8(2)	1990
Oleaceae	7(1)	1983	Simaroubaceae	2(1)	1963
Oliniaceae	4	1978	Smilacaceae	12(2)	2009
Onagraceae	4	1978	Solanaceae	8(4)	2005
Opiliaceae	2(1)	1963	Sonneratiaceae	4	1978
Orchidaceae	11(1)	1995	Sphenocleaceae	7(1)	1983
Orchidaceae	11(2)	1998	Sterculiaceae	1(2)	1961
Orobanchaceae			Strelitziaceae	13(4)	2010
see Scrophulariaceae	8(2)	1990	Taccaceae		
Oxalidaceae	2(1)	1963	see Dioscoreaceae	12(2)	2009
Palmae	13(2)	2010	Tecophilaeaceae	12(3)	2001
Pandanaceae	12(2)	2009	Tetragoniaceae	4	1978
Papaveraceae	1(1)	1960	Theaceae	1(2)	1961
Papilionoideae			Thymelaeaceae	9(3)	2006
see Leguminosae, Papilionoideae			Tiliaceae	2(1)	1963
Passifloraceae	4	1978	Trapaceae	4	1978
Pedaliaceae	8(3)	1988	Turneraceae	4	1978
Periplocaceae			Typhaceae	13(4)	2010
see Asclepiadaceae	-	-	Ulmaceae	9(6)	1991
Philesiaceae			Umbelliferae	4	1978
see Behniaceae	13(1)	2008	Urticaceae	9(6)	1991
Phormiaceae			Vacciniaceae		
see Hemerocallidaceae	12(3)	2001	see Ericaceae	7(1)	1983
Phytolaccaceae	9(1)	1988	Vahliaceae	4	1978
Piperaceae	9(2)	1997	Valerianaceae	7(1)	1983
Pittosporaceae	1(1)	1960	Velloziaceae	12(2)	2009
Plantaginaceae	9(1)	1988	Verbenaceae	8(7)	2005
Plumbaginaceae	7(1)	1983	Violaceae	1(1)	1960
Podostemaceae	9(2)	1997	Viscaceae	9(3)	2006
Polygalaceae	1(1)	1960	Vitaceae	2(2)	1966
Polygonaceae	9(3)	2006	Xyridaceae	13(4)	2010
Pontederiaceae	13(2)	2010	Zannichelliaceae	12(2)	2009
Portulacaceae	1(2)	1961	Zingiberaceae	13(4)	2010
Potamogetonaceae	12(2)	2009	Zosteraceae	12(2)	2009
Primulaceae	7(1)	1983	Zygophyllaceae	2(1)	1963
Proteaceae	9(3)	2006			